Nitrones, Nitronates and Nitroxides

THE CHEMISTRY OF FUNCTIONAL GROUPS

A series of advanced treatises under the general editorship of Professor Saul Patai

The chemistry of alkenes (2 volumes)
The chemistry of the carbonyl group (2 volumes)
The chemistry of the ether linkage
The chemistry of the amino group
The chemistry of the nitro and nitroso groups (2 parts)
The chemistry of carboxylic acids and esters
The chemistry of the carbon–nitrogen double bond
The chemistry of amides
The chemistry of the cyano group
The chemistry of the hydroxyl group (2 parts)
The chemistry of the azido group
The chemistry of the acyl halides
The chemistry of the carbon–halogen bond (2 parts)
The chemistry of the quinonoid compounds (2 volumes, 4 parts)
The chemistry of the thiol group (2 parts)
The chemistry of the hydrazo, azo and azoxy groups (2 parts)
The chemistry of amidines and imidates
The chemistry of cyanates and their thio derivatives (2 parts)
The chemistry of diazonium and diazo groups (2 parts)
The chemistry of the carbon–carbon triple bond (2 parts)
The chemistry of ketenes, allenes and related compounds (2 parts)
The chemistry of the sulphonium group (2 parts)
Supplement A: The chemistry of double-bonded functional groups (2 parts)
Supplement B: The chemistry of acid derivatives (2 parts)
Supplement C: The chemistry of triple-bonded functional groups (2 parts)
Supplement D: The chemistry of halides, pseudo-halides and azides (2 parts)
Supplement E: The chemistry of ethers, crown ethers, hydroxyl groups and their sulphur analogues (2 parts)
Supplement F: The chemistry of amino, nitroso and nitro compounds and their derivatives (2 parts)
The chemistry of the metal–carbon bond (4 volumes)
The chemistry of peroxides
The chemistry of organic selenium and tellurium compounds (2 volumes)
The chemistry of the cyclopropyl group (2 parts)
The chemistry of sulphones and sulphoxides

Updates

The chemistry of α-haloketones, α-haloaldehydes and α-haloimines
Crown ethers (in press)
The formation of carbon–halogen bonds (in press)

$$>C{=}N(\rightarrow O)<\qquad >C{=}N(\rightarrow O)O^- \qquad >N{-}\dot{O}$$

Nitrones, nitronates and nitroxides

By

ELI BREUER

The Hebrew University of Jerusalem

HANS GÜNTER AURICH

Phillips-Universität Marburg

ARNOLD NIELSEN

Michelson Laboratory, China Lake, California

Edited by

SAUL PATAI

and

ZVI RAPPOPORT

The Hebrew University of Jerusalem

Updates from the Chemistry of Functional Groups

1989

JOHN WILEY & SONS

CHICHESTER · NEW YORK · BRISBANE · TORONTO · SINGAPORE

An Interscience® Publication

Library of Congress of Cataloging-in-Publication Data:

Breuer, Eli.
Nitrones, nitronates, and nitroxides/by Eli Breuer, Hans Günter Aurich, Arnold Nielsen; edited by Saul Patai and Zvi Rappoport.
p. cm.—(The Chemistry of functional groups)
'An Interscience publication.'
ISBN 0 471 91709 5
1. Nitroxides. I. Aurich, Hans Günter, 1932–
II. Nielsen, Arnold. III. Title. IV. Series.
QD305.N8B83 1989
546'.711—dc19

88–17388
CIP

British Library Cataloguing in Publication Data:

Breuer, Eli
Nitrones, Nitronates and Nitroxides.—
(The chemistry of functional groups).
1. Nitroxides. Chemical reactions.
Catalysis
I. Title II. Aurich, Hans Günter
III. Nielsen, Arnold IV. Series
546'. 711595
ISBN 0 471 91709 5

Typeset by Thomson Press (India) Limited, New Delhi and
Printed in Great Britain by the Bath Press.

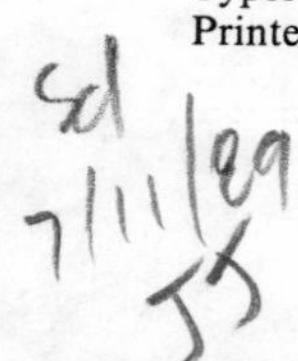

List of contributors

H. G. Aurich — Fachbereich Chemie, Phillips-Universität Marburg, Marburg, FRG.

E. Breuer — School of Pharmacy, Faculty of Medicine, The Hebrew University of Jerusalem, Jerusalem 91904, Israel.

A. T. Nielsen — Michelson Laboratory, Naval Weapon Center, China Lake, California, USA.

Contents

Foreword

This is the second volume published in the new series entitled 'Updates from the Chemistry of Functional Groups'.

As in the other volumes of this series, our intention is to present selected chapters in the original form in which they were published, together with an extensive updating of the subject, if possible by the authors of the original chapters.

The present volume contains the chapter 'Nitronic acids and esters' by A. T. Nielsen, first published in 1969 in *The chemistry of the nitro and nitroso groups*, edited by H. Feuer. This is followed by the chapter 'Nitrones and nitronic acid derivatives: their structure and their roles in synthesis' by E. Breuer, published originally in *Supplement F: The chemistry of amino, nitroso and nitro compounds and their derivatives* (Ed. S. Patai, 1982). Both the chapters mentioned above are updated by E. Breuer in Chapter 3 of the present book, which contains also some sections on material which has not been discussed previously in the main series.

This is followed by the chapter on 'Nitroxides' by H. G. Aurich, originally also published in *Supplement F*, and by an 'Appendix to Nitroxides' (Chapter 5), again written by H. G. Aurich.

An additional volume in the same series, now in the press will deal in a similar manner with crown ethers, based on the original chapters published in 1980 in *Supplement E*, and updated to account for the rapid development of this subject. Several other volumes are in the course of active preparation.

We will be grateful to readers who would call our attention to mistakes and/or omissions in this and other volumes of the series.

JERUSALEM
October 1988

SAUL PATAI
ZVI RAPPOPORT

Foreword

Nitrones, Nitronates and Nitroxides
Edited by S. Patai and Z. Rappoport

CHAPTER 1

Nitronic acids and esters

ARNOLD T. NIELSEN
Michelson Laboratory, Naval Weapons Center, China Lake, California

I. INTRODUCTION

The nitronic acids, or *aci*-nitro compounds, $R^1R^2C{=}NO_2H$, are an important group of organic acids. They may be characterized as rather unstable substances and good oxidizing agents, as are their esters. Most are relatively weak acids (pK_a^{Aci} 2–6) resembling carboxylic acids in acid strength. They are somewhat unique among organic acids of this strength in that their chemistry is closely linked with that of a stable tautomeric form, the parent nitroalkane. The nitro and *aci* forms share a common anion. This relationship, fundamental to nitronic acid chemistry, is illustrated with phenylnitromethane[1,2,3]. The equilibria of equation (1) illustrate the phenomenon of *aci*-nitro tautomerism.

$$C_6H_5CH_2NO_2 \underset{}{\overset{-H^+}{\rightleftharpoons}} C_6H_5CH{=}N\genfrac{}{}{0pt}{}{O^-}{O} \underset{}{\overset{H^+}{\rightleftharpoons}} C_6H_5CH{=}N\genfrac{}{}{0pt}{}{OH}{\to O} \quad (1)$$

B.p. 100° (8 mm)		M.p. 84°
Phenylnitromethane	Phenylmethanenitronate	Phenylmethanenitronic acid
Nitro compound	Nitronate anion	Nitronic acid (*aci* form)

The nomenclature of nitronic acids has often been a matter of discussion[4,5]. The term nitronic acid (*nitronsäure*) was introduced by Bamberger[6]. However, use of the prefix *aci* before the nitro compound name, a concept introduced by Hantzsch a few years later[7], achieved much wider use for many years. The general term *aci* was taken to mean the tautomeric, more acidic form of a *pseudo* acid. A *pseudo* acid is one whose proton is removed slowly[7]; a nitroalkane is a *pseudo* acid[8]. The prefix *iso*[9], and sometimes incorrectly *pseudo*, before the nitro compound name to indicate the nitronic acid form were widely employed for many years (*iso*nitro listings are found in *Chemical Abstracts* through 1946).

The nitronic acid naming system is now more widely accepted[5,10]. As pointed out originally by Bamberger[6], and later by Hass[5], it follows more closely the systematic naming of other organic acids and derivatives. It also clearly recognizes the identity of these substances as genuine organic acids. The naming of derivatives such as salts, esters, and anhydrides follows systematically. The *aci*-nitro term employed only as a prefix[11,12] does not readily adapt itself to naming these derivatives.

The nitronic acid function is $=NO_2H$; as a suffix it is called nitronic acid. If a prefix is required *aci*-nitro may be used[11,12]. For example, butane-2-nitronic acid = 2-*aci*-nitrobutane. However, the prefix nomenclature should be avoided with nitronic acids, as it is with carboxylic acids.

It is suggested that the two possible types of nitronic acids, $RCH=NO_2H$ and $R^1R^2CH=NO_2H$, derived from primary and secondary nitroalkanes, be designated *primary* and *secondary* nitronic acids, respectively. Some examples of nitronic acid nomenclature follow.

$CH_3CH=NO_2H$	Ethanenitronic acid (*primary*)
$(CH_3)_2C=NO_2H$	Propane-2-nitronic acid (*secondary*)
$n\text{-}C_3H_7CH=NO_2^-Na^+$	Sodium butane-1-nitronate
$C_6H_5CH=NO_2CH_3$	Methyl phenylmethanenitronate
(cyclohexane ring)$=NO_2Et$	Ethyl cyclohexanenitronate
$CH_3CH=N(\rightarrow O)OCOCH_3$	Acetic ethanenitronic anhydride
$CH_2=N(\rightarrow O)Cl$	Methanenitronyl chloride

The first preparation of a nitronic acid was apparently made by Konowalow (1893)[13] who isolated diphenylmethanenitronic acid, mp 90°, and described it as a very unstable substance, decomposing at room temperature. The phenomenon of *aci*-nitro tautomerism in solution was discovered by Holleman (1895)[14], who observed conductometrically the isomerization of 3-nitrophenylmethanenitronic acid into the nitro form. Hantzsch (1896)[1] was first to prepare both forms of a single nitro compound (phenylnitromethane and phenylmethanenitronic acid) and to recognize their tautomeric relationship.

The first preparation of a nitronic ester appears to be that of Nef (1894)[15], who synthesized, isolated, and recognized the unstable substance $H_2NCOC(CN)=NO_2C_2H_5$.

The question of nitronic acid structure was incompletely resolved and a subject of some debate and discussion for nearly 50 years after the discovery of these substances. The now accepted structure **1** was originally proposed by Nef[15] and employed by Bamberger[6]. An alternate oxazirane structure (**2**) was proposed by Hantzsch[1].

Reports that certain nitronate salts possess optical activity[16,17],

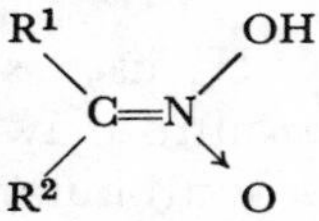

$$\begin{array}{ccc} R^1 & & OH \\ & C{=}N & \\ R^2 & & \searrow O \end{array} \qquad \begin{array}{ccc} R^1 & & \\ & C\text{——}NOH & \\ R^2 & O & \end{array}$$

(1) 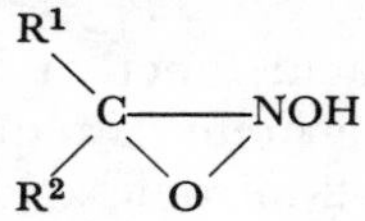 (2)

supporting structure **2**, were later shown to be in error[18–20]. Nitronate salts prepared from optically active nitroalkanes are, in fact, optically inactive[19]. The strong $\pi - \pi^*$ ultraviolet absorption of nitronic acids, esters, and salts supports structure **1**. Oxaziranes do not exhibit strong ultraviolet absorption[21]. Unsuccessful attempts have been made to prepare substances having structure **3**, with substituents directly attached to nitrogen[22].

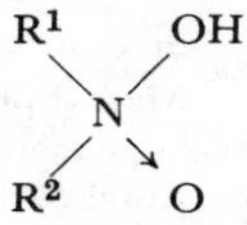

$$\begin{array}{ccc} R^1 & & OH \\ & N & \\ R^2 & & \searrow O \end{array}$$

(3)

Cis-trans isomerism might be observed, at least in the solid state, with unsymmetrically substituted nitronic acids, analogous to the *syn* and *anti* forms of oximes. This type of isomerism has been demonstrated for nitronic esters[23], nitrones[24], and oximes[25,26], but not for nitronic acids. It is possible that the rather stable unsymmetrical nitronic acids of wide melting range described by Hodge[27] are mixtures of *cis* and *trans* isomers. In solution in protic solvents such isomers would, of course, lose their identity rather rapidly.

II. *ACI*-NITRO TAUTOMERISM

A. Introduction

Study of the phenomenon of *aci*-nitro tautomerism has been, and remains, important to the development of acid-base catalysis and proton-transfer theory. The process in neutral or basic medium consists principally of equilibria involving nitronic acid (*aci* form), nitroalkane, and a common nitronate anion (equations 2, 3).

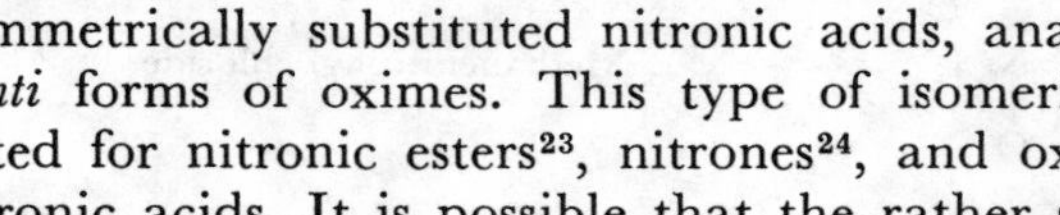

$$B + \begin{array}{ccc} R^1 & & OH \\ & C{=}N & \\ R^2 & & \searrow O \end{array} \underset{k_{-1}}{\overset{k_1}{\rightleftharpoons}} \begin{array}{ccc} R^1 & & O^- \\ & C{=}N & \\ R^2 & & O \end{array} + BH^+ \qquad (2)$$

Nitronic acid (*aci*) Nitronate

$$\mathrm{BH^+} + \begin{matrix} R^1 \\ \diagdown \\ C{=}N \\ \diagup \\ R^2 \end{matrix} \begin{matrix} O^- \\ \\ \\ \\ O \end{matrix} \underset{k_2}{\overset{k_{-2}}{\rightleftharpoons}} \begin{matrix} R^1 \\ \diagdown \\ CHNO_2 \\ \diagup \\ R^2 \end{matrix} + \mathrm{B} \qquad (3)$$

Nitronate Nitro

The acid-base equilibria of these equations define, to a close approximation, the ionization constants of nitronic acids ($K_a^{Aci} = k_1/k_{-1}$) and nitroalkanes ($K_a^{\mathrm{Nitro}} = k_2/k_{-2}$) where B is a solvent molecule (water, ethanol). Tautomerization involving ketonitronic acids is discussed in section III.C.4.

Many kinetic studies of *aci*-nitro tautomerism, employing diverse methods, have been made. The earliest studies[14,28,34], which employed conductivity measurements, made use of the much greater conductivity of the *aci* form, due to its greater dissociation into nitronate anion. Other methods have taken advantage of various properties of the *aci* form not observed in the nitro form, such as: rapid reaction with bromine (titration)[35–37], failure to be reduced polarographically[38–49], and strong ultraviolet absorption[50–53]. These studies have included rate measurements of forward and reverse processes.

In retrospect, an interesting aspect of many rate studies was the failure, even until recent times, to recognize the hybrid structure of the nitronate anion intermediate[29,30,38]. Confusion exists in many earlier papers relating to nitronate anions having different 'structures' (with negative charge on carbon or oxygen[38], or possessing optical activity[16]), and giving validity to these structures in kinetic and mechanistic expressions[38]. Present theory allows only *one* structure for the mesomeric nitronate anion, with negative charge delocalized principally on oxygen.

B. *Tautomerization of Nitronic Acids to Nitroalkanes*

The tautomerization of a nitronic acid to its parent nitroalkane (equations 2 and 3) proceeds essentially to completion for most simple nitroalkanes because of the relatively weaker acidity of a nitroalkane compared to its corresponding nitronic acid (*aci* form). The kinetics of tautomerization of the *aci* to the nitro form has been studied extensively, principally in water solvent. Early conductometric studies of Holleman[14,54] and Hantzsch[32,33] showed the process to be very rapid.

Proton removal from nitronic acid oxygen (k_1) has been measured for phenylmethanenitronic acid ($k_1 = 4.14 \times 10^{-5}$ l/mole-sec in

99.5% water, 0.5% ethanol at 25°). From the ionization constant in water, $K_a^{Aci} = 1.3 \times 10^{-4}$ [34,39,55], the reverse process, protonation on oxygen, may be calculated to be much faster ($k_{-1} = 3.2 \times 10^{-1}$ l/mole-sec).

Many more measurements have been made of the overall rate of tautomerization of the nitronic acid to the nitro form (protonation on carbon). One may assume a steady-state expression and define this rate as $K_a^{Aci}\ k_{-2}$ since k_{-2} is slow compared to k_{-1}. Table 1 summarizes the available rate data. From the K_a^{Aci} values the rate of protonation on carbon, k_{-2}, may be calculated.

The mechanism of the tautomerization process expressed in equations (2) and (3) requires that C-protonation occurs on the intermediate nitronate anion, not on the nitronic acid or some other species. In agreement with this postulate is the fact that the tautomerization rate is accelerated in slightly basic solution, inhibited in acid solution[51,65]. Strong nitronic acids which are highly ionized, such as bromomethanenitronic, tautomerize extremely rapidly[32,47]. Highly hindered acids of the type $R_3CC(C_6H_5){=}NO_2H$ do not tautomerize at a measurable rate in acid solution, but require a basic catalyst to increase the concentration of nitronate ion, thus permitting tautomerization to occur readily[65].

Tautomerization does occur in acid solution[44], but strong acid suppresses ionization and usually favors other reactions such as the Nef. Armand[47] found acids $XCH{=}NO_2H$, $X_2C{=}NO_2H$, and $CH_3CX{=}NO_2H$ (X = Cl, Br) to tautomerize rapidly and completely without undergoing Nef reaction, even at low pH. On the other hand, with *d*-2-nitrooctane acid-catalyzed Nef reaction (in *N* hydrochloric acid at 100°) to form 2-octanone occurred more rapidly than tautomerization since it was observed that the recovered unreacted nitro compound retained all its optical activity[50].

A different mechanism involving direct proton transfer to carbon from a nitronic acid intermediate (rather than a nitronate anion) is involved in the much-studied dark reaction of tautomerization of compounds such as arylmethanenitronic acids to the nitro compound (e.g. **5** → **4**)[66–72] (equation 4). Nitronic acids identical with those formed photochemically can also be formed by acidification of the alkali metal salts[73]. The presence of a nitro group *ortho* of the methylene group is required for the photochromic transformation[74–78]. The pyridyl group in **4** may be replaced by phenyl[75], alkyl[78], or hydrogen[71]. The reaction is not limited to solutions, but occurs also in the solid state where it was first observed[67,79,80].

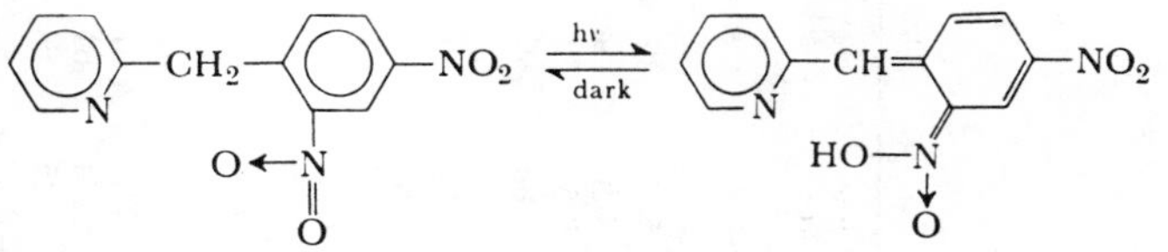

$$\text{(4)} \underset{\text{dark}}{\overset{h\nu}{\rightleftharpoons}} \text{(5)} \qquad (4)$$

(**4**) Pale yellow

(**5**) Deep blue
λ_{max}^{EtOH} 540 mμ

The direct intramolecular proton transfer mechanism involving a nitronic acid is supported by several facts, in addition to the requirement of an *ortho* nitro group. The dark reaction rate [*Aci* (**5**)→ Nitro (**4**)] is accelerated in acid solution[81]. It is also faster in an aprotic solvent such as isooctane than in a protic solvent (ethanol) by a factor of ca 10^4 [66]. The rate is strongly *accelerated* by electron-releasing groups; replacing NO_2 by NH_2 in **5** increases the rate ca 10^4 [82]. A large negative entropy (−45–50 eu) supports a rigid transition state[75,82]. The intramolecular nature of the process was demonstrated in a very simple system. Deuterium is incorporated into *o*-nitrotoluene (**6**)—but not *p*-nitrotoluene under the same conditions—when irradiated in deuterium oxide-dioxane[71,76–78]. Ionization of the nitronic acid (**7**) provides a facile deuterium exchange mechanism (equation 5).

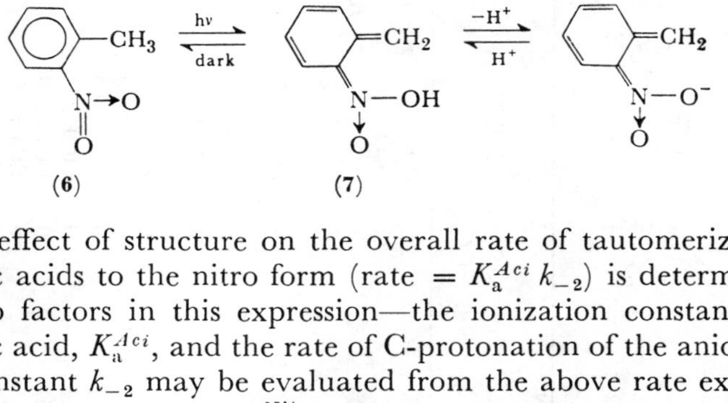

$$\text{(6)} \underset{\text{dark}}{\overset{h\nu}{\rightleftharpoons}} \text{(7)} \underset{H^+}{\overset{-H^+}{\rightleftharpoons}} \text{anion} \qquad (5)$$

(**6**) (**7**)

The effect of structure on the overall rate of tautomerization of nitronic acids to the nitro form (rate $= K_a^{Aci}\, k_{-2}$) is determined by the two factors in this expression—the ionization constant of the nitronic acid, K_a^{Aci}, and the rate of C-protonation of the anion (k_{-2}). The constant k_{-2} may be evaluated from the above rate expression if K_a^{Aci} is known, or, if K_a^{Nitro} and the uncatalyzed rate of proton removal from the nitroalkane (k_2) are known, k_{-2} may be calculated from the equilibrium expression $K_a^{\text{Nitor}} = k_2/k_{-2}$. The available data are summarized in Table 1. Structural factors which affect K_a^{Aci} and k_{-2} are germane to the general problem of anion stability.

Inductive effects may be compared by examining those nitronic acids not exhibiting large resonance or other anion-stabilizing

TABLE 1. Rates of tautomerization of nitronic acids to nitroalkanes.

$$\underset{R^2}{\overset{R^1}{>}}C{=}NO_2H \underset{k_{-1}}{\overset{k_1}{\rightleftharpoons}} \underset{R^2}{\overset{R^1}{>}}C{=}NO_2^- + H^+ \underset{k_2}{\overset{k_{-2}}{\rightleftharpoons}} \underset{R^2}{\overset{R^1}{>}}CHNO_2$$

$$K_a^{Aci} = k_1/k_{-1}$$

No.	Nitronic acid	Temp. (°C)	Solvent aqueous	$K_a^{Aci} \times 10^4$	pK_a^{Aci}	k_{-2} (l/mole-min)	log k_{-2}	$K_a^{Aci} k_{-2}$ (l/mole-min)	log $K_a^{Aci} k_{-2}$ + 2	Ref.
1.	$O_2NCH{=}NO_2H$	25	H_2O	140	1.86	1.9×10^5	5.28	2600	5.42	32, 56
2.	$CH_2{=}NO_2H$	25	H_2O	5.6	3.25	4.1×10^4	4.61	23	3.36	29–32, 56
3.	$CH_3CH{=}NO_2H$	25	H_2O	0.40	4.4	9.0×10^2	2.95	0.036	0.55	32, 56, 57
4.	$CH_3CH_2CH{=}NO_2H$	25	EtOH, 85%	0.25[a]	4.6[a]	4.5×10^3[a]	3.65[a]	0.076	0.88	36, 42, 52, 58–60
5.	$CH_3(CH_2)_5C(CH_3){=}NO_2H$	25	EtOH, 85%	(0.05)[b]	(5.3)[b]	(6×10^2)[b]	(2.8)[b]	0.0031	−0.51	50, 52
6.	$2\text{-}BrC_6H_4CH{=}NO_2H$	0	EtOH, 50%	—	—	—	—	0.042	0.62	61
7.	$3\text{-}BrC_6H_4CH{=}NO_2H$	0	EtOH, 50%	—	—	—	—	0.028	0.45	61
8.	$4\text{-}BrC_6H_4CH{=}NO_2H$	0	EtOH, 50%	—	—	—	—	0.018	0.25	61
9.	$2\text{-}ClC_6H_4CH{=}NO_2H$	0	EtOH, 50%	—	—	—	—	0.032	0.50	61
10.	$3\text{-}ClC_6H_4CH{=}NO_2H$	0	EtOH, 50%	—	—	—	—	0.025	0.40	61
11.	$4\text{-}ClC_6H_4CH{=}NO_2H$	0	EtOH, 50%	—	—	—	—	0.014	0.15	61, 62
12.	$2\text{-}O_2NC_6H_4CH{=}NO_2H$	0	EtOH, 50%	—	—	—	—	(0.15)[b]	1.18	61, 63
13.	$3\text{-}O_2NC_6H_4CH{=}NO_2H$	0	EtOH, 50%	—	—	—	—	0.092	0.96	61, 63
14.	$4\text{-}O_2NC_6H_4CH{=}NO_2H$	0	EtOH, 50%	—	—	—	—	0.11	1.04	61, 63
15.	$C_6H_5CH{=}NO_2H$	0	EtOH, 50%	—	—	—	—	0.0056	−0.25	61, 63
		0	H_2O	—	—	—	—	0.0050	−0.30	34, 36, 39,
		25	H_2O	1.3	3.9	(1×10^3)[b]	(3.0)[b]	(0.125)[b]	(1.1)[b]	53, 55, 64

16.	$O_2NCH_2CH_2CH{=}NO_2H$	25	EtOH, 85%	32	2.49	1.7×10^2	2.23	0.55	1.74	52
17.	$O_2NCH_2(CH_2)_2CH{=}NO_2H$	25	EtOH, 85%	2	3.70	4.6×10^2	2.66	0.093	0.97	52
18.	$O_2NCH_2(CH_2)_3CH{=}NO_2H$	25	EtOH, 85%	1.6	3.80	1.28×10^3	3.11	0.204	1.31	52
19.	$O_2NCH_2(CH_2)_4CH{=}NO_2H$	25	EtOH, 85%	1.4	3.85	1.14×10^3	3.06	0.159	1.20	52
20.	$O_2NCH(CH_2)_2C{=}NO_2H$ CH_2OH CH_2OH	25	EtOH, 85%	2.8	3.55	1.6×10^2	2.20	0.044	0.64	52
21.	$O_2NCH(CH_2)_3C{=}NO_2H$ CH_2OH CH_2OH	25	EtOH, 85%	0.22	4.66	2.95×10^3	3.47	0.065	0.81	52
22.	$O_2NCH(CH_2)_2C{=}NO_2H$ CH_2OCH_3 CH_2OCH_3	25	EtOH, 85%	1.0	4.00	1.64×10^3	3.22	0.164	1.21	52
23.	$HO_2N{=}CHCH_2CH{=}NO_2H$[c]	25	EtOH, 85%	112	1.95	1.28×10^3	3.11	14.5	3.16	52
24.	$HO_2N{=}CH(CH_2)_2CH{=}NO_2H$[c]	25	EtOH, 85%	5.0	3.30	1.86×10^3	3.28	0.93	1.97	52
25.	$HO_2N{=}CH(CH_2)_3CH{=}NO_2H$[c]	25	EtOH, 85%	3.5	3.46	2.91×10^3	3.46	1.02	2.01	52
26.	$HO_2N{=}CH(CH_2)_4CH{=}NO_2H$[c]	25	EtOH, 85%	2.8	3.55	1.04×10^3	3.02	0.29	1.46	52
27.	$HO_2N{=}C(CH_2)_2C{=}NO_2H$[c] CH_2OH CH_2OH	25	EtOH, 85%	7.1	3.15	3.56×10^3	3.55	2.53	2.40	52
28.	$HO_2N{=}C(CH_2)_3C{=}NO_2H$[c] CH_2OH CH_2OH	25	EtOH, 85%	0.5	4.30	9.4×10^3	3.97	0.47	1.67	52
29.	$HO_2N{=}C(CH_2)_2C{=}NO_2H$[c] CH_2OCH_3 CH_2OCH_3	25	EtOH, 85%	2.5	3.60	11.3×10^3	4.06	2.82	2.45	52

[a] Data in water (reference 42).
[b] Values in parentheses are estimated; cf. reference 56.
[c] Data not corrected for statistical factor.

effects. Of these, the strongest acids tautomerize at the fastest rate as shown in Figure 1. The acids which tautomerize at the fastest rate also have the fastest C-protonation rate (k_{-2}) as shown in

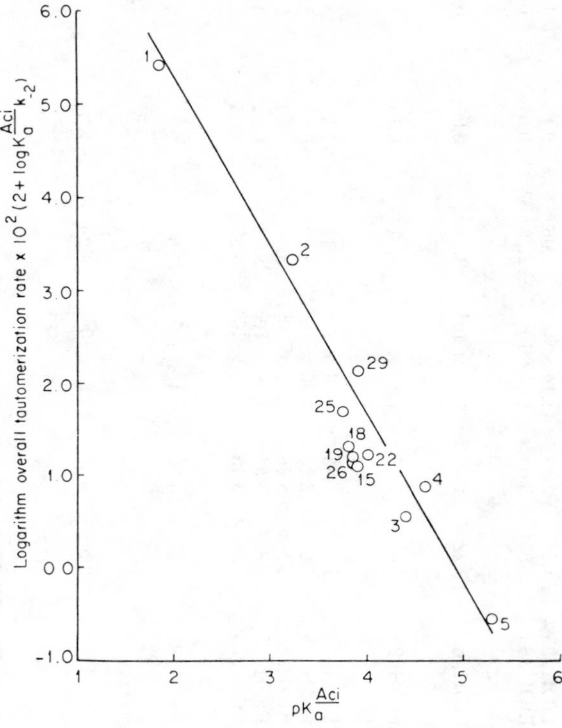

FIGURE 1. Plot of logarithm of overall rate of tautomerization of nitronic acids to nitroalkanes $\times 10^2$ ($2 + \log K_a^{Aci} k_{-2}$) vs. pK_a^{Aci}; data at 25° in Table 1. Data for bisnitronic acids have been corrected for statistical factor.

Figure 2. Hantzsch[32] reported bromomethanenitronic and nitromethanenitronic acids to tautomerize at rates much too rapid to measure compared to methanenitronic, the weaker acid. Armand[47] observed only instantaneous tautomerization, even at low pH (no Nef reaction), with the strong acids $X_2C{=}NO_2H$, $XCH{=}NO_2H$ and $CH_3CX{=}NO_2H$ (X = Cl, Br). *m*-Nitrophenylmethanenitronic acid tautomerizes to the nitro form ca 20 times faster than the weaker phenylmethanenitronic acid.

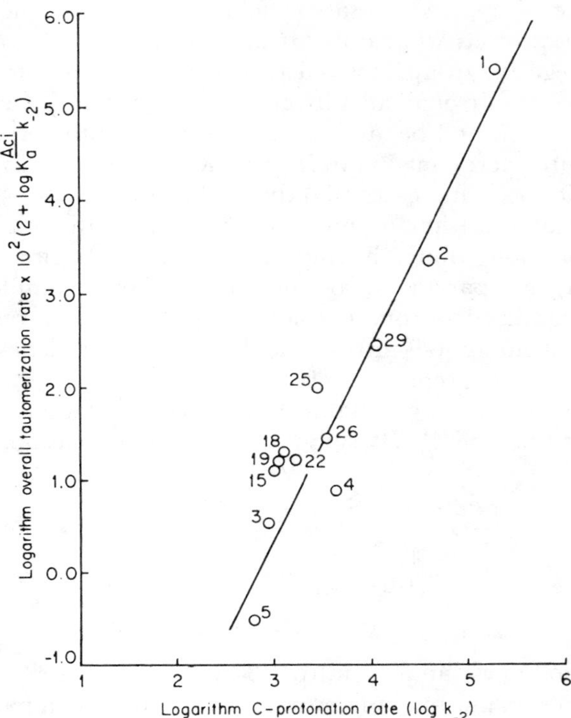

FIGURE 2. Plot of logarithm of overall rate of tautomerization of nitronic acids to nitroalkanes × 10^2 $(2 + \log K_a^{Aci}\ k_{-2})$ vs. logarithm of C-protonation rate ($\log k_{-2}$;) data at 25° in Table 1.

Inductive effects exhibit predictable behavior in relation to ionization constants, which affect tautomerization rates. Electron-withdrawing groups such as halogen and nitro increase K_a^{Aci} (see Table 5 in section II.D), and tautomerization rate. Electron-releasing groups such as methyl decrease K_a^{Aci} and tautomerization rate. For example, ethanenitronic acid tautomerizes ca 10^3 times more slowly than the stronger methanenitronic acid; α-phenylpropanenitronic acid tautomerizes more slowly than α-phenylethanenitronic acid[83].

On the other hand, the increase in C-protonation rate caused by substitution of electronegative groups on the nitronate carbon (Figure 2) is opposite to what one might expect. An electron-withdrawing group lowers electron density at the nitronate carbon and should slow C-protonation. One possible explanation may be the presence of a polar, negative field very close to the nitronate carbon; thus protons are readily attracted to the site. However, when the electronegative group is moved away from the nitronate carbon, as in the series of ω-nitronalkanenitronic acids $O_2N(CH_2)_nCH{=}NO_2H$, one observes expected behavior[52]. The overall tautomerization rate decreases with increasing chain length; K_a^{Aci} decreases also, and the C-protonation rate increases slightly (Table 1).

Slow tautomerization of nitronate anions to the nitro form is observed for those anions having ground state energies very much lower than the parent nitro compound. For example, certain resonance-stabilized nitronate anions may show this property. The nitro forms of fluorene-9-nitronic acid (**8**)[84,85] and indene-1-nitronic acid (**9**)[86] can be prepared only in aprotic solvents[85a]. Fluorene-9-nitronic acid (**8**) is very stable[84]. 2,4-Cyclopentadiene-1-nitronic acid is very unstable[86a]. Its sodium salt on acidification, produces

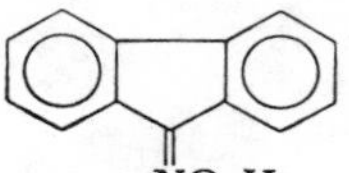

NO_2H NO_2H

(**8**) 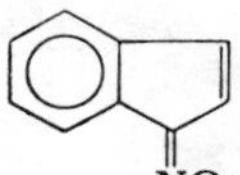 (**9**)

a black polymer and 1-nitro-1,3-cyclopentadiene[86a]. Phenylmethanenitronic and methanenitronic acids are of comparable acid strength (pK_a^{Aci} 3.9 and 3.25, respectively), but phenylmethanenitronic acid tautomerizes approximately 100 times more slowly (Table 1). Resonance-stabilized nitronic acids which tautomerize slowly do so principally because the C-protonation rate is relatively slow. There is a rather large activation energy barrier between the anion and the nitro form.

Resonance stabilization of a nitronate anion often increases acid strength which may sometimes actually account for a slight enhancement of overall tautomerization rate. For example, a 20% increase in tautomerization rate is observed for 4-nitrophenylmethanenitronic acid relative to the 3-nitro isomer[61]; the 4-bromo and 4-chloro isomers tautomerize more slowly than their corresponding *meta* isomers[61]. However, the most usual consequence of significant nitronate anion stabilization relative to the nitro form appears to be a decrease, rather than an increase, in overall tautomerization rate.

Stabilization of nitronate anions by hydrogen bonding also results in an increase in K_a^{Aci} and a decrease in C-protonation rate. The methylol derivative **10** is a somewhat stronger acid than the corresponding methyl ether (**11**)[52]. But, the *overall* tautomerization rate ($K_a^{Aci}\,k_{-2}$) of the hydrogen-bonded acid (**10**) is about 4 times slower, despite the greater acidity. This is due to the slower C-protonation

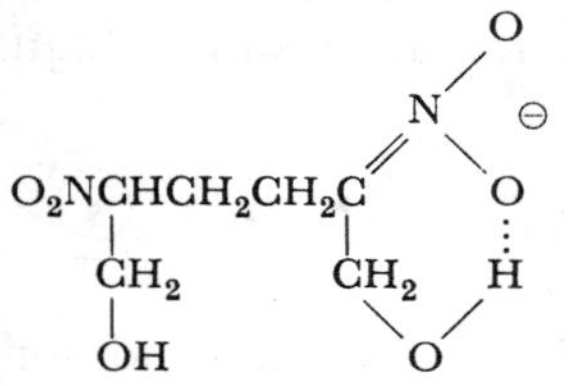

(**10**)

pK_a^{Aci} 3.55

$k_{-2} = 1.6 \times 10^{-2}\ \text{min}^{-1}$

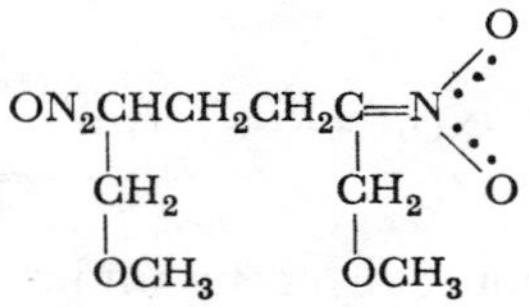

(**11**)

pK_a^{Aci} 4.00

$k_{-2} = 16.4 \times 10^{-2}\ \text{min}^{-1}$

rate (10 times slower). A similar effect is observed with the butane-1,4-bisnitronic acid derivatives of **10** and **11** (data in Table 1). The failure of certain hydroxynitronic acids (including **10**) to undergo a Nef reaction may be caused by their hydrogen-bonded, stabilized nitronate anions (see discussion in section III.C.1.a).

Rapid tautomerization of nitronate anions to the nitro form is observed with anions which do not differ much in ground state energy from the parent nitro compound. The nitronic acid forms are extremely unstable and are rarely isolated. Conjugated anions derived from conjugated nitroolefins, nitrodienes, and nitro aromatic compounds protonate rapidly. C-protonation usually occurs most rapidly at the terminal position of these nitronate anions and often favors the conjugated nitroolefin product under kinetic control. For example, anion **12** rapidly protonates at the terminal 3-position to yield 1-nitrocyclohexene (**13**) exclusively (equation 6) apparently without intervention of non-conjugated 3-nitrocyclohexene[58]; this protonation is very rapid and occurs at about the same rate as that for propane-1-nitronate to 1-nitropropane[52,58]. Conjugated nitronate anion **14** protonates at the terminal 5-position to yield only nitroolefin **15**[58] (equation 7).

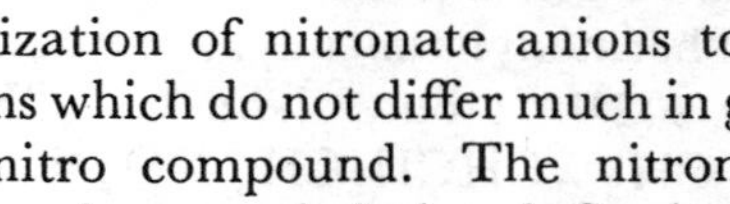

$$\mathbf{12}\ (=NO_2^-) \xrightarrow{H^+} \mathbf{13}\ (NO_2) \qquad (6)$$

(**12**) (**13**)

$$CH_2{=}CHCH{=}CHC(CH_3){=}NO_2^- \xrightarrow{H^+} CH_3CH{=}CHCH{=}C(CH_3)NO_2 \qquad (7)$$

(14) (15)

Protonation of nitronates derived from aromatic nitro compounds occurs very rapidly, also at a terminal position (equation 8).

$$^-O{-}C_6H_4{-}NO_2 \longleftrightarrow O{=}C_6H_4{=}NO_2^- \xrightarrow{H^+} HO{-}C_6H_4{-}NO_2 \qquad (8)$$

Protonation can occur under kinetic control to yield some of the unconjugated nitroolefin, as in the protonation of nitronate anion **16**; a 1:1 mixture of olefins **17** and **18** is produced[10] (equation 9).

$$CH_2{=}C(CH_3)CH{=}NO_2^- \xrightarrow{H^+} CH_2{=}C(CH_3)CH_2NO_2 + (CH_3)_2C{=}CHNO_2 \qquad (9)$$

(16) (17) (18)

The equilibrium composition of nitroolefins is a matter of interest and has received some study[10,87–91]. As with olefinic ketones and other unsaturated carbonyl compounds the α,β-unsaturated isomer is usually favored, and the position of equilibrium is affected by structure and solvent[87]. Nitrobenzenes and olefins such as **13** and **15** are favored. Thermodynamically, conjugated isomer **18** is favored over **17** by 4:1[10,87]; but, olefin $CH_3CH{=}C(CH_3)NO_2$ is favored 100% over $CH_2{=}CHCH(CH_3)NO_2$ [87].

Exceptional behavior is exhibited by certain olefins substituted with a nitromethyl group, CH_2NO_2 (note **17**, above). The β,γ-isomer is often favored at equilibrium as, for example, with olefins **19**[10,88], **20**[89,90], and **21**[91]. The product composition of nitroolefins

1-(nitromethyl)cyclohexene (19) $\qquad (O_2NCH_2)(CH_2{=})C{-}C({=}CH_2)(CH_2NO_2)$ (20) $\qquad CH_2{=}CHCH_2NO_2$ (21)

derived from the nitronate anions of **19**, **20**, and **21** under kinetic control is not known.

Slow tautomerization is observed with certain nitronic acids highly substituted about the nitronate carbon. Quantitative data are limited. Octane-2-nitronic acid tautomerizes 25 times more slowly than propane-1-nitronic acid[52]. Nitronic acids such as **22**[65] tautomerize much more slowly than phenylmethanenitronic acid[1].

These results indicate that a slow rate of C-protonation of the

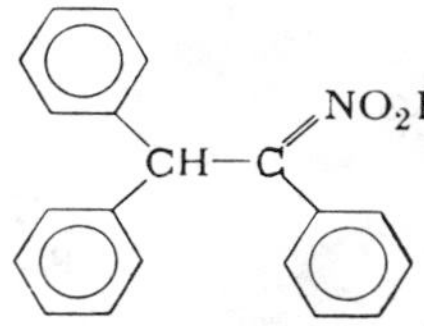

(22)

nitronate anion is believed to be a factor, a result of poor solvation about the nitronate carbon.

The stereochemistry of *aci*-nitro tautomerism has been studied in a few systems. In simple monocyclic nitronate anions with vicinal substituents one observes kinetic preference for protonation from the least hindered side; this result leads to the least stable product (steric approach control). Protonation of 2-phenylcyclohexanenitronate ion (**23**) leads to 98% *cis*-2-phenyl-1-nitrocyclohexane

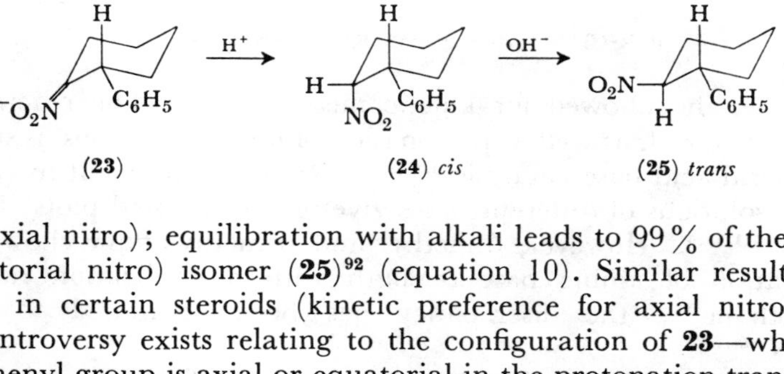

$$\textbf{(23)} \xrightarrow{H^+} \textbf{(24)}\ cis \xrightarrow{OH^-} \textbf{(25)}\ trans \qquad (10)$$

(**24**, axial nitro); equilibration with alkali leads to 99% of the *trans* (equatorial nitro) isomer (**25**)[92] (equation 10). Similar results are found in certain steroids (kinetic preference for axial nitro)[93,94]. (A controversy exists relating to the configuration of **23**—whether the phenyl group is axial or equatorial in the protonation transition state[92,95,96].) In cyclohexane systems, in the absence of vicinal substituents, there is a kinetic[92a] as well as equilibrium preference for equatorial nitro (ca 80–90%) over axial[93,94,97–104]. For example, 4-*t*-butylcyclohexanenitronate protonates under kinetic control to yield 76% *trans*-4-*t*-butylnitrocyclohexane (equatorial nitro)[92a]. Equilibration of 1,3- and 1,4-dinitrocyclohexanes in ethanolic sodium bicarbonate leads to the diequatorial isomers (ca 80–90%)[103].

In the rigid bicyclic system **26**, protonation with dilute acetic acid occurs with 84–97% kinetic preference for the most stable, *trans* product (**27**)[105,106] (equation 11). Proton approach is *cis* to the vicinal R group (product development control). In this more rigid system, eclipsing of incipient nitro and neighboring R group results in a higher energy transition state than in the more flexible cyclohexane ring system.

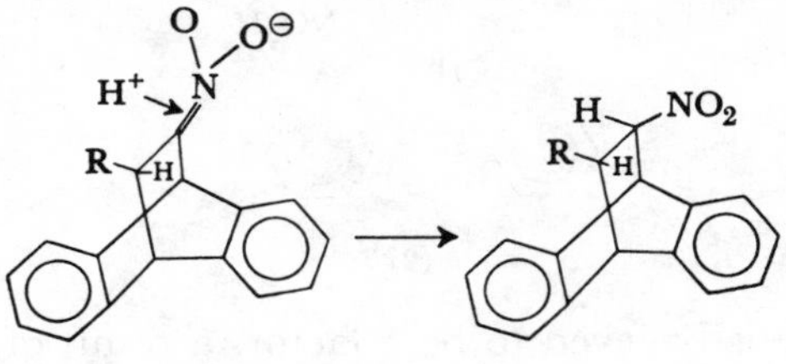

(11)

(**26**) R = CH_3, C_6H_5 (**27**) *trans*

C. Proton Removal from Nitroalkanes

Removal of a proton from a nitroalkane produces a nitronate anion (reverse of equation 3) (equation 12). This reaction occurs in the reverse of *aci*-nitro tautomerism and has been studied extensively; it proceeds at a convenient rate at ordinary temperatures

$$R^1R^2CHNO_2 + B \underset{k_{-2}}{\overset{k_2}{\rightleftharpoons}} R^1R^2C{=}NO_2^- + BH^+ \qquad (12)$$

and is easily followed for kinetic measurements. The reaction of proton removal, as well as protonation of nitronate anions, is subject to general acid-base catalysis[40,107–110]. Rate measurement in various buffer solutions of different bases gives good Brönsted plots (Figure 3)[40,109,111,111a]. However, dimethyl and trimethylamine show poor correlation of amine base strength with neutralization rates of nitroethane[112]; the discrepancy has been attributed to steric factors[111a,112].

The rates of uncatalyzed proton removal from various nitroalkanes determined in water solvent are summarized in Table 2. Also included are rates of reprotonation (C-protonation) and ionization constants (K_a^{Nitro}). As noted by Bell (reference 107, p. 160) an expected linear relationship is observed between acid strength of the nitroalkane and rate of proton removal (Figure 4). The nitroalkanes of greatest acidity ionize most rapidly.

The energy of activation for proton removal with water as a base is ca. 20–23 kcal/mole[57,119]. A relatively large negative entropy of activation is observed (−19–24 eu for nitroalkanes[28,42,57,112,119]). For other bases (amines, hydroxide, and acetate ion) the energy of activation is less (12–16 kcal/mole) and the entropy of activation

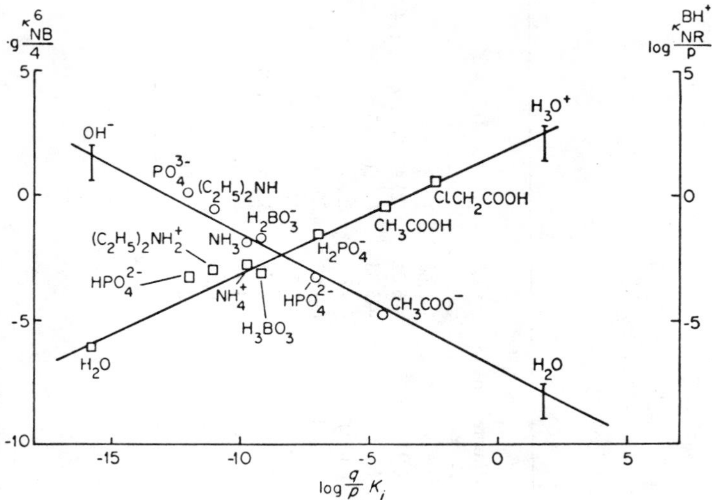

FIGURE 3. Dissociation and recombination rates of the nitro-form of 1-nitropropane as a function of the dissociation constant of the acting acid or of the acid conjugated with the acting base: □ recombination of nitro-form, ○ dissociation of the nitro-form. [Reproduced, by permission, from ref. 40.]

more positive (−7–16 eu)[92a,111,112,120]. These data agree with a slow proton removal process requiring a large amount of solvent reorganization in the transition state[121].

Base-catalyzed proton removal from nitroalkanes is an important reaction (equation 13). In kinetic studies solutions of sodium hydroxide in water, aqueous dioxane or aqueous ethanol have been fre-

$$R^1R^2CHNO_2 + OH^- \xrightarrow{k_3} R^1R^2C{=}NO_2^- + H_2O \quad (13)$$

quently employed. Reaction rate data are summarized in Table 3. Hantzsch[32] made the first studies with nitroethane employing a conductometric method.

Junell[37,130–132], and later Pedersen[133], made the first thorough kinetic investigations of proton removal from nitroalkanes employing acetate catalyst and bromine titration of the nitronate anion formed. They demonstrated that the bromine and chlorine reaction rates are identical, that the rate-limiting step is proton removal, and that the reaction is first order in nitro compound and base. Pearson and Dillon showed bromine and iodination rates to be identical[57].

Deuterium substitution in nitroalkanes results in a slower reaction rate; for hydroxide ion an isotope effect of 7.4–10.3 is observed[111,120].

TABLE 2. Rates of ionization of nitroalkanes in water at 25°.

$$B + \underset{R^2}{\overset{R^1}{>}}CHNO_2 \underset{k_{-2}}{\overset{k_2}{\rightleftharpoons}} \underset{R^2}{\overset{R^1}{>}}C{=}NO_2 + BH^+$$

Nitroalkane	$K_a^{Nitro}(k_2/k_{-2})$	pK_a^{Nitro}	k_{-2} (l/mole-min)	k_2 (l/mole-min)	$\log k_2$	Ref.
CH_3NO_2	6.1×10^{-11}	10.21	4.1×10^4	2.5×10^{-6}	−5.60	28, 37, 56, 59, 113
$CH_3CH_2NO_2$	3.2×10^{-9}	8.5	9.0×10^2	2.9×10^{-6}	−5.54	32, 47, 56, 59, 60, 114
$CH_3CH_2CH_2NO_2$	1.05×10^{-9}	8.98	4.5×10^3	4.7×10^{-6}	−5.33	42, 60, 115
$O_2N(CH_2)_6NO_2$	$(10^{-10})^a$	$(10.3)^b$	1.0×10^3	(10^{-7})	(−7)	52, 116
$CH_3(CH_2)_5CH(CH_3)NO_2$	$(10^{-10})^a$	(10)	(6×10^2)	(6×10^{-8})	(−7.2)	50, 52, 116
CH_2BrNO_2	5×10^{-9}	8.3	1.6×10^5	8×10^{-4}	−3.1	37, 47, 56
$CH_2(NO_2)_2$	2.7×10^{-4}	3.57	1.9×10^5	50	1.7	32, 56, 117, 118
$CH_3CHClNO_2$	1.3×10^{-7}	6.9	2.6×10^2	3.4×10^{-5}	−4.47	47, 56
$CH_3COCH_2NO_2$	2.3×10^{-5}	4.64	8.0×10^6	2.2	0.35	56
$C_6H_5CH_2NO_2$	1.6×10^{-7}	6.8	1.9×10^4	3×10^{-3}	−2.5	34, 38, 39, 53, 55

[a] Values in parentheses are estimated from available data in references cited.

[b] Corrected for statistical factor.

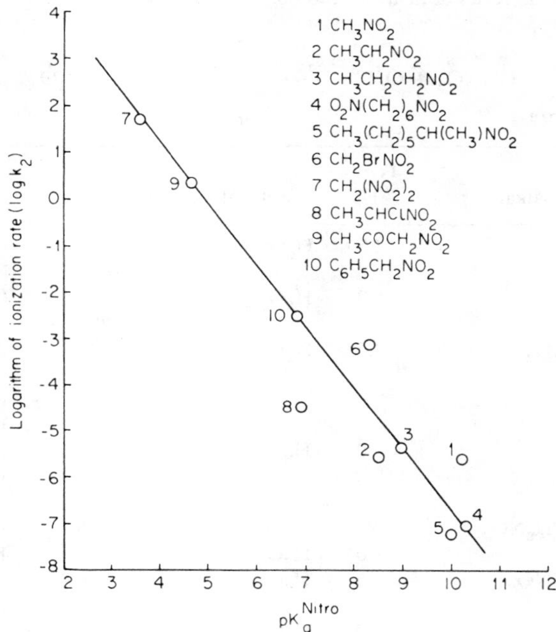

FIGURE 4. Plot of logarithm of ionization rate ($\log k_2$) of nitroalkanes in water at 25° vs. pK_a^{Nitro}; data at 25° in Table 2.

The isotope effect varies with the pK_a difference of the reacting systems[111,134–136]. The rate of neutralization of nitroalkanes is faster in D_2O[127] and in aprotic solvents[121,126] than in water. Hindrance in the attacking base (e.g. collidine) results in a relatively slower rate of proton removal[111,112,135–138].

The effect of structure on rates of proton removal may be seen from the data in Tables 2 and 3, and Figure 4. It has been pointed out by Bordwell[92a] that the transition state for proton abstraction resembles the ground state rather than nitronate ion; inductive and steric effects are the important factors which affect rate of proton removal. Thus electron-withdrawing groups such as nitro enhance the rate[32,57,139]. Rates of neutralization of substituted phenylnitromethanes indicate a 50-fold rate acceleration by *p*-nitro (relative to hydrogen)[120]. The substitution by electron-releasing *p*-methyl decreases the rate by one-half[120]. Bulky substituents about the acidic proton decrease the rate; 2-nitrobutane reacts ca. 25 times more

TABLE 3. Rates of neutralization of nitroalkanes by hydroxide ion.

$$R^1R^2CHNO_2 + OH^- \xrightarrow{k_3} R^1R^2C{=}NO_2^- + H_2O$$

No.	Nitroalkane	Temp. (°C)	Solvent	k_3 (l/mole-min)	Ref.
1.	CH_3NO_2	25	H_2O	1026	110, 122
				1600	111
		0	H_2O	173	123
				237	28
2.	$CH_3CH_2NO_2$	25	H_2O	236	122
				312	111
				336	124
				354	110, 125, 126
		0	H_2O	35	123
				37.5	127
				39	28
3.	$CH_3CH_2CH_2NO_2$	25	H_2O	195	110, 122
		0	H_2O	29	123
4.	$(CH_5)_2CHNO_2$	25	H_2O	16.4	110, 122
				19	111
		0	H_2O	2	28, 123
		25	CH_3OH^a	19.2	92a
5.	n-$C_3H_7CH_2NO_2$	25	H_2O	192	122
6.	$CH_3CH_2CH(CH_3)NO_2$	25	H_2O	8.8	122
7.	$C_6H_5CH_2CH(CH_3)NO_2$	25	CH_3OH^a	27.6	92a
8.	cyclopropyl-NO_2	0	1:1 Dioxane-H_2O	No reaction	120
9.	cyclobutyl-NO_2	0	1:1 Dioxane-H_2O	165	120
10.	cyclopentyl-NO_2	0	1:1 Dioxane-H_2O	39.8	120
11.	cyclohexyl-NO_2	28	1:1 Dioxane-H_2O	78.3	120
		0	1:1 Dioxane-H_2O	7.62	120
		25	H_2O	21.4	111, 128
		25	CH_3OH^a	19.8	92a
12.	t-C_4H_9-cyclohexyl-NO_2	25	CH_3OH^a	*cis* 60 *trans* 12	92a
13.	2-(C_6H_4-4-Cl)cyclohexyl-NO_2	25	CH_3OH^a	*cis* 114 *trans* 0.56	92a

(*continued*)

TABLE 3—*continued*

No.	Nitroalkane	Temp. (°C)	Solvent	k_3 (l/mole-min)	Ref.
14.	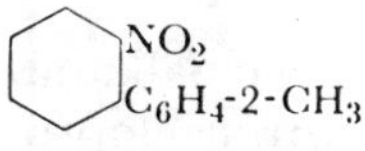 NO_2, C_6H_4-2-CH_3	25	CH_3OH[a]	*cis* 60 *trans* 0.28	92a
15.	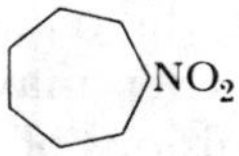 NO_2	0	1:1 Dioxane-H_2O	20.6	120
16.	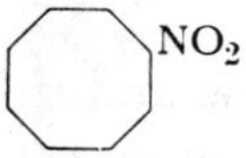 NO_2	0	1:1 Dioxane-H_2O	16.0	120
17.	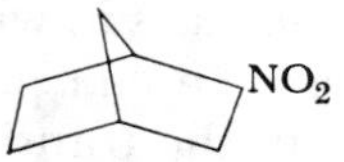 NO_2	0	1:1 Dioxane-H_2O	*endo* 134 *exo* 6.8	120
18.	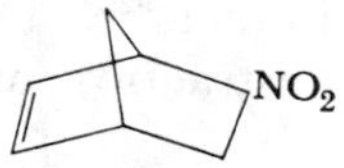 NO_2	0	1:1 Dioxane-H_2O	*endo* 210 *exo* 7.86	120
19.	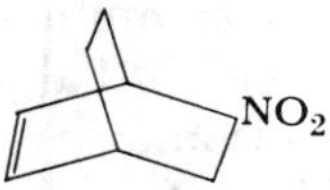 NO_2	0	1:1 Dioxane-H_2O	*endo* 50.6 *exo* 47.2	120
20.	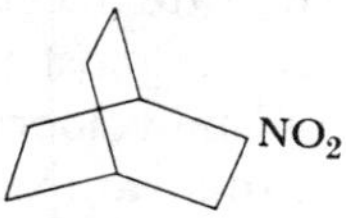 NO_2	0	1:1 Dioxane-H_2O	37.9	120
21.	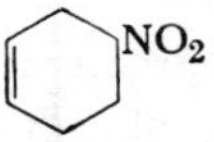 NO_2	0	1:1 Dioxane-H_2O	21.4	120
22.	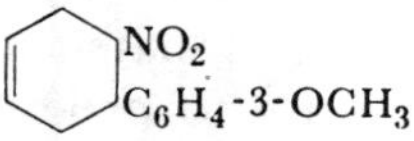 NO_2, C_6H_4-3-OCH_3	25	EtOH, 37%	(0.022)[b,c]	129
23.	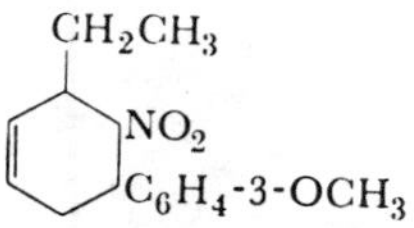 CH_2CH_3, NO_2, C_6H_4-3-OCH_3	25	EtOH, 37%	0.0069[b]	129
24.	3-$O_2NC_6H_4CH_2NO_2$	0	1:1 Dioxane-H_2O	2110	120
25.	4-$O_2NC_6H_4CH_2NO_2$	0	1:1 Dioxane-H_2O	4560	120
26.	$C_6H_5CH_2NO_2$	0	1:1 Dioxane-H_2O	92.8	120
27.	4-$CH_3C_6H_4CH_2NO_2$	0	1:1 Dioxane-H_2O	51.9	120

[a] Sodium methoxide base.

[b] Stereochemistry not established.

[c] Estimated from data in 27% EtOH reported in ref. 129.

slowly than nitroethane[122]. 3-Ethyl-5-(3-methoxyphenyl)-4-nitrocyclohexene[129] (Table 3, no. 23) is extremely unreactive compared to the less substituted parent 4-nitrocyclohexene itself (no. 21); stereochemistry, undetermined, is also important here.[92a].

The rate of proton removal from C_3–C_8 nitrocycloalkanes (in 1:1 dioxane-water at 0°) has been studied[120]. Nitrocyclopropane does not react, whereas nitrocyclobutane reacts fastest; the reaction rate order is: $C_4 > C_5 > C_7 > C_8 > C_6 \gg C_3$. The order may be explained in terms of relative ring strain and steric repulsions of ring hydrogens in the reactant.

The stereochemistry of proton removal has been examined for substituted nitrocyclohexanes[92a]. The *cis* isomers react more rapidly than the *trans*. This effect is large in 2-arylnitrocyclohexanes in which the *cis* isomers react ca. 200 times faster. The rate enhancement effect has been ascribed in part to a relief of strain in the transition state proceeding from the axial nitro in the *cis*-isomers[92a]. There appears also to be a slight steric preference for abstraction of an equatorial hydrogen.

The stereochemistry of proton removal has been examined for nitrobicycloheptanes, -heptenes, -octanes, and -octenes (Table 3, nos. 17–20)[120]. Proton removal is fastest for the *endo* isomers, where attack occurs at the least hindered *exo* hydrogen. The rate acceleration for the *endo* isomers relative to *exo* is greatest for the most rigid bicycloheptene compound (26-fold). In the less rigid bicyclooctene series (no. 19) both isomers react at essentially the same rate which is nearly that observed with nitrobicyclooctane (no. 20).

Proton removal from nitroalkanes is catalyzed by acids[37,50,51,133,140]. This process resembles the acid-catalyzed enolization of aldehydes and ketones. The mechanism probably involves a protonated intermediate which is attacked by solvent or other base to produce a

$$H^+ + (R^1)(R^2)CHNO_2 \rightleftharpoons (R^1)(R^2)CHNO_2H^+ \tag{14}$$

$$H_2O + (R^1)(R^2)CHNO_2H^+ \rightleftharpoons (R^1)(R^2)C{=}NO_2H + H_3O^+ \quad \text{slow} \tag{15}$$

$$(R^1)(R^2)C{=}NO_2H \longrightarrow \text{Products} \tag{16}$$

nitronic acid directly (equations 14, 15)[51,140a]. For acid-catalyzed halogenations of nitroalkanes the rate is independent of the halogen employed, or halogen concentration, and is first order in nitro compound[37]. The proton removal step (equation 15) is slow compared to the oxygen protonation equilibrium (equation 14) and reaction of nitronic acid to form products (equation 16). The proton removal step is quite slow compared to the base-catalyzed process. Its rate is close to the auto-dissociation rate in water (Table 2). Table 4 summarizes the rate data obtained in *N* hydrochloric acid solution.

TABLE 4. Rates of acid-catalyzed reactions of nitro compounds in *N* hydrochloric acid[51]

Compound	Reaction	Temp. (°C)	Pseudo first order rate constant $\min^{-1} \times 10^5$
Nitromethane	Bromination[37]	60	14
2-Nitropropane	Bromination[37]	60	3.0
2,5-Dinitro-1,6-hexanediol (**28**)	Isomerization[51]	60	20[a]
2,5-Dinitro-1,6-hexanediol (**28**)	Isomerization[51]	100	200[a]
2-Nitrooctane	Hydrolysis[50]	100	1.2[b]
Nitromethane	Bromination[37]	35	0.9
Bromonitromethane	Bromination[37]	35	240
Dibromonitromethane	Bromination[37]	35	12,000

[a] Value corrected for reaction at one asymmetric center.
[b] Solvent "50%" ethanol, *N* hydrogen chloride.

Many reactions of nitroalkanes in acid solution require an initial acid-catalyzed proton removal to produce a nitronic acid. Examples are formation of carboxylic[140a,141–143] and hydroxamic acids[153] from primary nitroalkanes, and the acid-catalyzed Nef reaction[52,142]. The epimerization of low-melting 2,5-dinitro-1,6-hexanediol (**28**) to the high-melting form (**29**) is acid-catalyzed. The bromination of **28** and **29** to yield **30** is acid-, as well as base-catalyzed (equation 17).

$$\underset{\textbf{(28)}\ \text{M.p. } 112\text{–}113^\circ}{HOCH_2\underset{NO_2}{\underset{|}{C}}HCH_2CH_2\underset{NO_2}{\underset{|}{C}}HCH_2OH} \xrightarrow{H^+} \underset{\textbf{(29)}\ \text{M.p. } 164\text{–}165^\circ}{HOCH_2\underset{NO_2}{\underset{|}{C}}HCH_2CH_2\underset{NO_2}{\underset{|}{C}}HCH_2OH}$$

$$\xrightarrow[H^+]{Br_2} \underset{\textbf{(30)}\ \text{M.p. } 111\text{–}112^\circ}{HOCH_2\overset{Br}{\overset{|}{\underset{NO_2}{\underset{|}{C}}}}CH_2CH_2\overset{Br}{\overset{|}{\underset{NO_2}{\underset{|}{C}}}}CH_2OH} \quad (17)$$

Epimerization does not occur at a measurable rate in acid solution with the bisphenylurethane or bismethyl ether derivatives of **28**; the corresponding derivatives of **29** are not formed under conditions whereby **28** → **29**. The epimerization **28** → **29** does *not* occur at a measurable rate in pure water in the absence of added acid or basic catalysts. No formaldehyde is produced during the epimerization. A similar epimerization reaction is observed with the next higher homolog, 2,7-dinitro-1,7-heptanediol[51]. The epimerization is believed to be favored over the competing Nef reaction due to stabilization of the nitronate anion (e.g. **31**) by hydrogen bonding which slows hydrolysis, and a relatively rapid C-protonation rate[52].

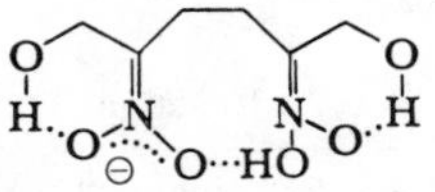

(31)

D. Ionization Constants of Nitronic Acids and Nitroalkanes

The ionization constants of nitronic acids may be expressed as

$$K_a^{Aci} = \frac{[H^+][A^-]}{[Aci]}$$

where $[Aci]$ = the concentration of the nitronic acid and $[A^-]$ is the concentration of nitronate anion[57,59]. Fewer values of K_a^{Aci} have been determined than K_a^{Nitro}. A complication in measurement lies in tautomerization to the more weakly acidic nitro form; but, by extrapolation of measurements back to zero time this error may be eliminated. Methods of determination include conductivity measurements[28], polarography[55], and potentiometric titrations[59]. Known values of K_a^{Aci} and K_a^{Nitro} are summarized in Table 5. Most nitronic acids are much stronger acids than the parent nitroalkanes (usually $pK_a^{\mathrm{Nitro}} - pK_a^{Aci} = 2\text{–}5$). However, this ratio narrows as acidity increases as shown in Figure 5. For very strong acids having $pK_a \leqq 0$, such as nitroform, $HC(NO_2)_3$, $pK_a^{Aci} \cong pK_a^{\mathrm{Nitro}}$.

Substitution affects acidity of nitronic acids in the manner observed in carboxylic acids[152].

There is evidence for intramolecular hydrogen bonding in 1,3-propanebisnitronic acid (**32**) and 1,4-butanebisnitronic acid[52]. A large $K_a^{\mathrm{I}}/K_a^{\mathrm{II}}$ ratio is observed, similar to the values found for *cis*-caronic and dipropylmalonic acids (**33** and **34** respectively). For the

TABLE 5. Ionization constants of nitronic acids and nitroalkanes in water at 25°.

Nitroalkane	pK_a^{Aci} [a]	pK_a^{Nitro} [a]	Ref.
	A. Mononitroalkanes		
CH_3NO_2	3.25	10.21	47, 56, 59, 60, 113, 130, 132
$CH_3CH_2NO_2$	4.4	8.5	28, 31, 47, 52, 56, 59, 60, 114, 130, 132
$CH_3CH_2CH_2NO_2$	4.6	8.98	40, 59, 60, 115
$(CH_3)_2CHNO_2$	5.1	7.68	47, 59, 60, 116
$CH_3(CH_2)_3NO_2$	—	10	116
$CH_3CH_2CH(CH_3)NO_2$	—	9.4	116
Cyclohexyl-NO_2	6.35	8.3	109, 128
$C_6H_5CH_2NO_2$	3.9	6.8	34, 36, 38, 39, 53, 55, 64
$CH_3(CH_2)_5CH(CH_3)NO_2$	(5.3)[b]	(10)[b]	50, 52, 116
	B. α,ω-Dinitroalkanes		
$(O_2N)_2CHCH_2CH(NO_2)_2$	—	I 1.11 (20°) II 4.96 (20°)	115, 117
$O_2N(CH_2)_3NO_2$	I (1.95)[b] II 8.40	—	52
$O_2N(CH_2)_4NO_2$	I (3.30)[b] II 8.30	—	52
$O_2N(CH_2)_5NO_2$	I 3.46 II 7.57	—	52
$O_2N(CH_2)_6NO_2$	I 3.55 II 4.80	(10)[b]	52
$O_2NCH(CH_2OH)(CH_2)_2CH(CH_2OH)NO_2$	I (3.15)[b] II 9.17	—	52
$O_2NCH(CH_2OH)(CH_2)_3CH(CH_2OH)NO_2$	I 4.30 II 8.45	—	52
$O_2NCH(CH_2OCH_3)(CH_2)_2CH(CH_2OCH_3)NO_2$	I (3.60)[b] II 8.35	—	52
	C. 1,1-Dinitroalkanes		
$BrCH(NO_2)_2$	—	3.47	144
$ClCH(NO_2)_2$	—	3.80	144
$FCH(NO_2)_2$	—	7.70 (20°)	118, 145
$CH(NO_2)_3$	—	0.06 0.17 (20°)	56, 115, 144, 146–148
$NCCH(NO_2)_2$	—	−6.23	144

(continued)

TABLE 5—*continued*

Nitroalkane	$pK_a^{Aci\,a}$	$pK_a^{Nitro\,a}$	Ref.
		C. 1,1-Dinitroalkanes	
$CH_2(NO_2)_2$	1.86	3.57	32, 56, 115, 117, 118, 145, 146, 148, 148a
$CH_3CH(NO_2)_2$	4.0	5.13	113–115, 144, 146, 148a 148–150
$CH_3CH_2CH(NO_2)_2$	4.1	5.6	114, 115, 148, 148a, 150
$HOCH_2CH_2CH(NO_2)_2$		4.44	156
$CH_3C(NO_2)_2CH_2CH(NO_2)_2$	—	1.36 (20°)	115, 117
$CH_3(CH_2)_2CH(NO_2)_2$	—	5.20	115, 148, 148a, 150
$(CH_3)_2CHCH(NO_2)_2$	—	6.75	148a 150,
$CH_3(CH_2)_3CH(NO_2)_2$	—	5.4	115, 148, 148a, 150, 151
$CH_3(CH_2)_4CH(NO_2)_2$	—	5.42 (20°)	115, 148a, 150, 151
$CH_3(CH_2)_8CH(NO_2)_2$	—	5.48 (20°)	115, 148a, 150, 151
$C_6H_5CH(NO_2)_2$	—	3.71	148a
		D. α-Substituted Mononitroalkanes	
Cl_2CHNO_2	—	5.99	118
F_2CHNO_2	—	12.4	149
$ClFCHNO_2$	—	10.14	118
$BrCH_2NO_2$	—	8.2 (21°)	47
$ClCH_2NO_2$	—	7.20	47, 118
CF_3CHFNO_2	—	9.1	149
$CF_3CH_2NO_2$	—	7.4	149
$CH_3CHBrNO_2$	—	7.3 (21°)	47
$CH_3CHClNO_2$	—	6.8 (21°)	47
$H_2NCOCHClNO_2$	—	3.50	118
$H_2NCOCHFNO_2$	—	5.89	118
$H_2NCOCH_2NO_2$	—	5.18	118
$CH_3COCH_2NO_2$	—	5.1	56
$C_2H_5O_2CCHClNO_2$	—	4.16	118
$C_2H_5O_2CCHFNO_2$	—	6.28	118
$C_2H_5O_2CCH_2NO_2$	—	5.75	56, 113, 118
$C_2H_5O_2CCH(CH_3)NO_2$	—	7.4	113, 116
$C_2H_5O_2CCH(C_2H_5)NO_2$	—	7.6	116
$C_2H_5O_2CCH(i\text{-}C_3H_7)NO_2$	—	9.0	116
$C_6H_5COCH_2NO_2$	2.2	—	32
$C_2H_5O_2CCH(n\text{-}C_5H_{11})NO_2$	—	7.7	116
$C_2H_5O_2CH(C_6H_5)NO_2$	—	6.9	116

[a] Average of values reported at 25°, neglecting divergent values.
[b] Estimated values.

longer chain 1,6-hexanebisnitronic acid $K_a^I/K_a^{II} = 17.7$, indicating

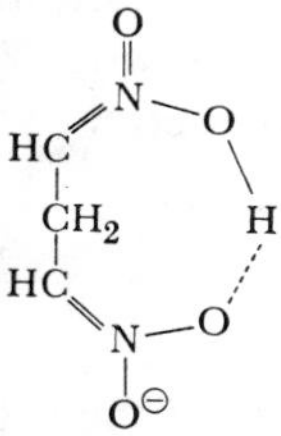

	(32)	(33)	(34)
$K_a{}^I/K_a{}^{II}$	2.9×10^6	9.3×10^5	2.8×10^5

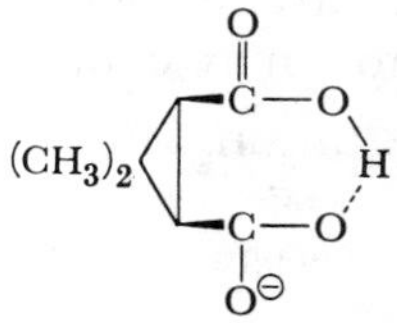

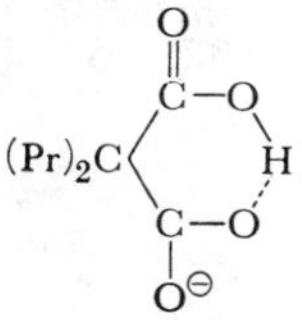

no intramolecular hydrogen bonding in the mononitronate anion.

The ionization constants of nitroalkanes may be expressed as

$$K_a^{Nitro} = \frac{[H^+][A^-]}{[Nitro]}$$

where [Nitro] = the concentration of the nitroalkane and $[A^-]$ is the concentration of nitronate anion. A complication lies in the fact that the nitronate anion is in equilibrium with the more strongly acidic nitronic acid. One observes an apparent ionization constant, $K_a^{App.}$ defined as

$$K_a^{App.} = \frac{[H^+][A^-]}{[Nitro] + [Aci]} \cong K_a^{Nitro}$$

where [*Aci*] = the concentration of nitric acid[30,132]. If K_a^{Aci} is known K_a^{Nitro} can be calculated from $K_a^{App.}$. For most nitroalkanes in solution the concentration of undissociated nitronic acid is very small and the [*Aci*] term may be neglected. Thus $K_a^{App.}$ is very nearly equal to K_a^{Nitro}. Values of pK_a^{Nitro} are summarized in Table 5; most are determined conductometrically or spectrophotometrically, some by titration. Ionization constants of nitroalkanes are solvent dependent[153].

Mononitroalkanes are relatively strong *pseudo* acids, $pK_a^{Nitro} = 7\text{–}10$. They are stronger than most monocarbonyl compounds (acetone $pK_a = 20$[154]), but are comparable to 1,3-diketones (acetylacetone $pK_a = 9$[155]). Nitromethane appears to be the weakest acid (pK_a^{Nitro} 10.2).

Electron-withdrawing groups such as halogen (Cl, Br) and nitro are inductively acid-strengthening[156]. However, *alpha*-fluorine substitution is decidedly acid-weakening. This unusual effect has been attributed to stablilizing no-bond resonance in the nitroalkane[118].

	$ClCH_2NO_2$	Cl_2CHNO_2	$ClFCHNO_2$
pK_a^{Nitro}	7.20	5.99	10.14

Electron-releasing alkyl groups, it appears, can also be acid-strengthening when α-substitution occurs (on the carbon bearing the nitro group) in mononitroalkanes. Compare 1- and 2-nitropropane (**35, 36**). Two factors may contribute to this effect which is

	$CH_3CH_2CH_2NO_2$	$(CH_3)_2CHNO_2$
	(**35**)	(**36**)
pK_a^{Nitro}	8.98	7.68

associated with a relatively slower C-protonation rate of the nitronate ion derived from **36**. One is a stabilization of the nitronate anion by hyperconjugation[157]. Another may be a steric effect related to poor solvation about the nitronate carbon. On the other hand, β-alkyl substitution is acid-weakening (1-nitrobutane, pK_a^{Nitro} 10), the expected result of electron-releasing bulky alkyl substitution.

In 1,1-dinitroalkanes and α-nitroesters both α- and β-alkyl substitution are acid-weakening (this effect is also observed in the carboxylic acids[152]). Dinitromethane (pK_a^{Nitro} 3.57) is the strongest acid of the 1,1-dinitroalkane series. The 1,1-dinitro-*n*-alkanes (C_2–C_9) all have similar, but smaller, ionization constants (pK_a^{Nitro} 5.2–5.7). An interesting compound is 1,1-dinitro-2-methylpropane (**38**). The weaker acidity of this more hindered β-alkyl substituted compound is probably associated with a relatively slower rate of proton removal (compare the unbranched isomer 1,1-dinitrobutane **37**). The α-alkyl-α-nitroester $C_2H_5O_2CCH(CH_3)NO_2$, pK_a^{Nitro} 7.4,

	$CH_3CH_2CH_2CH(NO_2)_2$	$(CH_3)_2CHCH(NO_2)_2$
	(**37**)	(**38**)
pK_a^{Nitro}	5.20	6.75

is weaker than the unbranched homolog $C_2H_5O_2CCH_2NO_2$, pK_a^{Nitro} 5.75[56,113,116,118]. More rate data are needed to supplement the available pK_a measurements. A quantitative correlation of structure with pK_a^{Nitro} has been reported[158].

III. NITRONIC ACIDS

A. Preparation of Nitronic Acids

Several methods are available for preparation of nitronic acids. All depend on oxygen-protonation of a nitronate anion as the final step. Certainly the most convenient and frequently employed method is acidification of a nitronate salt (equation 18). The procedure often involves preparation of a sodium or potassium salt by neutralization

$$R^1R^2C{=}NO_2^-Na^+ + HCl \xrightarrow{0-5^\circ} R^1R^2C{=}NO_2H + NaCl \qquad (18)$$

of a nitroalkane with aqueous alkali at 0–25°. Acidification of the salt usually proceeds best with an excess of a strong mineral acid, such as hydrochloric, keeping the temperature at 0–5° [1,13]. A low temperature is required to minimize Nef and other decomposition reactions. Use of a weak acid (acetic or carbonic) is preferred for C-protonation to regenerate a nitroalkane from its nitronate salt[13,159,160]. The weak acid permits a slightly acidic buffered solution having a relatively high concentration of nitronate ion needed for C-protonation to the nitroalkane. An excess of a strong mineral acid results in a strongly acidic solution having a low concentration of nitronate ion which inhibits C-protonation to the nitroalkane.

An interesting example of nitronic acid preparation is found in the acidification of a Meisenheimer-type salt[161–168]. Trinitrotoluene (**39**) reacts with potassium methoxide to form the thermodynamically stable potassium salt **40**[161,169]. Acidification of this salt with hydrogen chloride at −5° is reported to produce nitronic acid **41**,

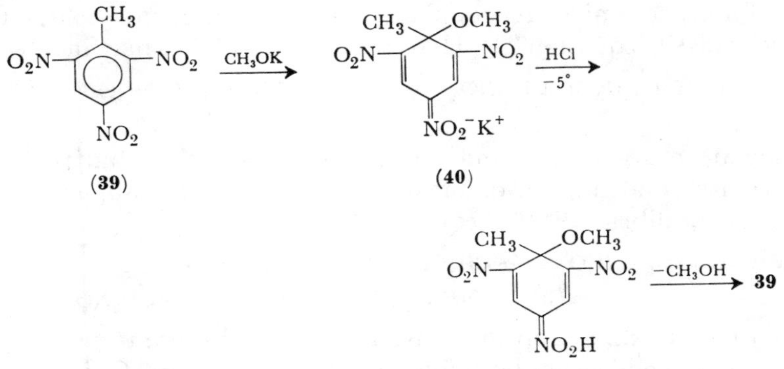

$$\text{(39)} \xrightarrow{CH_3OK} \text{(40)} \xrightarrow[-5^\circ]{HCl} \text{(41)} \xrightarrow{-CH_3OH} \textbf{39} \qquad (19)$$

(**39**): 2,4,6-$(O_2N)_3C_6H_2CH_3$; (**40**): 1-CH_3-1-OCH_3-2,6-$(NO_2)_2$-cyclohexadienylidene-4-$NO_2^-K^+$; (**41**): 1-CH_3-1-OCH_3-2,6-$(NO_2)_2$-cyclohexadienylidene-4-NO_2H

described as a dark red solid which explodes on heating[170] (equation 19). Many examples are known of reactions to form adducts like **40**[168,171–174]. Usually upon acidification of these substances a nitroaromatic compound is produced immediately since the nitronic acid decomposes so rapidly (**41** → **39**)[73,171–175].

Salts other than those of the alkali metals have been employed for preparing nitronic acids. Reaction of lead α-cyanophenylmethanenitronate with hydrogen sulfide produced the nitronic acid[176] (equation 20). Ammonium salts of nitronic acids on standing

$$(C_6H_5\underset{\displaystyle CN}{\underset{|}{C}}{=}NO_2^-)_2Pb^{++} + H_2S \xrightarrow{Et_2O} 2\ C_6H_5\underset{\displaystyle CN}{\underset{|}{C}}{=}NO_2H + PbS \qquad (20)$$

may evolve ammonia and produce a nitronic acid[177] (equation 21).

$$2,4,6\text{-}(CH_3)_3C_6H_2\text{—}CH{=}NO_2^-NH_4^+ \xrightarrow{25^\circ} 2,4,6\text{-}(CH_3)_3C_6H_2\text{—}CH{=}NO_2H + NH_3 \qquad (21)$$

Another important general method for preparation of nitronic acids involves addition of anions to nitroolefins. Alkyl and aryl Grignard reagents add to α-nitrostilbenes to form hindered nitronic acids in high yield[65] (equations 22 and 23). Addition of nitroform to

$$C_6H_5MgBr + C_6H_5CH{=}C(C_6H_5)NO_2 \xrightarrow{Et_2O} \underset{(90\%)}{(C_6H_5)_2CHC(C_6H_5){=}NO_2H} \qquad (22)$$

$$CH_3MgI + (C_6H_5)_2C{=}C(C_6H_5)NO_2 \xrightarrow{Et_2O} (C_6H_5)_2C(CH_3)C(C_6H_5){=}NO_2H \qquad (23)$$

nitroolefins is a similar reaction and provides excellent yields of nitronic acids[178] (equation 24). Hydration of nitroolefins may involve

$$(NO_2)_3CH + CH_2{=}C(CH_3)NO_2 \longrightarrow \underset{(95\%)}{(NO_2)_3CCH_2C(CH_3){=}NO_2H} \qquad (24)$$

intermediate nitronate ion and nitronic acid formation, but retrograde Henry condensation or nitroalcohol formation is apparently favored at equilibrium[179,179a] (equation 25).

$$CH_3CH{=}CHNO_2 + H_2O \longrightarrow CH_3CHOHCH{=}NO_2H \longrightarrow$$
$$CH_3CHOHCH_2NO_2 \longrightarrow CH_3CHO + CH_3NO_2 \qquad (25)$$

Oxidation of oximes appears to be a most useful route to nitronic acids. The method has not been fully developed, however. Oxidation of acetophenone and propiophenone oximes by Caro's acid (peroxymonosulfuric acid) leads to nitronic acids (not isolated) at room temperature[83,180] (equation 26). On warming, these nitronic acids rapidly tautomerize to the nitro form[83]. The use of other oxidizing

$$C_6H_5C(C_2H_5){=}NOH + H_2SO_5 \longrightarrow C_6H_5C(C_2H_5){=}NO_2H + H_2SO_4 \qquad (26)$$

agents, including dinitrogen tetroxide[83,181,182], manganese dioxide in acetic acid[183], peroxytrifluoroacetic acid[184], and nitric acid[184a], for conversion of oximes to nitroalkanes, probably proceeds through a nitronic acid intermediate. Action of powerful oxidizing agents (peroxytrifluoroacetic acid) on hindered oximes should yield nitronic acids directly.

Photochemical conversion of nitroalkanes to nitronic acids is the subject of a recent patent[185] (equation 27). The reaction has been

$$RCH_2NO_2 \xrightarrow{h\nu} RCH{=}NO_2H \qquad (27)$$

studied extensively with a limited group of compounds, the pyridylnitrophenylmethanes; nitronic acids derived from these compounds are unstable and tautomerize very rapidly to the nitro form (see section II.B)[73,186]. The scope of photochemical generation of nitronic acids has yet to be determined.

Thermal generation of a nitronic acid intermediate **42** has been postulated in the conversion of an *o*-nitrobiphenyl into the phenanthridine **43**[187,188] (equation 28).

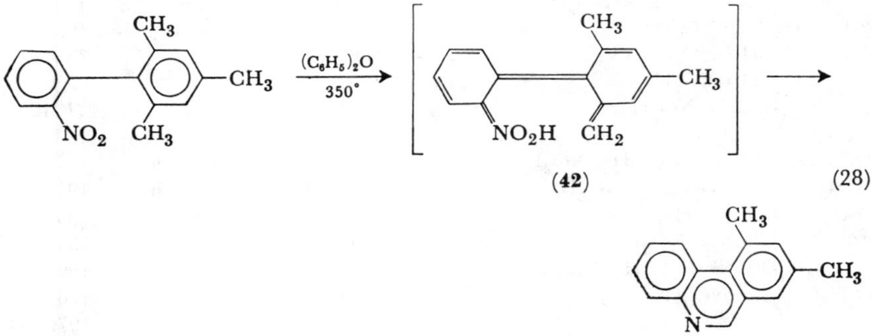

CH3, NO2, CH3, CH3 $\xrightarrow[350°]{(C_6H_5)_2O}$ [CH3, CH3, NO2H, CH2] → CH3, CH3, N (28)

(**42**)

(**43**) 32%

A nitronic ester has been converted into a nitronic acid. Potassium fluorene-9-nitronate and *t*-butyl bromide form *t*-butyl ester **44** in ethanol solution[84]. Ester **44** decomposes on standing at room temperature to form fluorene-9-nitronic acid (**8**) and butylene (equation 29). Application of this reaction to sodium phenylmethane-

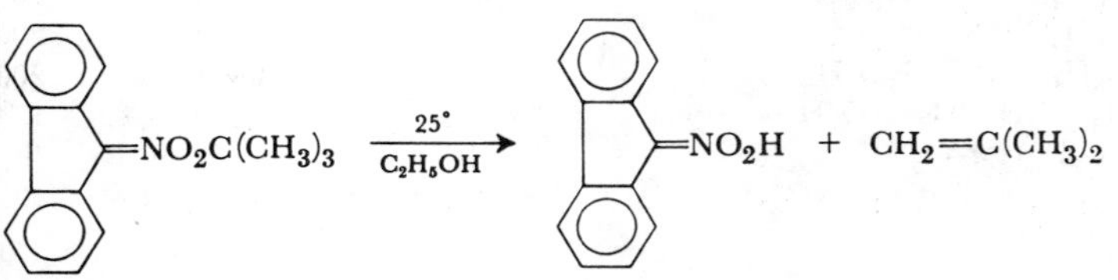

$=NO_2C(CH_3)_3 \xrightarrow[C_2H_5OH]{25°} =NO_2H + CH_2{=}C(CH_3)_2 \qquad (29)$

(**44**) (**8**)

nitronate, however, led to phenylnitromethane rather than the nitronic acid[84]. Hydrolysis of nitronic esters under mild conditions

TABLE 6. Properties of nitronic acids.

Nitronic acid	M.p. (°C)	Half-life (approx.) at 25°[a]	Ref.
$(O_2N)_2C{=}NO_2H$	50	few min	146, 190
$CH_3CH{=}NO_2H$	liquid	few min	32
$(O_2N)_3CCH_2C(CH_3){=}NO_2H$	91–91.5	2–3 h	178
$(NC)_2C{=}CHCH{=}NO_2H$	liquid	few h	191
$(O_2N)_3CCH_2C(C_2H_5){=}NO_2H$	70.5–71	2–3 h	178
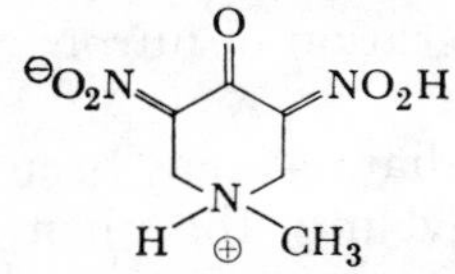 O; $^{\ominus}O_2N{=}$ … ${=}NO_2H$; N; H $\oplus$ CH_3	119	few days	192
$(O_2N)_3CCH_2C(n\text{-}C_3H_7){=}NO_2H$	85–85.5	2–3 h	178
$(O_2N)_3CCH_2C(i\text{-}C_3H_7){=}NO_2H$	93–93.5	2–3 h	178
$2\text{-}BrC_6H_4CH{=}NO_2H$	100	several h	193
$4\text{-}BrC_6H_4CH{=}NO_2H$	89–90	12 h	33
$4\text{-}ClC_6H_4CH{=}NO_2H$	64	2 days	62
$4\text{-}O_2NC_6H_4CH{=}NO_2H$	91	1 day	32, 194
$C_6H_5CH{=}NO_2H$	84	few days	1
$C_2H_5O_2CC(CN){=}CHCH{=}NO_2H$	liquid	several h	191
$(CH_3O_2C)_2C{=}CHCH{=}NO_2H$	liquid	several h	191
$(i\text{-}C_3H_7)_2C{=}NO_2H$	69–70	1 day	195
$2\text{-}BrC_6H_4C(CN){=}NO_2H$	51–52	1 week	193
$4\text{-}BrC_6H_4C(CN){=}NO_2H$	64	1 day	196
$C_6H_5C(CN){=}NO_2H$	39–40	few h	176, 197, 198
$C_6H_5C(CH_3){=}NO_2H$	45	few min	13, 83, 180
$4\text{-}CH_3OC_6H_4CH{=}NO_2H$	65–70	few h	199
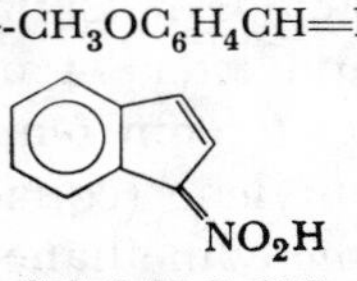 ${=}NO_2H$ (indene)	50	few h	86
$2\text{-}CH_3C_6H_4C(CN){=}NO_2H$	—	few h	200
$3\text{-}CH_3C_6H_4C(CN){=}NO_2H$	—	few h	200
$4\text{-}CH_3C_6H_4C(CN){=}NO_2H$	—	few h	200
$3,5\text{-}(CH_3)_2C_6H_3CH{=}NO_2H$	63	few min	13, 159
$C_6H_5C(C_2H_5){=}NO_2H$	—	few min	13, 83
$(C_2H_5O_2C)_2C{=}CHCH{=}NO_2H$	b.p. 130–140/0.2 mm[b]	few days	191
$C_6H_5C(CO_2C_2H_5){=}NO_2H$	liquid	few days	197
$2,4,5\text{-}(CH_3)_3C_6H_2CH{=}NO_2H$	102–110	few h	177
$C_6H_5C(i\text{-}C_3H_7){=}NO_2H$	54	few h	13

(continued)

TABLE 6—*continued*

Nitronic acid	M.p. (°C)	Half-life (approx.) at 25°[a]	Ref.
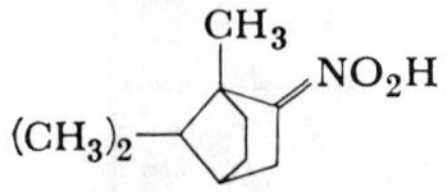 CH_3 / NO_2H / $(CH_3)_2$ (bicyclic structure)	74; 83	few min	201, 202
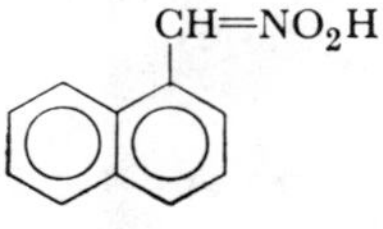 $CH{=}NO_2H$ (1-naphthyl structure)	—	few h	200
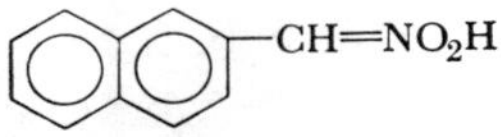 (2-naphthyl structure)—$CH{=}NO_2H$	—	few h	200
2,3,4,5-$(CH_3)_4C_6HCH{=}NO_2H$	—	few h	177
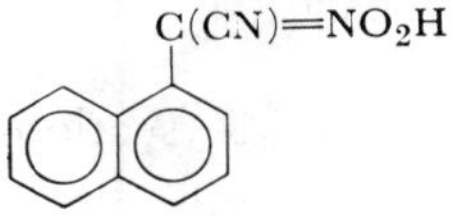 $C(CN){=}NO_2H$ (1-naphthyl structure)	—	few h	200
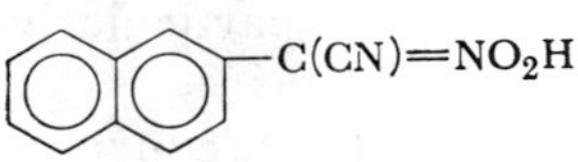 (2-naphthyl structure)—$C(CN){=}NO_2H$	—	few h	200
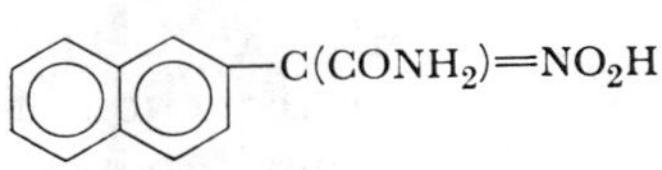 (2-naphthyl structure)—$C(CONH_2){=}NO_2H$	155–156	several h	200
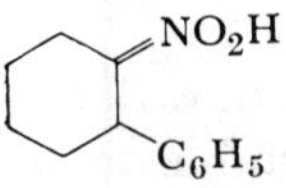 NO_2H / C_6H_5 (cyclohexane structure)	84–86	1 day	92
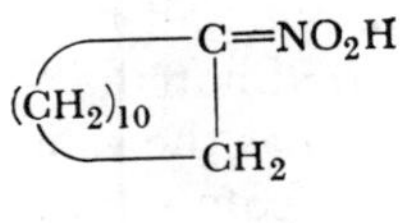 $(CH_2)_{10}$ / $C{=}NO_2H$ / CH_2 (ring structure)	85–86	several days	203
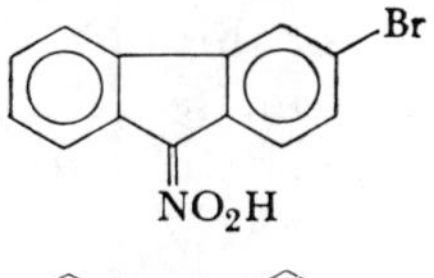 Br / NO_2H (fluorene structure)	132	several weeks	204
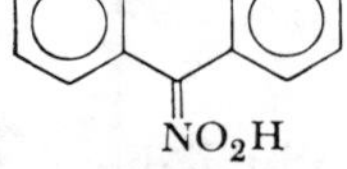 NO_2H (fluorene structure)	145–150; 132–135	several weeks	84, 85
$(C_6H_5)_2C{=}NO_2H$	90	few h	33
2-ClC_6H_4(4-BrC_6H_4)$CHC(CH_3){=}NO_2H$	101–118	>6 months	27

(*continued*)

TABLE 6—*continued*

Nitronic acid	M.p. (°C)	Half-life (approx.) at 25°[a]	Ref.
$4\text{-ClC}_6\text{H}_4(4\text{-BrC}_6\text{H}_4)\text{CHC(CH}_3)\text{=NO}_2\text{H}$	48–65	4–20 h	27
$2\text{-ClC}_6\text{H}_4(4\text{-IC}_6\text{H}_4)\text{CHC(CH}_3)\text{=NO}_2\text{H}$	116–130	>6 months	27
$2\text{-ClC}_6\text{H}_4(4\text{-ClC}_6\text{H}_4)\text{CHC(CH}_3)\text{=NO}_2\text{H}$	75–80	>6 months	27
$(2\text{-ClC}_6\text{H}_4)_2\text{CHC(CH}_3)\text{=NO}_2\text{H}$	42–55	4–20 h	27
$(\text{C}_6\text{H}_5)_2\text{CHC(CH}_3)\text{=NO}_2\text{H}$	70–77	4–20 h	27
$2\text{-ClC}_6\text{H}_4(4\text{-CH}_3\text{C}_6\text{H}_4)\text{CHC(CH}_3)\text{=NO}_2\text{H}$	89–94	8 days	27
$4\text{-ClC}_6\text{H}_4[2,4\text{-(CH}_3)_2\text{C}_6\text{H}_3]\text{CHC(C}_2\text{H}_5)\text{=NO}_2\text{H}$	47–70	4–20 h	27
(fluorene structure with COC_6H_5 substituent and $=\text{NO}_2\text{H}$ at C-9)	80–84	several weeks	205

[a] Approximate time for undiluted sample to liquefy or exhibit evident decomposition.
[b] About 90% of the sample is reported to decompose during the distillation.

may be expected to yield products other than nitronic acids (section IV.C.1). Thus, the conversion of nitronic esters to nitronic acids appears to be a reaction of limited scope and utility, particularly since most esters are prepared from nitronate salts.

Fluorene-9-nitronic acid (**8**) may also be prepared by reduction of 9-bromo or 9-iodo-9-nitrofluorene with potassium iodide[84]. This reaction should be considered unique, however, since reduction of other 1-bromo-1-nitroalkanes with mild reducing agents produces nitroalkanes[189].

Table 6 lists most of the known, isolable nitronic acids in order of molecular formula. Usually only solids are sufficiently stable to be isolated in pure form. Melting points and approximate half-lives are given where this information is available. The approximate half-life (measured at room temperature) is arbitrarily chosen as the time required for liquefaction or evident decomposition of a compound. Since no standardized procedure has been employed for this measurement, the times listed are necessarily very approximate. The information should prove useful, however, in correlating structure with stability.

B. Physical Properties of Nitronic Acids

Ionization constants of nitronic acids are summarized in Table 5, melting points in Table 6.

The ultraviolet spectra of nitronic acids resemble closely the spectra of nitronate anions (salts) and nitronic esters[206]. A strong $\pi - \pi^*$ band ($\varepsilon \cong 10000$) is found near 220–230 mμ for simple aliphatic nitronic acids measured in ethanol or water[47,52]. The corresponding nitronate anions absorb at nearly the same wavelength (usually ca. 10 mμ higher, depending on structure) with approximately equal extinction coefficients[47,52,207,208]. Absorption spectra of nitronate anions have been published elsewhere[120,209,210]. Olefinic or aromatic ring conjugation produces the expected bathochromic shift in absorption maximum wave length [$C_6H_5CH{=}NO_2$ $\lambda_{max}^{H_2O}$ 284 mμ (ε 20000); $C_6H_5CH{=}NO_2^-Na^+$ $\lambda_{max}^{H_2O}$ 294 mμ (ε 25000)][53,211–214]. The nitronic acids produced by irradiation of pyridyl and phenylnitrophenylmethanes are highly colored with strong absorption bands near 580–700 mμ (see section II.B)[74,75]; in this group of compounds the corresponding nitronate salts absorb at wave lengths ca. 10 mμ lower (ethanol solvent)[75].

The infrared spectra of nitronic acids are characterized by C=N absorption near 1620–1680 cm^{-1} [150,178,206]. This absorption is in the region of oxime C=N absorption, 1640–1684 cm^{-1} [215]. Conjugation shifts the absorption to slightly lower frequencies; fluorene-9-nitronic acid absorbs at 1652 cm^{-1} [211]. Nitronic esters absorb intensely in the region 1610–1660 cm^{-1} (C=N). Nitronate salts absorb at much lower frequencies, as one observes with carboxylate salts[215]. Sodium alkanenitronate salts reveal a C=N band in the region 1587–1605 cm^{-1} [216].

The infrared absorption of nitronic acids in the OH stretching region resembles that of carboxylic acids. A free OH stretching band is absent[211]. One observes the broad absorption band envelope in the region 2500–3000 cm^{-1} characteristic of associated weak acids[206,211].

Few nmr spectra of nitronic acids have been reported[217–219]. From nmr, infrared, and ultraviolet spectra measurements it was concluded that the imine **45a** exists as the nitronic acid **45c**, rather than nitro-olefin **45b**, in methanol or deuteriochloroform; in the latter solvent an AB quartet was observed at τ 2.38, 3.27[217].

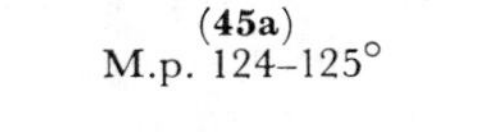

$C_6H_4(COCH_3)(N{=}CHCH_2NO_2)$

(45a)
M.p. 124–125°

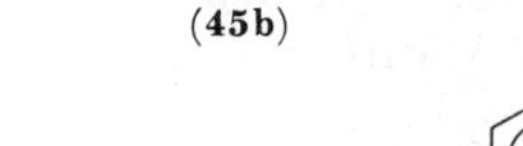

$C_6H_4(COCH_3)(NHCH{=}CHNO_2)$

(45b)

$C_6H_4(COCH_3)(N{=}CHCH{=}NO_2H)$

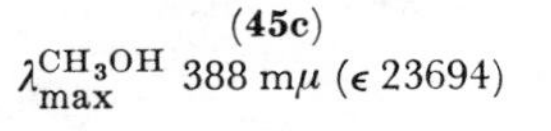

(45c)
$\lambda_{max}^{CH_3OH}$ 388 mμ (ϵ 23694)

C. Reactions of Nitronic Acids

Nitronic acids are quite reactive. The C=N bond undergoes many addition reactions. Nitronic acids are also good oxidizing agents and are readily reduced to oximes. They participate in autooxidation-reduction reactions. The reactions which are discussed in this section are principally those of nitronic acids and nitronate anions in *acid* solution. Reactions of nitronate salts, or of nitronate ions in basic solution, with a few exceptions, are not discussed. Formation of nitronic esters and anhydrides is described in following sections.

1. Addition reactions of nitronic acids

Two distinct patterns of addition are evident: (*a*) In acid solution a protonated nitronic acid (**46**) adds nucleophiles such as halide and hydroxide ion; simultaneously the nitronic group ultimately

$$R^1R^2C{=}\overset{+}{N}(OH)_2 + Nu^- \longrightarrow R^1R^2C(NO)(Nu) + H_2O \qquad (30)$$

(**46**) $Nu^- = OH^-(H_2O),\ Cl^-,\ Br^-,\ I^-$

becomes nitroso in the product (equation 30). Alternatively, in strongly acidic solution a dehydration to a nitrile oxide may precede addition of water to form a hydroxamic acid. (*b*) Nitronate anions exist in acid solution, although they are present in much lower concentration than in basic solution. They undergo addition with electrophiles such as nitrosonium, and nitronium ions, halogens, and hypohalogen acids. The nitronate group becomes nitro in the product (equation 31).

$$R^1R^2C{=}NO_2^- + E^+ \longrightarrow R^1R^2C(NO_2)(E) \qquad (31)$$

$E^+ = NO^+,\ NO_2^+\ (N_2O_4),\ Cl^+\ (Cl_2,\ HOCl),\ Br^+\ (Br_2,\ HOBr),\ CH_2O$

a. Nucleophilic addition. Nucleophilic additions occur on a protonated nitronic acid. Addition of water to nitronic acids is one of the most important nucleophilic addition reactions. It is a useful route to aldehydes and ketones (Nef reaction), hydroxamic acids, and carboxylic acids.

The Nef reaction[15] is important synthetically. It has been reviewed[4,220] and its mechanism studied[46,47,209,221–226]. The reaction involves treatment of a nitronate salt or nitronic acid with aqueous acid; in effect, it is the acid-catalyzed hydrolysis of a nitronic acid. The mechanism may be expressed by the equations 32, 33, 34, and 35[209,222,227]. The details of the decomposition of the hydrated

$$R^1R^2C{=}N(\rightarrow O)OH + H^+ \rightleftharpoons \underset{(\mathbf{46})}{R^1R^2C{=}\overset{\oplus}{N}(OH)_2} \longleftrightarrow R^1R^2\overset{\oplus}{C}{-}N(OH)_2 \quad \text{(fast)} \qquad (32)$$

$$\underset{(\mathbf{46})}{R^1R^2C{=}\overset{\oplus}{N}(OH)_2} + H_2O \rightleftharpoons \underset{(\mathbf{47})}{R^1R^2C(OH){-}\overset{\oplus}{N}H(OH)_2} \quad \text{(slow)} \qquad (33)$$

$$\underset{(\mathbf{47})}{R^1R^2C(OH){-}\overset{\oplus}{N}H(OH)_2} \longrightarrow R^1R^2C{=}O + H_2O + H^+ + HNO \qquad (34)$$

$$\underset{(\mathbf{47})}{R^1R^2C(OH){-}\overset{\oplus}{N}H(OH)_2} \longrightarrow \underset{(\mathbf{48})}{R^1R^2C(OH){-}N{=}O} + H_2O + H^+ \qquad (35)$$

intermediate **47** are not completely understood. The formation of the blue color which frequently accompanies the Nef reaction may be explained by formation of the hydroxynitroso compound **48**[209]. The initially formed nitrogenous product of the reaction is believed to be the unstable intermediate nitroxyl (HNO) which forms nitrous oxide (equation 36).

$$2\,HNO \longrightarrow H_2O + N_2O \qquad (36)$$

A direct acid-catalyzed Nef reaction is possible without starting with a nitronate salt or a nitronic acid[50,51,141,228]. 2-Octanone has been obtained directly from *d*-2-nitrooctane by refluxing with aqueous hydrochloric acid[50]. The recovered nitro compound retained all its optical activity indicating, in this example, that the Nef reaction was faster than tautomerization of the intermediate nitronic acid, formed by an acid-catalyzed process (equation 37).

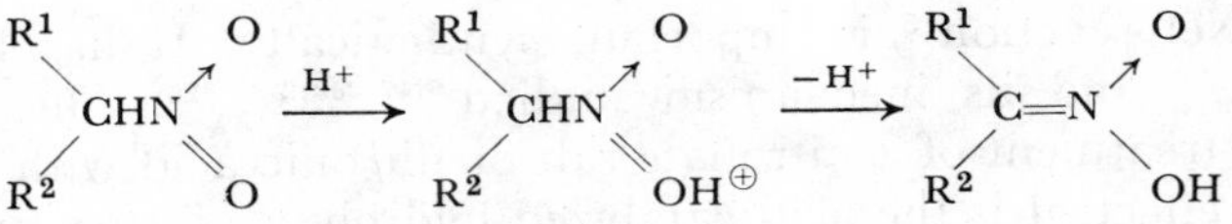

$$\begin{matrix}R^1 \\ \quad\diagdown \\ \quad CHN\diagup^{\nearrow O} \\ \quad\diagup\diagdown\!\!\diagdown O \\ R^2\end{matrix} \xrightarrow{H^+} \begin{matrix}R^1 \\ \quad CHN(\rightarrow O)(=OH^{\oplus}) \\ R^2\end{matrix} \xrightarrow{-H^+} \begin{matrix}R^1 \\ \quad C{=}N(\rightarrow O)(OH) \\ R^2\end{matrix} \qquad (37)$$

(49)

Studies have been made to determine optimum conditions for securing high yields of aldehydes and ketones in the Nef reaction[46,47,105,223]. The reaction is pH dependent. A low pH (0.1–1) favors Nef reaction over tautomerization which occurs more readily at pH 3–5 (see Table 7)[47]. A very low pH (as with 85 % sulfuric acid) favors hydroxamic acid formation, possibly proceeding by a different mechanism[222].

The yields of aldehydes and ketones on Nef hydrolysis vary (0–100 %) and depend on the structure of the nitronic acid (Table 8), as well as on pH (Table 7). Tautomerization to the parent nitro compound is an important competing reaction, although other reactions can occur[65,240]. Simple, unsubstituted aliphatic nitronic acids readily undergo the Nef reaction[32,160]. However, branching near the nitronate carbon, which hinders attack, decreases yields[221,229,231,232]. Compounds of structure $Ar_3CCH(Ar)NO_2$ fail to undergo the Nef reaction[65].

Factors which stabilize nitronate anions (and nitronic acids) inhibit the Nef reaction[32]. These include resonance stablization, presence of electron-withdrawing groups, and hydrogen bonding. *p*-Nitrophenylnitromethane[222] and 1,1,1,2,2,3,3-heptafluoro-5-nitropentane[233] fail to undergo the Nef reaction. Nitrodesoxyinsitols fail to undergo the Nef reaction[237,238] and may be recovered unchanged; stabilization of the nitronate anion by hydrogen bonding **49** has been suggested to explain this result[51]. Homoallylic resonance in the

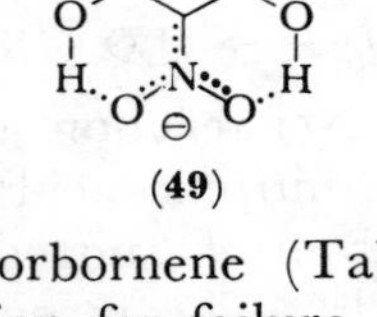

[Structure: cyclohexane ring bearing OH, HO, OH, and two O–H groups hydrogen-bonded (H···O) to the oxygens of a nitronate $N^{\ominus}$ group]

(49)

nitronic acid of 5-nitronorbornene (Table 8, no. 20) has been suggested as an explanation for failure of the Nef reaction[240,241]; however, the nitro compound is not recovered[240].

Ring strain in the transition state leading to a product having an exocyclic double bond is probably the explanation for failure of the

TABLE 7. Products of nitronic acid decomposition in water at various pH. (0.1 M solutions of nitronic acid in buffer solutions at 21°)[47]

A. Ethanenitronic acid

pH	$C_2H_5NO_2$	CH_3CHO	$CH_3CH{=}NOH$	$CH_3C(NO_2){=}NOH$	NO_2^-
3.4	100	0	0	0	0
2.9	85	7	3	4	0
2.07	59	28	6	6	0
1.50	27	61	6	6	0
1.18	6	82	6	5	0
0.45	0	98	0	0	0

B. Propane-2-nitronic acid

pH	$(CH_3)_2CHNO_2$	$(CH_3)_2CO$	$(CH_3)_2C{=}NOH$	$(CH_3)_2C(NO)(NO_2)$	NO_2^-
5.4	100	0	0	0	0
5	85	7–8	7–8	0	7–8
4.25	44	20	19	15	4
3.75	27	26	25	20	5
3.10	10	30	30	29	0
2.5	3	32	31	32	0
2.2	0	33	32	33	0
2	0	39	32	29	0
1.5	0	49	28	22	0
1.15	0	80	12	7	0
0.50	0	100	0	0	0

C. Cyclohexanenitronic acid

pH	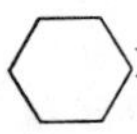 $C_6H_{11}NO_2$ (cyclohexyl–NO_2)	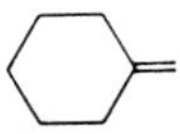 $C_6H_{10}{=}O$	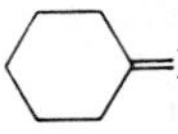 $C_6H_{10}{=}NOH$	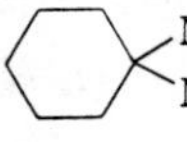 $C_6H_{10}(NO)(NO_2)$	NO_2^-
4.8	100	0	0	0	0
4.15	43	20	21	14	7
3.05	6	32	31	31	0
2.40	0	38	31	30	0
1.50	0	70	24	5	0
1	0	84	15	1	0
0.15	0	100	0	0	0

TABLE 8. Syntheses and attempted syntheses of carbonyl compounds by the Nef reaction.

No.	Nitroalkane	Yield carbonyl compound (%)	Ref.
1.	CH_3NO_2	100	47
2.	$CH_3CH_2NO_2$	77	47, 229
3.	$CH_3CH_2CH_2NO_2$	80	47, 229
4.	$(CH_3)_2CHNO_2$	84	47, 229
5.	$CH_3(CH_2)_3NO_2$	85	229
6.	$CH_3CH_2CH(CH_3)NO_2$	82	229
7.	$(CH_3)_2CHCH_2NO_2$	32	229
8.	$(CH_3)_3CCH_2NO_2$	very low	230, 231
9.	$(CH_3)_2C(CH_2NO_2)_2$[a]	0	232
10.	$CF_3CF_2CF_2CH_2CH_2NO_2$	0	233
11.	$CH_3CH_2CH(CH_2OH)NO_2$	50	179
12.	$(CH_3)_2CHCH(OH)CH_2NO_2$	0	179
13.	$-NO_2$	56	209
14.	$(C_6H_5)_2$ NO_2 C_6H_5	very low	234, 235
15.	$-NO_2$	89	209
16.	$-NO_2$	85–97	47, 209
17.	$-NO_2$ $-C_6H_5$	88	236
18.	OH, HO, NO_2, HO, OH, OH	0	237, 238
19.	$-NO_2$	80	226
20.	$-NO_2$	0[b]	221, 226, 236, 239–241
21.	$-NO_2$	68	242

(continued)

TABLE 8—*continued*

No.	Nitroalkane	Yield carbonyl compound (%)	Ref.
22.	$4\text{-}O_2NC_6H_4CH_2NO_2$	0	22
23.	$(C_6H_5)_2CHCH(C_6H_5)NO_2$[c]	94	65
24.	$(C_6H_5)_2C(CH_3)CH(C_6H_5)NO_2$[c]	0[d]	65

[a] Monosodium salt employed.

[b] [structure: bicyclic O–NH lactam fused to cyclopentene] was obtained in 42% yield with 9.2% aqueous HCl at −20 to −10° [240].

[c] Nitronic acids employed rather than salts.

[d] C_6H_5 [isoxazoline ring] $(C_6H_5)_2$ was obtained in 70% yield with methanolic hydrogen chloride.

Nef reaction with certain strained nitrocycloalkanes. This effect may explain the failure of 5-nitronorbornene to undergo Nef reaction in contrast to the behavior of 5-nitrobicyclo[2,2,2]-2-octene (Table 8, no. 21). An example of the effect of ring strain coupled with steric hindrance is shown by 1-nitro-2,3,3-triphenylcyclobutane (no. 14) which undergoes the Nef reaction in very low yield[234,235]. Nitrocyclobutane provides a lower yield (56%) of ketone than nitrocyclopentane (89%) and nitrocyclohexane (85–97%)[47,209].

Reaction of concentrated sulfuric acid (85–100%) with salts of primary nitroalkanes leads to hydroxamic acids[243–245]. The subject has been reviewed[246] and the mechanism discussed[222,240,247,248]. Direct conversion of nitroalkanes to hydroxamic acids has been observed in concentrated sulfuric acid[143,244] (equation 38). Nitronic

$$CH_3CH_2CH_2NO_2 \xrightarrow{H_2SO_4} \underset{44\%}{CH_3CH_2CONHOH} \qquad (38)$$

acid should be considered a primary reaction intermediate[244,249,250]. The nitronic acid would be protonated as in the first step of the Nef reaction. Two mechanisms may be considered[222]:

(1) *Nitrile oxide mechanism*[240]. In strong acid solution—in contrast to dilute aqueous acid used in the Nef reaction—dehydration of intermediate **46** might be expected to be favored over hydration. The resulting nitrile oxide intermediate **50** could rehydrate to

produce hydroxamic acid (equation 39). The nitrile oxide mechanism is favored to explain the cleavage and/or rearrangement of

$$\underset{(46)}{\mathrm{R(H)C{=}\overset{\oplus}{N}(OH)_2}} \longrightarrow \underset{(50)}{\mathrm{RC{\equiv}NOH^{\oplus}}} + \mathrm{H_2O}$$

$$\underset{(50)}{\mathrm{RC{\equiv}NOH^{\oplus}}} + \mathrm{H_2O} \longrightarrow \mathrm{H^+} + \underset{(51)}{\mathrm{RC(OH){=}NOH}} \rightleftharpoons \underset{(52)}{\mathrm{RC({=}O)NHOH}} \qquad (39)$$

α-nitro ketones[247]. Of two tautomeric forms, the hydroxyamide form **52** is usually favored over the oxime **51** at equilibrium[251].

(2) *Nitroso alcohol mechanism*[222,252]. A less favored mechanism, since water concentration is so low, involves initial hydration of **46** to the Nef intermediate **47**, followed by dehydration to the nitroso alcohol **48** and tautomerization to the hydroxamic acid (equation

$$\underset{(47)}{\mathrm{R(H)C(OH){-}\overset{\oplus}{N}H(OH)_2}} \xrightarrow{-\mathrm{H_3O^+}} \underset{(48)}{\mathrm{R(H)C(NO)(OH)}} \longrightarrow \underset{(52)}{\mathrm{RC({=}O)NHOH}} \qquad (40)$$

40). Argument against this mechanism is the fact that hydroxamic acids are not usually formed under Nef conditions where dilute acid is employed.

The effect of structure on yields of hydroxamic acid has been studied[222]. In contrast to behavior observed in the Nef reaction, electron-withdrawing groups facilitate hydroxamic acid formation. In 85% sulfuric acid solvent the yield from *p*-nitrophenylnitromethane is 86%; from 1-nitropropane, 28% (equation 41).

$$\mathrm{O_2N{-}C_6H_4{-}CH{=}NO_2H} \xrightarrow{85\%\ \mathrm{H_2SO_4}} \underset{86\%}{\mathrm{O_2N{-}C_6H_4{-}C({=}O)NHOH}} \qquad (41)$$

Carboxylic acids and hydroxylamine are formed by treatment of primary nitroalkanes with concentrated mineral acids[140a,141–143,243,253–263]. Nitronate salts may also be employed. The reaction is sometimes called the Victor Meyer reaction after its discoverer and developer (1873–1876)[141,142,243,253,254]. The reaction has synthetic utility both for preparation of carboxylic acids[264,265] and hydroxylamine[142,266–269].

Reaction conditions for carboxylic acid formation are somewhat more vigorous than those required for hydroxamic acid formation. Solutions of nitroalkane in 85% sulfuric acid, or concentrated hydrochloric acid-acetic acid, are heated under reflux for several hours (equation 42). Yields of both acid and hydroxylamine are often

$$CH_3CH_2NO_2 \xrightarrow[\text{Reflux 8 h}]{85\%\ H_2SO_4} \underset{88\%}{CH_3CO_2H} + \underset{86\%}{NH_3OH^+, HSO_4^-} \tag{42}$$

high (80–90%)[143]. Hydroxamic acids may be isolated when lower temperatures or shorter reaction times are employed[143,261]. It seems reasonable that hydroxamic acids are intermediates in the reaction. Thus the mechanism would involve the steps for hydroxamic acid formation, followed by acid-catalyzed hydrolysis of the hydroxamic acid[140a,270] (equation 43),

$$RC(=O)NHOH + H^+ \longrightarrow RC(=O)OH + \overset{\oplus}{N}H_3OH \tag{43}$$

One commercially feasible hydroxylamine synthesis employs 1,2-dinitroethane which forms oxalic acid[267–269]. Another process employs nitromethane[142,255,256,258,266,271].

An extension of the reaction to secondary nitroalkanes permits preparation of amides by including azide as a reactant[272,273] (equation 44).

$$n\text{-}Pr_2C{=}NO_2^-Na^+ \xrightarrow{H_2SO_4,\ NaN_3} \underset{62\%}{n\text{-}PrCONHPr\text{-}n} \tag{44}$$

Hydrogen halides add to nitronic acids[105]. With nitronate salts in ether solvent the blue α-halonitroso product [203,249] may occasionally be isolated as a colorless dimer[189,249,274] (equation 45). (The α-

$$CH_3CH{=}NO_2^-Na^+ \xrightarrow{HCl,\ Et_2O} \underset{\text{Blue oil}}{CH_3CH(Cl)NO} \longrightarrow \underset{\substack{\text{Colorless} \\ \text{M.p. } 65^\circ}}{CH_3CH(Cl)N(\rightarrow O){=}N(\rightarrow O)CH(Cl)CH_3} \tag{45}$$

$$CH_3CH{=}NOH \xrightarrow{Cl_2} CH_3CH(Cl)NO$$

halonitroso compounds are also readily prepared by halogenation of oximes[189,274,274a]).

The reaction mechanism is believed to depart from the protonated nitronic acid intermediate **46** common to nucleophilic addition reactions of nitronic acids. Addition of hydrogen chloride to form

53, followed by dehydration, produces the nitroso product **54** (equation 46),

$$\underset{(\mathbf{46})}{R^1R^2C{=}\overset{\oplus}{N}(OH)_2} + HCl \longrightarrow \underset{(\mathbf{53})}{R^1R^2C(Cl){-}\overset{\oplus}{N}H(OH)_2}$$

$$\underset{(\mathbf{53})}{R^1R^2C(Cl){-}\overset{\oplus}{N}H(OH)_2} \longrightarrow \underset{(\mathbf{54})}{R^1R^2C(Cl)NO} + H_3O^+ \qquad (46)$$

The stereochemistry of this addition has been examined (equation 47)[105]. The potassium salt of **55** was treated with hydrogen chloride in ether at 0°. The resulting nitroso group appears *trans* to the adjacent R group (phenyl, methyl) in product **56** in the kinetically controlled process. Interestingly, **56a** could not be prepared by chlorination of the required oxime[105].

$$\mathbf{55} \xrightarrow[\text{Et}_2\text{O, 0°}]{-\text{H}_3\text{O}^+} \mathbf{56} \qquad (47)$$

(**55a**) R = CH_3 (**56a**) (78%, R = CH_3)
(**55b**) R = C_6H_5 (**56b**) (23%, R = C_6H_5)

The α-halonitroso compounds derived from primary nitronic acids tautomerize readily to form hydroxamic acid chlorides (α-chlorooximes)[248,250,274,274a,275]. The blue phenylchloronitrosomethane **57** is observed in solution when phenylmethanenitronic

$$C_6H_5CH{=}NO_2H \xrightarrow[\text{Et}_2\text{O}]{\text{HCl}} \underset{\substack{(\mathbf{57})\ \text{Blue} \\ \text{(not isolated)}}}{C_6H_5CH(Cl)NO} \longrightarrow \underset{\substack{(\mathbf{58})\ \text{Colorless} \\ \text{M.p. 50–51°}}}{C_6H_5C(Cl){=}NOH} \qquad (48)$$

acid is treated with hydrogen chloride in ether; it has not been isolated, however, and rearranges to colorless benzohydroxamic acid chloride (**58**)[249,274,275] (equation 48). The ammonium salt of ethyl nitroacetate forms ethyl chlorooximinoacetate (**59**)[250] (equation 49).

$$C_2H_5O_2CCH{=}NO_2^-, NH_4^+ \xrightarrow[Et_2O]{HCl} C_2H_5O_2C\underset{\displaystyle Cl}{\underset{|}{C}}HNO \longrightarrow C_2H_5O_2C\underset{\displaystyle Cl}{\underset{|}{C}}{=}NOH \quad \textbf{(59)} \qquad (49)$$

Reaction of hydrogen bromide with nitronic acids to form α-bromonitrosoalkanes appears not to have been reported. The less stable α-bromonitrosoalkanes are prepared by bromination of oximes[189,274].

Hydrogen iodide (aqueous) reduces nitronic acids to oximes and is the basis for a quantitative analytical determination; the liberated iodine is titrated with sodium thiosulfate[276] (equation 50).

$$R_2C{=}NO_2H + 2\,HI(aq.) \longrightarrow R_2C{=}NOH + H_2O + I_2 \qquad (50)$$

The α-halonitrosoalkanes may be oxidized to α-halonitroalkanes with peroxytrifluoroacetic acid[189].

b. Electrophilic addition. Electrophilic additions to nitronic acids occur on the nitronate anion. The following discussion is limited to reactions of nitronic acids in acid solution. With a few exceptions the many reactions of nitronate salts and nitronate anions in basic solution—Michael addition and aldol-type condensations, for example—will be discussed in the second volume of this treatise.

Halogens, or hypohalogen acids, add readily to nitronic acids to yield α-halonitroalkanes[170,201] (equation 51). The mechanism

$$C_6H_5CH{=}NO_2H + Cl_2 \longrightarrow C_6H_5CHClNO_2 + HCl \qquad (51)$$

involves addition of halogen to a nitronate anion (equation 52).

$$R^1R^2C{=}N(\rightarrow O)OH \rightleftharpoons R^1R^2C{=}N(\dot{\vdots}O^-)(\dot{\vdots}O) + H^+$$

$$R^1R^2C{=}N(O^-)O + X_2 \text{ (or XOH)} \longrightarrow R^1R^2C(X){-}NO_2 + X^- \text{ (or } OH^-) \qquad (52)$$

$$X = Cl, Br, I$$

Addition of iodine monochloride (ICl) produces the iodo compound, $R^1R^2CINO_2$, and chloride ion[209]. It is to be noted that α-halonitroalkanes are most conveniently prepared by halogenation of nitronate salts.

Reaction of nitrous acid with nitronic acids and salts yields pseudonitroles (blue)[105,201,277–279]. These pseudonitroles derived from primary nitronic acids isomerize very readily to nitrolic acids[170,259,277–279]. The reaction was discovered by Victor Meyer (1873)[278,279] who first prepared acetonitrolic acid and dimethyl pseudonitrole by addition of dilute acid to a mixture of nitroalkane salt and potassium nitrite (equations 53 and 54). The blue pseudonitroles may be isolated in the solid state or in solution; often they are isolated as colorless crystalline dimers. On melting, the dimers form blue liquids containing pseudonitrole monomer (equation 54).

$$CH_3CH{=}NO_2^-Na^+ + KNO_2 \xrightarrow{H^+} CH_3C(NO_2){=}NOH \tag{53}$$

Acetonitrolic acid
M.p. 81–82°

$$(CH_3)_2C{=}NO_2^-Na^+ + KNO_2 \xrightarrow{H^+} (CH_3)_2C(NO)NO_2 \underset{\text{Heat}}{\rightleftharpoons} (CH_3)_2C(NO_2)N(\rightarrow O){=}N(\rightarrow O)C(CH_3)_2NO_2 \tag{54}$$

Dimethyl pseudonitrole; Dimer, M.p. 76°

The reaction mechanism for pseudonitrole formation very likely involves nitronate anion, rather than nitronic acid, in a reaction with nitrosonium ion (NO^+) or N_2O_4[105] (equation 55).

$$HNO_2 + H^+ \longrightarrow NO^+ + H_2O$$

$$R^1R^2C{=}N(O)O^- + NO^+ \ (\text{or } N_2O_4) \longrightarrow R^1R^2C(NO)NO_2 \ (\text{Blue})$$

$$\xrightarrow{\text{When } R^2 = H} R^1C(NO_2){=}NOH \ (\text{Colorless})$$

$$\xrightarrow{\text{When } R^1, R^2 \neq H} R^1R^2C(NO_2)N(\rightarrow O){=}N(\rightarrow O)CR^1R^2(NO_2) \ (\text{Dimer (Colorless)}) \tag{55}$$

The stereochemistry of pseudonitrole formation has been examined in the system shown in equation 56[105]. Addition of NO^+ to **60** occurs *cis* to the R group (C_6H_5, CH_3) leading to a *trans* arrangement of R

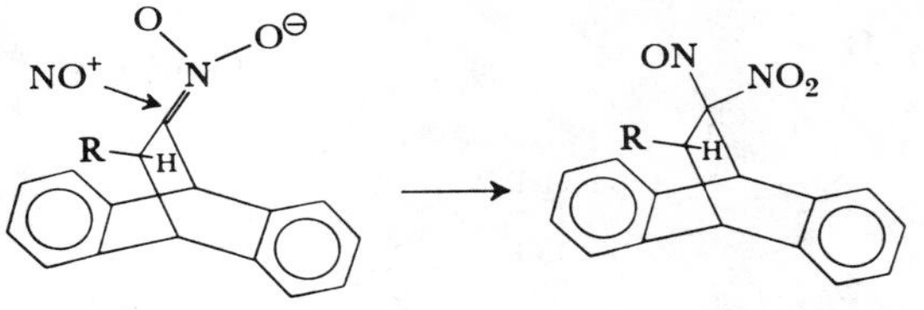

(56)

(**60a**) R = CH_3 (**61a**) (83%, R = CH_3)
(**60b**) R = C_6H_5 (**61b**) (58%, R = C_6H_5)

and nitro (product development control) in the product **61**; repulsion of R and incipient NO_2 in the transition state may account for the result.

Since nitrous acid may be formed from nitrite on acidification of nitronate salts (the nitrite formed *in situ* by air oxidation of nitronates[280] or autooxidation-reduction of nitronic acids[47]), pseudonitrole and nitrolic acid formation is a side reaction which often results on regeneration of nitroalkanes from their salts[160,180,226,249,281–283]. This reaction is favored by use of aged nitronate solutions at 0–25°, and by slow addition of the nitronate salt to the acid[226,283]. For example, sodium bicyclo[2,2,1]heptane-2-nitronate when added slowly to aqueous hydrochloric acid at room temperature leads to a 20% yield of pseudonitrole dimer **62**[226]. Conditions have been developed for

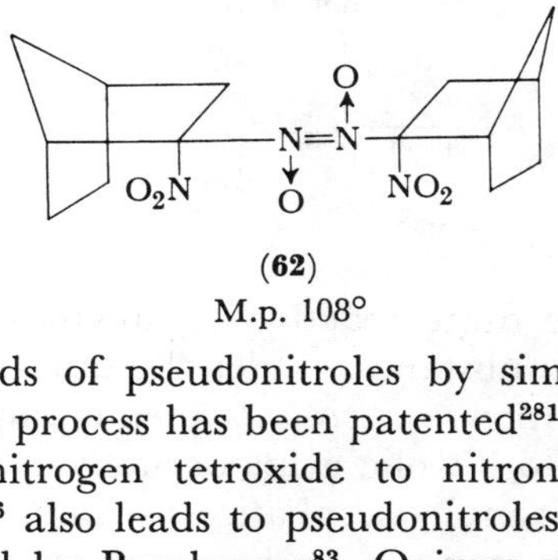

(**62**)
M.p. 108°

securing high yields of pseudonitroles by simple acidification of nitronate salts; the process has been patented[281–283].

Addition of dinitrogen tetroxide to nitronic acids[181,182,284] or nitronate salts[285,286] also leads to pseudonitroles or nitrolic acids, a reaction discovered by Bamberger[83]. Oximes also may be used as reactants[83,181,182,285]. Excess N_2O_4 converts phenylpseudonitrole into α,α-dinitrotoluene[181,182,284] (equations 57–59).

$$C_6H_5C(CH_3){=}NO_2^-Na^+ + N_2O_4 \xrightarrow[Et_2O]{} C_6H_5C(CH_3)(NO_2)NO + NaNO_3 \qquad (57)$$

$$C_2H_5O_2CCH{=}NO_2^-Na^+ + N_2O_4 \xrightarrow[Et_2O]{} C_2H_5O_2C(NO_2)C{=}NOH + NaNO_3 \qquad (58)$$

$$C_6H_5CH{=}NO_2H \xrightarrow{N_2O_4,\ CCl_4,\ 0^\circ} C_6H_5CH(NO)NO_2$$

$$C_6H_5CH{=}NOH \xrightarrow{N_2O_4,\ CCl_4,\ 0^\circ} C_6H_5CH(NO)NO_2$$

$$C_6H_5CH(NO)NO_2 \longrightarrow C_6H_5CH(NO_2)\overset{O}{\underset{}{\uparrow}}N{=}\underset{\downarrow O}{N}CH(NO_2)C_6H_5 \qquad (59)$$

$$C_6H_5CH(NO)NO_2 \xrightarrow{N_2O_4} C_6H_5CH(NO_2)_2$$

Attempts to convert 1,1-dinitroalkanes into 1,1-dinitro-1-nitrosoalkanes by this method have led to other products[181]. Potassium 1-nitroethanenitronate forms acetonitrolic acid, presumably due to hydrolysis by adventitious water of the observed blue intermediate, 1,1-dinitro-1-nitrosoethane (**63**)[286] (equation 60). α-Nitrophenylmethanenitronate forms trinitromethylbenzene by oxidation of the intermediate nitroso compound **64**[181,284] (equation 61).

$$CH_3C(NO_2){=}NO_2^-K^+ \xrightarrow[Et_2O]{N_2O_4} CH_3C(NO)(NO_2)NO_2 \xrightarrow{H_2O} CH_3C(NO_2){=}NOH \qquad (60)$$

(**63**) Blue
(not isolated)

$$C_6H_5C(NO_2){=}NO_2^-K^+ \xrightarrow[Et_2O]{N_2O_4} C_6H_5C(NO)(NO_2)NO_2 \xrightarrow{N_2O_4} C_6H_5C(NO_2)(NO_2)NO_2 \qquad (61)$$

(**64**)

Pseudonitroles are quite reactive compounds and may become converted into other substances under the usual preparation conditions. The isomerization of primary pseudonitroles (those having an *alpha* hydrogen) to nitrolic acids occurs very readily in the aliphatic series. No simple aliphatic pseudonitrole of structure $RCH(NO)NO_2$ appears to have been described. Phenyl pseudonitrole, $C_6H_5CH(NO)NO_2$, has been prepared and isolated as its dimer which was found to be unstable at room temperature, decomposing in one day, and melting with explosive decomposition[181]. With aqueous alkali it readily isomerizes to benzonitrolic acid[181]. Dimers derived from disubstituted pseudonitroles, $R^1R^2C(NO)NO_2$, are stable[279,282]. Benzonitrolic acid decomposes on standing in nitrous acid solution, or on warming, to yield diphenylfuroxan

(**65**)[277]; benzonitrile oxide may be an intermediate (equation 62).

$$C_6H_5\underset{\underset{NO}{|}}{C}HNO_2 \xrightarrow{OH^-} C_6H_5\underset{\underset{NO_2}{|}}{C}{=}NOH$$

$$C_6H_5\underset{\underset{NO_2}{|}}{C}{=}NOH \xrightarrow[\text{Warm}]{-HNO_2} [C_6H_5C{\equiv}NO] \longrightarrow C_6H_5\text{-(furoxan ring: N-O-N}\rightarrow\text{O)-}C_6H_5 \qquad (62)$$

The nitration of dipotassium tetranitroethane with mixed acid (5–70°) to produce hexanitroethane in 90% yield should be considered an electrophilic addition of nitronium ion to a bisnitronate ion[287] (equation 63).

$$\ominus\,(O_2N)_2\ddot{C}{-}\ddot{C}(NO_2)_2\,\ominus + 2\,NO_2^+ \xrightarrow{(H_2SO_4,\ HNO_3)} \underset{\text{M.p. }142^\circ}{\underset{(90\%)}{(O_2N)_3C{-}C(NO_2)_3}} \qquad (63)$$

The conversion of the trinitromethyl substituted nitronic acid **66** into a bisdinitromethyl derivative **67** is an interesting rearrangement[288,289]. The reaction, carried out in ethanol with potassium acetate[288] or ammonium hydroxide[289], may involve intramolecular or intermolecular nitration of the intermediate nitronate ion by the ω-trinitromethyl group (equation 64).

$$\underset{(\mathbf{66})}{(O_2N)_3CCH_2\underset{\underset{R}{|}}{C}{=}NO_2H} \xrightarrow[\text{EtOH}]{\text{KOAc}} \underset{(\mathbf{67})}{(O_2N)_2CHCH_2\underset{\underset{R}{|}}{C}(NO_2)_2} \qquad (64)$$

$R = H,\ CH_3,\ C_2H_5$

Nitronic acids, nitronate salts, and certain nitroalkanes react with diazonium salts to form α-nitroaldehyde hydrazones in high yield[6,170,259,290–299]. The reaction was discovered by Victor Meyer[290,291]. It involves an addition to a nitronate anion (equation 65). The

$$C_6H_5CH{=}NO_2H \rightleftharpoons C_6H_5CH{=}NO_2^- + H^+$$

$$C_6H_5CH{=}NO_2^- + C_6H_5N_2^+ \longrightarrow C_6H_5\underset{\underset{NO_2}{|}}{C}{=}NNHC_6H_5 \quad \text{M.p. } 101.5\text{–}102^\circ \qquad (65)$$

$$CH_3CH{=}NO_2^-Na^+ + C_6H_5N_2^+ \longrightarrow CH_3\underset{\underset{NO_2}{|}}{C}{=}NNHC_6H_5 \quad 92\%;\ \text{M.p. } 141\text{–}142^\circ$$

α-nitroaldehyde hydrazones react with diazomethane to produce orange-red methyl esters of phenylazonitronic acids (see section IV.A.1)[6,295–299].

Phenyl isocyanate reacts with arylmethanenitronic acids to produce diphenylurea[33] and unidentified oils[32,201].

2. Oxidation and reduction reactions

The reaction of nitronic acids with oxidizing agents has not been extensively studied. Bornane-2-nitronic acid is oxidized to camphor with permanganate[201,230] (equation 66). Oxidation of 2-(nitromethyl)-alkanenitronic acids (formed *in situ* from the sodium salts) occurs

$$\text{Bornane-2-nitronic acid } (=NO_2H) \xrightarrow{KMnO_4} \text{camphor } (=O) \qquad (66)$$

smoothly with nitric acid (but not sulfuric acid) to yield 2-nitromethyl-alkanoic acids[232] (equation 67). The hindered nitronic acid,

$$(CH_3)_2C(CH_2NO_2)CH{=}NO_2H \xrightarrow[90\text{–}95^\circ]{HNO_3} (CH_3)_2C(CH_2NO_2)CO_2H \quad 66\% \qquad (67)$$

$(C_6H_5)_2CHC(C_6H_5){=}NO_2H$, was found to be inert to sodium peroxide or ozone[65]. The oxidation of secondary nitronate anions by oxygen in basic solution to yield ketones[280] is not observed in acid solution.

Reduction of nitronic acids to oximes occurs readily in excellent yields, with a wide variety of reducing agents (Table 9). The polarographic half-wave potential at pH 0 for reduction of propane-2-nitronic acid to acetoxime is -0.9 v; at pH 2, $E_{\frac{1}{2}} = 1.05$ v[44]

$$(CH_3)_2C{=}NO_2H + 2\,H^+ + 2e^- \longrightarrow (CH_3)_2C{=}NOH + H_2O \qquad (68)$$

(equation 68). Reduction of cyclohexanenitronic acid to cyclohexanone oxime (70–80 % yield) has been accomplished with several reducing agents (see Table 9 (equation 69)). Hydroxylamine reduces cyclododecanenitronic acid to the oxime in 91 % yield[203].

$$C_6H_{10}{=}NO_2H + H_2S \longrightarrow C_6H_{10}{=}NOH + H_2O + S \quad 77\% \qquad (69)$$

TABLE 9. Reduction of nitronic acids to oximes.

Nitronic acid	Reducing agent	Yield oxime (%)	Ref.
$CH_3CH{=}NO_2H$	HI	100[a]	276
$CH_3CH_2CH{=}NO_2H$	HI	100[a]	276
$(CH_3)_2C{=}NO_2H$	HI	100[a]	276
cyclohexylidene$=NO_2H$	HI	100[a]	276
	H_2S	77	300, 301
	$H_2S_2O_3$	80	302
	NH_2OH	—	303
	NH_4Cl, CH_3OH	70	304
$4\text{-}BrC_6H_4CH{=}NO_2H$	NaHg	—	33
$C_6H_5CH{=}NO_2H$	NaHg	—	33
$(CH_2)_{10}$ ring: $-C{=}NO_2H$, $-CH_2$	NH_2OH	91	203
$(C_6H_5)_2CHC(C_6H_5){=}NO_2H$	AlHg	—	65

[a] Quantitative analytical method; product not isolated.

Phenylmethanenitronic acid is reduced to the oxime with zinc and alkali and sodium amalgam[33]; aluminum amalgam has also been used for nitronic acid reduction[65]. Complete reduction of a nitronic acid to the corresponding amine appears not to have been reported; catalytic (Pt) hydrogenation should effect it.

The mechanism of nitronic acid reduction may involve a radical-chain process initiated by electron transfer between nitronic acid and nitronate anion (equation 70)[305]. Reduction of the radical-anion intermediate **69** could include steps 71–74. These suggested steps include dissociation of **69** into oximate ion **70** and hydroxyl radical (**71**) (equation 71), followed by reduction of hydroxyl by hydrogen

$$R_2C{=}NO_2H + R_2C{=}NO_2^- \longrightarrow \underset{(\mathbf{68})}{R_2\dot{C}NO_2} + \underset{(\mathbf{69})}{R_2\dot{C}N\begin{matrix} O^- \\ OH \end{matrix}} \qquad (70)$$

$$\underset{(\mathbf{69})}{R_2\dot{C}N\begin{matrix} O^- \\ OH \end{matrix}} \longrightarrow \underset{(\mathbf{70})}{R_2C{=}NO^-} + \underset{(\mathbf{71})}{HO^\cdot} \qquad (71)$$

$$\underset{(\mathbf{71})}{HO^\cdot} + H_2S \longrightarrow H_2O + \underset{(\mathbf{72})}{HS^\cdot} \qquad (72)$$

$$R_2\dot{C}N\begin{matrix}O^-\\ \diagup\\ \diagdown\\ OH\end{matrix} + H_2S \longrightarrow R_2C{=}NO^- + H_2O + HS^\cdot \quad (73)$$

(69) (70) (72)

$$R_2C{=}NO_2^- + HS^\cdot \longrightarrow R_2\dot{C}N\begin{matrix}O^-\\ \diagup\\ \diagdown\\ OH\end{matrix} + S \quad (74)$$

(69)

sulfide to hydrosulfide radical (**72**) (equation 72). Alternatively, a concerted reaction of **69** with hydrogen sulfide (equation 73) may be more likely to occur in the presence of the reducing agent. Finally, another electron exchange (equation 74) would regenerate **69** and continue the chain.

An interesting and complex reaction exhibited by nitronic acids is an autooxidation-reduction[13,27,47,48,65,84,203,204,226,296,306,307]. This process has been observed in solution and in the solid state. Oxime is a characteristic product. Other products are ketones, substituted 1,2-dinitroethanes, nitrolic acids, nitrous acid, and oxides of nitrogen.

In solution the reaction is dependent on the structure of the nitronic acid and on the pH. It has been studied quantitatively in dilute aqueous solution by Armand[47,48] (Table 7, section III.C.1.a., summarizes some of the data). For example, cyclohexanenitronic acid at pH 2.4 produces the following products[47] (equation 75). Tautomerization to nitrocyclohexane is important only at higher pH.

$$C_6H_{10}{=}NO_2H \xrightarrow[\text{pH 2.4}]{H_3O^+} C_6H_{10}{=}O + C_6H_{10}{=}NOH + C_6H_{10}(NO)(NO_2) \quad (75)$$

38% 31% 30%

At lower pH (<1) one obtains only the Nef product, cyclohexanone. It is to be noted that oxime is not a Nef product. The autooxidation-reduction reaction is catalyzed by acids. At certain acid concentrations (pH 2–4) it competes with tautomerization and Nef reaction. Secondary nitronic acids undergo the reaction much more readily than primary.

A mechanism for the reaction is suggested by the facts above. Nitronic acids are very easily reduced to oximes and are present in unprotonated form in rather high concentration at pH 2–4. The initial step(s) is the Nef hydrolysis (equation 76). The reducing agent is believed to be the Nef hydrolysis product, nitroxyl, or its equivalent (equation 77); *cf.* the mechanism of oxime formation (equations 70–74). The stoichiometry of the process which indicates that

$$R^1R^2C{=}N(\rightarrow O)OH \xrightarrow{H_3O^+} R^1R^2C{=}O + H^+ + HNO + H_2O \qquad \text{Nef (76)}$$

$$R^1R^2C{=}N(\rightarrow O)OH + HNO \longrightarrow R^1R^2C{=}NOH + HNO_2 \qquad \text{Oxidation-reduction (77)}$$

$$R^1R^2C{=}N(O^{\ominus})O + NO^+ \longrightarrow R^1R^2C(NO)(NO_2) \qquad \text{Nitrosation (78)}$$

oxime and pseudonitrole form in equal amounts, and in yields always less than that of ketone, suggests an immediate reaction of nitrous acid with remaining nitronic acid (equation 78). A relatively slow tautomerization of nitronic acid to nitroalkane (observed with secondary nitronic acids) evidently favors the oxidation process.

The conversion of nitronic acids into oximes by boiling in ethanol solvent[84,203] could involve decomposition of an ethyl nitronate (see section IV.C.2), as well as autooxidation-reduction.

Decomposition of nitronic acids occurs in the absence of solvents[13,27,65,296], but this process has received no systematic study. Bamberger[296] and Konowalow[13] observed the facile formation of benzophenone and its oxime from diphenylmethanenitronic acid. The decomposition, which is believed to include autooxidation-reduction, often leads to oxime, ketone, and oxides of nitrogen. Tautomerization also occurs. The decomposition is accelerated by traces of water. It is inhibited by accumulation of bulky groups about the nitronate carbon (see half-lives listed in Table 6, section III.A) which also inhibits Nef hydrolysis.

An interesting intramolecular oxidation-reduction is observed with the very hindered nitronic acid **73**, which does not tautomerize to nitroalkane nor undergo the Nef reaction. In methanolic hydrogen chloride **73** forms the 3,4,4-triphenyl-2-isoxazoline (**74**) by participation of the neighboring alkyl group, CH_2R[65] (equation 79).

$$(C_6H_5)_2C(CH_2R)C(C_6H_5){=}NO_2H \xrightarrow[CH_3OH]{HCl} \text{(74)} + H_2O \qquad (79)$$

(**73**) $R = H, CH_3$ (**74**) 70%

Bimolecular coupling products (1,2-dinitroethanes such as **76**) are obtained from fluorene-9-nitronic acid (**75**) and its ring-substituted derivatives by warming in ethanol[84,205,308–310]. Fluorenone oxime (**77**) is also formed in 26% yield[84] (equation 80). Although other nitronic acids have not been observed to undergo this reaction, nitronate salts can form bimolecular coupling products on oxidation[311,312]. Dimer **76** is also prepared in quantitative yield

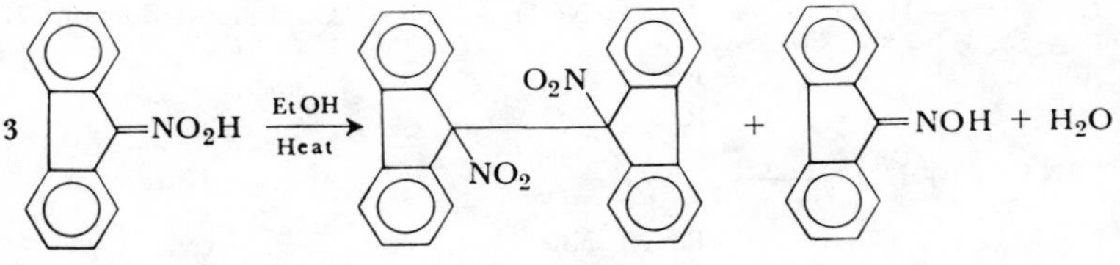

$$3\ (\text{fluorene})\!=\!NO_2H \xrightarrow[\text{Heat}]{\text{EtOH}} (\text{9-nitrofluoren-9-yl})_2 + (\text{fluorene})\!=\!NOH + H_2O \qquad (80)$$

(**75**) (**76**) (**77**)

by reaction of the potassium salt of **75** with iodine[313], or by heating 9-iodo-9-nitrofluorene[314]. It may also be prepared in 71% yield by electrolysis of the **75** salt[313].

Formation of dimer **76** is believed to involve the spontaneously initiated process leading to radicals **68**, **69** (equation 70)[305,312,314–316]. The following equations (81–86) are suggested to explain dimerization

$$R_2\dot{C}NO_2 + R_2C{=}NO_2^- \longrightarrow R_2C(NO_2)\text{—}CR_2NO_2^{\dot{-}} \qquad (81)$$

(**68**) (**78**)

$$R_2C(NO_2)\text{—}CR_2NO_2^{\dot{-}} + R_2C{=}NO_2H \longrightarrow R_2C(NO_2)\text{—}CR_2(NO_2) + R_2\dot{C}N(O^-)(OH) \qquad (82)$$

(**78**) (**79**) (**69**)

$$R_2\dot{C}N(O^-)(OH) \longrightarrow R_2C{=}NO^- + HO^{\cdot} \qquad (83)$$

(**69**) (**70**) (**71**)

$$HO^{\cdot} + R_2C{=}NO_2^- \longrightarrow OH^- + R_2\dot{C}NO_2 \qquad (84)$$

(**71**) (**68**)

$$R_2\dot{C}N(O^-)(OH) + R_2C{=}NO_2H \longrightarrow R_2C{=}NO^- + R_2\dot{C}NO_2 + H_2O \quad (85)$$

(**69**) (**70**) (**68**)

$$R_2C{=}NO^- + H^+ \longrightarrow R_2C{=}NOH \quad (86)$$

(**70**)

and oxime formation. Formation of a radical anion intermediate (**78**) (equation 81) would be favored over a reaction between radicals **68** and **69**[312,315]. An exchange would lead to dimer **79** (equation 82). Oxime formation is explained by the sequence suggested for nitronic acid reduction (equation 83) involving dissociation of **69** into oximate ion **70** and hydroxyl radical **71**. Radical **68** is regenerated by electron exchange between hydroxyl radical and nitronate ion (equation 84). Alternatively, a direct electron transfer between **69** and nitronic acid could lead to the same result (equation 85). Finally, protons made available by required ionization of the nitronic acid can produce oxime (equation 86).

The decomposition of *p*-bromophenylcyanomethanenitronic acid (**80**) into dimeric products, in the solid state or in benzene solution, may be a radical process[196]. Gentle heating leads to a 1-nitro-1,2-dicyano derivative **81** (equation 87); a 1,2-dinitroethane derivative was not isolated[196]. Prolonged heating or a slightly higher temperature leads to a 1,2-dicyanostilbene (**82**). *o*-Bromophenylcyanomethanenitronic acid behaves similarly[193]. In dilute aqueous sulfuric

$$p\text{-BrC}_6H_4C(CN){=}NO_2H \xrightarrow[C_6H_6,\ \text{warm}]{-HNO_2} p\text{-BrC}_6H_4C(NO_2)(CN){-}CH(CN)C_6H_4Br\text{-}p \quad (87)$$

(**80**) (**81**) 30%, M.p. 130–134°

81 $\xrightarrow{C_6H_6,\ \text{Reflux}-HNO_2}$ **82**; **80** $\xrightarrow{H_2SO_4,\ \text{aq.},\ 25°}$ **82**

$$p\text{-BrC}_6H_4C(CN){=}C(CN)C_6H_4Br\text{-}p \quad (88)$$

(**82**) M.p. 214–215°

acid solution at room temperature **80** is converted quantitatively into the dicyanostilbene derivative **82**[196,197] (equation 88). The mechanism may involve combination of spontaneously initiated radical and radical anion, with loss of nitrite from the initially formed adduct **83** (equation 89).

$$\mathrm{ArC(CN){=}NO_2H + ArC(CN){=}NO_2^- \longrightarrow Ar\dot{C}(CN)NO_2 + Ar\dot{C}(CN)N(O^-)OH}$$

$$\mathrm{Ar\dot{C}(CN)NO_2 + Ar\dot{C}(CN)N(O^-)OH \longrightarrow ArC(CN)(NO_2)C(Ar)(CN)N(O^-)OH \; \textbf{(83)}} \qquad (89)$$

$$\mathbf{83} \longrightarrow \mathrm{ArC(CN)(NO_2)CH(CN)Ar + NO_2^-}$$

Heating strongly in aqueous alkali converts the nitronate salt **84** into an unsubstituted stilbene **85**[196] (equation 90). Certain other aryl nitronate salts behave similarly[193,317].

$$\underset{\textbf{(84)}}{p\text{-}\mathrm{BrC_6H_4C(CN){=}NO_2^-Na^+}} \xrightarrow[\text{150–160°, 5–6 h}]{\text{NaOH, aq.}} \underset{\textbf{(85)}\ 70\text{–}80\%}{p\text{-}\mathrm{BrC_6H_4CH{=}CHC_6H_4Br}\text{-}p} \qquad (90)$$

3. Reactions of α-halonitronic acids

The α-halonitronic acids, $\mathrm{RC(X){=}NO_2H}$ (X = halogen), are somewhat unique in the ease with which they undergo displacement of halide ion by various nucleophiles. Reactions of these substances are believed to occur with a nitronic acid, rather than a nitronate intermediate. Examples of such reactions include the ter Meer[318] hydrolysis to carboxylic acids[319], and coupling to 1,2-dinitroethylenes[313,320]. α-Halonitronic acids are reactive and attempts to isolate them have failed[32]. They are readily reduced to halide and nitronate anion[321].

The ter Meer reaction[318] involves a displacement of halide by weakly nucleophilic nitrite ion[322]. For example, 1,1,4,4-tetranitrobutane (**86**) may be prepared by reaction of the dipotassium salt of 1,4-dibromo-1,4-dinitrobutane with potassium nitrite[323] (equation

$$\mathrm{K^{+}\,{}^{-}O_2N{=}C(Br)CH_2CH_2C(Br){=}NO_2^-K^+ + 2\ KNO_2 \longrightarrow}$$

$$\mathrm{K^{+}\,{}^{-}O_2N{=}C(NO_2)CH_2CH_2C(NO_2){=}NO_2^-K^+ + 2\ KBr}$$

$$\xrightarrow{\mathrm{HCl,\ Et_2O}} \underset{\textbf{(86)}}{\mathrm{(O_2N)_2CHCH_2CH_2CH(NO_2)_2}} \qquad (91)$$

91). The mechanism is depicted as a displacement on the α-halonitronic acid[322] (equation 92).

$$\underset{\displaystyle |\atop Br}{RC}{=}NO_2^- \underset{}{\overset{H^+}{\rightleftharpoons}} \underset{\displaystyle |\atop Br}{RC}{=}NO_2H \xrightarrow[-Br^-]{NO_2^-} \underset{\displaystyle |\atop NO_2}{RC}{=}NO_2H \overset{-H^+}{\rightleftharpoons} \underset{\displaystyle |\atop NO_2}{RC}{=}NO_2^- \xrightarrow{H^+} \underset{\displaystyle |\atop NO_2}{RCHNO_2} \quad (92)$$

The mechanism of α-chloronitroalkane hydrolysis has been studied[319,324]. The reaction is interpreted as a displacement of chloride, by water, from an α-chloronitronic acid intermediate (equation 93).

$$CH_3\underset{\displaystyle |\atop Cl}{C}{=}NO_2H + H_2O \xrightarrow{-Cl^-} CH_3\underset{\displaystyle |\atop OH}{C}{=}NO_2H \xrightarrow{H_3O^+} CH_3CO_2H + N_2O \quad (93)$$

Halonitromethanenitronic acids are unstable[32,325,326]. They undergo a complex rearrangement to dihalodinitromethanes[325] (equation 94).

$$Br\underset{\displaystyle |\atop NO_2}{C}{=}NO_2^-K^+ \xrightarrow{H_2SO_4} Br\underset{\displaystyle |\atop NO_2}{C}{=}NO_2H \longrightarrow Br\overset{\displaystyle Br\atop |}{\underset{\displaystyle |\atop NO_2}{C}}NO_2 \quad (94)$$

1,2-Dinitroethylene coupling products of α-halonitronic acids apparently are formed by displacement of halide—by attack of nitronate anion on an α-halonitronic acid[320]. For example, 1,2-dinitro-2-butene may be formed in 36% yield from 1-chloro-1-nitroethane by treatment with ca. one mole-equivalent of aqueous sodium hydroxide solution at 10–15° (pH ca. 9) (equation 95).

$$CH_3\underset{\displaystyle |\atop Cl}{C}{=}NO_2H + CH_3\underset{\displaystyle |\atop Cl}{C}{=}NO_2^- \longrightarrow HO_2N{=}\underset{\displaystyle |\atop CH_3}{C}{-}\overset{\displaystyle CH_3\atop |}{\underset{\displaystyle |\atop Cl}{C}}NO_2 + Cl^-$$

$$HO_2N{=}\underset{\displaystyle |\atop CH_3}{C}{-}\overset{\displaystyle CH_3\atop |}{\underset{\displaystyle |\atop Cl}{C}}NO_2 \xrightarrow{-H^+} {}^-O_2N{=}\underset{\displaystyle |\atop CH_3}{C}{-}\overset{\displaystyle CH_3\atop |}{\underset{\displaystyle |\atop Cl}{C}}NO_2 \xrightarrow{-Cl^-} \underset{\displaystyle 36\%}{O_2N\underset{\displaystyle |\atop CH_3}{C}{=}\overset{\displaystyle CH_3\atop |}{C}NO_2} \quad (95)$$

Coupling of α-halonitroalkanes and nitronate salts can lead to 1,2-dinitroethanes[84,310,313,327,328] (equation 96). This reaction may be conducted *in situ* by treating nitronate salts with iodine[84,310,313,327]. Such reactions apparently involve nitronate ions rather than nitronic

$$2\,C_6H_5CH{=}NO_2^-Na^+ + I_2 \longrightarrow C_6H_5CH(NO_2)CH(NO_2)C_6H_5 + 2\,NaI \qquad (96)$$

M.p. 155°
(low melting isomer)[327]

acids. They may proceed by displacement[84] or radical-anion[84,314] mechanisms.

4. Reactions of ketonitronic acids

Several α- and γ-ketonitronic acids have been described and their somewhat unique chemistry is discussed in this section. Ketonitronic acids are usually prepared by acidification of their alkali metal salts; the salts of α-nitroketones are conveniently prepared by the alkaline nitration of ketones[329,330]. α-Nitroketones may be prepared by reaction of α-bromoketones with silver nitrite[331] or sodium nitrite[332].

For all α-ketonitronic acids there exist three possible tautomeric forms: *aci* or α-ketonitronic acid (**87**), keto (**88**), and enol (**89**) (equation 97).

$$R^1C({=}O)C(R^2){=}NO_2H \rightleftharpoons R^1C({=}O)CH(R^2)NO_2 \rightleftharpoons R^1C(OH){=}C(R^2)NO_2 \qquad (97)$$

(**87**) *Aci* α-Ketonitronic acid — (**88**) Keto — (**89**) Enol

Compounds representing each of the three forms have been reported. In solution the three forms can exist in tautomeric equilibrium with the common anion, $R^1COC(R^2){=}NO_2^-$. The interconversion is catalyzed by bases and acids.

The relative concentration of **87**, **88**, and **89** may be measured by ultraviolet, infrared, and nmr spectroscopy[330,333,334], and with the aid of bromine titration[36]. A complication lies in the presence of the fourth species, the common anion, particularly in protic solvents. The composition of the equilibrium mixture is solvent dependent (Table 10)[330]. The keto form of α-nitroketones (**88**) is favored in polar protic solvents such as ethanol as one observes with β-dicarbonyl compounds[338]. Enol form **89** is favored more in aprotic solvents such as carbon tetrachloride or hexane suggesting an intramolecularly hydrogen-bonded form[330]. Its concentration usually remains low, however.

The amount of α-ketonitronic acid (**87**) present in these equilibria is believed to be quite small. However, freshly prepared solutions, obtained by acidification of salts of α-ketonitronic acids, probably contain relatively high concentrations of nitronic acid form; on standing the nitroketone usually results[329].

The acyclic α-ketonitronic acids derived from α-nitroacetophenone and α-nitroacetone were studied earlier by Hantzsch[32,339,340] and by others[36,249,331]. Other examples have been studied more recently[333,341]. The nitronic acid appears to be the least favored form at equilibrium. The keto form predominates (ca. 99%) in protic and aprotic solvents for aromatic and aliphatic acyclic nitro ketones.

The properties of alicyclic α-ketonitronic acids (**91**) generally resemble those of their acyclic counterparts. Alkali salts of alicyclic α-ketonitronic acids (**90**) have been prepared frequently since they are readily available by alkaline nitration of ketones[329,330]. An oil is obtained by acidification of the C_6 potassium salt (**90**, $n = 4$) with dilute sulfuric acid at 0°; the oil, possibly a mixture of tautomers

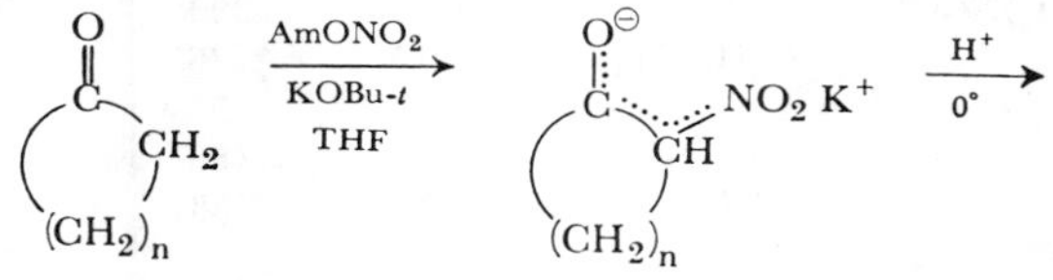

$$\text{cyclic ketone } C(=O)CH_2(CH_2)_n \xrightarrow[\text{KOBu-}t,\ \text{THF}]{\text{AmONO}_2} C(O^{\ominus})\cdots CH \cdots NO_2\ K^+ \ (\mathbf{90}) \xrightarrow[0°]{H^+}$$

$n = 4$–8; 10

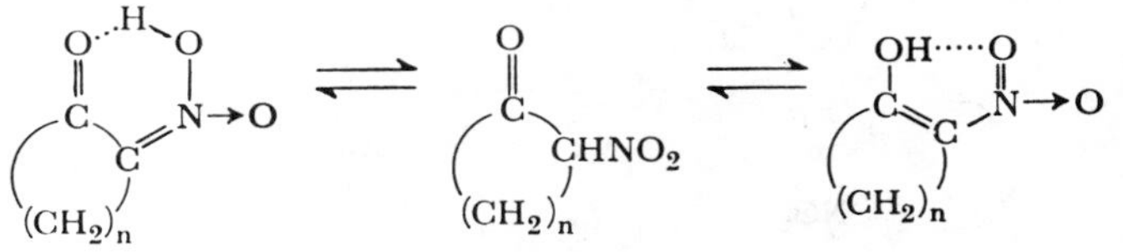

$$(\mathbf{91}) \rightleftharpoons (\mathbf{92})\ C(=O)CHNO_2(CH_2)_n \rightleftharpoons (\mathbf{93})\ C(OH)=C-N(\rightarrow O)O \qquad (98)$$

91, **92**, and **93** (equation 98), slowly crystallizes on standing to form α-nitrocyclohexanone (**92**, $n = 4$; m.p. 39.5–40°)[329]. Solutions of the freshly prepared oil give a red color with ferric chloride and are acidic.

In solution in aprotic solvents alicyclic α-nitrocycloalkanones appear to exist to some extent in the enol form **93**[334,342]. However, in protic solvents the keto form is strongly favored (Table 10). Ring size affects the enol content in carbon tetrachloride solution[330]; C_6, C_8, and C_{10} α-nitrocycloalkanones have higher enol contents than C_7, C_9, and C_{12} homologs.

Table 10. Equilibrium composition of nitroketones.

Nitroketone	Solvent	% Keto	Ref.
$CH_3CH_2CH_2COCH_2NO_2$	CCl_4	100	333
$CH_3CH_2COCH(CH_3)NO_2$	CCl_4	100	333
$CH_3COC(CH_3)_2NO_2$	CCl_4	100	333
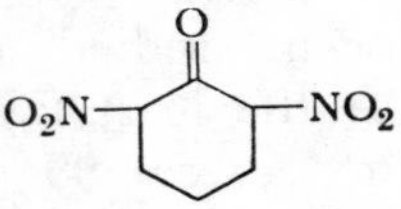 O, O_2N, NO_2	CH_2Cl_2	25	335
	$(CD_3)_2SO_2$	100	335
	EtOH	100	335
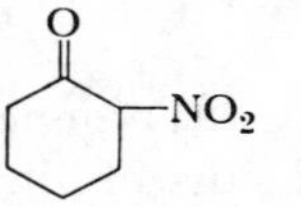 O, NO_2	CCl_4	50	330
	$CDCl_3$	69.4	333
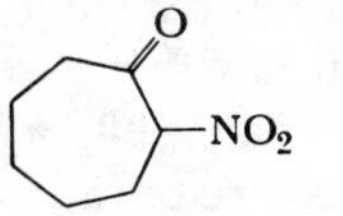 O, NO_2	CCl_4	100	330
$CH_3CH_2CH_2COCH(C_2H_5)NO_2$	CCl_4	100	330
$C_6H_5COCH_2NO_2$	$C_6H_5CH_3$	89.7	36
	C_2H_5OH	94.7	36
	CH_3OH	97.2	36
	$CDCl_3$	100	333
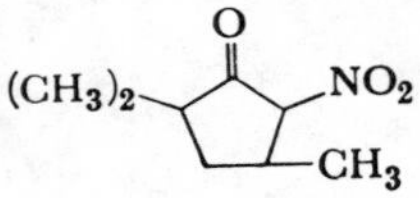 O, $(CH_3)_2$, NO_2, CH_3	CCl_4	100	330
O, NO_2	CCl_4	70	330
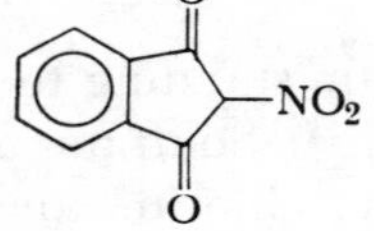 O, O, NO_2	H_2O	2	336
	EtOH	10	336
	CH_3CO_2H	12	336
	Et_2O	62	336
	C_6H_6	90	336
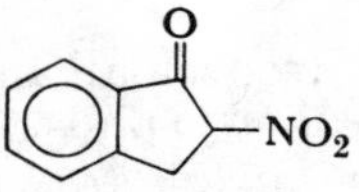 O, NO_2	n-C_6H_{14}	10	330
	CCl_4	70	330
	EtOH	70	330
$C_6H_5COCH(CH_3)NO_2$	neat	100	333
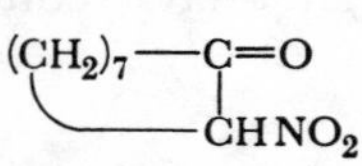 $(CH_2)_7$—C=O, CHNO$_2$	CCl_4	90	330

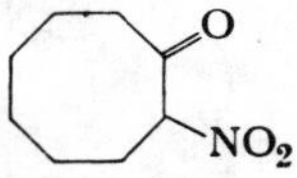

(*continued*)

TABLE 10—*continued*

Nitroketone	Solvent	% Keto	Ref.
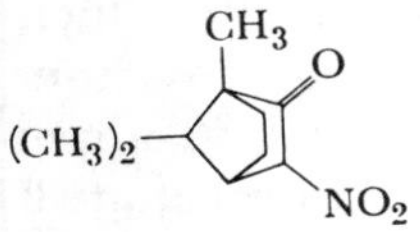 CH_3, O, $(CH_3)_2$, NO_2	C_6H_6 D_2O	93 100	337 337a
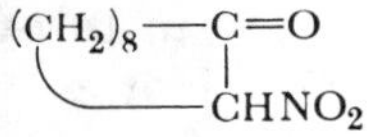 $(CH_2)_8$—C=O, $CHNO_2$	CCl_4	60	330
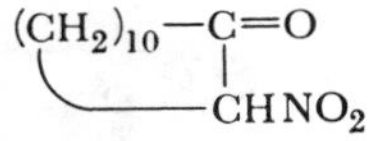 $(CH_2)_{10}$—C=O, $CHNO_2$	CCl_4	100	330
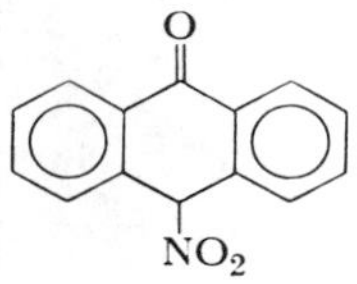 O, NO_2	EtOH	97	35
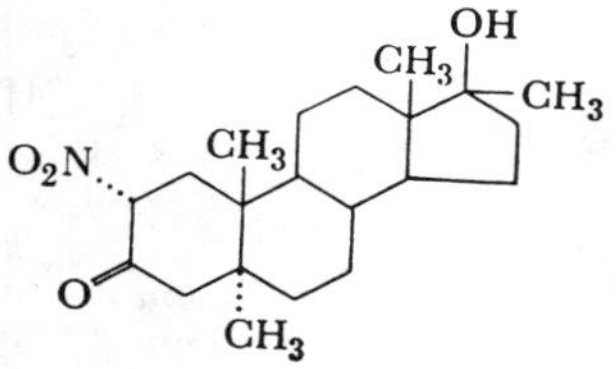 OH, CH_3, CH_3, O_2N, CH_3, O, CH_3	$CDCl_3$	50	334

Spectroscopic evidence strongly supports the presence of the enol **93** rather than nitronic acid form **91** in aprotic solvents[330]. For α-nitrocyclohexanone a sharp OH peak at $-3.6\ \tau$ (CCl_4), intensity $\simeq 0.5$ proton, is observed in the nmr spectrum of the equilibrated mixture. In addition there appears a weak band at 1613 cm^{-1} in the neat sample (possible C=C), and NO_2 bands at 1550 and 1515 cm^{-1} representing unconjugated and conjugated nitro groups, respectively. Other α-nitrocycloalkanones (C_7–C_{12}) have similar spectra. The carbonyl band appears in *reduced* intensity near 1720–1740 cm^{-1} in carbon tetrachloride solution indicating presence of α-nitroketone **92** rather than α-ketonitronic acid **91**; also the latter might be expected to have a carbonyl band near 1639 cm^{-1} as found in the salts **90.** No sharp OH stretching bands are found in the infrared spectra; only very broad bands, characteristic of more associated protons, occur. The ultraviolet spectrum of α-nitrocyclohexanone in carbon tetrachloride reveals a strong band at 320 mμ (ε 3970)[330]. Other α-nitrocycloalkanones (C_7–C_{12}) have similar ultraviolet spectra (λ_{max} 320–370 mμ; ε_{max} 1700–4000), which could be assigned to the enol form[330]. Alkali salts of α-nitroketones also have strong bands at ca. 340 mμ (ε 12000) in absolute ethanol[330].

Comparison of the ultraviolet spectra of the α-nitroketone **94** with that of its nitronic ester **98** and enol ether **97** (prepared by reaction of **94** with diazomethane) indicate very little keto form to exist in 96% ethanol solution[343–345] (equations 99, 100). However, the relative concentrations of enol **95** and nitronic acid **96** cannot be

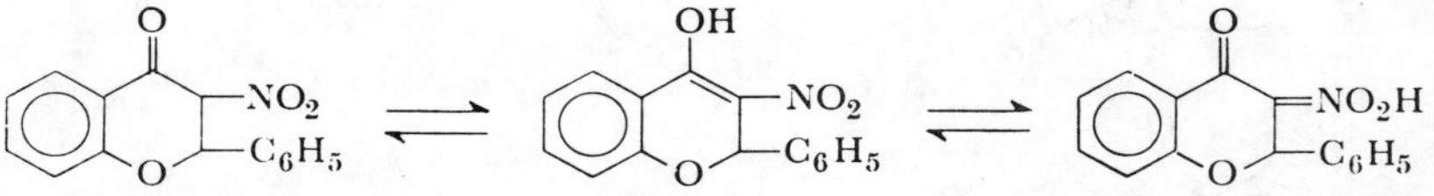

(**94**) ⇌ (**95**) ⇌ (**96**) (99)

(**94, 95, 96**): λ_{max}^{EtOH} 255(8450), 328 (4970), and 378 (11,100) mμ

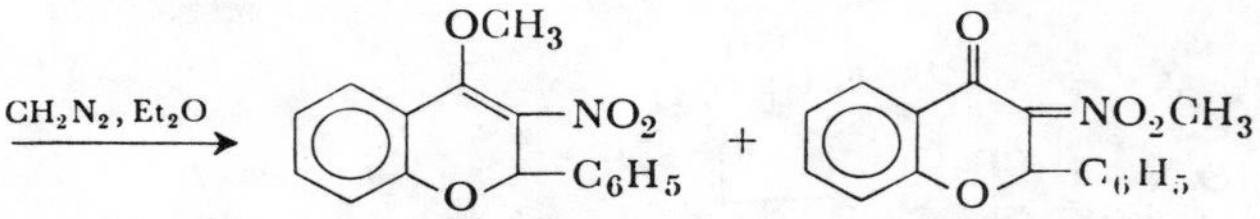

$\xrightarrow{CH_2N_2,\ Et_2O}$ (**97**) 30% + (**98**) 63% (100)

(**97**): λ_{max}^{EtOH} 285 mμ (5320)

(**98**): λ_{max}^{EtOH} 310 mμ (7950), 358 mμ (3740)

determined from the ultraviolet spectral data alone. 2-Nitro-1-tetralone (**99**) exhibits behavior different than that of **94**. It exists in the enol form in hexane, but in ethanol the keto form predominates[330]. The band at 370 mμ is not found in 2-bromo-2-nitro-1-tetralone or 1-tetralone and is believed to be characteristic of the enol form **100** rather than the nitronic acid, **101**[330,334] (equation 101),

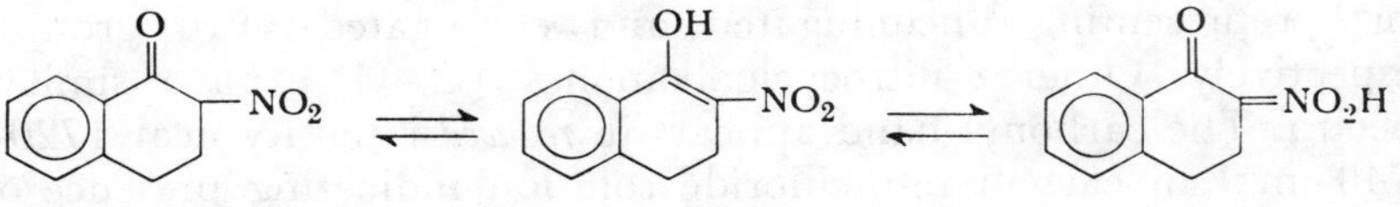

(**99**) ⇌ (**100**) ⇌ (**101**) (101)

(**99, 100, 101**): λ_{max}^{EtOH} 370 mμ (294); λ_{max}^{Hexane} 370 mμ (10,800)

The formation of oxindigos (e.g. **102**) by heating acidified α-ketonitronate salts suggests a radical-anion coupling reaction characteristic of nitronic acids whose anions are resonance stabilized[332,346] (equation 102).

The enol form **103** of 2-nitro-1-indanone (**104**) described by earlier workers[347,348] has recently been shown to have the isomeric

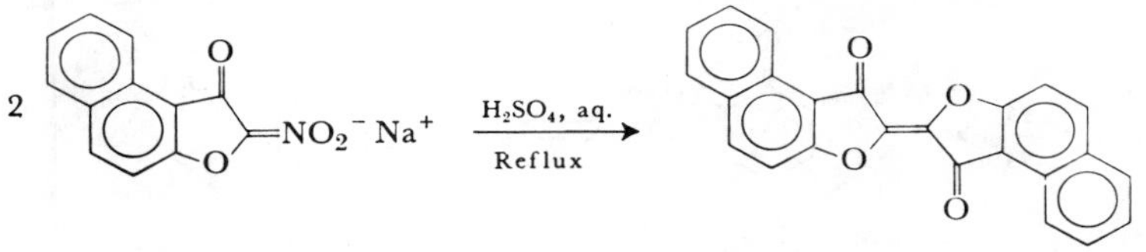

$$2\ \text{(naphthofuranone)}{=}NO_2^-\,Na^+ \xrightarrow[\text{Reflux}]{H_2SO_4,\ aq.} \textbf{(102)} \qquad (102)$$

nitroolefin structure **105**[349,350]. The substance is prepared by condensation of *o*-phthalaldehyde with nitromethane.

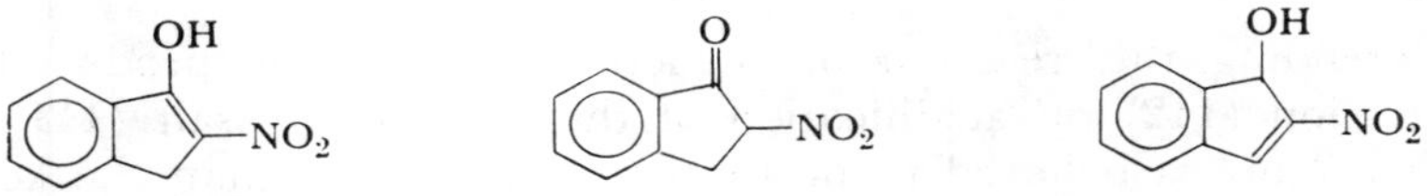

(**103**) (**104**) (**105**)

α-Nitrocamphor and certain bromo and chloro derivatives have been studied extensively[337,351–362]. With *pseudo* bromonitrocamphor (**106**) two forms have been isolated, m.p. 108° and 142°; the higher melting form is said to be a nitronic acid[337,353,355]. However, the various crystalline so-called *aci* forms could also be epimeric nitro

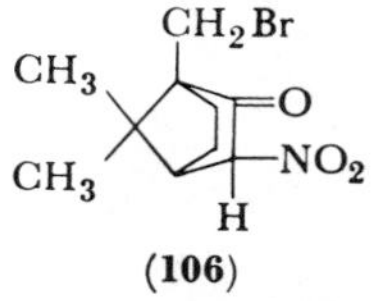

(**106**)

ketones[362]. The mutarotation of α-nitrocamphor is catalyzed by acids and bases[357,362]; general acid catalysis is observed[362].

Salts of alicyclic α,α′-dinitronic acids (e.g. **107**) have been prepared from cycloalkanones by reaction with alkyl nitrates and alkali alkoxides[192,329,335,363]. Acidification of salt **107** produces an oil described as a bisnitronic acid **108**[329], but which may contain ketone **109**, enol **110**, as well as a nitronic acid derived from **110.** On standing, the oil slowly crystallizes forming ketone **109.** In methylene chloride solution **109** appears to exist 75% as the enol form **110,** but in ethanol solely as ketone **109**[335] (equation 103).

$$\text{cyclohexanone} \xrightarrow[\text{THF}]{\text{2 KOBu-}t,\ \text{2 AmONO}_2} \underset{(\mathbf{107})}{K^{+}\,{}^{-}O_2N{=}\text{C}_6\text{-one}{=}NO_2^{-}\,K^{+}} \xrightarrow{H^{+}} \underset{(\mathbf{108})}{HO_2N{=}\text{C}_6\text{-one}{=}NO_2H} \quad (103)$$

$$\xrightarrow[\text{24 h}]{25^{\circ}} \underset{\substack{(\mathbf{109})\\ \text{M.p. } 110.5^{\circ}}}{O_2N\text{-C}_6\text{-one-}NO_2} \rightleftharpoons \underset{(\mathbf{110})}{O_2N\text{-C}_6(OH)\text{-}NO_2}$$

Dipotassium cyclopentanone-2,5-*bis*nitronate on acidification gave crystals (not identified) which decomposed readily to produce an oil[329].

Extending this reaction to N-methyl-4-piperidone produced a zwitterion (**112**) on acidification of the bisnitronate salt (**111**)[192]. Its structure is believed to be the enol **112**, rather than the keto-nitronic acid **113** since its spectra are different from the disalt **111**. The carbonyl band of **111** at 1600 cm^{-1} (Nujol) is not found in **112**, and **112** has a broad OH band near 2750 cm^{-1} and an ultraviolet band at 364 mμ not found in **111** (equation 104).

$$\underset{(\mathbf{111})}{K^{+}\,{}^{-}O_2N{=}\ldots{=}NO_2^{-}\,K^{+}\ (N\text{-}CH_3)} \xrightarrow{2\,H^{+}} \underset{\substack{(\mathbf{112})\\ \text{M.p. } 119^{\circ}}}{{}^{\ominus}O_2N{=}\ldots(OH)\ldots NO_2\ (\overset{\oplus}{N}(H)CH_3)} \quad \underset{(\mathbf{113})}{{}^{\ominus}O_2N{=}\ldots(O)\ldots{=}NO_2H\ (\overset{\oplus}{N}(H)CH_3)} \quad (104)$$

$\lambda_{max}^{1:1\ EtOH\text{-}H_2O}$ 234 mμ (4960)
308 mμ (4080)
390 mμ (10,300)

$\lambda_{max}^{1:1\ EtOH\text{-}H_2O}$ 304 mμ (2240)
364 mμ (3440)
415 mμ (4150)

Alicyclic α,α'-diketonitro compounds **114–118** have been prepared[54,336,339,364–376]. These substances are rather strong acids and in solution in protic solvents exist largely as resonance stabilized nitronate anions. The 2-nitro-1,3-cyclohexanedione derivatives **114a** and **114b**, have been described as nitroenols **115a,b** in the solid state[374] (equation 105). The strong absorption bands at 293 mμ (ε 5000) and at 296 mμ (ε 7000) of chloroform solutions of **114a**

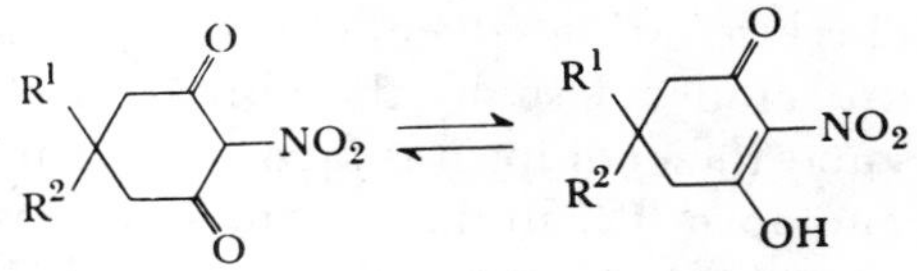

(105)

(**114a**) $R^1 = R^2 = CH_3$ (**115a,b**)
(**114b**) $R^1 = H$; $R^2 = C_6H_5$

and **114b**, respectively, suggest a high concentration of enol in the aprotic solvent. 2-Nitrodimedone (**114a**) is a relatively weak acid ($pK_a^{Nitro} \simeq 3$) compared to nitrobarbituric acid (**116a**) and 2-nitro-1,3-indanedione (**118**, $pK_a^{Nitro} < 0$)[54,339,365,369]. Nitrobarbituric acid (**116a**) and its dimethyl derivative (**116b**, m.p. 152°) were

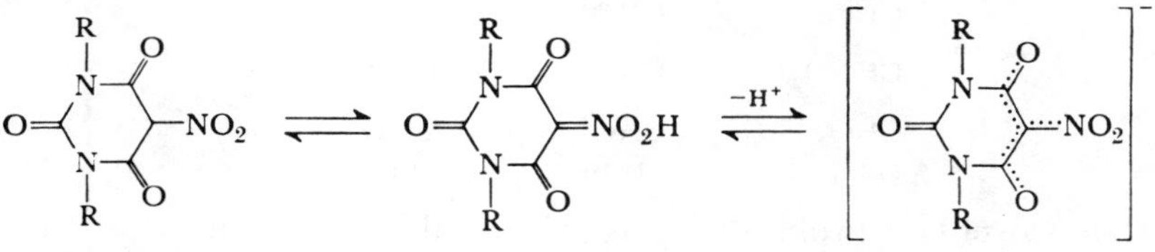

(**116a**) R = H (**117**) (106)
(**116b**) $R = CH_3$
M.p. 152°

prepared and studied by Holleman[54] who described them as nitronic acids **117** (equation 106).

2-Nitro-1,3-indanedione (**118**), a much studied compound[366–376], is a strong acid, comparable to hydrochloric[365,369]. It cannot be acetylated[365] and in water exists only 2% in the nitrodiketo form **118**, or 98% as forms **119–121** by bromine titration[336]; in benzene

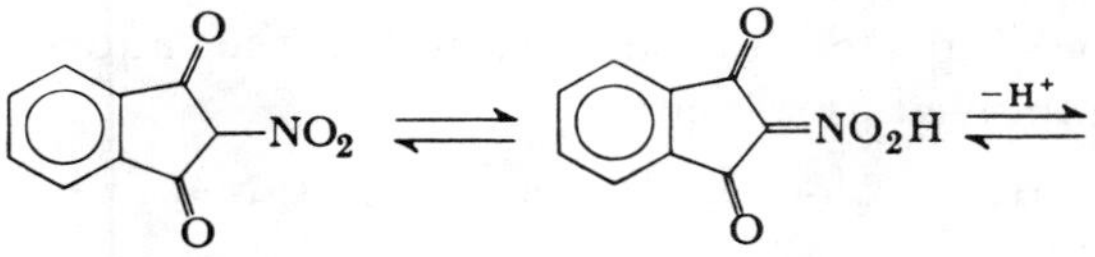

(**118**) (**119**)

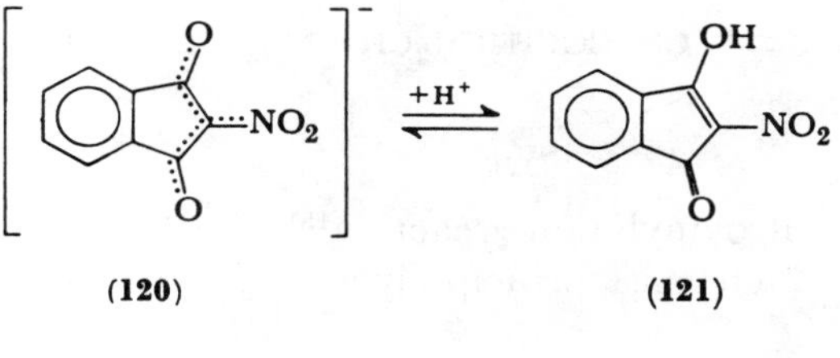

(107)

(**120**) (**121**)

solution, however, it exists 90% in the nitrodiketo form **118** (equation 107). This interesting solvent effect is opposite to that found for all other α-nitroketones. Usually the nitroketone form is favored in protic solvents and the enol form is favored in aprotic solvents. It appears that the enol form **121** in this unique system is not important in either protic or aprotic solvents; the ionic form **120** is evidently very important in protic solvents.

Nitromalonaldehyde (**122, 123**) is an unstable substance, m.p. 50–51° [377]. It is prepared by acidification of its silver salt with ethereal hydrogen chloride[377]. In water it produces a yellow, strongly acidic solution, but it decomposes rapidly in this solvent to yield 1,3,5-trinitrobenzene and formic acid. It is soluble in benzene and may be crystallized from ligroin. In the solid state and in

$$\begin{array}{ccc} \mathrm{CHO} & \mathrm{CHOH} & \mathrm{CHO} \\ | & \| & | \\ \mathrm{CHNO_2} & \mathrm{CNO_2} & \mathrm{C{=}NO_2H} \\ | & | & | \\ \mathrm{CHO} & \mathrm{CHO} & \mathrm{CHO} \\ (\mathbf{122}) & (\mathbf{123}) & (\mathbf{124}) \end{array}$$

aprotic solvents it probably exists principally as the enol **123**. Very little nitronic acid **124** would be present in protic solvents. Rather, one would find principally the anion since this is a strong acid ($pK_a^{Nitro} \cong 0$). The sodium salt is nearly colorless in the solid state and relatively stable[378,379]. Aqueous solutions of nitromalonaldehyde, however, are colored yellow.

γ-Ketonitronic acids are known. Some are readily prepared by Michael addition of vinyl ketones to suitable nitroalkanes[375,380,381]. The slow rate of reduction of the carbonyl group of **125** with sodium borohydride may be explained by formation of the cyclic *pseudo* ester **127** from γ-ketonitronate anion **126**[381] (equation 108). Some keto-nitronic acid might be expected to be present in the aqueous methanol which was employed as solvent.

$$\mathrm{CH_3COCH{=}CH_2 + CH_2(NO_2)_2 \longrightarrow CH_3COCH_2CH_2CH_2(NO_2)_2}\quad(\mathbf{125})$$

$$\mathbf{125} \rightleftharpoons \underset{(\mathbf{126})}{\mathrm{CH_3COCH_2CH_2\overset{NO_2}{\overset{|}{C}}H{=}NO_2^-}} \rightleftharpoons \underset{(\mathbf{127})}{\text{[cyclic: } \mathrm{CH_3}, \mathrm{HO}\text{–C–O–N(}\rightarrow\mathrm{O)}{=}\mathrm{C(NO_2)}\text{ ring]}} \qquad (108)$$

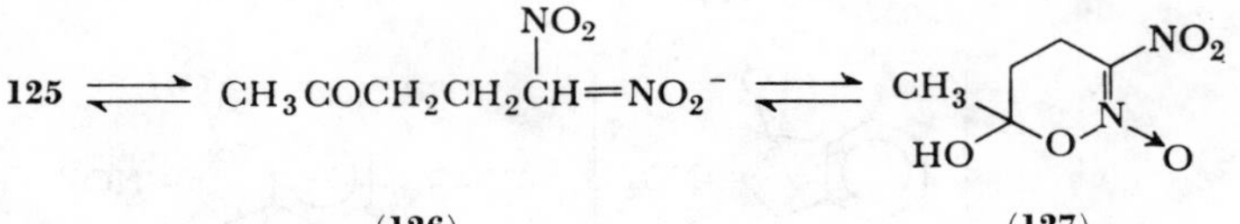

The nitrovinylation reaction[191,213,382] leads to 4-keto-1-nitro-olefins which exist principally in the nitronic acid form in protic solvents, and principally in the nitro form in aprotic solvents[213,219]. The compound **128** in methylene chloride solution has the characteristic phenyl vinyl ketone absorption (e.g. $C_6H_5COCH{=}CHCH_3$

$$C_6H_5COCH_3 + (CH_3)_2NCH{=}CHNO_2 \xrightarrow[\text{2. HCl}]{\text{1. EtOK, EtOH; }-(CH_3)_2NH}$$

$$\underset{(\mathbf{128})}{C_6H_5COCH{=}CHCH_2NO_2} \rightleftharpoons \underset{(\mathbf{129})}{C_6H_5COCH{=}CHCH{=}NO_2H} \quad (109)$$

$\lambda_{max}^{CH_2Cl_2}$ 258 mμ (10,900) $\lambda_{max}^{CH_3OH}$ 400 mμ (6600)

$\lambda_{max}^{CH_3OH}$ 256 mμ, ε_{max} 17,400) and absorbs at much lower wave length than the nitronic acid **129**, with its extended conjugation, which is formed in methanol (equation 109). Unlike compound **125** the carbonyl group in **128** can be easily reduced to hydroxyl with sodium borohydride[213].

The benzoylindenenitronic acid **130** is known and is prepared by acylation of potassium indene-1-nitronate[86] (equation 110).

(110)

$NO_2^{\ominus}$ ⟷ NO_2 (⊖) $\xrightarrow{C_6H_5COCl}$ COC_6H_5, NO_2H

(**130**) 77%
M.p. 121°

The tautomers of *o*- and *p*-nitrophenols are α- and γ-ketonitronic acids, respectively[383,384]. Their yellow color in basic solution in protic solvents is due to nitronate anions. The yellow color was observed by Hantzsch to remain momentarily on acidification of these salts due to formation of the relatively weak nitronic acids **131**, **132**[383,384] (equation 111); esters of these acids have been prepared[384].

OH, NO_2 $\underset{}{\overset{-H^+}{\rightleftharpoons}}$ O, $=NO_2^-$ $\overset{H^+}{\rightleftharpoons}$ O, $=NO_2H$

Colorless Yellow (**131**) Yellow (111)

OH, NO_2 $\overset{-H^+}{\rightleftharpoons}$ O, NO_2^- $\overset{H^+}{\rightleftharpoons}$ O, NO_2H

Colorless Yellow (**132**) Yellow

Tautomerization of the yellow nitronic acids **131**, **132** to the colorless nitrophenols occurs very rapidly.

Anthrone-10-nitronic acid (**134**) is an interesting γ-ketonitronic acid[35,385,386]. Two tautomeric forms have been reported—nitro ketone **133** and nitrophenol **135**. Colorless 10-nitroanthrone (**133**) is prepared by nitration of anthracene in acetic acid. It forms a deep red potassium salt when treated with potassium hydroxide solution.

HNO_3, HOAc, 60°

(**133**) Colorless
80%; M.p. 137°; 148°

H_2O, 25° | 1. KOH 2. Dil. H_2SO_4, 0°

(112)

(**135**) R = H; yellow; unstable
(**136**) R = OAc; yellow; m.p. 182°
(**137**) R = $O_2CC_6H_5$; yellow; m.p. 238°

(**134**) Red
M.p. 80–85°

The salt reacts with cold dilute sulfuric acid to yield carmine-red crystals of the ketonitronic acid **134**[386] (equation 112). The nitronic acid is quite stable and may be stored for months in a desiccator. It melts unsharply, ca. 80–85°, with decomposition, ultimately forming anthraquinone on continued heating[386]. On standing with water or dilute acids it is converted into the nitro form **133** with formation of some anthraquinone[386]. The nitronic acid or its potassium salt can be brominated to form colorless 10-bromo-10-nitroanthrone, m.p. 116° [386]. Reaction of the silver salt of **134** with methyl iodide produced a resin from which only anthraquinone could be isolated[386]. The red potassium nitronate salt may be acetylated or benzoylated to produce yellow 10-nitro-9-anthryl acetate (**136**) or benzoate, **137**[35]. 10-Nitro-9-anthrol (**135**, an unstable yellow substance, isolable only at −5°) was prepared by Hantzsch[385] by acidification of the ammonium salt of **134** with hydrogen chloride in ether at Dry Ice temperature; evaporation of the yellow solution gave **135**. On warming to room temperature **135** immediately produced the red nitronic acid **134**.

IV. NITRONIC ACID ESTERS

A. Preparation of Nitronic Acid Esters

The various methods available for preparing the rather unstable nitronic acid esters are presented in this section. Acyclic and cyclic esters (e.g. the 2-isoxazoline *N*-oxides) are considered separately. Certain side reactions which defeat the syntheses are discussed, including C-alkylation of nitronates.

1. Acyclic nitronic acid esters

Three principal methods are available for preparation of acyclic nitronic esters: (*a*) alkylation of sodium or potassium nitronate salts, (*b*) alkylation of silver nitronate salts, and (*c*) reaction of nitroalkanes and nitronic acids with diazomethane.

The O-alkylation of sodium and potassium nitronate salts has been examined extensively[5,23,84,275,387–393]. The initial product is a nitronic ester (**138**; equation 113), which may decompose easily under the reaction conditions to form an oxime and an aldehyde or ketone (equation 114). C-Alkylation may also occur (equation 115). The course of the reaction depends on the structures of the

$$R^1R^2C{=}NO_2^-Na^+ + R^3R^4CHX \longrightarrow \underset{\substack{\textbf{(138)}\\ \text{Nitronic ester}}}{R^1R^2C{=}NO_2CHR^3R^4} + NaX \quad (113$$

$$\underset{\textbf{(138)}}{R^1R^2C{=}NO_2CHR^3R^4} \longrightarrow R^1R^2C{=}NOH + R^3R^4C{=}O \quad (114)$$

$$R^1R^2C{=}NO_2^-Na^+ + R^3R^4CHX \longrightarrow R^1R^2C(NO_2)CHR^3R^4 + NaX \quad (115)$$

nitronate salt and alkylating agent. Alkylating agents of various types have been employed, including alkyl fluoroborates[23,390,391], sulfates[275,394–396], and halides[65,196,387–389,392,393,396,397]. Nitronic esters prepared by this method are listed in Table 11 (method A).

O-Alkylation with alkyl fluoroborates is the best method when alkali metal nitronate salts are employed[23,390,391]. In a procedure developed by Kornblum and coworkers[23,391], nearly quantitative yields are obtained in a rapid reaction at 0° (equation 116).

$$R^1R^2C{=}NO_2^-Na^+ + (EtO)_3BF_4 \xrightarrow{0^\circ} \underset{\substack{75\text{–}95\%\\ R^1,\, R^2\, =\, \text{alkyl}}}{R^1R^2C{=}NO_2Et} + Et_2O + NaBF_4 \quad (116)$$

TABLE 11. Synthesis and properties of acyclic nitronic esters.

Nitronic ester	Synthetic method[a]	Yield (%)	M.p., (°C)	Half-life (approx.)[b] at 25°	Ref.
$(NO_2)_2C{=}NO_2CH_3$	A,C	—	dec.[c]	few min	398, 399
$O_2NCH{=}NO_2CH_3$	C	—	dec.[c]	few min	400
$NCC(NO_2){=}NO_2CH_3$	B	58	62–64	—	401, 402
$(NO_2)_2C{=}NO_2C_2H_5$	A,C	—	dec.[c]	few min	398
$H_2NCOCH{=}NO_2CH_3$	B	—	112°	—	403
$CH_3CH{=}NO_2CH_3$	A	90–95	liquid	3–24 h	23
$H_2NCOCH{=}NO_2C_2H_5$	B	30	114	several h	403
$CH_3C(NO_2){=}NO_2C_2H_5$	B	—	dec.	few h	15
$CH_3CH{=}NO_2C_2H_5$	A	94	liquid	1 day	23
$NCC(CONH_2){=}NO_2C_2H_5$	B	—	—	few h	15, 404–406
$C_2H_5O_2CCH{=}NO_2CH_3$	B,C	—	b.p. 84°/2.5 mm	—	394, 407
$CH_3CH_2CH{=}NO_2C_2H_5$	A	79	liquid	1 day	23
$(CH_3)_2C{=}NO_2C_2H_5$	A	75–80	liquid	few h	23
$(NC)_2C{=}CHCH{=}NO_2CH_3$	B	65	105	—	191
$(CH_3O_2C)_2C{=}NO_2CH_3$	C	90	68	—	394, 408
$C_2H_5O_2CCH{=}NO_2C_2H_5$	B	40	b.p. 81°/3 mm	—	407
$H_2NCOCH{=}NO_2C_3H_7\text{-}n$	B	32	107	—	403, 409
$n\text{-}C_3H_7CH{=}NO_2C_2H_5$	A	90–95	liquid	1 day	23
$C_2H_5C(CH_3){=}NO_2C_2H_5$	A	90–95	liquid	5 min	23
2,6-dinitro-4-(methoxy-aci-nitro)cyclohexadienone: $O{=}C_6H_2(NO_2)_2{=}NO_2CH_3$ (structure: NO_2, NO_2, $O{=}$, $=NO_2CH_3$)	B	—	40–42	few min	384

$=NO_2CH_3$, O	B	—	−5	few min	384
O $=$ $=NO_2CH_3$	B	—	—	few min	384
$(NO_2)_2C{=}NO_2CH_2C_6H_4$-$NO_2$-4	B	—	—	few min	410
NO_2, O $=$ $=NO_2C_2H_5$, NO_2	B	1.5	50–52	few min	384
4-$BrC_6H_4CH{=}NO_2CH_3$(*'trans'*)	C	80	66.5–67.5	2 weeks	23, 394
4-$BrC_6H_4CH{=}NO_2CH_3$(*'cis'*)	C	20	—	few h	23
4-$O_2NC_6H_4CH{=}NO_2CH_3$(*'trans'*)	C	67	118–120	several weeks	23
4-$O_2NC_6H_4CH{=}NO_2CH_3$(*'cis'*)	C	33	—	several h	23
$=NO_2C_2H_5$, O_2N, O	B	—	—	few min	410
$C_6H_5CH{=}NO_2CH_3$	A,C	—	liquid	3–4 h	394, 395
$C_6H_5N_2CH{=}NO_2CH_3$	C	17	54.5	1 week	297
$H_2NCOCH{=}NO_2C_5H_{11}$-*n*	B	28	100	—	403
$(CH_3O_2C)_2C{=}CHCH{=}NO_2CH_3$	B	70	87	—	191

(*continued*)

TABLE 11—*continued*

Nitronic ester	Synthetic method[a]	Yield (%)	M.p., (°C)	Half-life (approx.)[b] at 25°	Ref.
(cyclohexylidene)$=NO_2C_2H_5$	A	100	liquid	1–5 min	23
$2\text{-}BrC_6H_4C(CN){=}NO_2CH_3$	B	—	104–105	few h	193
$4\text{-}BrC_6H_4C(CN){=}NO_2CH_3$	B	70–80	110	several h	196
$2,4,6\text{-}Cl_3C_6H_2N_2C(CH_3){=}NO_2CH_3$	C	29	89–90	—	299
$C_6H_5C(CN){=}NO_2CH_3$	B	50–70	41–42	several h	194, 411, 412
$2,4\text{-}Cl_2C_6H_3N_2C(CH_3){=}NO_2CH_3$	C	40	110–111	—	299
$C_6H_5COCH{=}NO_2CH_3$	C	—	liquid	—	394
$4\text{-}BrC_6H_4CH{=}NO_2C_2H_5$ ('*trans*'/'*cis*' = 3/1; mixture)	A	95	35–40	1–3 h	23
$4\text{-}ClC_6H_4N_2C(CH_3){=}NO_2CH_3$	C	—	112	—	298
$4\text{-}CH_3C_6H_5SO_2CH{=}NO_2CH_3$	C	—	liquid	—	394
$4\text{-}O_2NC_6H_4CH{=}NO_2C_2H_5$('*trans*')	A	74	100–101	several weeks	23
$4\text{-}O_2NC_6H_4CH{=}NO_2C_2H_5$('*cis*')	A	18	—	few h	23
$C_6H_5N_2C(CH_3){=}NO_2CH_3$	C	65	71.5–72	several h	6, 298
$4\text{-}BrC_6H_4COCH{=}CHCH{=}NO_2CH_3$	C	35	128	—	213
$C_6H_5C(CN){=}CHCH{=}NO_2CH_3$	C	38	68	—	191
$C_6H_5COCH{=}CHCH{=}NO_2CH_3$	C	28	118	—	213
(1-indanone-2-ylidene)$=NO_2CH_3$	C	40	132	—	213
(1-tetralone-4-ylidene)$=NO_2C_2H_5$	B	—	—	few min	384

$4\text{-}CH_3OC_6H_4COCH{=}CHCH{=}NO_2CH_3$	C	32	111–118	—	213
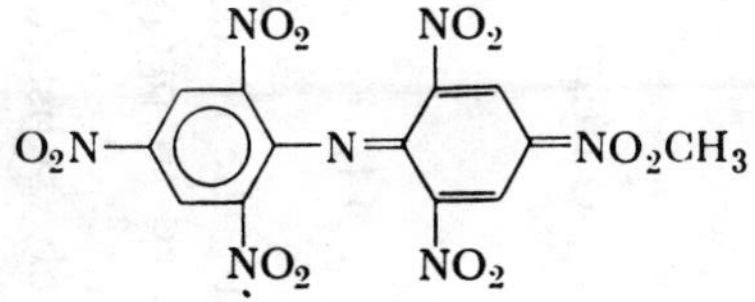 O_2N–(2,6-$(NO_2)_2C_6H_2$)–$N{=}$(2,6-$(NO_2)_2$-cyclohexadienylidene)$=NO_2CH_3$	B	poor	140–141	few h	413
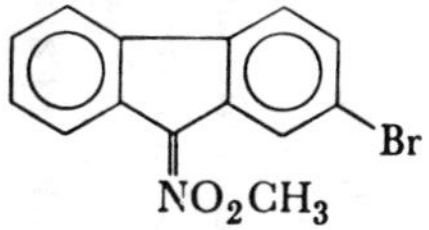 2-Br-fluorenylidene $=NO_2CH_3$	B	—	72–80	few min	204
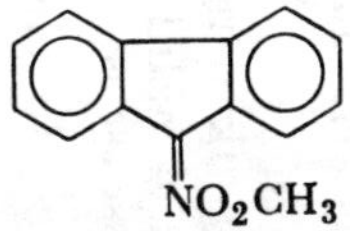 fluorenylidene $=NO_2CH_3$	A	—	84	—	275
$C_6H_5N_2C(C_6H_5){=}NO_2CH_3$	C	17	92	—	295
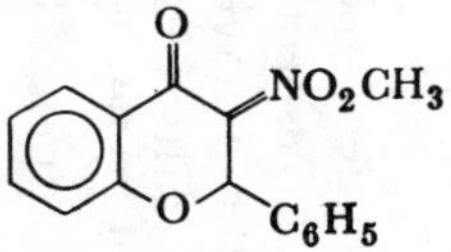 2-C_6H_5-chroman-4-one-3-ylidene $=NO_2CH_3$	C	63	—	—	344
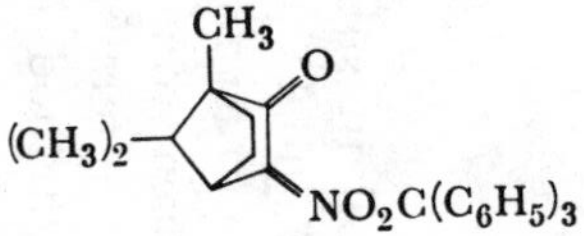 $(CH_3)_2$, CH_3 bicyclic ketone $=NO_2C(C_6H_5)_3$	A	44	149–152	—	414

[a] Methods of synthesis: A. Alkylation of sodium or potassium nitronate salt. B. Alkylation of silver nitronate salt. C. Diazomethane reaction with nitroalkane or nitronic acid.

[b] Very approximate time for half of undiluted sample to decompose.

[c] Isolated in solution only. Ester decomposes on attempted isolation by removal of solvent.

Several ethyl alkanenitronates have been prepared by this method.

However, when this reaction is applied at 50–70° to sodium cyclopentane- and cyclohexanenitronates and propane-2-nitronate, the esters are not obtained[390]. Oximes and *N*-alkyl oximes result (equation 117). Esters derived from simple secondary nitroalkanes are less stable than those from primary[23].

$$\text{C}_6\text{H}_{10}\text{=NO}_2^-\text{Na}^+ + (\text{EtO})_3\text{BF}_4 \xrightarrow[-\text{Et}_2\text{O};\ -\text{NaBF}_4]{50-70^\circ} \underset{50-74\%}{\text{C}_6\text{H}_{10}\text{=NOH}} + \underset{12-18\%}{\text{C}_6\text{H}_{10}\text{=NOEt}} \quad (117)$$

Methyl sulfate has been successfully employed as an alkylating agent for the preparation of certain nitronic esters, in unstated yields[275,394,395] (equation 118). In the reaction of methyl sulfate with cyclohexanenitronate, cyclohexanone oxime was formed[396].

$$C_6H_5CH{=}NO_2^-Na^+ + (CH_3)_2SO_4 \longrightarrow C_6H_5CH{=}NO_2CH_3 + Na_2SO_4$$

$$\text{(fluorenylidene)}{=}NO_2^-K^+ + (CH_3)_2SO_4 \longrightarrow \text{(fluorenylidene)}{=}NO_2CH_3 + Na_2SO_4 \quad (118)$$

Alkylation of sodium or potassium nitronates with alkyl halides has not yet produced a nitronic ester by any reported procedure[65,196,387,389,392,393,396,397]. Oximes, aldehydes, ketones, or C-alkyl products result[314,393] (equation 119). These reactions are discussed

$$(CH_3)_2C{=}NO_2^-Na^+ + ArCH_2Br \xrightarrow[25-80^\circ]{EtOH} \underset{68-77\%}{ArCHO} + (CH_3)_2C{=}NOH$$

$$C_6H_5CH{=}NO_2^-Na^+ + p\text{-}O_2NC_6H_4CH_2Cl \longrightarrow \underset{37\%}{C_6H_5CH(NO_2)CH_2C_6H_4NO_2\text{-}p} + C_6H_5CH{=}NOH + p\text{-}O_2NC_6H_4CHO \quad (119)$$

in detail in subsequent sections (*cf.* Tables 13 and 14, sections IV.A.2 and IV.C.1, respectively).

In contrast to the behavior of alkali nitronate salts, the reaction of silver nitronate salts with alkyl halides can be used to prepare nitronic esters[15,191,193,194,196,306,384,386,387,401,403–408,410–413]. As a synthetic method it appears limited to electronegatively substituted salts. Methyl and other alkyl iodides have been employed more frequently than bromides or chlorides (ether solvent) to produce nitronic esters (Table 11, method B) (equation 120). Silver picrate

$$(CH_3O_2C)_2C{=}CHCH{=}NO_2Ag + CH_3I \longrightarrow (CH_3O_2C)_2C{=}CHCH{=}NO_2CH_3 + AgI \quad 80\%$$

$$\underset{\displaystyle CN}{4\text{-}BrC_6H_4\overset{}{C}}{=}NO_2Ag + CH_3I \longrightarrow \underset{\displaystyle CN}{4\text{-}BrC_6H_4C}{=}NO_2CH_3 + AgI \quad 70\text{–}80\% \tag{120}$$

$$H_2NCOCH{=}NO_2Ag + RI \longrightarrow H_2NCOCH{=}NO_2R + AgI \quad 28\text{–}32\%$$

$$R = Et, Pr, Am$$

led to a very low yield of the unstable methyl or ethyl ester, **139**[384,386] (equation 121). As with alkali nitronate salts, certain silver nitronates react with alkyl halides to produce C-alkylation products and/or

$$O{=}C_6H_2(NO_2)_2{=}NO_2Ag + RX \longrightarrow O{=}C_6H_2(NO_2)_2{=}NO_2R + AgX \tag{121}$$

$$R = CH_3, Et \qquad X = Br, I \qquad \textbf{(139)}$$

oximes, oxime ethers, and aldehydes or ketones[15,407,410,415–418] (equation 122).

$$CH_3C(NO_2){=}NO_2Ag + CH_3I \longrightarrow CH_3C(NO_2)_2CH_3 + CH_3C(NO_2){=}NOCH_3 + CH_2O \tag{122}$$

Methyl nitronic esters are conveniently obtained by reaction of nitronic acids with diazomethane in ether solvent (*direct method*). More conveniently, certain acidic nitroalkanes may be used as reactant (*indirect method*). The diazomethane reaction has the advantage that mild conditions may be employed, and C-alkylation products are not obtained. Oximes can result, however, by decomposition of the ester[394]. Nitronic esters prepared by reactions of diazomethane are listed in Table 11, method C.

The *direct method* which requires a nitronic acid reactant has seldom been employed[23,191,394], but would appear to be potentially quite useful. It is applicable at low temperature, and, although not yet exploited, should be applicable to those nitronic acids derived from more weakly acidic nitroalkanes which do not react with diazomethane. 4-Nitro, 4-bromophenyl- and phenylmethanenitronic acid have been converted into their methyl esters; the parent nitroalkanes also react to give the same product[23,394] (equation 123).

$$4\text{-}O_2NC_6H_4CH{=}NO_2H + CH_2N_2 \longrightarrow 4\text{-}O_2NC_6H_4CH{=}NO_2CH_3 + N_2$$

$$C_6H_5C(CN){=}CHCH{=}NO_2H + CH_2N_2 \longrightarrow \underset{38\%}{C_6H_5C(CN){=}CHCH{=}NO_2CH_3} + N \qquad (123)$$

The *indirect method* is quite effective with negatively substituted nitroalkanes ($pK_a^{Nitro} <$ ca. 8)[23,213,344,394,419,420,420a]. α- and γ-Nitro ketones and carboxylic esters react, but some enol methyl ether formation may be expected to result with some of these compounds (see section III.C.4)[344] (equation 124). More weakly acidic nitroalkanes ($pK_a^{Nitro} >$ ca. 8) such as nitromethane, 2-nitrobornane, and 3-phenyl-1-nitropropane do not react with diazomethane[23,421].

$$4\text{-}BrC_6H_4CH_2NO_2 + CH_2N_2 \longrightarrow 4\text{-}BrC_6H_4CH{=}NO_2CH_3 + N_2$$

$$C_2H_5O_2CCH_2NO_2 + CH_2N_2 \longrightarrow C_2H_5O_2CCH{=}NO_2CH_3 + N_2 \qquad (124)$$

$$C_6H_5COCH{=}CHCH_2NO_2 + CH_2N_2 \xrightarrow{THF} \underset{(28\%)}{C_6H_5COCH{=}CHCH{=}NO_2CH_3} + N$$

The deeply red-colored α-arylazonitronic esters may be prepared by reaction of diazomethane with aldehyde 1-nitrohydrazones. Several of these esters have been prepared by Bamberger and coworkers (see Table 11)[6,295–299]. The orange-red hydrazones are readily prepared in high yield by reaction of nitronate salts with diazonium salts at 0° (see section III.C.1.b)[293–296]. Some oxime formation accompanies formation of these unstable esters[298] (equation 125).

$$C_6H_5NHN{=}\underset{\displaystyle CH_3}{\underset{|}{C}}NO_2 + CH_2N_2 \xrightarrow[\text{Et}_2\text{O, Several days}]{0°}$$

$$\underset{65\%}{C_6H_5N{=}N\underset{\displaystyle CH_3}{\underset{|}{C}}{=}NO_2CH_3} + \underset{20\%}{C_6H_5N{=}N\underset{\displaystyle CH_3}{\underset{|}{C}}{=}NOH} + \underset{6\%}{C_6H_5N{=}N\underset{\displaystyle CH_3}{\underset{|}{C}}{=}NOCH_3} \qquad (125)$$

A special method of synthesis of nitronic esters derived from 2,6-di-*t*-butyl-4-nitrophenol employs a trialkyl phosphite and ethyl acrylate reacting at room temperature without a solvent (equation 125*a*)[421a].

$$t\text{-}C_4H_9\text{-}C_6H_2(OH)(NO_2)\text{-}C_4H_9\text{-}t + (RO)_3P + H_2C{=}CHCO_2C_2H_5 \longrightarrow$$

$$t\text{-}C_4H_9,\ C_4H_9\text{-}t\ \text{quinone nitronic ester } (O{=}C_6H_2{=}N(\rightarrow O)OR) + (RO)_2\overset{O}{\overset{\|}{P}}CH_2CH_2CO_2C_2H_5 \quad (125a)$$

28–56%

$R = CH_3, C_2H_5, i\text{-}C_3H_7$

A new, special method of ester preparation is said to involve addition of iodo- or bromotrinitromethane to olefins in a solvent such as dimethyl sulfoxide[399,422–423a,b], e.g. ethylene and iodotrinitromethane yields **140** (equation 126).

$$CH_2{=}CH_2 + IC(NO_2)_3 \longrightarrow \underset{\mathbf{(140)}}{ICH_2CH_2O_2N{=}C(NO_2)_2} \quad (126)$$

However, recent evidence has shown that the reaction in dimethyl sulfoxide does not lead to simple addition compounds of structure **140**, but rather to compounds of sulfonium structure **141**[423a]. The structure of **141** was proved by synthesis. Dimethyl sulfoxide was

$$\underset{\mathbf{(141)}}{CH_3\text{—}\underset{CH_3}{\overset{CH_3}{S}}\text{—}O\text{—}\underset{O^-}{\overset{+}{N}}{=}C(NO_2)_2} \rightleftharpoons CH_3O\text{—}\underset{CH_3}{\overset{CH_3}{S^+}} + {}^-O\text{—}\underset{O^-}{\overset{+}{N}}{=}C(NO_2)_2$$

methylated with dimethyl sulfate to compound **142** (equation 127).

$$(CH_3)_2SO + (CH_3)_2SO_4 \xrightarrow{\text{room temp.}} \underset{\mathbf{(142)}}{CH_3O\overset{+}{S}(CH_3)_2;\ CH_3OSO_3^-} \quad (127)$$

Addition of potassium trinitromethane dissolved in dimethoxyethane gave **141** (equation 128).

$$\mathbf{142} + K^+C(NO_2)_3^- \longrightarrow \mathbf{141} + CH_3OSO_3K \quad (128)$$

Nitronic esters have not been prepared by reaction of a nitronic acid with an alcohol.

Finally, a comparison of C- and O-alkylation of nitronate salts should be considered at this point[423c]. The extent of C- and O-alkylation is known to depend on three principal factors: (*a*) the nature of the leaving group in the alkylating agent, (*b*) the structure of the alkylating agent, and (*c*) the structure of the nitronate anion. Other factors include nature of the cation and solvent, reaction temperature, and solubility of reactants and products.

The alkylation of alkali nitronate salts has been studied extensively with substituted benzyl alkylating agents (Table 12)[5,314,388,393,397,424–428a,b]. When O-alkylation is the sole reaction, the yield is independent of the nature of the leaving group. However, 2- and 4-nitro substituted benzyl alkylating agents are notably exceptional in their behavior. The extent of C- and O-alkylation of lithium or sodium propane-2-nitronate salts by 2-O_2N- and 4-$O_2NC_6H_4CH_2X$ does depend on the nature of the leaving group X. Here, O-alkylation is favored by the best leaving groups.

The extent of C- and O-alkylation depends on the structure of the alkylating agent. 2- and 4-nitrobenzyl chloride[427], and 2,4-dinitrobenzyl chloride[387] effect principally C-alkylation. However, 3-nitrobenzyl chloride and other benzyl halides effect O-alkylation only[393].

It is to be expected that yields of C- and O-alkylation products would depend on the structure of the nitronate anion. For example, with 4-nitrobenzyl chloride yields of C-alkylation products are: $CH_3CH{=}NO_2^-$ (24%) and $(CH_2)_5C{=}NO_2^-$ (62%)[397].

Evidence for radical anion intermediates has been obtained by esr measurements in the C-alkylation with 4-nitrobenzyl chloride[314,424,427]. The mechanism in this particular case is considered to be an exchange leading to a radical anion (**143**) and a nitro radical (**144**), followed by loss of chloride ion and coupling in a chain process (equations 129–132)[314,427].

$$O_2NC_6H_4CH_2Cl + (CH_3)_2C{=}NO_2^- \longrightarrow \underset{(\mathbf{143})}{O_2NC_6H_4CH_2Cl^{\dot{-}}} + \underset{(\mathbf{144})}{(CH_3)_2NO_2\cdot} \quad (129)$$

$$O_2NC_6H_4CH_2Cl^{\dot{-}} \longrightarrow O_2NC_6H_4CH_2\cdot + Cl^- \quad (130)$$

$$O_2NC_6H_4CH_2\cdot + (CH_3)_2C{=}NO_2^- \longrightarrow O_2NC_6H_4CH_2C(CH_3)_2NO_2^{\dot{-}} \quad (131)$$

$$O_2NC_6H_4CH_2C(CH_3)_2NO_2^{\dot{-}} + O_2NC_6H_4CH_2Cl \longrightarrow O_2NC_6H_4CH_2C(CH_3)_2NO_2 + O_2NC_6H_4CH_2Cl^{\dot{-}} \quad (132)$$

Supporting this mechanism is the finding that addition of 1,4-dinitrobenzene to this system as an electron scavenger increases the extent of O-alkylation from 6 to 88%[424].

TABLE 12. Effect of substituents and leaving group on the extent of C- and O-alkylation of substituted benzyl alkylating agents reacting with sodium and lithium propane-2-nitronate salts.

$$ArCH_2X + (CH_3)_2C{=}NO_2^-M^+ \xrightarrow{-MX}$$
$$M = Li, Na$$

$$\underset{\text{C-Alkylation Product}}{ArCH_2C(CH_3)_2NO_2} + \underbrace{ArCHO + (CH_3)_2C{=}NOH}_{\text{O-Alkylation Products}}$$

Ar	X	%C-Alkylation	%O-Alkylation[a]	Ref.
4-$O_2NC_6H_4$	$^+N(CH_3)_2$	93[b]	0	424
		90[c]	0	425
	$C_6Cl_5CO_2$	93[b]	0	424, 426
	Cl	92[b]	6	424, 426
		83[c]	1	5, 388
	OTos	40[b]	32	424, 426
	Br	20[b]	60	422, 426
	I	8[b]	86	424, 426
2-$O_2NC_6H_4$	Cl	46[c]	30	388
	Cl	31[b]	52	424
	Br	1[b]	98	424
3-$O_2NC_6H_4$	Cl	0[b]	82	424
		—[c]	73	388
	Br	0[b]	80	424
	I	0[b]	84	424
2,4-$(O_2N)_2C_6H_3$	Cl	33[c]	—	387, 388
C_6H_5	Cl	0[b]	82–84	424
		0[c]	73	5
	OTos	0[b]	82–84	424
	Br	0[b]	82–84	424
		0[c]	73	393
	I	0[b]	82–84	424
4-NCC_6H_4	Br	0[c]	70	5, 393
4-$CH_3O_2CC_6H_4$	Br	0[c]	72	5, 393
4-$CH_3COC_6H_4$	Br	0[c]	77	5
4-$(CH_3)_3N^+I^-C_6H_4$	I	0[c]	68	5
4-BrC_6H_4	Br	0[c]	75	5, 393
4-$CF_3C_6H_4$	Br	0[c]	77	5, 393
4-$CH_3C_6H_4$	Br	0[c]	70	5
2-$CH_3C_6H_4$	Br	0[c]	68–73	393

[a] Determined by yield of aldehyde, ArCHO, or corresponding acid, $ArCO_2H$.
[b] Lithium propane-2-nitronate in dimethylformamide at 0°.
[c] Sodium propane-2-nitronate in ethanol at 25–80°.

Other C-alkylations with various alkylating agents are known (equations 133–136)[328,428a,b,429–431]. In these examples, in contrast to the 4-$O_2NC_6H_4CH_2X$ example, good leaving groups (Br^-, I^-,

$$\text{C}_7\text{H}_7^+\text{Br}^- + (CH_3)_2C{=}NO_2^-Na^+ \xrightarrow[-NaBr]{CH_3OH} \text{C}_7\text{H}_7\text{(H)}C(CH_3)_2NO_2 \quad (133)^{429}$$

68%

$$(C_6H_5)_2I^+, OTos^- + R^1R^2C{=}NO_2^-, Na^+ \xrightarrow[\text{—NaOTos}]{DMF} C_6H_5I + C_6H_5C(R^1R^2)NO_2 \quad (134)^{430}$$

58–69%

R^1,R^2 = H, alkyl, cycloalkyl

$$(CH_3)_2C(X)NO_2 + (CH_3)_2C{=}NO_2^-, Na^+ \xrightarrow{\text{—NaX}} (CH_3)_2C(NO_2)C(NO_2)(CH_3)_2 \quad (135)^{328}$$

9%, X = Cl
29%, X = Br
43%, X = I

$$\text{3-}(TosOCH_2CH_2)\text{-1-nitrocyclohexane} \xrightarrow[-NaOTos]{NaH,\ DMF} \text{1-nitrobicyclo[3.2.1]octane} \quad (136)^{431}$$

40% (vpc)

$OTos^-$) *favor* C-alkylation. Recently, the reaction of equation 135 has been shown to proceed by a radical-anion chain mechanism[314,428a].

The mechanism of silver dinitromethanenitronate alkylation with alkyl halides has been studied in acetonitrile at 25° [401,410,415,416,418]. Overall third-order kinetics are observed in a mechanism which involves dinitromethanenitronate anion[418] (equation 137). With the silver salt C-alkylation is limited to primary halides, including allyl

$$CH_3I + (O_2N)_2C{=}NO_2Ag \xrightarrow{CH_3CN} CH_3C(NO_2)_3 + AgI \quad (137)$$

51%

and benzyl halides (28–52% yield)[401,418,432]. 2-Bromopropane and other secondary, as well as tertiary halides are believed to undergo O-alkylation; the resulting nitronic esters decompose to yield alkyl nitrate as the principal anomalous product[418].

Silver nitrocyanomethanenitronate gave a 58% yield of nitronic ester by O-alkylation with methyl iodide (equation 138), but could also be C-alkylated with *t*-butyl and allyl bromides[418] (equation 139).

$$\underset{\displaystyle \text{CN}}{\text{O}_2\text{NC}}\text{=NO}_2\text{Ag} + \text{CH}_3\text{I} \longrightarrow \underset{\displaystyle \text{CN}}{\text{O}_2\text{NC}}\text{=NO}_2\text{CH}_3 + \text{AgI} \quad (58\%) \qquad (138)$$

$$\underset{\displaystyle \text{CN}}{\text{O}_2\text{NC}}\text{=NO}_2\text{Ag} + (\text{CH}_3)_3\text{CBr} \longrightarrow \underset{\displaystyle \text{CN}}{(\text{O}_2\text{N})_2\text{C}}\text{C}(\text{CH}_3)_3 + \text{AgBr} \quad (17\%) \qquad (139)$$

C-Alkylation is observed with silver and mercury phenylmethanenitronate and silver α-cyanophenylmethanenitronate with diphenylmethyl bromide and trityl chloride[327,387,411,417,433] (equations 140, 141).

$$\text{C}_6\text{H}_5\text{CH=NO}_2\text{Hg} + (\text{C}_6\text{H}_5)_3\text{CCl} \longrightarrow \text{C}_6\text{H}_5\overset{\displaystyle \text{NO}_2}{\text{C}}\text{HC}(\text{C}_6\text{H}_5)_3 + \text{HgCl} \quad (33\text{–}40\%) \qquad (140)$$

$$\text{C}_6\text{H}_5\underset{\displaystyle \text{CN}}{\text{C}}\text{=NO}_2\text{Ag} + (\text{C}_6\text{H}_5)_2\text{CHBr} \longrightarrow$$

$$\text{C}_6\text{H}_5\overset{\displaystyle \text{NO}_2}{\underset{\displaystyle \text{CN}}{\text{C}}}\text{CH}(\text{C}_6\text{H}_5)_2 \; (10\text{–}18\%) + \text{C}_6\text{H}_5\underset{\displaystyle \text{CN}}{\text{C}}\text{=NOH} \; (50\%) + (\text{C}_6\text{H}_5)_2\text{C=O} \qquad (141)$$

In these particular reactions O-alkylation predominates with the silver salts leading to the nitronic ester decomposition products oxime and ketone.

A generalization for the direction of alkylation of ambident anions has been presented by Kornblum[434]. It states that the greater the S_N1 character of the transition state, the greater the preference for bonding to the most electronegative atom in the ambient anion. Because of the instability of nitronic esters it is difficult to test this generalization using the available data on C- vs O-alkylation of nitronates[418,438]. O-Alkylation products (nitronic esters) are thermodynamically less stable than the corresponding C-alkylation products. Also, several routes for ester decomposition are available making it difficult to assess the extent of O-alkylation. With one exception only those nitronic esters having O-*n*-alkyl groups are sufficiently stable to be isolable (Table 11); those having all other types of groups, including secondary and tertiary O-alkyl groups, decompose. On the other hand, stable C-alkylation products have been obtained (usually in low to moderate yields) with a variety of alkylating agents, including those with primary, secondary, and tertiary alkyl groups. Thus, in making predictions in thermodynamic terms about the extent of C- vs O-alkylation of nitronates, one needs knowledge of the thermodynamic stability of the products and their potential routes of decomposition—as well as a good material balance of reaction products.

2. Cyclic nitronic acid esters

Cyclic nitronic esters corresponding to the lactones in the carboxylic acid series are known. Only one type has been investigated extensively, the 2-isoxazoline *N*-oxides, **145** (Table 13). The cyclic esters, unlike the acyclic, are relatively stable, crystalline solids.

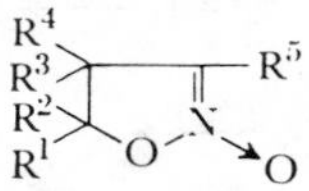

R^4, R^3, R^2, R^1, R^5; O, N, →O

(**145**)

They are good oxidizing agents, the *N*-oxo group being easily removed.

Available synthetic routes to 2-isoxazoline *N*-oxides proceed from either 3-halo-1-nitroalkanes or 1,3-dinitroalkanes. The methods were discovered and developed by Kohler and coworkers[26,438,441,444]. Usually the starting compound is a Michael-addition derived product. The O-alkylation cyclization reaction to produce ester employs one mole-equivalent of a base such as potassium hydroxide, potassium acetate, or diethylamine. For example, benzal malonic ester (**146**) adds phenylnitromethane to produce the Michael adduct **147**. Bromination of **147** leads to the 3-bromo-1-nitroalkane **148**, which upon refluxing with ethanolic potassium acetate for 1 h produces isoxazoline *N*-oxide **149** in 90% yield (equation 142).

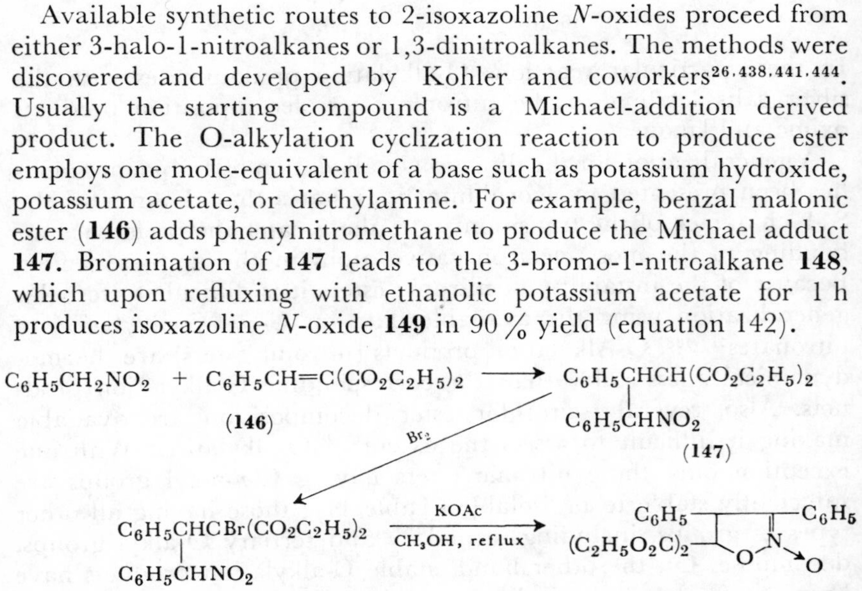

$C_6H_5CH_2NO_2$ + $C_6H_5CH{=}C(CO_2C_2H_5)_2$ (**146**) ⟶ $C_6H_5CHCH(CO_2C_2H_5)_2$ / $C_6H_5CHNO_2$ (**147**)

$\xrightarrow{Br_2}$ $C_6H_5CHCBr(CO_2C_2H_5)_2$ / $C_6H_5CHNO_2$ (**148**) $\xrightarrow[CH_3OH,\ reflux]{KOAc}$ C_6H_5, C_6H_5, $(C_2H_5O_2C)_2$; O, N, →O (**149**) 90% (142)

The reaction has been extended to 3-bromo-1,1-dinitroalkanes (Table 13)[436]. 3-Bromo-1,1-dinitropropane (**150**) reacts with potassium acetate in water to precipitate 3-nitro-2-isoxazoline-2-oxide (**151**)[436] (equation 143).

$$BrCH_2CH_2CH(NO_2)_2 \xrightarrow[H_2O,\ 25°]{KOAc} \text{3-nitro-2-isoxazoline-2-oxide} \quad (143)$$

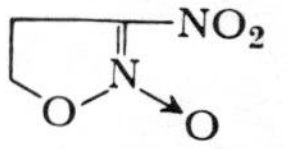

(**150**) (**151**) 84%; M.p. 96.5°

Use of a 1,3-dinitropropane for cyclic ester synthesis is illustrated with 1,3-dinitro-1,2,3-triphenylpropane (**152**), prepared by Michael addition of phenylnitromethane to α-nitrostyrene[441,444a]. One equivalent of sodium methoxide converts the dinitro compound to 3,4,5-triphenyl-2-isoxazoline-2-oxide (**153**) (equation 144).

$$C_6H_5CH_2NO_2 + C_6H_5CH{=}C(NO_2)C_6H_5 \xrightarrow{NaOCH_3}$$

$$C_6H_5CH(CH(NO_2)C_6H_5)(C_6H_5CHNO_2) \xrightarrow{NaOCH_3} \text{3,4,5-triphenyl-2-isoxazoline-2-oxide}$$

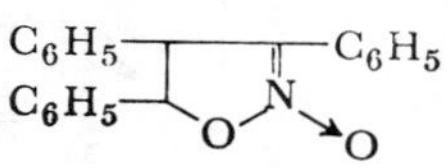

(**152**) (**153**) (144)

The mechanism of the isoxazoline synthesis from 3-bromoalkane-1-nitronates is reasonably a bromide displacement by nitronate oxygen. The mechanism departing from 1,3-dinitroalkanes has been shown not to involve nitroolefin intermediate **154**[444a]. The reaction occurs by intramolecular displacement of nitrite ion from the mononitronate anion **155** (equation 145)[444a].

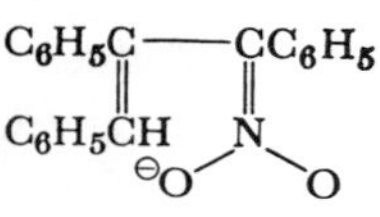

$$C_6H_5CH{=}C(C_6H_5)C(C_6H_5){=}N(O^-){\to}O \quad C_6H_5CH(NO_2)CH(C_6H_5)C(=NO_2^-)C_6H_5 \xrightarrow{-NO_2^-} \mathbf{153} \quad (145)$$

(**154**) (**155**)

Cyclic nitronic esters with six-membered rings are available from 4-keto-1-nitroalkanes. As discussed in section III.C.4, 4-keto-1,1-dinitropentane is believed to cyclize in solution, although the *pseudo* ester **127** was not isolated.

Michael addition of a 1-nitroalkene to cyclohexane-1,3-dione anion (**156**) (sodium methoxide catalyst) leads to the bicyclic nitronic ester **160**[446] (equation 146).

TABLE 13. Synthesis and properties of cyclic nitronic esters (2-isoxazoline *N*-oxides).

Reactant(s)	Base	Nitronic ester	M.p. (°C)	Yield (%)	Ref.
		A. Synthesis from 3-halo-1-nitroalkanes			
$BrCH_2CH_2CH(NO_2)_2$	KOAc	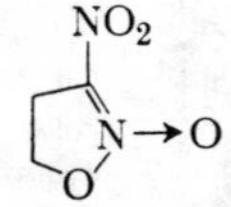 NO_2; N→O; O	96.5	84	436
$BrCH_2CH(CH_3)CH(NO_2)_2$	KOH	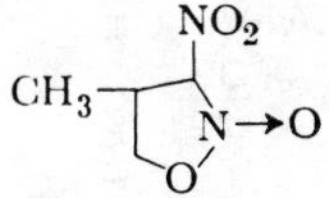 NO_2; CH_3; N→O; O	liquid	95	436
$(CH_3)_2C{=}NO_2^-Na^+$ *and* $CH_3C(NO_2){=}CHCH_2Cl$	none added	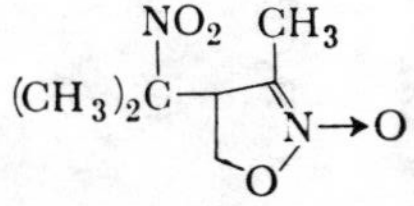 NO_2 CH_3; $(CH_3)_2C$; N→O; O	77–78	35–40	408
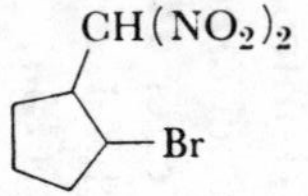 $CH(NO_2)_2$; Br	KOH	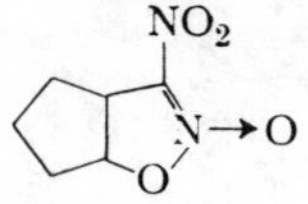 NO_2; N→O; O	83–83.5	32	436

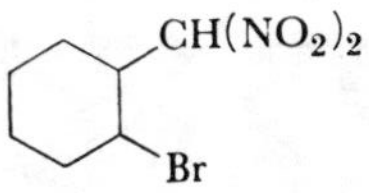 (cyclohexane ring) $CH(NO_2)_2$, Br	KOH	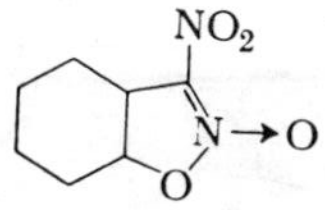 NO_2, $N\rightarrow O$, O (fused to cyclohexane ring)	94–95	45	436, 423b
$C_6H_5CH(CH_2Br)CH(NO_2)_2$	KOH	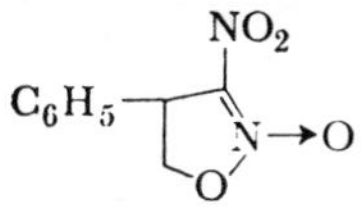 NO_2, C_6H_5, $N\rightarrow O$, O	80	75	436
$(EtO_2C)_2C(Br)CH(C_6H_5)CH(C_6H_5)NO_2$	KOAc	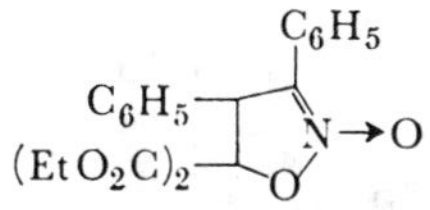 C_6H_5, C_6H_5, $(EtO_2C)_2$, $N\rightarrow O$, O	107–108	90	437, 438
$C_6H_5COCHBrCH(C_6H_5)CH(C_6H_5)NO_2$	KOAc	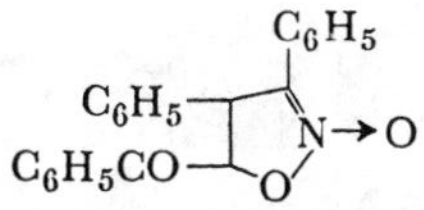 C_6H_5, C_6H_5, C_6H_5CO, $N\rightarrow O$, O	123	95	26
B. Synthesis from 1,3-dinitroalkanes					
$C_6H_5CH[CH(NO_2)CO_2Et]_2$	Et_2NH	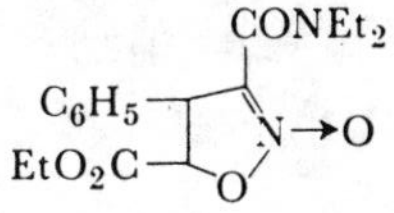 $CONEt_2$, C_6H_5, EtO_2C, $N\rightarrow O$, O	181	80[a]	439, 440

(*continued*)

TABLE 13—*continued*

Reactant(s)	Base	Nitronic ester	M.p. (°C)	Yield (%)	Ref.
$C_6H_5CH{=}C(NO_2)C_6H_5$ *and* $C_6H_5CH_2NO_2$	$NaOCH_3$	C_6H_5; C_6H_5; C_6H_5; $N{\rightarrow}O$; O	162	45	441
$C_6H_5CH[CH(NO_2)C_6H_5]NH(CH_2)_5$	none added	C_6H_5; C_6H_5; C_6H_5; $N{\rightarrow}O$; O	162	59	442
$C_6H_5CH{=}CHNO_2$	$(CH_2)_5NH$	C_6H_5; C_6H_5; C_6H_5; $N{\rightarrow}O$; O	162	45	442
$C_6H_5CH{=}C(C_6H_5)NO_2$ *and* 4-Br$C_6H_4CH_2NO_2$	$NaOCH_3$	C_6H_5; C_6H_5; 4-BrC_6H_4; $N{\rightarrow}O$; O	—	60 (total)	443
		C_6H_4-Br-4; C_6H_5; C_6H_5; $N{\rightarrow}O$; O	—		

[a] Eight other examples reported (30–90% yield) with C_6H_5 replaced by 4-$CH_3OC_6H_4$, 4-HOC_6H_5, 2-ClC_6H_4, 4-ClC_6H_4, 2-$O_2NC_6H_4$, 3-$O_2NC_6H_4$, 4-$O_2NC_6H_4$, 2-HO-naphthyl[440].

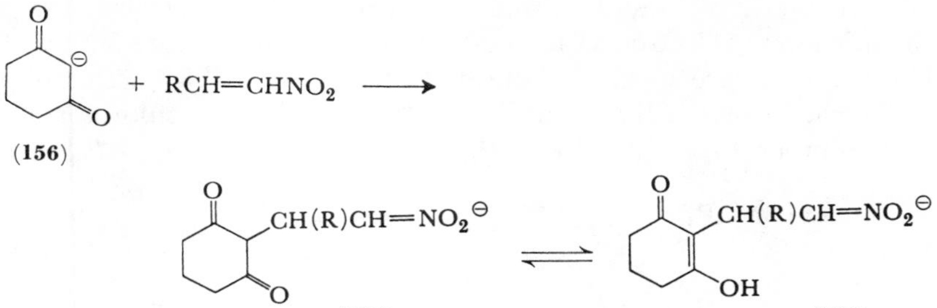

(156) + $RCH{=}CHNO_2$ ⟶

$CH(R)CH{=}NO_2^{\ominus}$ (157) ⇌ $CH(R)CH{=}NO_2^{\ominus}$ (158)

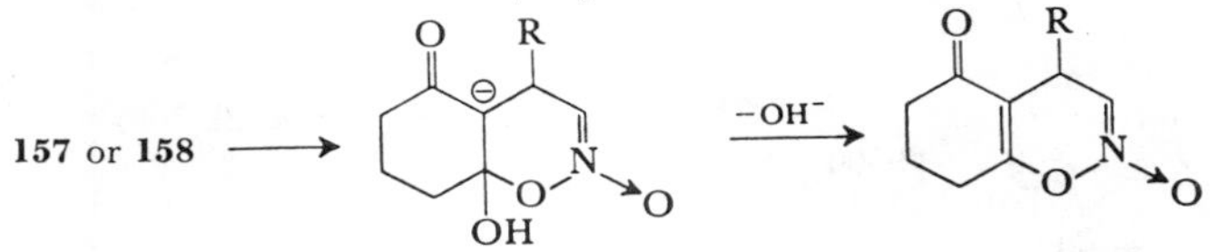

157 or **158** ⟶ (159) $\xrightarrow{-OH^-}$ (146)

(**160a**) R = CH_3 (52%); m.p. 164–166°
(**160b**) R = C_6H_5 (72%); m.p. 165–167°

Cyclic nitronic esters having four, seven, eight, and larger membered rings (homologs of 2-isoxazoline *N*-oxides) are unknown. Heterocyclic compounds such as isoxazole and oxadiazole *N*-oxides, etc., may be called cyclic nitronic esters[447,448].

Cyclic nitronic esters are not readily prepared by oxidation of the corresponding desoxy compounds (2-isoxazolines, for example[65])[437].

B. Physical Properties of Nitronic Acid Esters

Melting points of nitronic esters are listed in Tables 11 and 13.

The ultraviolet spectra of a few nitronic esters have been reported[23,401,402,409,449]. None representing the simple aliphatic type, $(alkyl)_2C{=}NO_2R$ are available, but these would be expected to have strong $\pi - \pi^*$ bands near 220–230 mμ like the parent nitronic acids (see section III.B). Several α-nitro acyclic and cyclic esters (e.g. **161**, **162**) have been examined in methylene chloride at 5° and in water solutions[402,449]. They all have strong absorption of ca. 315–320 mμ; the corresponding nitronate anions absorbed at wavelengths 25–50 mμ higher[402]. The extinction coefficient in the spectrum of unstable ethyl nitroformate (**161**) was obtained by extrapolation to zero time.

$(O_2N)_2C{=}NO_2C_2H_5$

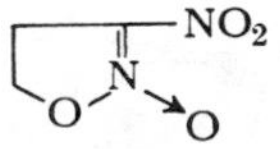

(**161**)
$\lambda_{max}^{CH_2Cl_2}$ 315 mμ (6000)

(**162**)
$\lambda_{max}^{H_2O}$ 320 mμ (8366)

The spectra of the high-melting, presumably *trans* isomers of nitronic esters **163a** and **164** reveal intense absorption near 300 mμ[23]. The solutions are unstable; the half-life of **164** in 95% ethanol at 19° is ca. 7 days[23]. The half-life of **163a** in deuteriochloroform at room temperature is ca. 2 days[23].

$4\text{-}O_2NC_6H_4(H)C{=}N(\rightarrow O)OR$ (*trans*)

(**163a**) R = C_2H_5; m.p. 100–101° λ_{max}^{EtOH} 240 mμ (9300) 337 mμ (17,600)

(**163b**) R = CH_3; m.p. 118–120°

$4\text{-}BrC_6H_4(H)C{=}N(\rightarrow O)OCH_3$ (*trans*)

(**164**) M.p. 66.5–67.5° λ_{max}^{EtOH} 288 mμ (32,700)

The ultraviolet spectra of α-keto nitronic esters (**165, 96**) have been reported; that of **165** resembles that of nitrone **166**[414].

CH$_3$; (CH$_3$)$_2$; O; NO$_2$C(C$_6$H$_5$)$_3$

(**165**) $\lambda_{max}^{Et_2O}$ 288 mμ (13,900)

O; NO$_2$CH$_3$; O; C$_6$H$_5$

(**96**) λ_{max}^{EtOH} 310 mμ (7950) 358 mμ (3740)

(CH$_3$)$_2$; O; $\overset{+}{N}$; O$^-$

(**166**) λ_{max}^{EtOH} 280 mμ (17,500)

The infrared spectra of nitronic esters reveal intense C=N absorption in the region 1610–1660 cm^{-1} [23,178,408]. This is within the C=N absorption region of nitronic acids (1620–1680 cm^{-1}) and oximes (1640–1685 cm^{-1}); see section III.B.1.(1).

Examination of the n.m.r. spectra of certain nitronic esters permits stereochemical assignments[23,213]. Like oximes, nitronic esters exist in two geometric forms. Two forms of ester **168** have been isolated by reaction of nitroolefin **167** (presumably *trans*) with diazomethane[213] (equation 147). Both produce the same oxime **169** by

$$4\text{-}BrC_6H_4CO(H)C{=}C(H)CH_2NO_2 \xrightarrow{CH_2N_2} 4\text{-}BrC_6H_4(H)C{=}C(H)CH{=}NO_2CH_3 \xrightarrow[\text{Heat}]{-CH_2O}$$

(**167**) M.p. 91°

(**168a**) 27%; M.p. 114°
(**168b**) 35%; M.p. 128°

$$\begin{array}{ccc} 4\text{-}BrC_6H_4 & & H \\ & C{=}C & \\ H & & CH{=}NOH \end{array} \qquad (147)$$

(**169**) M.p. 164°

(one isomer only from **168a** or **168b**)

heating in toluene[213] and are believed to be isomeric about the C=N bond, i.e. *cis* and *trans* forms of the nitronic ester.

The n.m.r. spectra of nitronic esters prepared from primary nitro compounds exhibit vinyl hydrogen peaks near 6.0 δ for alkyl R in $RCH{=}NO_2C_2H_5$, and near 7.0 δ (singlet) for aryl R ($CDCl_3$ solvent)[23]. A mixture of *cis* and *trans* isomers is indicated in crude ester samples by noticeable splitting of these peaks. For example, one isomer (probably *trans*) of methyl 4-nitrophenylmethanenitronate (**163b**), m.p. 118–120°, has a sharp vinyl singlet at 7.20 δ, whereas crude ester, m.p. 100–108°, clearly containing a mixture of *cis* and *trans* forms, exhibits vinyl singlets at 7.20 and 6.92 δ[23]. Similar observations were made with the 4-bromo compound **164.** Pure samples of the other (probably lower-melting, *cis*) isomers of **163b** and **164** were not isolated; these isomers were observed (by n.m.r.) to be much less stable than the *trans* isomers; the half-life of *cis*-**164** is estimated at ca. 40 min in deuteriochloroform at room temperature[23].

C. Reactions of Nitronic Acid Esters

Reactions of the rather unstable nitronic esters parallel those of nitronic acids. The same products often are formed from both substances under similar reaction conditions. Hydrolysis under acidic conditions can lead to Nef products or hydroxamic acids. Nitronic esters are good oxidizing agents, are easily reduced, and participate in auto-oxidation-reduction reactions, the most important of these being a disproportionation to an oxime and an aldehyde or ketone. Diene addition, a reaction not yet observed with nitronic acids, leads to 1,2-isoxazolidines with nitronic esters.

1. Hydrolysis of nitronic esters

It seems remarkable that hydrolysis of a nitronic ester to produce an isolable nitronic acid directly (equation 148) has never been observed. Surprising also is the fact that the reverse reaction, synthesis of a nitronic ester from an alcohol and a nitronic acid, has

$$R^1R^2C{=}NO_2R^3 + H_2O \rightleftharpoons R^1R^2C{=}NO_2H + R^3OH \quad (148)$$

not been found. It has been possible, however, to hydrolyze certain nitronic esters (those which form stabilized nitronate anions) to nitroalkanes and alcohols[384,394,401,409,414,421a]. Either acidic or basic catalysts have been employed (equations 149–152).

$$C_6H_5COCH{=}NO_2CH_3 + H_2O \xrightarrow{HCl} C_6H_5COCH_2NO_2 + CH_3OH \quad (149)^{394}$$

$$O{=}C_6H_2(NO_2)_2{=}NO_2C_2H_5 + H_2O \longrightarrow HO{-}C_6H_2(NO_2)_2{-}NO_2 + C_2H_5OH \quad (150)^{384}$$

$$(CH_3)_2[\text{bicyclic ketone, }CH_3]{=}NO_2C(C_6H_5)_3 + H_2O \xrightarrow[EtOH]{NaOH} (CH_3)_2[\text{bicyclic ketone, }CH_3]{-}NO_2 + (C_6H_5)_3COH \quad (151)^{414}$$

$$O_2NC(CN){=}NO_2CH_3 + H_2O \xrightarrow[50°]{NaOH} HC(NO_2)_2CN + CH_2O + NO_2^{\ominus} \quad (152)^{401}$$

Ethyl acetamidomethanenitronate reacts with ammonia or silver nitrate to form nitronate salts[403,407] (equations 153, 154).

$$H_2NCOCH{=}NO_2C_2H_5 + NH_3 \longrightarrow H_2NCOCH{=}NO_2^-NH_4^+ + C_2H_5OH \quad (153)$$

$$H_2NCOCH{=}NO_2C_2H_5 + AgNO_3 \longrightarrow H_2NCOCH{=}NO_2Ag + HNO_3 + C_2H_5OH \quad (154)$$

Acid hydrolysis of nitronic esters under Nef reaction conditions produces aldehydes and ketones (Table 14)[222]. With aliphatic nitronic esters the products and product yields are virtually the same as those obtained from the corresponding nitronic acids (generated from nitronate salts)[47,222,229]. For example, either sodium butane-2-nitronate or ethyl butane-2-nitronate produces 2-butanone in 81–82 % yield when treated with 4*N* sulfuric acid[222,229] (equation 155).

$$C_2H_5C(CH_3){=}NO_2C_2H_5 \xrightarrow[81\%]{dil.\ H_2SO_4} C_2H_5COCH_3 \xleftarrow[82\%]{dil.\ H_2SO_4} C_2H_5C(CH_3){=}NO_2^-Na^+ \quad (155)$$

TABLE 14. Reaction of nitronic esters and nitronate salts with sulfuric acid[222].

A. Reactions of nitronic esters with sulfuric acid

Nitronic ester	Products with $4N$ H_2SO_4 (%)	Products with $31N$ H_2SO_4 (%)
$4\text{-}BrC_6H_4CH{=}NO_2Et$	$4\text{-}BrC_6H_4CHO$ (65) $4\text{-}BrC_6H_4CONHOH$ (12)	$4\text{-}BrC_6H_4CONHOH$ (63)
$4\text{-}O_2NC_6H_4CH{=}NO_2Et$	$4\text{-}O_2NC_6H_4CHO$ (80–82) $4\text{-}O_2NC_6H_4CONHOH$ (6)	$4\text{-}O_2NC_6H_4CONHOH$ (98)
$CH_3CH{=}NO_2Et$	CH_3CHO (67)	$CH_3CONHOH$ (41)
$CH_3CH_2CH{=}NO_2Et$	CH_3CH_2CHO (good)	—
$(CH_3)_2C{=}NO_2Et$	$(CH_3)_2CO$ (72)	—
$CH_3CH_2CH_2CH{=}NO_2Et$	$CH_3CH_2CH_2CHO$ (good)	$CH_3CH_2CH_2CONHOH$ (42)
$CH_3CH_2C(CH_3){=}NO_2Et$	$CH_3CH_2COCH_3$ (81)	—

B. Reactions of nitronate salts with sulfuric acid

Nitronate salt	Products with $4N$ H_2SO_4 (%)	Products with $31N$ H_2SO_4 (%)
$4\text{-}BrC_6H_4CHNO_2^-Na^+$	$4\text{-}BrC_6H_4CH_2NO_2$ (90)[a]	$4\text{-}BrC_6H_4CONHOH$ (29)
$4\text{-}O_2NC_6H_4CH{=}NO_2^-Na^+$	$4\text{-}O_2NC_6H_4CH_2NO_2$ (93)	$4\text{-}O_2NC_6H_4CONHOH$ (86)
$CH_3CH{=}NO_2^-Na^+$	CH_3CHO (85) $CH_3CONHOH$ (2)[b]	$CH_3CONHOH$ (45)
$CH_3CH_2CH{=}NO_2^-Na^+$	CH_3CH_2CHO (80)[47]	—
$(CH_3)_2C{=}NO_2^-Na^+$	$(CH_3)_2C{=}O$ (84)[47]	—
$CH_3CH_2CH_2CH{=}NO_2^-Na^+$	$CH_3CH_2CH_2CHO$ (70) $CH_3CH_2CH_2CONHOH$ (4)[b]	$CH_3CH_2CH_2CONHOH$ (28)
$CH_3CH_2C(CH_3){=}NO_2^-Na^+$	$CH_3CH_2COCH_3$ (81)[229]	

[a] With 10% H_2SO_4.
[b] With 21% H_2SO_4.

On the other hand, 4-nitrophenylmethanenitronic acid does not undergo the Nef reaction (equation 156). Yet, its ethyl ester yields the Nef product, 4-nitrobenzaldehyde (80–82%) under the same conditions[222] (equation 157). A similar observation is made with 4-bromophenylmethanenitronic acid[222].

$$O_2NC_6H_4CH{=}NO_2^-H^+ \xrightarrow{\text{dil. } H_2SO_4} O_2NC_6H_4CH_2NO_2 \quad 93\% \tag{156}$$

$$O_2NC_6H_4CH{=}NO_2Et \xrightarrow{\text{dil. } H_2SO_4} O_2NC_6H_4CHO \quad 80\text{–}82\% \tag{157}$$

These results indicate a mechanism with aromatic nitronic esters (and probably with all nitronic esters) which does *not* involve a nitronic acid intermediate and does *not* involve prior hydrolysis of the ester to a nitronic acid. As pointed out by Kornblum[222], a protonated nitronic ester must be involved. Hydration of this species **170**, followed by loss of alcohol, would yield carbonyl compound by a mechanism (equations 158–160) like that of the Nef; compare section III.C.1a.

$$R^1R^2C{=}N(\rightarrow O)OC_2H_6 + H^+ \longrightarrow R^1R^2C{=}\overset{\oplus}{N}(OH)OC_2H_6 \;\; (\mathbf{170}) \tag{158}$$

$$R^1R^2C{=}\overset{\oplus}{N}(OH)OC_2H_6 \;(\mathbf{170}) + H_2O \longrightarrow R^1R^2C(OH){-}\overset{\oplus}{N}H(OH)OC_2H_5 \tag{159}$$

$$R^1R^2C(OH){-}\overset{\oplus}{N}H(OH)OC_2H_5 \longrightarrow R^1R^2C{=}O + C_2H_5OH + H^+ + HNO \tag{160}$$

The observation that nitroalkanes are not usually obtained by acid hydrolysis of nitronic esters suggests that the rate of acid hydrolysis of protonated ester **171** to nitronic acid (equation 161) is usually slower than the 'ester Nef' reaction (equations 159, 160). The acid-

$$R^1R^2C{=}N(\rightarrow O)\overset{\oplus}{O}(H)C_2H_5 \;(\mathbf{171}) + H_2O \rightleftharpoons R^1R^2C{=}N(\rightarrow O)\overset{\oplus}{O}H_2 + C_2H_5OH \tag{161}$$

catalyzed hydrolysis of α-ketonitronic esters to α-ketonitroalkanes (equations 149–151) may be particular and proceed by a mechanism involving a protonated carbonyl group; formation of a resonance stabilized nitronate anion $\left[R^1C(=O)-C(R^2)=N(\to O)-O^-\right]$, also facilitates these reactions.

Hydroxamic acids are formed in concentrated sulfuric acid (31*N*) from both aliphatic and aromatic nitronic esters (equation 162); yields (higher with aromatics) are comparable to those obtained from nitronic acids (*via* nitronate salts) (Table 14)[222]. However, in

$$\begin{aligned}&PrCH{=}NO_2Et \xrightarrow{\text{Concd } H_2SO_4} PrCONHOH \quad (42\%)\\&4\text{-}O_2NC_6H_4CH{=}NO_2Et \xrightarrow{\text{Concd } H_2SO_4} 4\text{-}O_2NC_6H_4CONHOH \quad (98\%)\end{aligned} \qquad (162)$$

dilute (4*N*) sulfuric acid solution a difference in behavior is noted between nitronic esters and nitronic acids in forming hydroxamic acids. In the more dilute acid aliphatic nitronic acids form hydroxamic acids, but aromatic ones do not. The opposite is true of the nitronic esters. In 4*N* sulfuric acid solution aliphatic nitronic esters do not, but aromatic nitronic esters do form hydroxamic acids.

This difference in behavior between nitronic acids and esters in yet another acid-catalyzed reaction suggests, as in the 'ester Nef' reaction (equations 158–160), a mechanism which does not require a nitronic acid intermediate. In a nitrile oxide mechanism (see section III.C.1) the protonated ester **170** could lose alcohol, ultimately forming a protonated nitrile oxide **172**, which hydrates to the hydroxamic acid (equation 163). The formation, in 4*N* sulfuric acid, of more hydroxamic acid from aliphatic nitronic acids than

$$\underset{\textbf{(170)}}{(R)(H)C{=}\overset{\oplus}{N}(OH)(OC_2H_6)} \longrightarrow \underset{\textbf{(172)}}{RC{\equiv}NOH^{\oplus}} + C_2H_5OH$$

$$\underset{\textbf{(172)}}{RC{\equiv}NOH^{\oplus}} + H_2O \longrightarrow H^+ + RCONHOH \qquad (163)$$

from aliphatic nitronic esters (which form aldehydes) is readily explained by assuming that **170** is more stable (forms nitrile oxide more slowly) than the corresponding intermediate, $RCH{=}\overset{\oplus}{N}(OH)_2$ (**173**), obtained from the acid. The assumed relatively greater stability of **170** over **173** also explains the observed formation of hydroxamic acids from aromatic nitronic esters, and the absence of formation of hydroxamic acids from the corresponding aromatic nitronic acids (which tautomerize to nitroalkanes) under the same conditions in $4N$ sulfuric acid.

A hydroxamic acid **175** has been prepared from a nitronic ester (**174**) in a basic medium; a nitrile oxide intermediate has been postulated for this reaction[414] (equation 164).

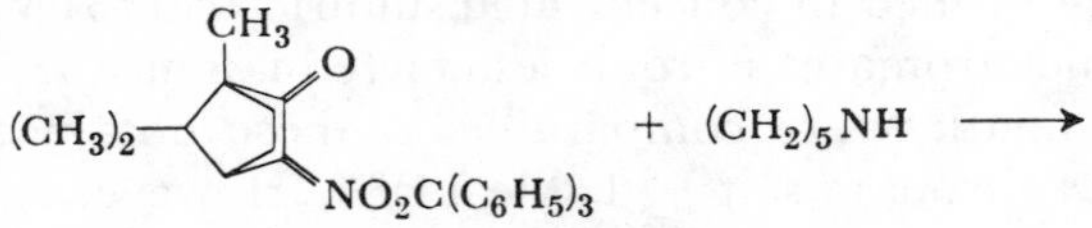

$$(\mathbf{174}) + (CH_2)_5NH \longrightarrow (\mathbf{175}) + (C_6H_5)_3COH \qquad (164)$$

Hydrogen chloride should add to nitronic esters—as it does to nitronic acids—to yield chloronitroso compounds or hydroxamic acid chlorides (section III.C.1a). Methyl dicarbomethoxymethanenitronate (**176**) produces oxides of nitrogen and a blue color (possibly compound **177**; not isolated) when treated with aqueous hydrogen chloride[394] (equation 165).

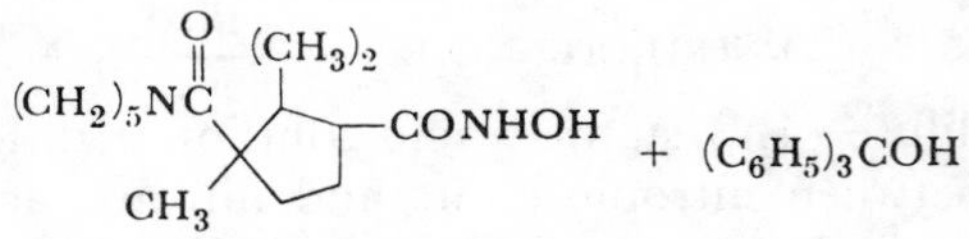

$$\underset{(\mathbf{176})}{(CH_3O_2C)_2C{=}NO_2CH_3} + HCl \longrightarrow \underset{\substack{(\mathbf{177})\\ \text{(not isolated)}}}{(CH_3O_2C)_2C(Cl)NO} + CH_3OH \qquad (165)$$

The addition of hydrogen chloride to ethyl carbethoxymethanenitronate leads to the hydroxamic acid chloride **181**. Addition probably proceeds, as with nitronic acids, through a protonated ester **178** and an HCl adduct **179** which loses ethanol to form a chloronitroso compound **180**. Rearrangement of **180** would yield the hydroxamic acid chloride product, **181** (equation 166).

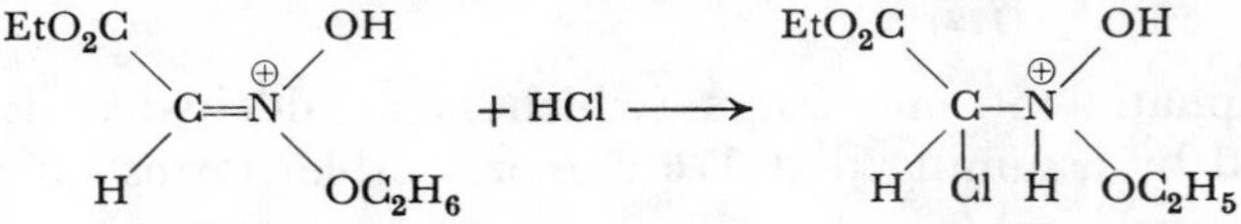

$$\underset{(\mathbf{178})}{EtO_2C(H)C{=}\overset{\oplus}{N}(OH)OC_2H_5} + HCl \longrightarrow \underset{(\mathbf{179})}{EtO_2C(H)(Cl)C{-}\overset{\oplus}{N}(H)(OH)OC_2H_5} \qquad (166)$$

$$\underset{(\mathbf{179})}{EtO_2C(H)(Cl)C{-}\overset{\oplus}{N}(H)(OH)OC_2H_5} \xrightarrow[-H^+]{-C_2H_5OH} \underset{(\mathbf{180})}{EtO_2C(H)(Cl)CNO} \longrightarrow \underset{(\mathbf{181})}{EtO_2C(Cl)C{=}NOH}$$

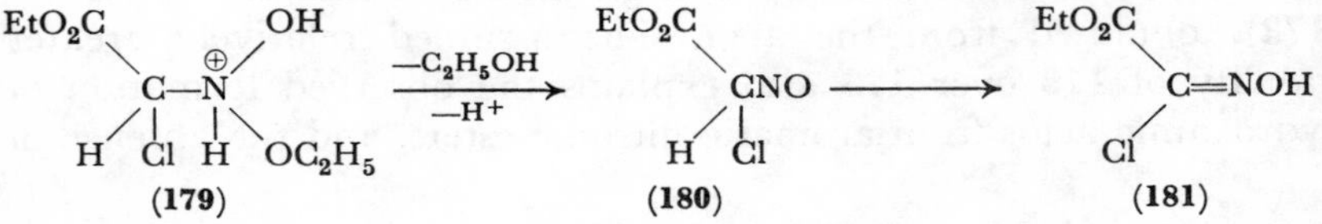

Since electrophilic additions to nitronic acids (section III.C.1b) appear to involve the nitronate anion in acid solution, no reactions of this type are to be expected for nitronic esters. None are found. Reactions of ethyl and methyl acetamidomethanenitronate with aqueous bromine are reported to produce 1-bromonitronic esters [$H_2NCOC(Br){=}NO_2R$; R = CH_3, C_2H_5], solids which decompose slowly at room temperature[403]. The possibility also exists, however, that these products are N-bromo derivatives, $BrNHCOCH{=}NO_2R$.

2. Oxidation and reduction reactions of nitronic acid esters

Nitronic esters, like nitronic acids, are good oxidizing agents. They are easily reduced to oximes. Only a few reducing agents have been employed in reactions with nitronic esters. Reaction of nitronic esters with added oxidizing agents apparently has not been studied.

Hydrogen iodide reduces nitronic esters to oximes with formation of iodine[394,412] (equation 167). The reaction may involve an addition

$$\underset{\displaystyle\text{CN}}{C_6H_5\overset{|}{C}}{=}NO_2CH_3 + 2\,HI \longrightarrow \underset{\displaystyle\text{CN}}{C_6H_5\overset{|}{C}}{=}NOH + CH_3OH + I_2 \qquad (167)$$

of hydrogen iodide, followed by loss of alcohol to yield a transient iodonitroso compound; reduction of this product by hydrogen iodide would yield the oxime. Unlike the reduction of nitronic acids with hydrogen iodide[276], the reaction rate is quite variable and is not cleanly quantitative. Arndt and Rose[394] observed that when esters are treated with concentrated hydriodic acid, the number of equivalents of iodine formed, and the rate of reaction, varied with the structure of the ester: 4-$CH_3C_6H_4SO_2CH{=}NO_2CH_3$ reacted exothermically to liberate 2–2.5 equivalents of iodine; $C_6H_5CH{=}NO_2CH_3$ reacted on warming to produce 0.5 equivalent, and $(CH_3O_2C)_2C{=}NO_2CH_3$ produced 4.36–4.38 equivalents. No iodine was formed when 4-$BrC_6H_4CH{=}NO_2CH_3$ was treated with cold colorless azeotropic hydroiodic acid[394].

Although reduction of a nitronic acid to an amine appears not to have been described, nitronic esters have been so reduced. Methyl α-cyanophenylmethanenitronate (**182**) is hydrogenated (platinum, acetic anhydride) completely to phenyl-1,2-diaminoethane, isolated as the bisacetyl derivative **183**[411] (equation 168).

$$\underset{\displaystyle\textbf{(182)}}{\underset{\displaystyle\text{CN}}{C_6H_5\overset{|}{C}}{=}NO_2CH_3} \xrightarrow[Ac_2O]{H_2,\,Pt} \underset{\displaystyle\textbf{(183)}}{\underset{\displaystyle\text{NHAc}}{C_6H_5\overset{|}{C}H}CH_2NHAc} \qquad (168)$$

Hydrogenation of keto ester **174** led to triphenyl carbinol and a mixture of epimeric amino alcohols **184**[414] (equation 169).

$$(CH_3)_2\text{-bicyclic}(CH_3)(=O)(=NO_2C(C_6H_5)_3)\ (\mathbf{174}) \xrightarrow{H_2,\ Pt} (CH_3)_2\text{-bicyclic}(CH_3)(OH)(NH_2)\ (\mathbf{184})\ \text{Epimers} + (C_6H_5)_3COH\ \ 86\% \quad (169)$$

An auto-oxidation-reduction reaction of nitronic esters is one of their most important and characteristic properties. It is the disproportionation of the nitronic ester to form an oxime and an aldehyde or ketone (Table 15) and is of synthetic utility for preparing each of these products. The reaction proceeds by heating in solution in various solvents, or, in the absence of solvents (equations 170–172). It appears to have been discovered by Nef (1894)[15,404]. Yields,

$$C_6H_5COCH{=}CHCH{=}NO_2CH_3 \xrightarrow[\text{Reflux 3–4 h}]{EtOH} C_6H_5COCH{=}CHCH{=}NOH + CH_2O \quad (170)^{213}$$
$$80\%$$

$$C_6H_5N{=}NC(CH_3){=}NO_2CH_3 \xrightarrow[\text{Reflux 3 min}]{H_2O} C_6H_5N{=}NC(CH_3){=}NOH + CH_2O \quad (171)^{298}$$
$$90\% \qquad 86\%$$

$$C_6H_5C(CN){=}NO_2CH_3 \xrightarrow[10\ h]{100^\circ} C_6H_5C(CN){=}NOH + CH_2O \quad (172)^{197}$$
$$50\%$$

although seldom reported, generally appear to be very good.

The reaction is not limited to the few types of stable nitronic esters. It is also observed with esters generated *in situ*. Alkylations of nitronic acids or nitroalkanes with diazomethane, alkyl fluoroborates, or alkyl halides can produce oximes. The synthetic value of the reaction applied to simple esters (methyl, ethyl) lies in oxime synthesis (e.g. equation 173)[419]; cf. Table 15 for other examples.

$$\text{10-nitroanthrone}\ (NO_2,\ O) \xrightarrow[Et_2O,\ C_6H_6]{CH_2N_2} \text{anthraquinone monoxime}\ (NOH,\ O) \quad (173)^{419}$$

Preparation of aldehydes and ketones from nitronic esters generated *in situ* is important synthetically[392,393,409,450]. An alkali or silver nitronate reacts with a primary or secondary alkyl halide in a solvent such as ethanol, usually at reflux temperature (equations 174–176).

$$C_6H_5C(CN){=}NO_2Ag + 4\text{-}O_2NC_6H_4CH_2Br \longrightarrow C_6H_5C(CN){=}NOH + 4\text{-}O_2NC_6H_4CHO \quad (174)^{387}$$

80% 80%

$$(CH_3)_2C{=}NO_2^-Na^+ + \text{3-bromocyclohexene} \longrightarrow (CH_3)_2C{=}NOH + \text{2-cyclohexen-1-one} \quad (175)^{429}$$

44%

$$\text{potassium fluorene-9-nitronate } (NO_2^-K^+) + BrCH_2COCH_3 \longrightarrow \text{fluorenone oxime (NOH)} + CH_3COCHO \quad (176)^{308}$$

97%

It is possible to examine the scope of this reaction since many examples are known (Table 16). The reaction has been developed as a synthetic method[392,393,450]. The customary procedure involves preparation of a nitronate salt from the nitroalkane and sodium ethoxide in ethanol, followed by addition of the halide. A short period of heating under reflux (1–3 h) is usually sufficient to complete the reaction. Yields of both carbonyl compound and oxime are usually excellent. Side reactions seem to present no difficulties and are seldom encountered. The range of structural variations allowed in nitro compound and halide is quite large. Primary and secondary nitroalkanes seem equally effective. Yields of aldehydes (from primary halides) appear to equal those of ketones from secondary halides. Aliphatic, alicyclic, and arylalkyl halides and nitronate salts have been employed with equal success. The method has been employed successfully in the synthesis of 1,2-dicarbonyl compounds from α-halo ketones[308,450]. Although alkali metal salts have been employed in most studies, silver salts are equally effective[15,387,417]. What appears to be the first example of the reaction, due to Nef[15], employed silver 1-nitroethanenitronate and ethyl iodide to yield acetaldehyde and α-nitroacetaldoxime.

The mechanism of the disproportionation of nitronic esters to carbonyl compounds has been studied[421a]. Generally, the reaction

TABLE 15. Decomposition of nitronic esters.

Nitronic ester	Solvent, Temp., (°C)	Products	Yield (%)	Ref.
$H_2NCOC(CN){=}NO_2C_2H_5$	H_2O, heat	$H_2NCOC(CN){=}NOH$	—	15, 405,
		CH_3CHO	—	406
$C_2H_5O_2CCH{=}NO_2CH_3$	none, 65°	$C_2H_5O_2CCH{=}NOH$	—	394
		CH_2O	—	
$(CH_3)_2C{=}NO_2C_2H_5$[a]	H_2O, 60–70°	$(CH_3)_2C{=}NOH$	20	390
		CH_3CHO	—	
cyclopentylidene$=NO_2C_2H_5$[b]	H_2O, 60–70°	cyclopentylidene$=NOH$	86	,390
		cyclopentylidene$=NOC_2H_5$[b]	5	
		CH_3CHO	—	
$4\text{-}BrC_6H_4CH{=}NO_2CH_3$	none, 80°	$4\text{-}BrC_6H_4CH{=}NOH$	—	394
		CH_2O	—	
	15% HCl, 100°	3-(4-BrC_6H_4)-5-(C_6H_4-Br-4)-1,2,4-oxadiazole	—	394
		$4\text{-}BrC_6H_4CO_2H$	—	
$C_6H_5CH{=}NO_2CH_3$	none, 25–100°	C_6H_5CHO	—	394
		CH_2O	—	
	15% HCl, 100°	3-C_6H_5-5-C_6H_5-1,2,4-oxadiazole	—	394
cyclohexylidene$=NO_2C_2H_5$[b]	H_2O, 50–60°	cyclohexylidene$=NOH$	51–79	390

		(cyclohexylidene)$=NOC_2H_5$[b]	7-19	
		CH_3CHO	—	
$2,4,6\text{-}Cl_3C_6H_2N_2C(CH_3){=}NO_2CH_3$	H_2O, 100°	$2,4,6\text{-}Cl_3C_6H_2N_2C(CH_3){=}NOH$	—	229
		CH_2O	—	
$2,4\text{-}Cl_2C_6H_3N_2C(CH_3){=}NO_2CH_3$	H_2O, 100°	$2,4\text{-}Cl_2C_6H_3N_2C(CH_3){=}NOH$	—	299
		CH_2O	—	
$4\text{-}BrC_6H_4CH{=}NO_2C_2H_5$	none, 25°	3-(4-BrC_6H_4)-5-(C_6H_4Br-4)-1,2,4-oxadiazole	60	23
		CH_3CHO	—	
$4\text{-}ClC_6H_4N_2C(CH_3){=}NO_2CH_3$	H_2O, 100°	$4\text{-}ClC_6H_4N_2C(CH_3){=}NOH$	—	298
		CH_2O	—	
2-quinoxalinyl-$CH{=}NO_2CH_3$	none, 120–140°	2-quinoxalinyl-$CH{=}NOH$	—	420
$C_6H_5C(CN){=}CHCH{=}NO_2CH_3$	toluene, 110°	$C_6H_5C(CN){=}CHCH{=}NOH$	58	191
		CH_2O	—	
$4\text{-}BrC_6H_4COCH{=}CHCH{=}NO_2CH_3$	EtOH, 80°	$4\text{-}BrC_6H_4COCH{=}CHCH{=}NOH$	70–80	213
		CH_2O	—	
$4\text{-}CH_3OC_6H_4COCH{=}CHCH{=}NO_2CH_3$	EtOH, 80°	$4\text{-}CH_3OC_6H_4COCH{=}CHCH{=}NOH$	70–80	213
		CH_2O	—	
2-Br-fluoren-9-ylidene$=NO_2CH_3$	EtOH, warm	2-Br-fluoren-9-ylidene$=NOH$	—	204

(continued)

TABLE 15–*continued*

Nitronic ester	Solvent, Temp., (°C)	Products	Yield (%)	Ref.
$C_6H_5C(CN){=}NO_2CH_2C_6H_5$	none, 25°	CH_2O $C_6H_5C(CN){=}NOH$ C_6H_5CHO	— — —	411
O, NO_2CH_3, O, C_6H_5	none, 100°	O, NOH, O, C_6H_5	25	344
$CH{=}NO_2CH_3$ | $C{=}NNHC_6H_5$ | $CH{=}NNHC_6H_5$	none, 120–130°	CH_2O $CH{=}NOH$ | $C{=}NNHC_6H_5$ | $CH{=}NNHC_6H_5$	— —	420
$NO_2CH_2C_6H_5$ [a]	EtOH, 80°	CH_2O NOH C_6H_5CHO	— 87 —	84

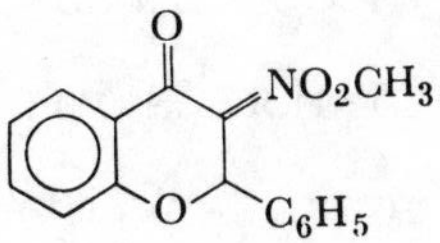

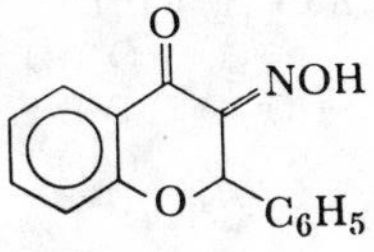

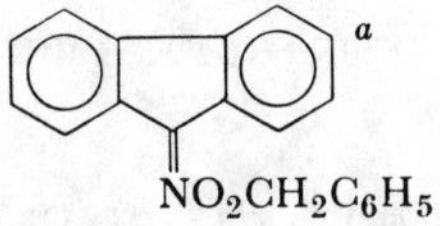

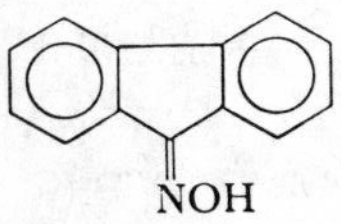

[a] Formed *in situ*.

[b] Formed *in situ* from sodium nitronates and $(EtO)_3BF_4^-$. The O-alkyl oxime products are believed to result from the alkylating agent present.

Table 16. Synthesis of aldehydes and ketones from *in situ* generated nitronic esters.

$$R^1R^2C{=}NO_2^-,\ M^+ + R^3R^4CHX \longrightarrow R^1R^2C{=}NOH + R^3R^4C{=}O + MX$$

Nitronate salt $R^1R^2C{=}NO_2^-$, M^+	Alkyl halide R^3R^4CHX; X =	Aldehyde or ketone, $R^3R^4C{=}O$	Yield (%)	Ref.
$(CH_3)_2C{=}NO_2^-Na^+$	Br	$HC{\equiv}CC(CH_3){=}CHCHO$	54	392
	Cl	cyclohexane-1,2-dione (structure)	30	450
	Br	$4\text{-}BrC_6H_4CHO$	70	5, 388, 393
	Cl	C_6H_5CHO	73	5, 388, 393, 450
	Br	$4\text{-}F_3CC_6H_4CHO$	77	5, 388, 393
	Br	$4\text{-}NCC_6H_4CHO$	70	5, 388, 393
	Br	$4\text{-}CH_3C_6H_4CHO$	70	5, 388
	Br	$4\text{-}CH_3O_2CC_6H_4CHO$	72	5, 388
	Br	$3,4\text{-}(CH_2O_2)C_6H_3CH_2COCH_3$	90	450
	Br	$OHCC(CH_3){=}CHC{\equiv}CCH{=}C(CH_3)CHO$	61	392
	I	$4\text{-}[(CH_3)_3N^+I^-]C_6H_4CHO$	68	5, 388
	Br	$(CH_3)_2C{=}CHCH_2CH_2C(CH_3){=}CHCHO$	75	392
	Br	2,2-dimethyl-6-methylcyclohexylidene with =CHCHO (structure: CH_3, CH_3, $=CHCHO$, $-CH_3$)	80	392

(*continued*)

Table 16—*continued*

Nitronate salt $R^1R^2C{=}NO_2^-$, M^+	Alkyl halide R^3R^4CHX; X =	Aldehyde or ketone, $R^3R^4C{=}O$	Yield (%)	Ref.
	Br	$n\text{-}C_9H_{19}CHO$	46	450
	Br	$(CH_3)_2C{=}C(CH_3)CH_2CH_2C(CH_3){=}CHCHO$	—	392
	Br	$n\text{-}C_{10}H_{21}CHO$	85	450
	Br	$CH_3COC(CH_3){=}CHC{\equiv}CCH{=}C(CH_3)COCH_3$	56	392
	Br	$(CH_3)_2C{=}CHCH_2CH_2C(CH_3){=}CHCH_2CH_2C(CH_3){=}CHCHO$	—	392
(cyclohexylidene)$=NO_2^-\,Na^+$	Cl	C_6H_5CHO	69	389
$C_6H_5CH{=}NO_2^-Na^+$	Cl	C_6H_5CHO	77	387
	Cl	$4\text{-}NCC_6H_4CHO$	—	387
	Cl	$2,4\text{-}(CH_3O)_2C_6H_3CHO$	—	387
$C_6H_5C(CN){=}NO_2Ag$	Br	$(C_6H_5)_2C{=}O$	(Very good)	417
$C_6H_5C(CN){=}NO_2^-Na^+$	Cl	C_6H_5CHO	—	387
(fluorenylidene)$=NO_2^-\,K^+$	I	CH_2O	—	84
	Br	CH_3CHO	—	84
	Br	CH_3COCHO	97	308
	Br, I	$(CH_3)_2C{=}O$	—	84
	Cl	$C_2H_5O_2CCHO$	—	84
	Br	(cyclohexanone)$=O$	—	84
	Br	$(C_2H_5O_2)_2C{=}O$	—	84
	Br	C_6H_5COCHO	84	308

appears to proceed more rapidly, and in higher yield, in a solvent than without a solvent. A slightly basic medium (pH 7–9) favors the reaction. A strongly basic medium is avoided to minimize self-condensation of the aldehyde or ketone products. Heating a nitronic ester with acids leads to a different reaction, formation of an alcohol rather than aldehydes or ketones (section IV.C.1). These observations agree with a mechanism, suggested by Kornblum, involving base attack at the α-carbon of the alkyl group[23] (equations 177–179).

$$\mathrm{B{:}} + \mathrm{R^1R^2C{=}N(\rightarrow O)OCHR^3R^4} \longrightarrow \mathrm{R^1R^2C{=}N(\rightarrow O)\overset{\ominus}{O}CR^3R^4} + \mathrm{BH^+} \qquad (177)$$

$$\mathrm{R^1R^2C{=}N(\rightarrow O)\overset{\ominus}{O}CR^3R^4} \longrightarrow \mathrm{R^1R^2C{=}NO^-} + \mathrm{R^3R^4C{=}O} \qquad (178)$$

$$\mathrm{R^1R^2C{=}NO^-} + \mathrm{BH^+} \longrightarrow \mathrm{R^1R^2C{=}NOH} + \mathrm{B{:}} \qquad (179)$$

However, the rate of decomposition of alkyl 3,5-di-*t*-butyl-4-oxo-2,5-cyclohexadiene nitronates is not base-catalyzed, and a cyclic intramolecular decomposition mechanism has been suggested.[421a].

Variations of the nitronic ester disproportionation reaction can lead to products other than aldehydes, ketones, and oximes. Epoxides react with nitronates (lithium ethoxide catalyst) to produce oxime ethers (e.g. **185**) when an excess of epoxide is employed; the expected α-hydroxy aldehydes were not isolated[451] (equation 180). Amides have been prepared by reaction of nitroalkanes with amines[450] (equation 181).

$$\mathrm{C_2H_5C(CH_3){=}NO_2^-} + \underset{\text{excess}}{\mathrm{CH_2CHCH_3\ (epoxide, \ O)}} \xrightarrow[\text{EtOH}]{\text{LiOEt}} \underset{(\mathbf{185})\ 39\%}{\mathrm{C_2H_5C(CH_3){=}NOCH_2CHOHCH_3}} + \underset{\text{(not isolated)}}{\mathrm{CH_3CHOHCHO}} \qquad (180)$$

$$\mathrm{(CH_3)_2CHNO_2} + \mathrm{C_6H_5CH[N(CH_3)_2]_2} \xrightarrow[\text{Reflux}]{\text{EtOH}} \mathrm{(CH_3)_2C{=}NOH} + \underset{94\%}{\mathrm{C_6H_5CON(CH_3)_2}} + \mathrm{(CH_3)_2NH} \qquad (181)$$

Heating methyl 4-toluenesulfonylmethanenitronate produces 4-tolyl thiocyanate and carbon dioxide, but no formaldehyde[394] (equation

182). Heating methyl dicarbomethoxymethanenitronate with aqueous sodium hydroxide produced methanol, carbonate, and fulminate (isolated as the silver salt)[394] (equation 183)

$$4\text{-}CH_3C_6H_4SO_2CH{=}NO_2CH_3 \xrightarrow{95°} 4\text{-}CH_3C_6H_4SC{\equiv}N + CO_2 + 2\,H_2O \quad (182)$$

$$(CH_3O_2C)_2C{=}NO_2CH_3 \xrightarrow[\text{Heat}]{\text{Aq. NaOH}} CH_3OH + CO_3^{=} + C{\equiv}NO^- \quad (183)$$

The formation of stilbenes from what appear to be *in situ*-generated nitronic esters has been observed[84,411] (equations 184, 185).

$$2\ 4\text{-}BrC_6H_4CH{=}NO_2^-Na^+ + CH_3I \xrightarrow[\text{Heat}]{\text{EtOH}} 4\text{-}BrC_6H_4CH{=}CHC_6H_4\text{-}4\text{-}Br \quad (184)$$

$$2\ C_6H_5\underset{\displaystyle CN}{\underset{|}{C}}{=}NO_2^-Na^+ + (CH_3)_2SO_4 \xrightarrow[25°,\ 3\ \text{weeks}]{CH_3OH,\ NaOH} C_6H_5\underset{\displaystyle CN}{\underset{|}{C}}{=}\underset{\displaystyle CN}{\underset{|}{C}}C_6H_5 \quad (185)$$

Nitronic esters of tertiary alcohols cannot disproportionate to form aldehydes or ketones. Such nitronic esters are reported to form *in situ* in hot ethanol solution from potassium fluorene-9-nitronate (**186**) and tertiary bromides (*t*-butyl bromide, 2-bromo-2-phenylpropane, and triphenylchloromethane)[84,205]. These esters were not isolated. The products actually isolated are the same as those derived from fluorene-9-nitronic acid (see section III.C.2), namely fluorenone oxime (**77**) and the 1,2-dinitroethane **76** (equation 186). When *t*-butyl bromide was a reactant, isobutylene was isolated as the dibromide.

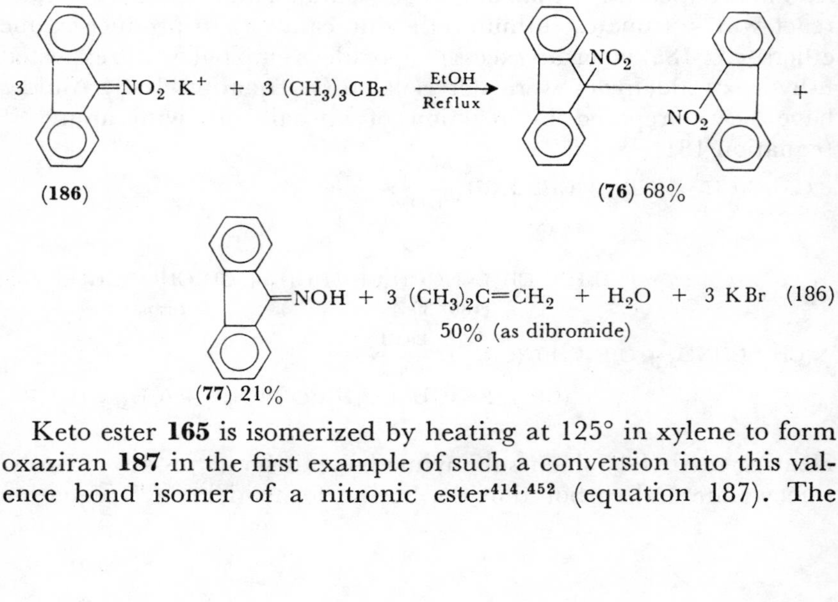

3 [fluorene]$=NO_2^-K^+$ + 3 $(CH_3)_3CBr$ $\xrightarrow[\text{Reflux}]{\text{EtOH}}$ [9,9'-bifluorenyl with NO_2 at each 9-position] +

(**186**) (**76**) 68%

[fluorene]$=NOH$ + 3 $(CH_3)_2C{=}CH_2$ + H_2O + 3 KBr (186)

50% (as dibromide)

(**77**) 21%

Keto ester **165** is isomerized by heating at 125° in xylene to form oxaziran **187** in the first example of such a conversion into this valence bond isomer of a nitronic ester[414,452] (equation 187). The conjugated carbonyl band of the nitronic ester at 1724 cm^{-1} is

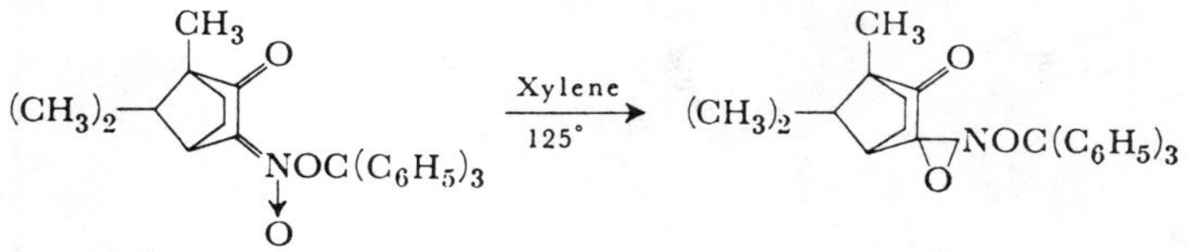

$$\text{(165)} \xrightarrow[125^\circ]{\text{Xylene}} \text{(187)} \qquad (187)$$

(**165**) M.p. 149–152°
$\nu_{C=O}^{cm^{-1}}$ 1724

(**187**) 33%;
M.p. 176–177°
$\nu_{C=O}^{cm^{-1}}$ 1751

shifted to 1751 cm^{-1} by the rearrangement. The extent to which oxaziran intermediates are involved in nitronic acid and ester chemistry is yet to be determined[453]. The conversion of the related nitrones into oxazirans is known; e.g. **188** → **189**[21,24,454,455] (equation 188).

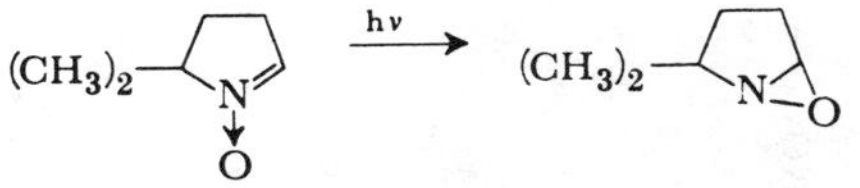

$$\text{(188)} \xrightarrow{h\nu} \text{(189)} \qquad (188)$$

(**188**) (**189**)

3. 1,3-Dipolar addition reactions of nitronic acid esters

Nitronic esters undergo 1,3-dipolar addition to olefins, a reaction discovered[456] and developed by Tartakovskii and co-workers[398,400,436,456–458]. The products are stable, often crystalline, 2-alkoxyisoxazolidines (Table 17). A wide variety of olefins have been added to methyl dinitromethanenitronate (**190**). The reactions are conducted under mild conditions—room temperature in methylene chloride or without solvent. Yields are good to excellent in most examples (equations 189, 190; Table 17).

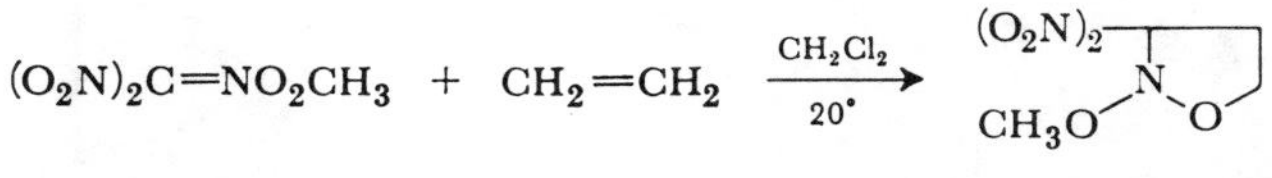

$$(O_2N)_2C{=}NO_2CH_3 + CH_2{=}CH_2 \xrightarrow[20^\circ]{CH_2Cl_2} \text{3,3-}(O_2N)_2\text{-2-}CH_3O\text{-isoxazolidine} \qquad (189)$$

(**190**)

73%; B.p. 66° (0.33 mm)

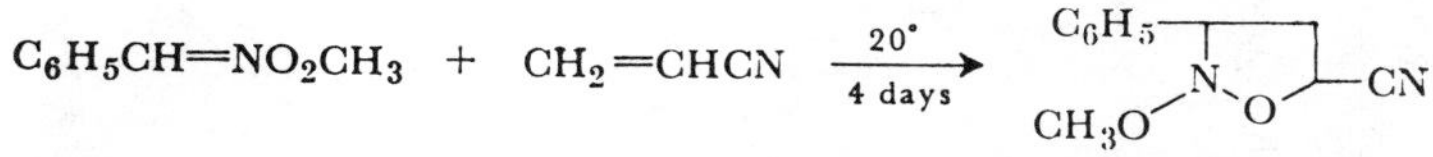

$$C_6H_5CH{=}NO_2CH_3 + CH_2{=}CHCN \xrightarrow[4\ \text{days}]{20^\circ} \text{3-}C_6H_5\text{-2-}CH_3O\text{-5-CN-isoxazolidine} \qquad (190)$$

46%; M.p. 96°

TABLE 17. 1,3-Dipolar additions of nitronic acid esters to olefins. Synthesis of isoxazolidines

Nitronic ester	Olefin	Product	Yield (%)	Ref.
$(NO_2)_2C{=}NO_2CH_3$[a]	$CH_2{=}CH_2$	$(NO_2)_2$, CH_3O–N–O isoxazolidine	73	398
	$CH_2{=}CHCH_3$	$(NO_2)_2$, CH_3O–N–O isoxazolidine, –CH_3	68	398
	$CH_2{=}CHCO_2H$	$(NO_2)_2$, CH_3O–N–O isoxazolidine, –CO_2H	91	398
	$CH_2{=}CHCH_2Cl$	$(NO_2)_2$, CH_3O–N–O isoxazolidine, –CH_2Cl	81	398, 456
	$CH_2{=}CHCH_2OH$	$(NO_2)_2$, CH_3O–N–O isoxazolidine, –CH_2OH	91	398
	$CH_2{=}CHCOCH_3$	$(NO_2)_2$, CH_3O–N–O isoxazolidine, –$COCH_3$	88	398
	$CH_2{=}CHCO_2CH_3$	$(NO_2)_2$, CH_3O–N–O isoxazolidine, –CO_2CH_3	27	398, 456

	$CH_2{=}CHOCOCH_3$	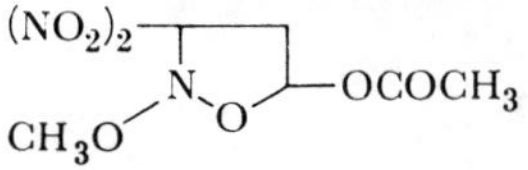 $(NO_2)_2$, N, O, $OCOCH_3$, CH_3O	93	398
		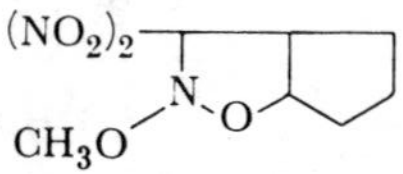 $(NO_2)_2$, N, O, CH_3O	76	398
$(NO_2)_2C{=}NO_2CH_3$[a]	$(CH_3)_2C{=}C(CH_3)_2$	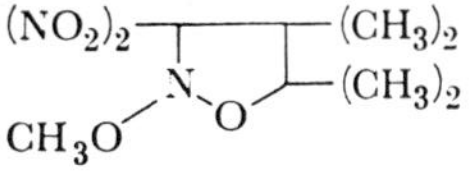 $(NO_2)_2$, $(CH_3)_2$, $(CH_3)_2$, N, O, CH_3O	20	398
	$CH_2{=}CHC_6H_5$	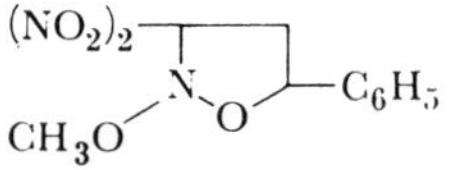 $(NO_2)_2$, N, O, C_6H_5, CH_3O	63	398
$O_2NCH{=}NO_2CH_3$[a]	$CH_2{=}CHCO_2CH_3$	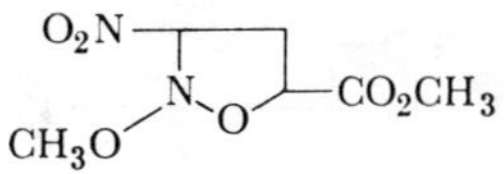 O_2N, N, O, CO_2CH_3, CH_3O	65	400
$(NO_2)_2C{=}NO_2CH_2CH_2I$[a]	$CH_2{=}CH_2$	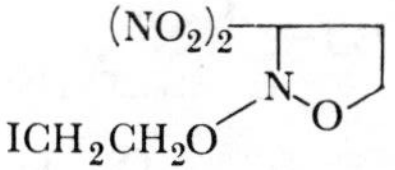 $(NO_2)_2$, N, O, ICH_2CH_2O	—	423
	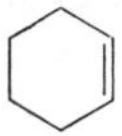	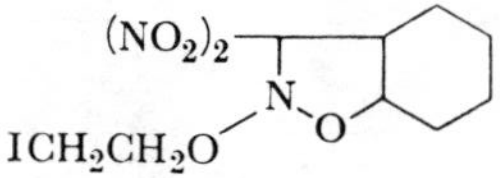 $(NO_2)_2$, N, O, ICH_2CH_2O	—	423
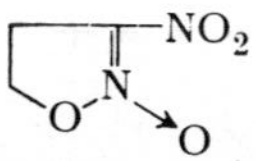 NO_2, N, O, O	$CH_2{=}CH_2$	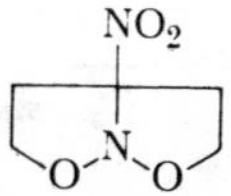 NO_2, O, N, O	—	456

(*continued*)

TABLE 17—*continued*

Nitronic ester	Olefin	Product	Yield (%)	Ref.
	$CH_2{=}CHCO_2CH_3$	NO_2, O, N, O, CO_2CH_3	—	456
$(NO_2)_2C{=}NO_2C_2H_5$[a]	$CH_2{=}CHCO_2CH_3$	$(NO_2)_2$, N, O, CO_2CH_3, C_2H_5O	38	398
CH_3, N, O, →O	$CH_2{=}CHC_6H_5$	CH_3, O, N, O, C_6H_5	—	456
$CH_3O_2CCH{=}NO_2CH_3$	$CH_2{=}CHCO_2CH_3$	CH_3O_2C, N, O, CO_2CH_3, CH_3O	74	458
$EtO_2CCH{=}NO_2CH_3$	$CH_2{=}CHCN$	EtO_2C, N, O, CN, CH_3O	76	458
	$CH_2{=}CHCH_2Cl$	EtO_2C, N, O, CH_2Cl, CH_3O	75	458
	$CH_2{=}CHCOCH_3$	EtO_2C, N, O, $COCH_3$, CH_3O	78	458

	$CH_2{=}CHCO_2CH_3$	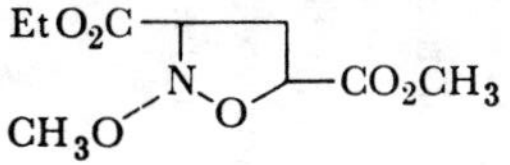 EtO_2C, N, O, CH_3O, CO_2CH_3	90	458
	$CH_2{=}CHC_6H_5$	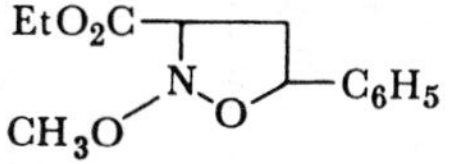 EtO_2C, N, O, CH_3O, C_6H_5	64	458
$C_6H_5CH{=}NO_2CH_3$	$CH_2{=}CHCN$	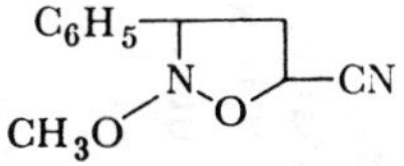 C_6H_5, N, O, CH_3O, CN	46	456, 457
	$CH_2{=}CHCO_2CH_3$	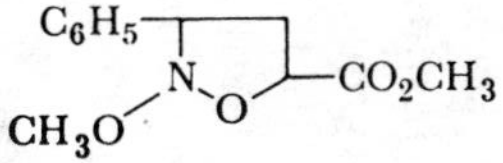 C_6H_5, N, O, CH_3O, CO_2CH_3	34	457
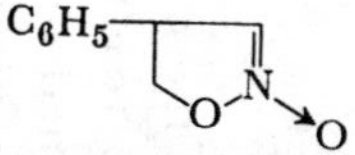 C_6H_5, N→O, O	$CH_2{=}CHC_6H_5$	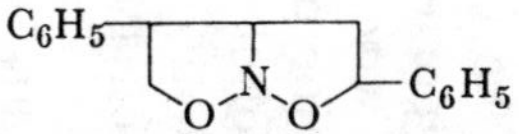 C_6H_5, O, N, O, C_6H_5	—	456

[a] Prepared *in situ*.

The reaction has been extended to cyclic nitronic esters as in the preparation of bicyclic **191**[436,456,458a] (equation 191).

$$\text{(cyclic nitronic ester)}-NO_2 + CH_2{=}CH_2 \longrightarrow \textbf{191} \qquad (191)$$

(**191**)

The scope of the reaction appears large, but has not been completely defined since relatively few nitronic esters have been employed. The required nitronic esters, some of which are very unstable (e.g. **190**), may be generated conveniently *in situ*. Thus, the reaction is not limited to the few stable nitronic esters. The reaction has failed in certain reported instances. The reaction of methyl phenylmethanenitronate with styrene failed to yield an isoxazolidine[457]. 2-Alkoxyisoxazolidine products having a hydrogen in the 3-position may eliminate alcohol to form a 2-isoxazoline; for example, in the reaction of methyl nitromethanenitronate with styrene the product was **192**[400] (equation 192).

$$O_2NCH{=}NO_2CH_3 + CH_2{=}CHC_6H_5 \longrightarrow$$

$$\text{3-}O_2N\text{-2-}CH_3O\text{-5-}C_6H_5\text{-isoxazolidine} \xrightarrow{-CH_3OH} \text{3-}O_2N\text{-5-}C_6H_5\text{-2-isoxazoline} \qquad (192)$$

(not isolated) (**192**) 34%

The structure of the adducts derived from unsymmetrical olefins has been established in a few instances[398,400,457]. Vinyl compounds ($CH_2{=}CHR$) examined thus far add so that the R group appears in the 5-position. Hydrolysis of **193** with dilute sulfuric acid led to β-benzoylacetic acid, thereby establishing the 3,5-relationship of phenyl and carbomethoxy[457] (equation 193).

$$\text{3-}C_6H_5\text{-2-}CH_3O\text{-5-}CO_2CH_3\text{-isoxazolidine} \xrightarrow[20^\circ]{20\%\ H_2SO_4} C_6H_5COCH_2CHOHCO_2H \qquad (193)$$

(**193**)

The addition of methyl phenylmethanenitronate (**194**) to benzaldoxime (**195**) led readily to 3,5-diphenyl-1,2,4-oxadiazole (**197**), which is also formed slowly from **194** on standing[457]. Since oximes form readily from nitronic esters (equation 194), the formation of oxadiazoles from certain nitronic esters on standing[23,457] appears to be simply an addition of product to reactant. The intermediate **196** demethanolates and dehydrates to yield the oxadiazole **197** (equation 195)

$$\underset{(\mathbf{194})}{C_6H_5CH{=}NO_2CH_3} \longrightarrow \underset{(\mathbf{195})}{C_6H_5CH{=}NOH} + CH_2O \qquad (194)$$

$$\underset{(\mathbf{194})}{C_6H_5CH{=}NO_2CH_3} + \underset{(\mathbf{159})}{C_6H_5CH{=}NOH} \longrightarrow \underset{(\mathbf{196})}{\left[\text{C}_6\text{H}_5\text{/NOH/CH}_3\text{O-N isoxazolidine}\text{-C}_6\text{H}_5\right]} \xrightarrow[-H_2O]{-CH_3OH} \underset{(\mathbf{197})}{\text{3,5-(C}_6\text{H}_5)_2\text{-1,2,4-oxadiazole}} \qquad (195)$$

The demethanolation of a 2-methoxyisoxazolidine has been shown to be acid-catalyzed[400], thus offering an explanation for the observed facile conversion of nitronic esters into oxadiazoles in hot hydrochloric acid[394] (equation 196). An alternate acid-catalyzed

$$2\ 4\text{-}BrC_6H_4CH{=}NO_2CH_3 \xrightarrow[\text{Reflux 5 min.}]{15\%\ HCl} \text{3-(4-BrC}_6\text{H}_4\text{)-5-(C}_6\text{H}_4\text{Br-4)-1,2,4-oxadiazole} \qquad (196)$$

mechanism could involve formation of a nitrile oxide intermediate from the nitronic ester (see section III.C.1.a), followed by addition of the oxime[400]. Diene addition of nitrile oxides to olefins has been reported[457,459] (equation 197).

$$ArCH{=}NO_2CH_3 \xrightarrow{H^+} ArC{\equiv}NO + CH_3OH$$

$$ArC{\equiv}NO + ArCH{=}NOH \longrightarrow \left[\text{Ar/NOH/Ar isoxazoline}\right] \xrightarrow{-H_2O} \text{3,5-Ar}_2\text{-1,2,4-oxadiazole} \qquad (197)$$

V. NITRONIC ACID DERIVATIVES OTHER THAN ESTERS

A. Nitronic Acid Salts

Salts of nitronic acids may be prepared by reaction of bases with nitronic acids. However, they are usually most readily prepared from nitroalkanes, employing a suitable solvent. Unlike most nitroalkanes, nitronic acids are soluble in sodium bicarbonate solution.

Many nitronate salts are shock sensitive explosives, and are particularly hazardous when anhydrous. The alkali metal salts are useful for purifying and isolating nitronic acids and nitroalkanes[1]. Properties of salts of polynitroalkanes have been reviewed[460].

Several metal cations have been employed in the preparation of nitronate salts. The sodium and potassium salts are the most common[32,198]; these are prepared by treatment of a nitroalkane with aqueous sodium or potassium hydroxide, or the ethanolic metal ethoxides (equation 198).

$$R^1R^2CHNO_2 + KOH \longrightarrow R^1R^2C{=}NO_2^-K^+ + H_2O \tag{198}$$

The usually colorless mononitronate salts often precipitate from cold solutions and may be isolated by filtration[339]. Alkali metal salts of 1,1-dinitroalkanes are yellow and often less soluble than mononitronates[194]. The potassium salts are less soluble than sodium salts.

Several heavy metal salts are known, the most common being silver, mercury, and copper[1,198,229,461–464]. These insoluble, largely covalent compounds may be prepared from the alkali salts by metathesis (equation 199)

$$R^1R^2C{=}NO_2^-Na^+ + AgNO_3 \longrightarrow R^1R^2C{=}NO_2Ag\downarrow + NaNO_3 \tag{199}$$

The silver salts are employed for nitronic ester synthesis (Table 11, method B, in section IV.A.1). The silver salt of nitroform exists in colorless and yellow modifications suggesting the possibility of CAg and OAg forms[147,415]. Insoluble mercury methanenitronate decomposes to form mercury fulminate[15,462] (equation 200).

$$2\,CH_2{=}NO_2^-Na^+ \xrightarrow{HgCl_2} (CH_2{=}NO_2)_2Hg \xrightarrow{-2\,H_2O} Hg(ON{\equiv}C)_2 \tag{200}$$

The qualitative test for nitronic acids employs aqueous ferric chloride. The resulting characteristic red-brown color[1,159] is probably that of a ferric salt, $(FeO_2N{=}CR^1R^2)^{++}$; *cf.* section VI.

Nitronate salts of weak bases such as ammonia and amines may be prepared from nitronic acids[177]. The reaction is conveniently conducted in ether solvent in which the salts are insoluble. On standing, the ammonium salts of weak nitronic acids liberate ammonia to regenerate the nitronic acid[32,177] (equation 201).

$$R^1R^2C{=}NO_2H + NH_3 \rightleftharpoons R^1R^2C{=}NO_2^-NH_4^+ \tag{201}$$

Nitroalkanes react directly with ammonia or amines[366,465]. The kinetics of this second-order process has been examined with nitroethane[112]. 2-Nitro-1,3-indanedione readily forms stable salts (**198**) useful for characterization of amines[366].

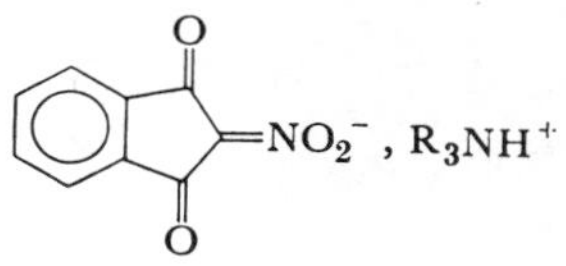

(**198**)

Nitronate salts are very useful and important reaction intermediates. They are employed in numerous reactions, either in solution or in suspension in anhydrous solvents. Typical are aldol-type condensation (Henry reaction), Michael addition, acylation, O- and C-alkylation, and halogenation (to form α-halonitroalkanes). Thermal decomposition of nitronate salts has been studied[422].

B. Nitronic Acid Anhydrides

Simple nitronic anhydrides—acyclic **199** or cyclic **201**—appear to be unstable compounds. The cyclic anhydrides **201** are known as furazan dioxides. Attempts to prepare them by oxidation of furazan oxides (**200**) have failed[466–468]. Evidence for a cyclic nitronic acid

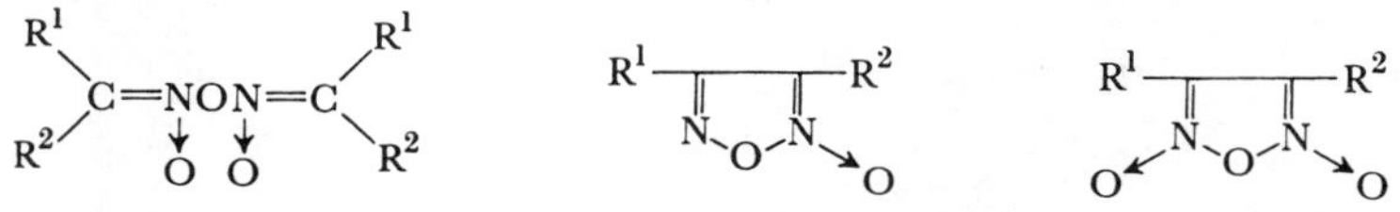

(**199**) (**200**) (**201**)

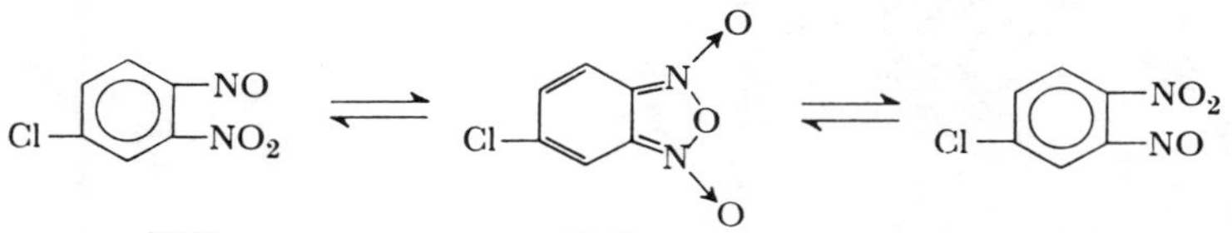

(**202**) (**203**) (**204**) (202)

anhydride **203** as a reaction intermediate is found in the facile interconversion of 4- and 5-chloro-2-nitronitrosobenzenes (**202** and **204**, respectively) (equation 202). Heating pure samples of either **202** or **204** at low concentration in refluxing tetrachloroethane gave a mixture of equal parts of the two isomers, regardless of the direction from which equilibrium was approached[466].

The known stable nitronic acid anhydrides are mixed anhydrides derived from *secondary* nitronic acids (with one exception) and carboxylic acids. These nitronic carboxylic acid anhydrides (Table 18) appear to be somewhat more stable than most nitronic esters.

TABLE 18. Synthesis of nitronic carboxylic acid anhydrides.

Nitro compound	Acylating agent	Anhydride	Yield (%)	Ref.
$(CH_3)_2C{=}NO_2^-Na^+$	$(CH_3CO)_2O$	$(CH_3)_2C{=}NO_2COCH_3$	17	470
	$(C_2H_5CO)_2O$	$(CH_3)_2C{=}NO_2COC_2H_5$	6	470
$(CH_3)_2CHNO_2$	$(CH_3CO)_2O$, $CH_3CO_2^-K^+$	$(CH_3)_2C{=}NO_2COCH_3$	8	470
	$(C_2H_5CO)_2O$, $C_2H_5CO_2^-K^+$	$(CH_3)_2C{=}NO_2COC_2H_5$	10	470
$C_6H_5CH{=}NO_2H$	$CH_2{=}C{=}O$	$C_6H_5CH{=}NO_2COCH_3$	—	275
$C_6H_5C(CN){=}NO_2Ag$	C_6H_5COCl	$C_6H_5C(CN){=}NO_2COC_6H_5$	85	411, 433
$C_6H_5C(CN){=}NO_2^-Na^+$	C_6H_5COCl	$C_6H_5C(CN){=}NO_2COC_6H_5$	—	411
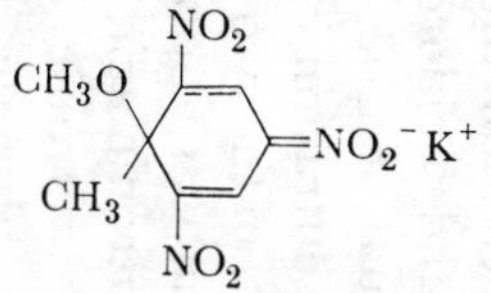 1-CH_3O-1-CH_3-2,6-$(NO_2)_2$-cyclohexadienylidene $=NO_2^-K^+$ (structure: NO_2, CH_3O, CH_3, NO_2, $=NO_2^-K^+$)	CH_3COCl	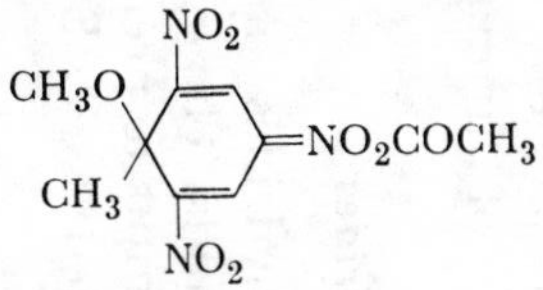 (structure: NO_2, CH_3O, CH_3, NO_2, $=NO_2COCH_3$)	—	170
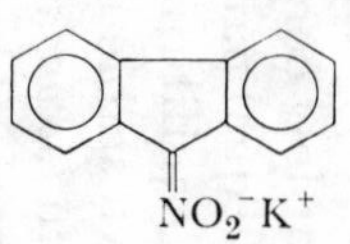 fluorenylidene $=NO_2^-K^+$ (structure)	CH_3COCl	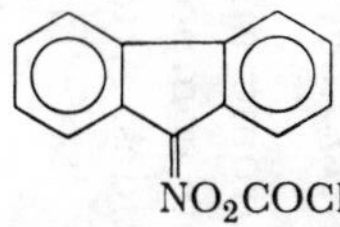 fluorenylidene $=NO_2COCH_3$ (structure)	—	275
	C_6H_5COCl	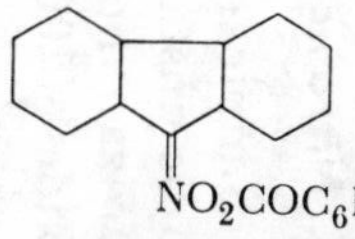 perhydrofluorenylidene $=NO_2COC_6H_5$ (structure)	—	275

They may be prepared by acylation of a secondary nitronate salt with an acid chloride, or anhydride[275,411,469–471]. Silver and alkali metal salts have been used (equations 203, 204).

$$\underset{\underset{\displaystyle CN}{|}}{C_6H_5C}{=}NO_2Ag + C_6H_5COCl \xrightarrow{C_6H_6} \underset{\underset{\displaystyle CN}{|}}{C_6H_5C}{=}\underset{\underset{\displaystyle O}{\downarrow}}{N}O\underset{\underset{\displaystyle O}{\|}}{C}C_6H_5 + AgCl \qquad (203)$$

85%
M.p. 116°

$$(CH_3)_2C{=}NO_2^-Na^+ + (CH_3CO)_2O \xrightarrow{Et_2O} (CH_3)_2C{=}\underset{\underset{\displaystyle O}{\downarrow}}{N}O\underset{\underset{\displaystyle O}{\|}}{C}CH_3 + CH_3CO_2^-Na^+ \qquad (204)$$

17%

The nitronate salt does not need to be prepared first. Treatment of a secondary nitroalkane with potassium acetate and an acid anhydride leads to a low yield of mixed anhydride[470]. Ketene may be used as in the preparation of **205**, the only known anhydride derived from a primary nitronic acid[275,471] (equation 205). This compound could not be prepared by reaction of sodium phenylmethanenitronate with acetyl chloride[470].

$$C_6H_5CH{=}NO_2H + CH_2{=}C{=}O \longrightarrow C_6H_5CH{=}\underset{\underset{\displaystyle O}{\downarrow}}{N}O\underset{\underset{\displaystyle O}{\|}}{C}CH_3 \qquad (205)$$

(**205**) M.p. 98°

The physical properties of nitronic carboxylic anhydrides have not been examined extensively. No spectra appear to have been recorded. The compounds which have been prepared (Table 18) are relatively stable, distillable liquids or crystalline solids.

Studies of reactions of nitronic carboxylic anhydrides are few. From what is known their reactions appear comparable to those of carboxylic anhydrides. However, a simple hydrolytic cleavage reaction to carboxylic and nitronic acids (or nitroalkanes) is not always found.

Two reaction patterns are distinguished. Anhydrides derived from *primary* nitronic acids rearrange with great ease to hydroxamic acid esters. Those derived from *secondary* nitronic acids cannot undergo this rearrangement, but form other products.

Reactions of primary nitronic carboxylic anhydrides prepared *in situ* can be observed by examining the reaction of salts of primary nitroalkanes with acid chlorides and anhydrides (equation 206). Nitronic carboxylic anhydrides have not been isolated from these reactions. Hydroxamic acid esters are the principal products[471–475].

The reaction was discovered by Kissel in 1882[472]. Both sodium

$$CH_3CH{=}NO_2^-Na^+ + CH_3COCl \longrightarrow CH_3CHNOCCH_3 + NaCl$$
$$\qquad\qquad\qquad\qquad\qquad\qquad \| \quad \|$$
$$\qquad\qquad\qquad\qquad\qquad\qquad O \quad O$$

$$C_6H_5CH{=}NO_2^-Na^+ + CH_3COCl \longrightarrow C_6H_5CNHOCC_6H_5 + NaCl \qquad (206)$$
$$\qquad\qquad\qquad\qquad\qquad\qquad \| \quad \|$$
$$\qquad\qquad\qquad\qquad\qquad\qquad O \quad O$$

ethanenitronate and sodium phenylmethanenitronate can yield dibenzohydroxamic acid with benzoyl chloride[62,472–475] (equations 207, 208).

$$C_6H_5CH{=}NO_2^-Na^+ + C_6H_5COCl \longrightarrow C_6H_5\underset{\|}{\overset{}{C}}NHO\underset{\|}{C}C_6H_5 + NaCl \qquad (207)$$
$$\qquad\qquad\qquad\qquad\qquad\qquad O \quad O$$

$$CH_3CH{=}NO_2^-Na^+ + C_6H_5COCl \longrightarrow$$
$$CH_3CNHOCC_6H_5 + C_6H_5CONHOCC_6H_5 + NaCl \qquad (208)$$
$$\| \quad \| \qquad\qquad\qquad \|$$
$$O \quad O \qquad\qquad\qquad O$$

When primary nitronate salts are treated with an excess of acylating agent, trisacylhydroxylamines result[245]. Acylation and transacylation of the hydroxamic ester are involved (equation 209).

$$RCH_2NO_2 \xrightarrow[CH_3CO_2^-Na^+]{(CH_3CO)_2O} (CH_3CO)_2NOCCH_3 + RCO_2^-Na^+ \qquad (209)$$
$$\qquad\qquad\qquad\qquad\qquad\qquad \|$$
$$\qquad\qquad\qquad\qquad\qquad\qquad O$$

76–79% (R = CH_3)
75–76% (R = C_2H_5)
6% (R = H)

Van Raalte observed that acyl exchange can occur when different aryl groups are present in the acid chloride and nitronate salt[62] (equation 210).

$$C_6H_5CH{=}NO_2^-Na^+ + 4\text{-}ClC_6H_4COCl \searrow$$
$$4\text{-}ClC_6H_4CNHOCC_6H_4\text{-}Cl\text{-}4 \qquad (210)$$
$$\| \quad \|$$
$$O \quad O$$
$$4\text{-}ClC_6H_4CH{=}NO_2^-Na^+ + C_6H_5COCl \nearrow$$

The mechanism of hydroxamic ester formation as suggested originally by Nef[473] probably involves initial formation of a nitronic carboxylic acid anhydride followed by a tautomeric rearrangement (equation 211). An oxaziran intermediate is possibly involved. The thermal rearrangement of oxazirans to amides has been studied[476]

$$R^1CH{=}\underset{\downarrow}{N}OCR^2 \longrightarrow \left[R^1\overset{H}{\overset{|}{C}}\text{——}NOCR^2 \right] \longrightarrow R^1\overset{H}{\overset{|}{C}}NOCR^2 \qquad (211)$$
$$O \;\; O \qquad\quad (\text{C–O–N ring}) \;\; O \qquad\quad O \;\; O$$

and is accelerated (relative to *N*-alkyl) by *N*-aryl substitution[477]. However, the actual conversion of a primary nitronic carboxylic anhydride into its isomeric hydroxamic ester has yet to be described.

Reactions of primary nitronic carboxylic anhydrides, prepared *in situ*, can also be examined by studying the reactions of primary nitronic acids, rather than the salts, with acid chlorides and anhydrides. Reaction of phenylmethanenitronic acid with acetyl or benzoyl chloride (or hydrogen chloride) leads to hydroxamic acid chloride **58**[249,275] (equation 212). In the presence of pyridine one obtains the benzoyl derivative **206**, which is also obtained from **58** under the same conditions[478] (equation 213).

$$C_6H_5CH{=}NO_2H + RCOCl \longrightarrow C_6H_5C(Cl){=}NOH + RCO_2H \qquad (212)$$

$R = CH_3, C_6H_5$ **(58)** M.p. 50–51°

$$C_6H_5CH{=}NO_2H + C_6H_5COCl \xrightarrow{C_5H_5N} C_6H_5C(Cl){=}NOCOC_6H_5 + H_2O \qquad (213)$$

(206)

The mechanism of this reaction may parallel the hydroxamic acid chloride forming reactions of nitronic acids (section III.C.1.a) and esters (section IV.C.1). Addition of hydrogen chloride to protonated anhydride **207**, followed by loss of water and a proton from adduct **208** would yield **206**; loss of benzoic acid would yield **58** (equation 214).

$$(C_6H_5)(H)C{=}\overset{\oplus}{N}(OH)(OCOC_6H_5)\ \textbf{(207)} \xrightarrow{HCl}$$

$$(C_6H_5)(H)(Cl)C{-}\overset{\oplus}{N}(H)(OH)(OCOC_6H_5)\ \textbf{(208)} \xrightarrow{-C_6H_5CO_2H,\ -H^+} C_6H_5C(Cl){=}NOH \qquad (214)$$

$$\textbf{(208)} \xrightarrow{C_5H_5N;\ -H_2O,\ -C_5H_5NH^{\oplus}} C_6H_5C(Cl){=}NOC(=O)C_6H_5\ \textbf{(206)}$$

Reactions of secondary nitronic carboxylic acid anhydrides can be examined readily because of the stability of these substances. The acyl function remains intact in reaction products. The nitronic acid function may be destroyed, however.

Hydrolysis of anhydride **209** with aqueous sodium hydroxide led to benzoic acid and α-cyanophenylnitromethane; reaction with phenylhydrazine led to hydrazide **210** (equations 215, 216)[411].

$$\underset{\textbf{(209)}}{C_6H_5\underset{\displaystyle CN}{\underset{|}{C}}{=}\overset{\displaystyle O}{\overset{\uparrow}{N}}OCOC_6H_5} \xrightarrow[2.\ H^+]{1.\ NaOH} C_6H_5\underset{\displaystyle CN}{\underset{|}{C}}HNO_2 + C_6H_5CO_2H \qquad (215)$$

$$\mathbf{209} \xrightarrow{C_6H_5NHNH_2} C_6H_5\underset{\displaystyle CN}{\underset{|}{C}}HNO_2 + \underset{\textbf{(210)}}{C_6H_5CONHNHC_6H_5} \qquad (216)$$

Hydrolysis of propane-2-nitronic acetic anhydride (**211**) by boiling water, with or without an acid catalyst, gave equal amounts of acetone and acetic acid in quantitative yield; nitrogen appeared as nitrous oxide and hydroxylamine (equation 217)[470]. In sodium hydroxide, hydrolysis of **211** occurred to give the same carbon products; nitrogen appeared as ammonia, nitrogen, and nitrous oxide, but no hydroxylamine was found (equation 218)[470].

$$\underset{\textbf{(211)}}{(CH_3)_2C{=}\underset{\displaystyle O}{\underset{\downarrow}{N}}OCOCH_3} \xrightarrow[\text{Reflux 15 min}]{H_2O\ \text{or}\ H_3O^+} \underset{100\%}{(CH_3)_2C{=}O} + \underset{100\%}{CH_3CO_2H} + N_2O + NH_2OH \qquad (217)$$

$$\mathbf{211} \xrightarrow[25°]{10\%H_2O,\ NaOH} (CH_3)_2C{=}O + CH_3CO_2^-Na^+ + N_2O + NH_3 + N_2 \qquad (218)$$

(209)

Ethanolysis of **211** in the presence of an acid catalyst, but not in ethanol alone, occurred to yield acetoxime, ethyl acetate, and a trace of acetone (equation 219)[470]. Aminolysis of **211** with aniline led to acetanilide (quantitative yield) as well as acetoxime and ammonia (equation 220)[470].

$$\mathbf{211} + C_2H_5OH \xrightarrow[\text{Heat}]{H^+} \underset{27\%}{(CH_3)_2C{=}NOH} + \underset{100\%}{CH_3CO_2C_2H_5} \qquad (219)$$

$$\mathbf{211} + C_6H_5NH_2 \xrightarrow[\text{Heat}]{} \underset{18\%}{(CH_3)_2C{=}NOH} + \underset{100\%}{CH_3CONHC_6H_5} + NH_3 \qquad (220)$$

The solvolysis reactions of nitronic carboxylic anhydrides would appear to require more than one mechanism. A simple cleavage to nitronic and carboxylic acids and derivatives (esters, salts) would explain most of the reactions. Alternative mechanisms may be involved, however. For example, acid-catalyzed hydrolysis of propane-2-nitronic anhydride (**211**) (equation 217) to yield acetone and nitrous oxide could be the result of Nef hydrolysis of the resulting propane-2-nitronic acid. However, as in the acid-catalyzed hydrolysis of nitronic esters to Nef products, the failure to isolate any nitroalkane (equation 217) as a product suggests an alternate mechanism. Like the 'ester-Nef' (section IV.C.1), an 'anhydride-Nef' is possible (equations 221, 222).

$$R^1R^2C{=}\overset{\oplus}{N}(OH)(O_2CR^3) + H_2O \longrightarrow R^1R^2C(OH){-}\overset{\oplus}{N}H(OH)(O_2CR^3) \qquad (221)$$

$$R^1R^2C(OH){-}\overset{\oplus}{N}H(OH)(O^2CR^3) \longrightarrow R^1R^2C{=}O + R^3CO_2H + HNO + H^+ \qquad (222)$$

The formation of acetoxime and ethyl acetate from anhydride **211** by acid-catalyzed ethanolysis (equation 219) could be a result of nitronic ester formation (**212**), followed by disproportionation to the oxime (equations 223, 224),

$$\underset{(\mathbf{211})}{(CH_3)_2C{=}\overset{\downarrow O}{N}OCOCH_3} + C_2H_5OH \longrightarrow \underset{(\mathbf{212})}{(CH_3)_2C{=}\overset{\downarrow O}{N}OC_2H_5} + CH_3CO_2C_2H_5 \qquad (223)$$

$$\mathbf{212} \longrightarrow (CH_3)_2C{=}NOH + CH_3CHO \qquad (224)$$

The base-catalyzed cleavage of anhydride **209** to α-cyanophenylnitromethane and benzoic acid (equation 215) appears as a simple cleavage to nitronate anion followed by tautomerization to the nitro compound. On the other hand, the base-catalyzed cleavage of anhydride **211** (equation 218) to acetone and ammonia, and the aminolysis of **211** (equation 220) to yield acetoxime and ammonia appear not to be nitronate anion reactions.

O-Alkyl oximes (e.g. **214**) were observed on attempted acetylation of potassium 1-nitromethanenitronate (**213**) and 1-nitroethanenitronate with acetyl chloride or acetyl nitrate (52% yield of **214**)[469] (equation 225). Some **214** (12%) was also formed in the reaction of

$$2\,CH_3\underset{\underset{NO_2}{|}}{C}{=}NO_2^-K^+ + CH_3COCl \longrightarrow$$

(**213**)

$$CH_3\underset{\underset{NO_2}{|}}{C}{=}NOC(NO_2)_2CH_3 + CH_3CO_2^-K^+ + KCl \quad (225)$$

(**214**) 31%

$$\mathbf{213} + C_6H_5COCl \longrightarrow \mathbf{214} + CH_3\underset{\underset{NO_2}{|}}{C}{=}NOCOC_6H_5 \quad (226)$$

(**215**)

213 with benzoyl chloride in which the benzoyl oxime **215** was produced[469] (equation 226). Formation of **214** suggests C-alkylation of an anhydride intermediate **216** by anion **217** (equation 227). Compound **214** has also been prepared from **217** and 1,1-dinitroethane[478a].

$$CH_3\underset{\underset{NO_2}{|}}{C}{=}\overset{\overset{O}{\uparrow}}{N}O\underset{\underset{O}{\|}}{C}R + CH_3\overset{\ominus}{C}(NO_2)_2 \longrightarrow \mathbf{214} \quad (227)$$

(**216**) (**217**)

C. Nitronic Acid Halides

There are two published accounts purporting to describe nitronic acid chlorides[32,479]. In both instances the unstable oils obtained are very poorly characterized substances. Hantzsch and Veit[32] treated phenylmethanenitronic acid with phosphorous pentachloride. The product was said to be acid chloride **218** (equation 228),

$$C_6H_5CH{=}NO_2H \xrightarrow{PCl_5} C_6H_5CH{=}N\begin{matrix}\nearrow O \\ \searrow Cl\end{matrix} \quad (228)$$

(**218**)

an exceptionally unstable oil. A more vigorous reaction occurred with 4-nitrophenylmethanenitronic acid to yield 4-nitrophenylnitromethane[32].

Reaction of nitroalkanes with picryl or *N*-(oxydichlorophosphino)pyridinium chloride at 80–120° produces low-boiling oils (distilled from the reaction mixture and condensed in cold traps) and said to be nitronic acid chlorides **219**[479]. The compounds have the following colors; a, colorless; b and c, faint greenish blue; d,

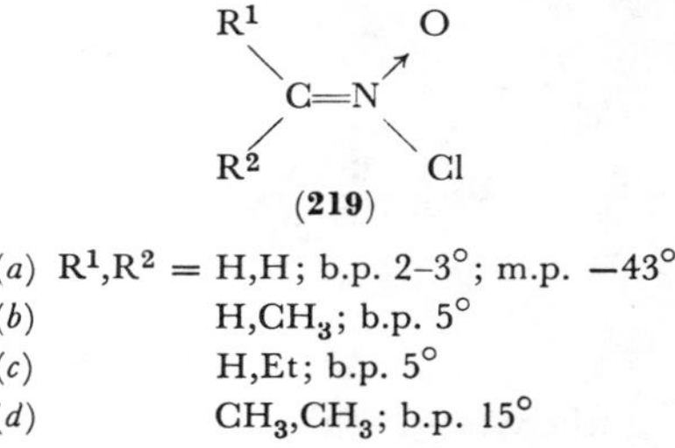

(219)

(*a*) R^1,R^2 = H,H; b.p. 2–3°; m.p. −43°
(*b*) H,CH_3; b.p. 5°
(*c*) H,Et; b.p. 5°
(*d*) CH_3,CH_3; b.p. 15°

intense blue. With aqueous or ethanolic ferric chloride no colors are developed with the compounds. The blue colors suggest the presence of nitroso compounds **220**. 2-Chloro-2-nitropropane is known as a blue liquid b.p. 70° (760 mm)[480]. However, chloronitrosoalkanes

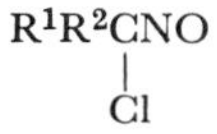

$R^1R^2C(Cl)NO$ **(220)** RC(Cl)=NOH **(221)**

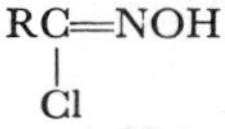

of the type RCH(Cl)NO are known only in solution and readily isomerize to colorless α-chlorohydroxamic acids **221**[274a]. It appears that the substances assigned structure **219** have not been adequately characterized, and an authentic nitronic acid chloride is yet to be described.

D. Nitronic Acid Amides

Nitronic acid amides are not well known. No substance having the hydrazone oxide structure **222** has been described. Nitronic acid amides of primary amides **223** would be azoxy tautomers **224**.

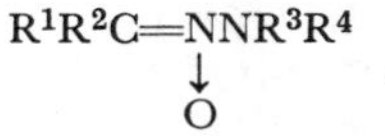

$R^1R^2C{=}N(\rightarrow O)NR^3R^4$

(222)

$R^1R^2C{=}N(\rightarrow O)NHR^3 \rightleftharpoons R^1R^2CHN(\rightarrow O){=}NR^3$

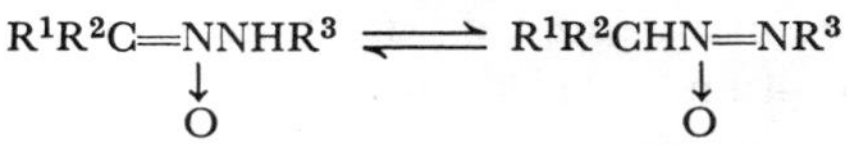

(223) **(224)**

A cyclic nitronic acid amide ('lactam') would be a pyrazole oxide **225** or pyrazoline oxide **226**. Examples of the former are the indazole oxides; e.g. **227**[481–484] (equation 229).

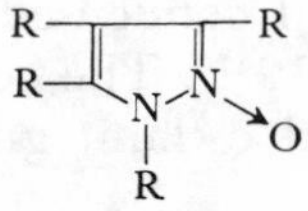

(225) 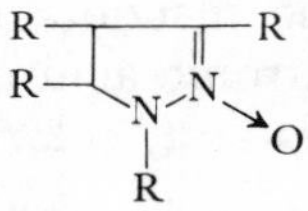 (226) (229)

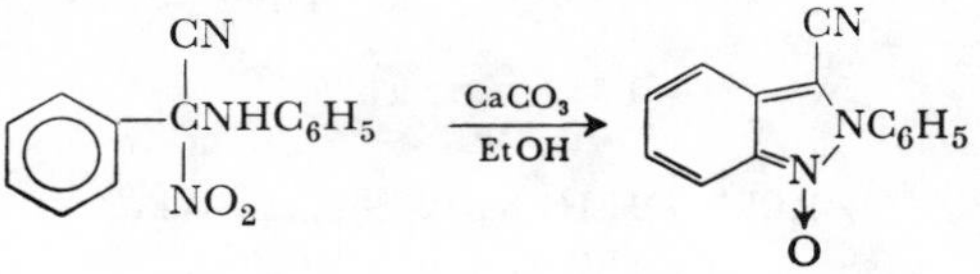

$$\text{C}_6\text{H}_5\text{C(CN)(NO}_2\text{)NHC}_6\text{H}_5 \xrightarrow[\text{EtOH}]{\text{CaCO}_3} $$

(227)

A nitronic acid imide would be a 2-*H*-1,2,3-triazole-1,3-dioxide (**228**); none has been described. However, 1-oxides such as **229** have been prepared[485–487].

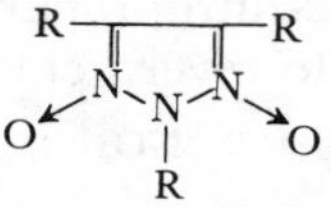

(228) 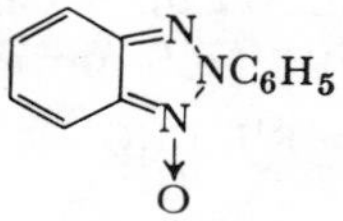 (229)

Nitronic acid amides could exist as reaction intermediates. Aminolysis of nitronic carboxylic anhydrides might involve a nitronic acid amide intermediate (equation 220).

Finally, nitrones (e.g. **231**) may be considered the 'ketones' of the nitronic acid series. They have not been prepared directly from nitronic acids, but are available from oxazirans (**230** → **231**)[24,454,477] (equation 230),

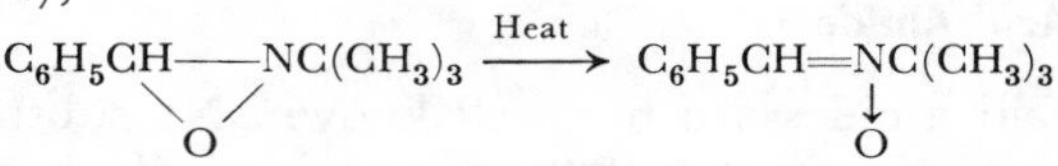

$$C_6H_5CH\text{—}NC(CH_3)_3 \xrightarrow{\text{Heat}} C_6H_5CH{=}NC(CH_3)_3 \quad (230)$$

(**230**) (**231**) 100%

VI. ANALYTICAL METHODS FOR NITRONIC ACIDS

Several methods are available for qualitative detection and quantitative determination of nitronic acids. Since tautomerization to the nitroalkane form often occurs readily in solution, the analytical method selected must consider this fact. The following tests do not apply to nitroalkanes. Some tests for nitronic acids are also applicable to nitronate anions.

The ferric chloride test used for enolic substances may be applied to nitronic acids. A red color usually develops when aqueous or alcoholic solutions of nitronic acids are treated with dilute ferric chloride solution[1,13]. Green[85] or brown[33] colors are also observed. This test, sometimes called the Konowalow reaction after its discoverer[13], was employed by the early workers in nitronic acid

Colors with ferric chloride

$CH_3CH{=}NO_2H$	$Br{-}C_6H_4{-}CH{=}NO_2H$	9-fluorenylidene $=NO_2H$
Red	Deep brown	Dark green

chemistry[1,13,159]. The color is probably due to a Fe^{III} nitronate salt, $Fe(O_2N{=}CR^1R^2)^{++}$, similar to the colored Fe^{III} phenolate salts, $Fe(OAr)^{++}$[488]. The test has been the basis for a quantitative colorimetric method[489].

Bromine titration of nitronic acids occurs rapidly and is the basis of the Kurt Meyer analysis[36] (equation 231). Bromine or ferric chloride may be used as an indicator. Iodine monochloride has been

$$R^1R^2C{=}NO_2H + Br_2 \longrightarrow R^1R^2C(Br)NO_2 + HBr \qquad (231)$$

employed for quantitative analysis of nitronic acids[209] (equation 232). The unreacted iodine monochloride, which is employed in excess, is allowed to react with *N,N,N′,N′*-tetramethyl-*p*-phenylenediamine to produce the intensely blue Würster radical cation **232**, which may be assayed spectrophotometrically (equation 233).

$$R^1R^2C{=}NO_2H + ICl \longrightarrow R^1R^2C(I)NO_2 + HCl \qquad (232)$$

$$ICl + 2\,(CH_3)_2N{-}C_6H_4{-}N(CH_3)_2 \longrightarrow 2\,(CH_3)_2\overset{\oplus}{N}{-}C_6H_4{-}N(CH_3)_2 + I^- + Cl^- \qquad (233)$$

(**232**) Blue

The oxidizing property of nitronic acids is the basis of an excellent quantitative method[276]. A mixture of potassium nitronate salt and potassium iodide is acidified. The hydrogen iodide reacts with the liberated nitronic acid to produce iodine and an oxime (equation 234). The iodine is then titrated with sodium thiosulfate employing starch indicator.

$$R^1R^2C{=}NO_2H + HI \longrightarrow R^1R^2C{=}NOH + I_2 \qquad (234)$$

Polarography has frequently been employed as a convenient method of quantitative analysis of nitronic acid-nitroalkane mixtures[38,44,47,110,321,490]. The nitronic acid form, as well as the nitronate anion, are not reduced polarographically at the dropping mercury electrode at the same voltage as nitroalkanes.

The greater acidity of nitronic acids (pK_a 3–6) over the parent nitroalkanes (pK_a 8–10) [ΔpK_a = ca. 3–6 (Figure 5)] is a property

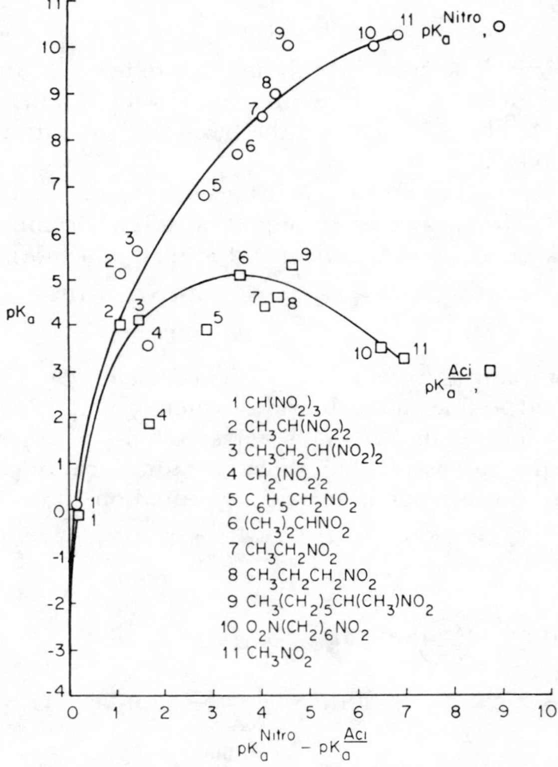

FIGURE 5. Plot of pK_a^{Aci} and pK_a^{Nitro} vs. $pK_a^{Nitro} - pK_a^{Aci}$. Data for nitronic acids and nitroalkanes in water at 25° in Table 5.

frequently employed for analysis. The nitronic acids are usually soluble in sodium bicarbonate solution[1]; most nitroalkanes are not (see pK_a values in Table 5, section II.D). Sodium hydroxide solution is not suitable for nitronic acid titration since nitroalkanes also react. 1,1-Dinitroalkanes and their nitronic acids are of nearly equal acid strength (pK_a 5–6); *cf.* Figure 5. Because of their relatively greater acidity, nitronic acids are better conductors in solution than nitroalkanes. The conductometric method has frequently been employed for quantitative analysis of those nitronic acids which have conductivities significantly greater than the corresponding nitroalkanes[28–31,112].

A most convenient analytical method, useful for rapid determination—as in kinetic studies—takes advantage of the strong characteristic nitronic acid $\pi - \pi^*$ ultraviolet absorption band in the ultraviolet region near 240 mμ[52,53]. Nitroalkanes do not absorb significantly in this region. Rapid reactions, such as *aci*-nitro tautomerization, may easily be followed by this spectrophotometric method[52].

VII. REFERENCES

1. A. Hantzsch and O. W. Schultze, *Chem. Ber.*, **29**, 699 (1896).
2. H. B. Hass and E. F. Riley, *Chem. Rev.*, **32**, 373 (1943).
3. P. A. S. Smith, *The Chemistry of Open-Chain Organic Nitrogen Compounds*, Vol. II, W. A. Benjamin, Inc., New York, 1966, pp. 391–454.
4. W. E. Noland, *Chem. Rev.*, **55**, 136 (1955).
5. H. B. Hass and M. L. Bender, *J. Am. Chem. Soc.*, **71**, 1767 (1949).
6. E. Bamberger, *Chem. Ber.*, **35**, 54 (1902).
7. A. Hantzsch, *Chem. Ber.*, **38**, 998 (1905).
8. A. Hantzsch, *Chem. Ber.*, **32**, 575 (1899).
9. A. F. Holleman, *Chem. Ber.*, **33**, 2913 (1900).
10. H. Shechter and J. W. Shepherd, *J. Am. Chem. Soc.*, **76**, 3617 (1954).
11. *The Naming and Indexing of Chemical Compounds from Chemical Abstracts*, *Chem. Abstr.*, **56**, 1 N (1962).
12. *Nomenclature of Organic Chemistry*, Definitive Rules for Sections A, B, and C, 2 Vols., International Union of Pure and Applied Chemistry Edition, Butterworths, London, 1965, 1966.
13. M. I. Konowalow, *Chem. Ber.*, **29**, 2193 (1896).
14. A. F. Holleman, *Rec. Trav. Chim.*, **4**, 121 (1895).
15. J. U. Nef, *Ann. Chem.* **280**, 264 (1894).
16. R. Kuhn and H. Albrecht, *Chem. Ber.*, **60**, 1297 (1927).
17. R. L. Shriner and J. H. Young, *J. Am. Chem. Soc.*, **52**, 3332 (1930).
18. N. Kornblum, N. M. Lichtin, J. T. Patton, and D. C. Iffland, *J. Am. Chem. Soc.*, **69**, 307 (1947).
19. N. Kornblum, J. T. Patton, and J. B. Nordmann, *J. Am. Chem. Soc.*, **70**, 746 (1948).
20. W. Theilacker and G. Wendtland, *Ann. Chem.*, **570**, 33 (1950).
21. E. Schmitz, *Adv. Heterocyclic Chem.*, **2**, 85 (1963).
22. F. Klages, R. Heisle, H. Sitz, and P. Specht, *Chem. Ber.*, **96**, 2387 (1963).
23. N. Kornblum and R. A. Brown, *J. Am. Chem. Soc.*, **86**, 2681 (1964).
24. G. R. Delpierre and M. Lamchen, *Quart. Rev.* (*London*), **19**, 329 (1965).
25. J. Meisenheimer and H. Meis, *Chem. Ber.*, **57**, 289 (1924).

26. E. P. Kohler, *J. Am. Chem. Soc.*, **46**, 1733 (1924).
27. E. B. Hodge, *J. Am. Chem. Soc.*, **73**, 2341 (1951).
28. S. H. Maron and V. K. LaMer, *J. Am. Chem. Soc.*, **60**, 2588 (1938).
29. S. H. Maron and V. K. LaMer, *J. Am. Chem. Soc.*, **61**, 692 (1939).
30. S. H. Maron and V. K. LaMer, *J. Am. Chem. Soc.*, **61**, 2018 (1939).
31. S. H. Maron and T. Shedlovsky, *J. Am. Chem. Soc.*, **61**, 753 (1939).
32. A. Hantzsch and A. Veit, *Chem. Ber.*, **32**, 607 (1899).
33. A. Hantzsch and O. W. Schultze, *Chem. Ber.*, **29**, 2251 (1896).
34. G. E. K. Branch and J. Jaxon-Deelman, *J. Am. Chem. Soc.*, **49**, 1765 (1927).
35. K. H. Meyer and A. Sander, *Ann. Chem.*, **396**, 133 (1913).
36. K. H. Meyer and P. Wertheimer, *Chem. Ber.*, **47**, 2374 (1914).
37. R. Junell, *Z. Physik. Chem.*, **A141**, 71 (1929).
38. V. M. Belikov, S. G. Mairanovskii, T. B. Korchemnaya, S. S. Novikov, and V. A. Klimova, *Izv. Akad. Nauk SSSR Ser. Khim.*, **1960**, 1675; *Chem. Abstr.*, **55**, 8325 (1961).
39. V. M. Belikov, S. G. Mairanovskii, T. B. Korchemnaya, and S. S. Novikov, *Izv. Akad. Nauk SSSR Ser. Khim.*, **1961**, 1108; *Chem. Abstr.*, **58**, 8880 (1963).
40. V. M. Belikov, S. G. Mairanovskii, T. B. Korchemnaya, and S. S. Novikov, *Izv. Akad. Nauk SSSR Ser. Khim.*, **1962**, 605.
41. V. M. Belikov, S. G. Mairanovskii, T. B. Korchemnaya, and V. P. Gul'tyai, *Izv. Akad. Nauk. SSSR Ser. Khim.*, **1964**, 439; *Chem. Abstr.*, **61**, 1741 (1964).
42. V. M. Belikov, T. B. Korchemnaya, S. G. Mairanovskii, and S. S. Novikov, *Izv. Akad. Nauk SSSR Ser. Khim.*, **1964**, 1599; *Chem. Abstr.*, **61**, 15953 (1964).
43. S. S. Novikov, V. M. Belikov, Yu. P. Egorov, E. N. Safonova, and L. V. Semenov, *Izv. Akad. Nauk SSSR Ser. Khim.*, **1959**, 1438; *Chem. Abstr.*, **53**, 21149 (1959).
44. J. Armand, *Compt. Rend.*, **254**, 2777 (1962).
45. J. Armand and P. Souchay, *Compt. Rend.*, **255**, 2112 (1962).
46. J. Armand, P. Souchay, and S. Deswarte, *Tetrahedron*, **20**, Suppl. 1, 249 (1964).
47. J. Armand, *Bull. Soc. Chim. France*, **1965**, 3246.
48. P. Souchay and J. Armand, *Compt. Rend.*, **253**, 460 (1961).
49. P. Souchay and S. Deswarte, *Compt. Rend.*, **255**, 688 (1962).
50. H. Feuer and A. T. Nielsen, *J. Am. Chem. Soc.*, **84**, 688 (1962).
51. H. Feuer and A. T. Nielsen, *Tetrahedron*, **19**, Suppl. 1, 65 (1963).
52. A. T. Nielsen and H. F. Cordes, *Tetrahedron*, **20**, Suppl. 1, 235 (1964).
53. W. Kemula, *Roczniki Chem.*, **35**, 1169 (1961); *Chem. Abstr.*, **56**, 6798 (1962).
54. A. F. Holleman, *Rec. Trav. Chim.*, **16**, 162 (1897).
55. S. G. Mairanovskii, V. M. Belikov, T. B. Korchemnaya, V. A. Klimova, and S. S. Novikov, *Izv. Akad. Nauk SSSR Ser. Khim.*, **1960**, 1785; *Chem. Abstr.*, **55**, 19832 (1961).
56. R. G. Pearson and R. L. Dillon, *J. Am. Chem. Soc.*, **75**, 2439 (1953).
57. R. G. Pearson and R. L. Dillon, *J. Am. Chem. Soc.*, **72**, 3574 (1950).
58. A. T. Nielsen, *J. Org. Chem.* **27**, 2001 (1962).
59. D. Turnbull and S. H. Maron, *J. Am. Chem. Soc.*, **65**, 212 (1943).
60. G. W. Wheland and J. Farr, *J. Am. Chem. Soc.*, **65**, 1433 (1943).
61. R. G. Cooke and A. K. Macbeth, *J. Chem. Soc.*, **1938**, 1024.
62. M. A. Van Raalte, *Rec. Trav. Chim.*, **18**, 378 (1899).
63. A. F. Holleman, *Rec. Trav. Chim.*, **15**, 365 (1896).
64. T. S. Patterson and A. McMillan, *J. Chem. Soc.*, **1908**, 1041.
65. E. P. Kohler and J. F. Stone, Jr., *J. Am. Chem. Soc.*, **52**, 761 (1930).

66. J. A. Sousa and J. Weinstein, *J. Org. Chem.*, **27,** 3155 (1962).
67. R. E. Hardwick, H. S. Mosher, and P. Passailaigue, *Trans. Faraday Soc.*, **56,** 44 (1960).
68. R. E. Hardwick and H. S. Mosher, *J. Chem. Phys.*, **36,** 1402 (1962).
69. H. S. Mosher, C. Souers, and R. Hardwick, *J. Chem. Phys.*, **32,** 1888 (1960).
70. A. Ficalbi, *Gazz. Chim. Ital.*, **93,** 1530 (1963).
71. H. Morrison and B. H. Migdalof, *J. Org. Chem.*, **30,** 3996 (1965).
72. H. S. Mosher, R. E. Hardwick, and D. Ben Hur, *J. Chem. Phys.*, **37,** 904 (1962).
73. H. Hiraoka and R. E. Hardwick, *Bull. Chem. Soc. Japan*, **39,** 380 (1966).
74. A. L. Bluhm, J. Weinstein, and J. A. Sousa, *J. Org. Chem.*, **28,** 1989 (1963).
75. A. L. Bluhm, J. A. Sousa, and J. Weinstein, *J. Org. Chem.*, **29,** 636 (1964).
76. G. Wettermark, *Nature*, **194,** 677 (1962).
77. J. D. Margerum, L. J. Miller, E. Saito, M. S. Brown, H. S. Mosher, and R. E. Hardwick, *J. Phys. Chem.*, **66,** 2434 (1962).
78. J. D. Margerum and R. G. Brault, *J. Am. Chem. Soc.*, **88,** 4733 (1966).
79. A. E. Tschitschibabin, B. Kuindshi, and S. W. Benewolenskaja, *Chem. Ber.*, **58,** 1580 (1925).
80. J. Weinstein, J. A. Sousa, and A. L. Bluhm, *J. Org. Chem.*, **29,** 1586 (1964).
81. G. Wettermark, *J. Am. Chem. Soc.*, **84,** 3658 (1962).
82. J. Weinstein, A. L. Bluhm, and J. A. Sousa, *J. Org. Chem.*, **31,** 1983 (1966).
83. E. Bamberger and R. Seligman, *Chem. Ber.*, **35,** 3884 (1902).
84. C. D. Nenitzescu and D. A. Isacescu, *Chem. Ber.*, **63,** 2484 (1930).
85. W. Wislicenus and M. Waldmüller, *Chem. Ber.*, **41,** 3334 (1908).
85a. R. C. Kerber and M. Hodos, *J. Org. Chem.*, **33,** 1169 (1968).
86. W. Wislicenus and K. Pfeilsticker, *Ann. Chem.*, **436,** 36 (1924).
86a. R. C. Kerber and M. J. Chick, *J. Org. Chem.*, **32,** 1329 (1967).
87. M. Bourillot, P. Rostaing, and G. Descotes, *Compt. Rend.*, **262** (C), 1080 (1966).
88. H. B. Fraser and G. A. R. Kon, *J. Chem. Soc.*, **1934,** 604.
89. E. S. Lipina, V. V. Perekalin, and Ya, S. Bobovich, *Dokl. Akad. Nauk SSSR*, **163,** 894 (1965); *Chem. Abstr.*, **63,** 13109 (1965).
90. E. S. Zonis, O. M. Lerner, and V. V. Perekalin, *Zh. Prikl. Khim.*, **34,** 711 (1961); *Chem. Abstr.*, **55,** 17476 (1961).
91. Yu. V. Baskov and V. V. Perekalin, *Zh. Obshch. Khim.*, **32,** 3106 (1962); *Chem. Abstr.*, **58,** 7816 (1963).
92. H. E. Zimmerman and T. E. Nevins, *J. Am. Chem. Soc.*, **79,** 6559 (1957).
92a. F. G. Bordwell and M. M. Vestling, *J. Am. Chem. Soc.*, **89,** 3906 (1967).
93. A. Bowers, M. B. Sanchez, and H. J. Ringold, *J. Am. Chem. Soc.*, **81,** 3702 (1959).
94. A. Bowers, L. C. Ibañez, and H. J. Ringold, *J. Am. Chem. Soc.*, **81,** 3707 (1959).
95. S. K. Malhotra and F. Johnson, *J. Am. Chem. Soc.*, **87,** 5493 (1965).
96. R. Caple and W. R. Vaughan, *Tetradedron Letters*, **1966,** 4067.
97. W. Hofman, L. Stefaniak, T. Urbański, and M. Witanowski, *J. Am. Chem. Soc.*, **86,** 554 (1964).
98. A. C. Huitric and W. F. Trager, *J. Org. Chem.*, **27,** 1926 (1962).
99. W. F. Trager and A. C. Huitric, *J. Org. Chem.*, **30,** 3257 (1965).
100. H. Feltkamp and N. C. Franklin, *J. Am. Chem. Soc.*, **87,** 1616 (1965).
101. N. C. Franklin and H. Feltkamp, *Tetrahedron*, **22,** 2801 (1966).
102. R. J. Ouellette and G. E. Booth, *J. Org. Chem.*, **31,** 587 (1966).
103. A. T. Nielsen, *J. Org. Chem.*, **27,** 1998 (1962).
104. T. Severin and R. Schmitz, *Chem. Ber.*, **95,** 1417 (1962).

105. W. E. Noland and R. Libers, *Tetrahedron*, **19**, Suppl. 1, 23 (1963).
106. W. E. Noland and J. M. Eakman, *J. Org. Chem.*, **26**, 4118 (1961).
107. R. P. Bell, *The Proton in Chemistry*, Cornell University Press, Ithaca, New York, 1959.
108. R. P. Bell and T. Spencer, *Proc. Roy. Soc. (London), Ser. A*, **251**, 41 (1959).
109. P. M. Zaitsev, Ya. I. Turyan, and Z. V. Zaitseva, *Kinetika i Kataliz*, **4**, 534 (1963); *Chem. Abstr.*, **59**, 15388 (1963).
110. E. W. Miller, A. P. Arnold, and M. J. Astle, *J. Am. Chem. Soc.*, **70**, 3971 (1949).
111. R. P. Bell and D. M. Goodall, *Proc. Roy. Soc. (London), Ser. A*, **294**, 273 (1966).
111a. M. J. Gregory and T. C. Bruice, *J. Am. Chem. Soc.*, **89**, 2327 (1967).
112. R. G. Pearson, *J. Am. Chem. Soc.*, **70**, 204 (1948).
113. H. Ley and A. Hantzsch, *Chem. Ber.*, **39**, 3149 (1906).
114. J. S. Belew and L. G. Hepler, *J. Am. Chem. Soc.*, **78**, 4005 (1956).
115. V. I. Slovetskii, A. A. Fainzilberg, and S. S. Novikov, *Izv. Akad. Nauk SSSR Ser. Khim.*, **1962**, 989; *Chem. Abstr.*, **58**, 5487 (1963).
116. N. Kornblum, R. K. Blackwood, and J. W. Powers, *J. Am. Chem. Soc.*, **79**, 2507 (1957).
117. S. S. Novikov, V. I. Slovetskii, S. A. Shevelev, and A. A. Fainzilberg, *Izv. Akad. Nauk SSSR Ser. Khim.*, **1962**, 598.
118. H. G. Adolph and M. J. Kamlet, *J. Am. Chem. Soc.*, **88**, 4761 (1966).
119. R. P. Bell and A. D. Norris, *J. Chem. Soc.*, **1941**, 855.
120. P. W. K. Flanagan, Ph.D. Thesis, Ohio State University, 1957; *Dissertation Abstr.*, **18**, 1980 (1958).
121. C. D. Ritchie and R. Uschold, *J. Am. Chem. Soc.*, **86**, 4488 (1964).
122. P. J. Elving and J. Lakritz, *J. Am. Chem. Soc.*, **77**, 3217 (1955).
123. J. Junell, *Arkiv Kemi, Mineral. Geol.*, **11B**, No. 34, 11 (1934).
124. R. P. Bell and J. C. Clunie, *Proc. Roy. Soc. (London), Ser. A*, **212**, 16 (1952).
125. R. P. Bell and M. H. Panckhurst, *J. Chem. Soc.*, **1956**, 2836.
126. P. Jones, J. L. Longridge, and W. F. K. Wynne-Jones, *J. Chem. Soc.*, **1965**, 3606.
127. W. F. K. Wynne-Jones, *J. Chem. Phys.*, **2**, 381 (1934).
128. Ya. I. Turyan, Yu. M. Tyurin, and P. M. Zaitsev, *Dokl. Akad. Nauk SSSR*, **134**, **1960**, 850; *Chem. Abstr.*, **55**, 6405 (1961).
129. G. W. Goward, C. E. Bricker, and W. C. Wildman, *J. Org. Chem.*, **20**, 378 (1955).
130. R. Junell, *Svensk. Kem. Tidskr.*, **46**, 125 (1934).
131. R. Junell, *Arkiv Kemi, Mineral. Geol.*, **11B**, No. 27, 1 (1934).
132. R. Junell, *Arkiv Kemi, Mineral. Geol.*, **11B**, No. 30, 1 (1934).
133. K. J. Pedersen, *Kgl. Danske Videnskab. Selskab., Mat. Fys. Medd.*, **12**, 3 (1932); *Chem. Abstr.*, **26**, 5825 (1932).
134. O. Reitz, *Z. Physik. Chem. (A)*, **176A**, 363 (1936).
135. E. S. Lewis and J. D. Allen, *J. Am. Chem. Soc.*, **86**, 2022 (1964).
136. L. Funderburk and E. S. Lewis, *J. Am. Chem. Soc.*, **86**, 2531 (1964); **89**, 2322 (1967).
137. J. A. Feather and V. Gold, *J. Chem. Soc.*, **1965**, 1752.
138. R. G. Pearson and F. V. Williams, *J. Am. Chem. Soc.*, **75**, 3073 (1953).
139. V. M. Belikov, S. G. Mairanovskii, T. B. Korchemnaya, and S. S. Novikov, *Izv. Akad. Nauk SSSR Ser. Khim.*, **1962**, 2103; *Chem. Abstr.*, **58**, 8645 (1963).
140. T. Kamentani, H. Sugahara, and H. Yagi, *J. Chem. Soc.*, 717 (1966).
140a. R. B. Cundall and A. W. Locke, *J. Chem. Soc. B*, **1968**, 98.
141. V. Meyer and J. Locher, *Chem. Ber.*, **8**, 219 (1875).
142. V. Meyer and J. Locher, *Ann. Chem.*, **180**, 163 (1876).

143. S. B. Lippincott and H. B. Hass, *Ind. Eng. Chem.*, **31,** 118 (1939).
144. T. N. Hall, *J. Org. Chem.*, **29,** 3587 (1964).
145. V. I. Slovetskii, L. V. Okhlobystina, A. A. Fainzilberg, A. I. Ivanov, L. I. Biryukova, and S. S. Novikov, *Izv. Akad. Nauk SSSR Ser. Khim.*, **1965,** 2063; *Chem. Abstr.*, **64,** 10469 (1966).
146. S. S. Novikov, V. I. Slovetskii, V. A. Tartakovskii, S. A. Shevelev, and A. A. Fainzilberg, *Dokl. Akad. Nauk SSSR*, **146,** 104 (1962); *Chem. Abstr.*, **58,** 3289 (1963).
147. A. Hantzsch and A. Rinckenberger, *Chem. Ber.*, **32,** 628 (1899).
148. V. I. Slovetskii, S. A. Shevelev, A. A. Fainzilberg, and S. S. Novikov, *Zh. Vses. Khim. Obshchestva im. D. I. Mendeleeva*, **6,** 599,707 (1961); *Chem. Abstr.*, **56,** 6712, 4986 (1962).
148a. M. E. Stizman, H. G. Adolph, and M. J. Kamlet, *J. Am. Chem. Soc.*, **90,** 2815 (1968).
149. I. L. Knunyants, L. S. German, and I. N. Rozhkov, *Izv. Akad. Nauk SSSR Ser. Khim.*, **1966,** 1062; *Chem. Abstr.*, **65,** 12087 (1966).
150. S. S. Novikov, V. M. Belikov, A. A. Fainzilberg, L. V. Ershova, V. I. Slovetskii, and S. A. Shevelev, *Izv. Akad. Nauk SSSR Ser. Khim.*, **1959,** 1855; *Chem. Abstr.*, **54,** 7337 (1960).
151. S. S. Novikov, V. I. Slovetskii, V. M. Belikov, I. M. Zavilovich, and L. V. Epishina, *Izv. Akad. Nauk SSSR Ser. Khim.*, **1962,** 520; *Chem. Abstr.*, **57,** 5349 (1962).
152. H. C. Brown, D. H. McDaniel, and O. Häflinger in *Determination of Organic Structures by Physical Methods* (Ed. E. A. Braude and F. C. Nachod), Academic Press, New York, 1955, p. 567.
153. V. I. Slovetskii, A. I. Ivanov, S. A. Shevelev, A. A. Fainzilberg, and S. S. Novikov, *Tetrahedron Letters*, **1966,** 1745.
154. R. P. Bell, *Trans. Faraday Soc.*, **39,** 253 (1943).
155. R. P. Bell, E. Gelles, and E. Möller, *Proc. Roy. Soc.* (*London*), *Ser. A*, **198,** 308 (1949).
156. V. I. Slovetskii, S. A. Shevelev, V. I. Erashko, L. I. Biryukova, A. A. Fainzilberg, and S. S. Novikov, *Izv. Akad. Nauk SSSR Ser. Khim.*, **1966,** 655; *Chem. Abstr.*, **65,** 3714 (1966).
157. H. M. E. Cardwell, *J. Chem. Soc.*, **1951,** 2442.
158. V. M. Belikov, A. Talvik, and T. B. Korchemnaya, *Reaktsionnaya Sposobnost Organ. Soedin.*, *Tartusk. Gos. Univ.*, **2,** 10 (1965); *Chem. Abstr.*, **63,** 17860 (1965).
159. M. I. Konowalow, *Chem. Ber.*, **28,** 1852 (1895).
160. N. Kornblum and G. E. Graham, *J. Am. Chem. Soc.*, **73,** 4041 (1951).
161. J. Meisenheimer, *Ann. Chem.*, **323,** 205 (1902).
162. C. L. Jackson and W. F. Boos, *Am. Chem. J.*, **20,** 444 (1898).
163. C. L. Jackson and F. H. Gazzolo, *Am. Chem. J.*, **23,** 376 (1900).
164. C. L. Jackson and R. B. Earle, *Am. Chem. J.*, **29,** 89 (1903).
165. R. Foster and D. L. Hammick, *J. Chem. Soc.*, **1954,** 2153.
166. T. Abe, T. Kumai, and H. Arai, *Bull. Chem. Soc. Japan*, **38,** 1526 (1965).
167. M. R. Crampton and V. Gold, *Chem. Commun.*, **1965,** 549.
168. J. F. Bunnett, *Quart. Rev.* (*London*), **12,** 1 (1958).
169. K. L. Servis, *J. Am. Chem. Soc.*, **89,** 1508 (1967).
170. A. Hantzsch and H. Kissel, *Chem. Ber.*, **32,** 3137 (1899).
171. J. B. Ainscough and E. F. Caldin, *J. Chem. Soc.*, **1956,** 2528.
172. A. Ya. Kaminskii and S. S. Gitis, *Zh. Obshch. Khim.*, **34,** 3743 (1964); *Chem. Abstr.*, **62,** 7664 (1965).
173. R. C. Farmer, *J. Chem. Soc.*, **1959,** 3425.
174. J. F. Bunnett and C. Bernasconi, *J. Am. Chem. Soc.*, **87,** 5209 (1965).

175. A. J. Boulton and D. P. Clifford, *J. Chem. Soc.*, **1965**, 5414.
176. W. Steinkopf, W. Malinowski, and A. Supan, *Chem. Ber.*, **44**, 2898 (1911).
177. R. Willstätter and A. Kubli, *Chem. Ber.*, **42**, 4152 (1909).
178. S. S. Novikov, K. K. Babievskii, and I. S. Korsakova, *Dokl. Akad. Nauk SSSR*, **125**, 560 (1959); *Chem. Abstr.*, **53**, 19849 (1959).
179. O. Convert and J. Armand, *Compt. Rend.*, **262**(C), 1013 (1966).
179a. G. A. Bonetti, C. B. DeSavigny, C. Michalski, and R. Rosenthal, *J. Org. Chem.*, **33**, 237 (1968).
180. E. Bamberger and R. Seligman, *Chem. Ber.*, **36**, 701 (1903).
181. L. I. Khmelnitskii, S. S. Novikov, and O. V. Lebedev, *Izv. Akad. Nauk SSSR Ser. Khim.*, **1961**, 477; *Chem. Abstr.*, **55**, 23389 (1961).
182. L. I. Khmelnitskii, S. S. Novikov, and O. V. Lebedev, *Zh. Obshch. Khim.*, **28**, 2303 (1958); *Chem. Abstr.*, **53**, 3111 (1959).
183. L. Canonica, *Gazz. Chim. Ital.*, **77**, 92 (1947).
184. W. D. Emmons and A. S. Pagano, *J. Am. Chem. Soc.*, **77**, 4557 (1955).
184(a). I. L. Knunyants, B. L. Dyatkin, E. P. Mochalina, and L. T. Lantseva, *Izv. Akad. Nauk SSSR Ser. Khim.*, **1966**, 164.
185. G. H. Dorion and K. O. Loeffler, *U.S. Pat.* 3,127,335 (1964); *Chem. Abstr.*, **60**, 14434 (1965).
186. I. Tanasescu, *Bull. Soc. Chim. France*, **39**, 1718 (1926).
187. G. Smolinsky and B. I. Feuer, *J. Org. Chem.*, **31**, 3882 (1966).
188. R. H. Smith and H. Suschitzky, *Tetrahedron*, **16**, 80 (1961).
189. A. T. Nielsen, *J. Org. Chem.*, **27**, 1993 (1962).
190. L. W. Andrew and D. L. Hammick, *J. Chem. Soc.*, **1934**, 244.
191. T. Severin, B. Brück, and P. Adhikary, *Chem. Ber.*, **99**, 3097 (1966).
192. H. Feuer and L. R. Blecker, unpublished results.
193. W. Wislicenus and M. Fischer, *Chem. Ber.*, **43**, 2234 (1910).
194. A. Hantzsch, *Chem. Ber.*, **40**, 1533 (1907).
195. A. T. Nielsen and E. N. Platt, unpublished results.
196. W. Wislicenus and H. Elvert, *Chem. Ber.*, **41**, 4121 (1908).
197. W. Wislicenus and A. Endres, *Chem. Ber.*, **35**, 1755 (1902).
198. H. V. Frost, *Ann. Chem.*, **250**, 163 (1889).
199. J. Meisenheimer and K. Weibezahn, *Chem. Ber.*, **54**, 3195 (1921).
200. W. Wislicenus and H. Wren, *Chem. Ber.*, **38**, 502 (1905).
201. M. O. Forster, *J. Chem. Soc.*, **77**, 251 (1900).
202. R. Schiff, *Chem. Ber.*, **13**, 1402 (1880).
203. L. I. Zakharkin and G. G. Zhigareva, *Izv. Akad. Nauk SSSR Ser. Khim.*, **1962**, 183; *Chem. Abstr.*, **57**, 11040 (1962).
204. J. T. Thurston and R. L. Shriner, *J. Am. Chem. Soc.*, **57**, 2163 (1935).
205. F. E. Ray and S. Palinchak, *J. Am. Chem. Soc.*, **62**, 2109 (1940).
206. S. S. Novikov, V. I. Slovetskii, A. A. Fainzilberg, S. A. Shevelev, V. A. Shlyapochnikov, A. I. Ivanov, and V. A. Tartakovskii, *Tetrahedron*, **20**, Suppl. 1, 119 (1964).
207. W. Kemula and W. Turnowska-Rubaszewska, *Roczniki Chem.*, **35**, 1169 (1961); *Chem. Abstr.*, **56**, 6798 (1961).
208. W. Kemula and W. Turnowska-Rubaszewska, *Roczniki Chem.*, **37**, 1597 (1963); *Chem. Abstr.*, **61**, 184 (1964).
209. M. F. Hawthorne, *J. Am. Chem. Soc.*, **79**, 2510 (1957).
210. F. T. Williams, Jr., P. W. K. Flanagan, W. J. Taylor, and H. Shechter, *J. Org. Chem.*, **30**, 2674 (1965).

211. J. Freeman and K. S. McCallum, *J. Org. Chem.*, **21,** 472 (1956).
212. A. Hantzsch and K. Voigt, *Z. Physik. Chem. (Leipzig)*, **79,** 592 (1912).
213. T. Severin and B. Brück, *Chem. Ber.*, **98,** 3847 (1965).
214. N. Zelinsky and N. Rosanoff, *Z. Physik. Chem. (Leipzig)*, **78,** 629 (1912).
215. L. J. Bellamy, *The Infra-red Spectra of Complex Molecules*, 2nd Edition, John Wiley and Sons, New York, 1958.
216. H. Feuer, C. Savides, and C. N. R. Rao, *Spectrochim. Acta.*, **19,** 431 (1963).
217. W. L. F. Armarego, T. J. Batterham, K. Schofield, and R. S. Theobald, *J. Chem. Soc.*, (*C*), **1966,** 1433.
218. D. Meuche and E. Heilbronner, *Helv. Chim. Acta*, **45,** 1965 (1962).
219. A. N. Campbell-Crawford, A. M. Gorringe, and D. Lloyd, *Chem. Ind. (London)*, **1966,** 1961.
220. J. C. Sowden, *Advances in Carbohydrate Chemistry*, Vol. 6, Academic Press, New York, 1951, pp. 291–318.
221. E. E. Van Tamelen and R. J. Thiede, *J. Am. Chem. Soc.*, **74,** 2615 (1952).
222. N. Kornblum and R. A. Brown, *J. Am. Chem. Soc.*, **87,** 1742 (1965).
223. V. Meyer, *Chem. Ber.*, **28,** 202 (1895).
224. L. C. Leitch, *Can. J. Chem.*, **33,** 400 (1955).
225. S. S. Nametkin, *J. Russ. Phys. Chem. Soc.*, **45,** 1414 (1913); *Chem. Abstr.*, **8,** 324 (1914).
226. W. C. Wildman and C. H. Hemminger, *J. Org. Chem.*, **17,** 1641 (1952).
227. P. Salomaa in *The Chemistry of the Carbonyl Group*, (Ed. S. Patai), Interscience Publishers, New York, 1966, pp. 205–206.
228. B. Weinstein and A. H. Fenselau, *J. Org. Chem.*, **27,** 4094 (1962).
229. K. Johnson and E. F. Degering, *J. Org. Chem.*, **8,** 10 (1943).
230. H. Schechter and F. T. Williams, Jr., *J. Org. Chem.*, **27,** 3699 (1962).
231. A. P. Howe and H. B. Hass, *Ind. Eng. Chem.*, **38,** 251 (1946).
232. R. A. Smiley and W. A. Pritchett, *J. Chem. Eng. Data*, **11,** 617 (1966).
233. D. J. Cook, O. R. Pierce, and E. T. McBee, *J. Am. Chem. Soc.*, **76,** 83 (1954).
234. O. L. Chapman and E. Hoganson, unpublished results; *cf.* E. Hoganson, Ph.D. Thesis, Iowa State University, 1965.
235. O. L. Chapman, A. A. Griswold, E. Hoganson, G. Lenz, and J. Reasoner, *Pure Appl. Chem.*, **9,** 585 (1964).
236. W. C. Wildman and R. B. Wildman, *J. Org. Chem.*, **17,** 581 (1952).
237. B. Iselin and H. O. L. Fischer, *J. Am. Chem. Soc.*, **70,** 3946 (1948).
238. J. M. Grosheintz and H. O. L. Fischer, *J. Am. Chem. Soc.*, **70,** 1479 (1948).
239. W. E. Parham, W. T. Hunter, and R. Hanson, *J. Am. Chem. Soc.*, **73,** 5068 (1951).
240. W. E. Noland, J. H. Cooley, and P. A. McVeigh, *J. Am. Chem. Soc.*, **81,** 1209 (1959).
241. M. Simonetta and S. Winstein, *J. Am. Chem. Soc.*, **76,** 18 (1954).
242. W. C. Wildman and D. R. Saunders, *J. Org. Chem.*, **19,** 381 (1954).
243. V. Meyer and C. Wurster, *Chem. Ber.*, **6,** 1168 (1873).
244. E. Bamberger and E. Rüst, *Chem. Ber.*, **35,** 45 (1902).
245. T. Urbański, *J. Chem. Soc.*, **1949,** 3374.
246. H. L. Yale, *Chem. Rev.*, **33,** 209 (1943).
247. T. Simmons and K. L. Kreuz, forthcoming publication.
248. A. Hassner and J. Larkin, *J. Am. Chem. Soc.*, **85,** 2181 (1963).
249. W. Steinkopf and B. Jürgens, *J. Prakt. Chem.*, [2] **84,** 686 (1911).
250. W. Steinkopf and B. Jürgens, *J. Prakt. Chem.*, [2] **83,** 453 (1911).
251. R. E. Plapinger, *J. Org. Chem.*, **24,** 802 (1959).
252. E. Bamberger, *J. Prakt. Chem.*, [2] **101,** 328 (1921).

253. V. Meyer, *Ann. Chem.*, **171,** 1 (1874).
254. V. Meyer, *Chem. Ber.*, **8**, 29 (1875).
255. R. Preibisch, *J. Prakt. Chem.*, [2] **7,** 480 (1873).
256. R. Preibisch, *J. Prakt. Chem.*, [2] **8,** 309 (1873).
257. A. Geuther, *Chem. Ber.*, **7,** 1620 (1874).
258. J. Donath, *Chem. Ber.*, **10,** 776 (1877).
259. J. Züblin, *Chem. Ber.*, **10,** 2083 (1877).
260. P. Alexéeff, *Bull. Soc. Chim. France*, **46,** 266 (1886).
261. S. Gabriel and M. Koppe, *Chem. Ber.*, **19,** 1145 (1886).
262. L. Henry, *Rec. Trav. Chim.*, **17,** 399 (1898).
263. R. A. Worstall, *Am. Chem. J.*, **21,** 218 (1899).
264. V. Perekalin and S. Sopova, *Zh. Obshch. Khim.*, **24,** 513 (1954); *Chem. Abstr.*, **49,** 6180 (1955).
265. B. F. Burrows and W. B. Turner, *J. Chem. Soc.* (*C*), **1966,** 255.
266. N. N. Mel'nikov, *Zh. Obshch. Khim.*, **4,** 1061 (1934); *Chem. Abstr.*, **29,** 3979 (1935).
267. N. Levy, C. W. Scaife, and A. E. Wilder-Smith, *J. Chem. Soc.*, **1946,** 1096.
268. T. Urbański and T. Dobosz, *Bull. Acad. Polon. Sci. Classe* (*III*), **5,** 541 (1957); *Chem. Abstr.*, **52,** 2738 (1958).
269. T. Urbański, *Tetrahedron*, **2,** 296 (1958).
270. D. C. Berndt and R. L. Fuller, *J. Org. Chem.*, **31,** 3312 (1966).
271. M. C. Sneed and R. C. Brasted, *Comprehensive Inorganic Chemistry*, Vol. 5, D. Van Nostrand, Princeton, N.J., 1956, p. 40.
272. L. G. Donaruma and M. I. Huber, U.S. Pat. 2,702,801 (1955); *Chem. Abstr.*, **50,** 1896 (1956).
273. L. G. Donaruma and M. I. Huber, *J. Org. Chem.*, **21,** 965 (1956).
274. O. Piloty and H. Steinbock, *Chem. Ber.*, **35,** 3101 (1902).
274a. G. Gasnati and A. Ricca, *Tetrahedron Letters*, **1967,** 327.
275. C. D. Nenitzescu and D. A. Isacescu, *Bull. Soc. Chim. Romania*, **14,** 53 (1932); *Chem. Abstr.*, **27,** 964 (1933).
276. V. A. Klimova and K. S. Zabrodina, *Izv. Akad. Nauk SSSR Ser. Khim.*, **1961,** 176; *Chem. Abstr.*, **55,** 18453 (1961).
277. H. Wieland and L. Semper, *Chem. Ber.*, **39,** 2522 (1906).
278. V. Meyer, *Chem. Ber.*, **6,** 1492 (1873).
279. V. Meyer, *Ann. Chem.*, **175,** 88 (1875).
280. G. A. Russell, *J. Am. Chem. Soc.*, **76,** 1595 (1954).
281. E. M. Nygaard, J. H. McCracken, and T. T. Noland, *U.S. Pat.* 2,370,185 (1945); *Chem. Abstr.*, **39,** 3551 (1945).
282. E. M. Nygaard, *U.S. Pat.* 2,401,267 (1946); *Chem. Abstr.*, **40,** 6092 (1946).
283. E. M. Nygaard and T. T. Noland, *U.S. Pat.* 2,401,269 (1946); *Chem. Abstr.*, **40,** 6093 (1946).
284. A. I. Titov and V. V. Smirnov, *Dokl. Akad. Nauk SSSR*, **83,** 243 (1952); *Chem. Abstr.*, **47,** 4298 (1953).
285. L. I. Khmelnitskii, S. S. Novikov, and O. V. Lebedev, *Izv. Akad. Nauk SSSR Ser. Khim.*, **1960,** 1783; *Chem. Abstr.*, **55,** 19833 (1961).
286. S. S. Novikov, O. V. Lebedev, L. I. Khmelnitskii, and Yu. P. Egorov, *Zh. Obshch. Khim.*, **28,** 2305 (1958); *Chem. Abstr.*, **53,** 4112 (1959).
287. W. Will, *Chem. Ber.*, **47,** 961 (1914).
288. S. S. Novikov, A. A. Fainzilberg, S. A. Shevelev, I. S. Korsakova, K. K. Babievskii, *Dokl. Akad. Nauk SSSR*, **1960,** 846; *Chem. Abstr.*, **54,** 20841 (1960).

289. S. S. Novikov, A. A. Fainzilberg, S. A. Shevelev, I. S. Korsakova, and K. K. Babievskii, *Dokl. Akad. Nauk SSSR*, **124**, 589 (1959); *Chem. Abstr.*, **53**, 11206 (1959).
290. V. Meyer and G. Ambühl, *Chem. Ber.*, **8**, 1073 (1875).
291. V. Meyer, *Chem. Ber.*, **9**, 384 (1876).
292. A. F. Holleman, *Rec. Trav. Chim.*, **13**, 405 (1894).
293. E. Bamberger, *Chem. Ber.*, **31**, 2626 (1898).
294. E. Bamberger, *Chem. Ber.*, **33**, 1781 (1900).
295. E. Bamberger, *Chem. Ber.*, **36**, 90 (1903).
296. E. Bamberger, O. Schmidt, and H. Levinstein, *Chem. Ber.*, **33**, 2043 (1900).
297. E. Bamberger and O. Schmidt, *Chem. Ber.*, **34**, 574 (1901).
298. E. Bamberger and J. Grob, *Chem. Ber.*, **35**, 67 (1902).
299. E. Bamberger and J. Frei, *Chem. Ber.*, **35**, 82 (1902).
300. Farbenfabriken Bayer, *Brit. Pat.* 684,369 (1952); *Chem. Abstr.*, **48**, 2095 (1954).
301. C. S. Coe and T. F. Doumani, *U.S. Pat.* 2,656,393 (1953); *Chem. Abstr.*, **48**, 10049 (1954).
302. A. A. Artem'ev, E. V. Genkina, A. B. Malimonova, V. P. Trofil'kina, and M. A. Isaenkova, *Zh. Vses. Khim. Obshchestva im. D. I. Mendeleeva*, **10**, 588 (1965); *Chem. Abstr.*, **64**, 1975 (1966).
303. M. A. Bakin, *U.S.S.R. Pat.* 173,777 (1965); *Chem. Abstr.*, **64**, 1980 (1966).
304. H. Welz and J. Weise, *Ger. Pat.* 837,692 (1952); *Chem. Abstr.*, **47**, 1729 (1953).
305. G. A. Russell, E. G. Janzen, and E. T. Strom, *J. Am. Chem. Soc.*, **86**, 1807 (1964).
306. F. Ratz, *Monatsh. Chem.*, **25**, 55 (1904).
307. O. v. Schickh, U.S. Pat. 2,712,032 (1955); Brit. Pat., 716,099 (1954); *Chem. Abstr.*, **50**, 13080 (1956).
308. D. A. Isacescu, *Bull. Soc. Chim. Romania*, **18A**, 63 (1936); *Chem. Abstr.*, **31**, 3036 (1937).
309. C. D. Nenitzescu and I. G. Dinulescu, *Izv. Akad. Nauk. SSSR Ser. Khim.*, **1958**, 1228; *Chem. Abstr.*, **53**, 5208 (1959).
310. C. Dale and R. L. Shriner, *J. Am. Chem. Soc.*, **58**, 1502 (1936).
311. H. Shechter and R. B. Kaplan, *J. Am. Chem. Soc.*, **75**, 3980 (1953).
312. G. A. Russell, A. J. Moye, E. G. Janzen, S. Mak, and E. R. Talaty, *J. Org. Chem.* **32**, 137 (1967).
313. C. D. Nenitzescu, *Chem. Ber.*, **62**, 2669 (1929).
314. G. A. Russell and W. C. Danen, *J. Am. Chem. Soc.*, **88**, 5663 (1966).
315. G. A. Russell and E. G. Janzen, *J. Am. Chem. Soc.*, **89**, 300 (1967).
316. B. C. Gilbert and R. O. C. Norman, *J. Chem. Soc.*, (*B*), **1966**, 722.
317. W. Wislicenus and A. Endres, *Chem. Ber.*, **36**, 1194 (1903).
318. E. ter Meer, *Ann. Chem.*, **181**, 1 (1876).
319. M. F. Hawthorne and R. D. Strahm, *J. Am. Chem. Soc.*, **79**, 3471 (1957).
320. E. M. Nygaard and T. T. Noland, U.S. Pat. 2,396,282 (1946); *Chem. Abstr.*, **40**, 3126 (1946).
321. J. Armand, *Bull. Soc. Chim. France*, **1966**, 543.
322. M. F. Hawthorne, *J. Am. Chem. Soc.*, **78**, 4980 (1956).
323. H. Feuer, C. E. Colwell, G. Leston, and A. T. Nielsen, *J. Org. Chem.*, **27**, 3598 (1962).
324. M. Simonetta and G. Favini, *Atti Accad. Nazl. Lincei. Mem., Classe Sci. Fis., Mat. Nat. Sez.*, **18**, 636 (1955); *Chem. Abstr.*, **50**, 4600 (1956).
325. R. A. Gotts and L. Hunter, *J. Chem. Soc.*, **125**, 442 (1924).
326. S. M. Losanitsch, *Chem. Ber.*, **15**, 471 (1882).

327. G. B. Brown and R. L. Shriner, *J. Org. Chem.*, **2**, 376 (1937).
328. L. W. Seigle and H. B. Hass, *J. Org. Chem.*, **5**, 100 (1940).
329. H. Wieland, P. Garbsch, and J. J. Chavan, *Ann. Chem.*, **461**, 295 (1928).
330. H. Feuer and P. M. Pivawer, *J. Org. Chem.*, **31**, 3152 (1966).
331. A. Lucas, *Chem. Ber.*, **32**, 600 (1899).
332. K. Fries and E. Pusch, *Ann. Chem.*, **442**, 272 (1925).
333. T. Simmons, R. F. Love, and K. L. Kreuz, *J. Org. Chem.*, **31**, 2400 (1966).
334. R. E. Schaub, W. Fulmor, and M. J. Weiss, *Tetrahedron*, **20**, 373 (1964).
335. H. Feuer in *Abstracts* 152*nd American Chemical Society Meeting*, New York, N.Y., Sept. 1966 (Paper No. S-157).
336. G. Vanags and J. Bungs, *Chem. Ber*, **75B**, 987 (1942).
337. T. M. Lowry, *J. Chem. Soc.*, **75**, 211 (1899).
337a. A. A. Griswold and P. S. Starcher, *J. Org. Chem.*, **30**, 1687 (1965).
338. G. S. Hammond in *Steric Effects in Organic Chemistry*, (Ed. M. S. Newman), John Wiley and Sons, New York, 1956, p. 442.
339. A. Hantzsch, *Chem. Ber.*, **40**, 1523 (1907).
340. A. Hantzsch, *Ann. Chem.*, **392**, 286 (1912).
341. R. G. Pearson, D. H. Anderson, and L. L. Alt, *J. Am. Chem. Soc.*, **77**, 527 (1955).
342. M. A. Matskanova and A. Arens, *Latvijas PSR Zinatnu Akad. Vestis, Kim. Ser.*, **1966**, 362; *Chem. Abstr.*, **65**, 15193 (1966).
343. G. I. Samokhvalov, M. K. Shakhova, M. I. Budagyants, A. Ya. Veinberg, L. V. Luk'yanova, and N. A. Preobrazhenskii, *Zh. Obshch. Khim.*, **31**, 1147 (1961); *Chem. Abstr.*, **55**, 24731 (1961).
344. M. K. Shakhova, M. I. Budagyants, G. I. Samokhvalov, and N. A. Preobrazhenskii, *Zh. Obshch. Khim.*, **32**, 2832 (1962); *Chem. Abstr.*, **58**, 12500 (1963).
345. M. Michalska, *Chem. Ind.* (*London*), **1966**, 628.
346. R. Stoermer and K. Brachmann, *Chem. Ber.*, **44**, 315 (1911).
347. J. Thiele and E. Weitz, *Ann. Chem.*, **377**, 1 (1910).
348. R. D. Campbell and C. L. Pitzer, *J. Org. Chem.*, **24**, 1531 (1959).
349. H. H. Baer, *Tetrahedron*, **20**, Suppl. 1, 263 (1964).
350. H. H. Baer and B. Achmatowicz, *J. Org. Chem.*, **29**, 3180 (1964).
351. P. Cazeneuve, *Compt. Rend.*, **108**, 243; 302 (1889).
352. P. Cazeneuve, *Bull. Soc. Chim. France*, [3] **1**, 558 (1889).
353. A. Lapworth and F. S. Kipping, *J. Chem. Soc.*, **69**, 304 (1896).
354. T. M. Lowry, *J. Chem. Soc.*, **73**, 986 (1898).
355. T. M. Lowry, *J. Chem. Soc.*, **83**, 953 (1903).
356. T. M. Lowry and W. Robertson, *J. Chem. Soc.*, **85**, 1541 (1904).
357. T. M. Lowry and E. H. Magson, *J. Chem. Soc.*, **93**, 107 (1908).
358. T. M. Lowry and C. H. Desch, *J. Chem. Soc.*, **95**, 807 (1909).
359. T. M. Lowry and V. Steele, *J. Chem. Soc.*, **107**, 1038 (1915).
360. T. M. Lowry and H. Burgess, *J. Chem. Soc.*, **123**, 2111 (1923).
361. A. Hantzsch and K. Voigt, *Chem. Ber.*, **45**, 85 (1912).
362. R. P. Bell and J. A. Sherred, *J. Chem. Soc.*, **1940**, 1202.
363. H. Feuer and C. Savides, *J. Am. Chem. Soc.*, **81**, 5826 (1959).
364. R. Andreasch, *Monatsh. Chem.*, **16**, 17 (1895).
365. G. Vanags, *Chem. Ber.*, **69**, 1066 (1936).
366. G. Vanags and A. Lode, *Chem. Ber.*, **70**, 547 (1937).
367. G. Vanags and A. Lode, *Chem. Ber.*, **71**, 1267 (1938).
368. G. Vanags and A. Dombrowski, *Chem. Ber.*, **75**, 82 (1942).

369. G. Vanags and E. Yu. Gudrinietze, *Zh. Obshch. Khim.*, **19**, 1542 (1949); *Chem. Abstr.*, **44**, 1087 (1950).
370. G. Vanags, J. Eiduss, and S. Gillers, *Latvijas PSR Zinatnu Akad. Vestis*, **1949**, 21; *Chem. Abstr.*, **48**, 446 (1954).
371. G. Vanags, Ya. J. Eiduss, and S. Gillers, *Dokl. Akad. Nauk SSSR*, **79**, 977 (1951); *Chem. Abstr.*, **49**, 4382 (1955).
372. L. P. Zalukaevs, *Zh. Obshch. Khim.*, **26**, 3125 (1956); *Chem. Abstr.*, **51**, 18052 (1957).
373. J. Stradins, O. Neilands, Ya. Freimanis, and G. Vanags, *Dokl. Akad. Nauk SSSR*, **129**, 594 (1959); *Chem. Abstr.*, **54**, 8753 (1960).
374. O. Neilands, J. Stradins, and G. Vanags, *Dokl. Akad. Nauk SSSR* **131**, 1084 (1960); *Chem. Abstr.*, **54**, 20911 (1960).
375. L. P. Zalukaevs and L. V. Klykova, *Zh. Obshch. Khim.*, **34**, 3821 (1964); *Chem. Abstr.*, **62**, 9077 (1965).
376. E. Gudriniece and G. Vanags, *Tetrahedron*, **20**, Suppl. 1, 33 (1964).
377. H. B. Hill and J. Torrey, Jr., *J. Am. Chem. Soc.*, **22**, 89 (1899).
378. T. L. V. Ulbricht and C. C. Price, *J. Org. Chem.*, **22**, 235 (1957).
379. P. E. Fanta in *Org. Syntheses*, Coll. Vol. **IV** (Ed. N. Rabjohn), John Wiley and Sons, New York, 1963, p. 844.
380. L. P. Zalukaevs and E. V. Vanags, *Zh. Obshch. Khim.*, **30**, 145 (1960); *Chem. Abstr.*, **54**, 24746 (1960).
381. H. Shechter, D. L. Ley, and L. Zeldin, *J. Am. Chem. Soc.*, **74**, 3664 (1952).
382. T. Severin and B. Brück, *Angew. Chem. Intern. Ed. Engl.*, **3**, 806 (1964).
383. A. Hantzsch, *Chem. Ber.*, **39**, 1084 (1906).
384. A. Hantzsch and H. Gorke, *Chem. Ber.*, **39**, 1073 (1906).
385. A. Hantzsch and A. Korcynski, *Chem. Ber.*, **42**, 1217 (1909).
386. J. Meisenheimer and E. Connerade, *Ann. Chem.*, **330**, 133 (1904).
387. L. Weisler and R. W. Helmkamp, *J. Am. Chem. Soc.*, **67**, 1167 (1945).
388. H. B. Hass and M. L. Bender, *J. Am. Chem. Soc.*, **71**, 3482 (1949).
389. K. Hamann and K. Bauer, *Ger. Pat.*, 825,547 (1951); *Chem. Abstr.*, **47**, 2204 (1953).
390. L. G. Donaruma, *J. Org. Chem.*, **22**, 1024 (1957).
391. N. Kornblum and R. A. Brown, *J. Am. Chem. Soc.*, **85**, 1359 (1963).
392. M. Montavon, H. Lindler, R. Marbet, R. Rüegg, G. Ryser, G. Saucy, P. Zeller, and O. Isler, *Helv. Chim. Acta*, **40**, 1250 (1957).
393. H. B. Hass and M. L. Bender in *Org. Syntheses*, Coll. Vol. IV, (Ed. N. Rabjohn), John Wiley and Sons, New York, 1963, p. 932.
394. F. Arndt and J. D. Rose, *J. Chem. Soc.*, **1935**, 1.
395. K. v. Auwers and R. Ottens, *Chem. Ber.*, **57**, 446 (1924).
396. R. E. McCoy and R. E. Gohlke, *J. Org. Chem.* **22**, 286 (1957).
397. H. B. Hass, E. J. Berry, and M. L. Bender, *J. Am. Chem. Soc.*, **71**, 2290 (1949).
398. V. A. Tartakovskii, I. E. Chlenov, G. V. Lagodzinskaya, and S. S. Novikov, *Dokl. Akad. Nauk SSSR*, **161**, 136 (1965); *Chem. Abstr.*, **62**, 14646 (1965).
399. K. Torssell, *Arkiv Kemi*, **23**, 537; 543 (1965).
400. V. A. Tartakovskii, I. E. Chlenov, N. S. Morozova, and S. S. Novikov, *Izv. Akad. Nauk SSSR Ser. Khim.*, **1966**, 370.
401. C. O. Parker, W. D. Emmons, A. S. Pagano, H. A. Rolewicz, and K. S. McCallum, *Tetrahedron*, **17**, 89 (1962).
402. V. I. Slovetskii, A. I. Ivanov, A. A. Fainzilberg, S. A. Shevelev, and S. S. Novikov, *Zh. Org. Khim.*, **2**, 937 (1966); *Chem. Abstr.*, **65**, 16827 (1966).
403. F. Ratz, *Monatsh. Chem.*, **26**, 1487 (1905).

404. J. U. Nef, *Ann. Chem.*, **280**, 331 (1894).
405. P. Seidel, *Chem. Ber.*, **25**, 431 (1892).
406. P. Seidel, *Chem. Ber.*, **25**, 2756 (1892).
407. W. Steinkopf, *Ann. Chem.* **434**, 21 (1923).
408. H. Shechter and F. Conrad, *J. Am. Chem. Soc.*, **76**, 2716 (1954).
409. W. Steinkopf, *J. Prakt. Chem.*, **81**, 97 (1910).
410. W. S. Reich, G. G. Rose, and W. Wilson, *J. Chem. Soc.*, **1947**, 1234.
411. J. T. Thurston and R. L. Shriner, *J. Org. Chem.*, **2**, 183 (1937).
412. F. Arndt, L. Loewe, and H. Isik, *Rev. Fac. Sci. Univ. Instanbul.* **2**, 139 (1937); *Chem. Abstr.*, **31**, 5774 (1937).
413. A. Hantzsch and S. Opolski, *Chem. Ber.*, **41**, 1745 (1908).
414. A. Young, O. Levand, W. K. H. Luke, and H. O. Larson, *Chem. Commun.*, **1966**, 230.
415. A. Hantzsch and K. S. Caldwell, *Chem. Ber.*, **39**, 2472 (1906).
416. A. Hantzsch, *Z. Anorg. Allgem. Chem.*, **209**, 213 (1932).
417. R. L. Shriner and G. B. Brown, *J. Org. Chem.*, **2**, 560 (1937).
418. G. S. Hammond, W. D. Emmons, C. O. Parker, B. M. Graybill, J. H. Waters, and M. F. Hawthorne, *Tetrahedron*, **19**, Suppl. 1, 177 (1963).
419. W. G. H. Edwards, *Chem. Ind. (London)*, **1951**, 112.
420. P. E. Fanta, R. M. W. Rickett, and D. S. James, *J. Org. Chem.*, **26**, 938 (1961).
420a. L. A. Cohen and W. M. Jones, *J. Am. Chem. Soc.*, **85**, 3397 (1963).
421. J. L. Heinke, *Chem. Ber.*, **31**, 1395 (1898).
421a. J. S. Meek and J. S. Fowler, *J. Org. Chem.*, **33**, 226 (1968).
422. K. Torssell and R. Ryhage, *Arkiv Kemi*, **23**, 525 (1965).
423. V. A. Tartakovskii, L. A. Nikonova, and S. S. Novikov, *Izv. Akad. Nauk SSSR Ser. Khim.*, **1966**, 1290; *Chem. Abstr.*, **65**, 16808 (1966).
423a. K. Torssell, *Acta Chem. Scand.*, **21**, 1 (1967).
423b. K. Torssell, *Acta Chem. Scand.*, **21**, 1392 (1967).
423c. V. I. Erashko, S. A. Shevelev, and A. A. Fainzilberg, *Uspekhi Khim.*, **35**, 1740 (1966).
424. R. C. Kerber, G. W. Urry, and N. Kornblum, *J. Am. Chem. Soc.*, **87**, 4520 (1965).
425. H. Shechter and R. B. Kaplan, *J. Am. Chem. Soc.*, **73**, 1883 (1951).
426. N. Kornblum, P. Pink, and K. V. Yorka, *J. Am. Chem. Soc.*, **83**, 2779 (1961).
427. N. Kornblum, R. E. Michel, and R. C. Kerber, *J. Am. Chem. Soc.*, **88**, 5661 (1966).
428. N. Kornblum and P. Pink, *Tetrahedron*, **19**, Suppl. 1, 17 (1963).
428a. G. A. Russell and W. C. Danen, *J. Am. Chem. Soc.*, **90**, 347 (1968).
428b. W. S. Saari, *J. Org. Chem.*, **32**, 4074 (1967).
429. M. Bersohn, *J. Am. Chem. Soc.*, **83**, 2137 (1961).
430. N. Kornblum and H. J. Taylor, *J. Org. Chem.*, **28**, 1424 (1963).
431. S. J. Eldridge, *Tetrahedron Letters*, **1965**, 4527.
432. P. Duden, *Chem. Ber.*, **26**, 3003 (1893).
433. H. Wieland and A. Höchtlen, *Ann. Chem.*, **505**, 237 (1933).
434. N. Kornblum, R. A. Smiley, R. K. Blackwood, and D. C. Iffland, *J. Am. Chem. Soc.*, **77**, 6269 (1955).
435. V. I. Erashko, S. A. Shevelev, and A. A. Fainzilberg, *Usp. Khim.*, **35**, 1740 (1966); *Chem. Abstr.*, **66**, 5143 (1967).
436. V. A. Tartakovskii, B. G. Gribov, I. A. Savost'yanova, and S. S. Novikov, *Izv. Akad. Nauk SSSR Ser. Khim.*, **1965**, 1644; *Chem. Abstr.*, **64**, 2080 (1966).
437. E. P. Kohler and G. R. Barrett, *J. Am. Chem. Soc.*, **48**, 1770 (1926).

438. E. P. Kohler and S. F. Darling, *J. Am. Chem. Soc.*, **52**, 1174 (1930).
439. A. Dornow and G. Wiehler, *Ann. Chem.*, **578**, 113 (1952).
440. A. Dornow and A. Frese, *Ann. Chem.*, **578**, 122 (1952).
441. E. P. Kohler and G. R. Barrett, *J. Am. Chem. Soc.*, **46**, 2105 (1924).
442. A. Dornow and F. Boberg, *Ann. Chem.*, **578**, 94 (1952).
443. E. P. Kohler and N. K. Richtmyer, *J. Am. Chem. Soc.*, **50**, 3092 (1928).
444. L. I. Smith, *Chem. Rev.*, **23**, 255 (1938).
444a. A. T. Nielsen and T. G. Archibald, *Tetrahedron Letters*, **1968**, 3375, and forthcoming publication.
445. A. T. Nielsen and W. G. Finnegan, *Tetrahedron*, **22**, 925 (1966).
446. H. Stetter and K. Hoehne, *Chem. Ber.*, **91**, 1344 (1958).
447. F. D. Chattaway and A. J. Walker, *J. Chem. Soc.*, **1925**, 2407.
448. N. K. Kochetkov and S. D. Sokolov, *Adv. Heterocyclic Chem.*, **2**, 365 (1963).
449. A. I. Ivanov, I. E. Chlenov, V. A. Tartakovskii, V. I. Slovetskii, and S. S. Novikov, *Izv. Akad. Nauk SSSR Ser. Khim.*, **1965**, 1491; *Chem. Abstr.*, **63**, 16152 (1965).
450. S. V. Lieberman, *J. Am. Chem. Soc.*, **77**, 1114 (1955).
451. G. B. Bachman and T. Hokama, *J. Am. Chem. Soc.*, **81**, 4223 (1959).
452. K. v. Auwers and L. Harres, *Chem. Ber.*, **62**, 2287 (1929).
453. W. D. S. Bowering, V. M. Clark, R. S. Thakur, and Lord Todd, *Ann. Chem.*, **669**, 106 (1963).
454. J. Hamer and A. Macaluso, *Chem. Rev.*, **64**, 473 (1964).
455. R. Bonnett, V. M. Clark, and A. Todd, *J. Chem. Soc.*, **1959**, 2102.
456. V. A. Tartakovskii, I. E. Chlenov, S. S. Smagin, and S. S. Novikov, *Izv. Akad. Nauk SSSR Ser. Khim.*, **1964**, 583; *Chem. Abstr.*, **61**, 4335 (1964).
457. V. A. Tartakovskii, S. S. Smagin, I. E. Chlenov, and S. S. Novikov, *Izv. Akad. Nauk SSSR Ser. Khim.*, **1965**, 552; *Chem. Abstr.*, **63**, 594 (1965).
458. V. A. Tartakovskii, I. E. Chlenov, S. L. Ioffe, G. V. Lagodzinskaya , and S. S. Novikov, *Zh. Org. Khim.*, **2**, 1593 (1966).
458a. V. A. Tartakovskii, A. A. Onishchenko, I. E. Chlenov, and S. S. Novikov, *Dokl. Akad. Nauk SSSR*, **167**, 844 (1966).
459. P. Rajagopalan and C. N. Talaty, *Tetrahedron Letters*, **1966**, 4537.
460. P. Noble, F. G. Borgardt, and W. L. Reed, *Chem. Rev.*, **64**, 19 (1964).
461. V. I. Slovetskii, V. A. Tartakovskii, and S. S. Novikov, *Izv. Akad. Nauk SSSR Ser. Khim.*, 1400 (1962); *Chem. Abstr.*, **58**, 2465 (1963).
462. V. Meyer and A. Rilliet, *Chem. Ber.*, **5**, 1030 (1872).
463. M. I. Konowalow, *J. Russ. Chem. Soc.*, **25**, 509 (1893); *Brit. Abstr.*, **I**, 277 (1894).
464. V. I. Slovetskii, A. I. Ivanov, S. A. Shevelev, A. A. Fainzilberg, and S. S. Novikov, *Zh. Org. Khim.*, **2**, 1445 (1966).
465. A. Dornow and H. Menzel, *Ann. Chem.*, **588**, 40 (1954).
466. F. B. Mallory, K. E. Schueller, and C. S. Wood *J. Org. Chem.*, **26**, 3312 (1961).
467. C. R. Kinney, *J. Am. Chem. Soc.*, **51**, 1592 (1929).
468. J. V. R. Kaufman and J. P. Picard, *Chem. Rev.*, **59**, 429 (1959).
469. L. A. Ustynyuk, S. A. Shevelev, and A. A. Fainzilberg, *Izv. Akad. Nauk SSSR Ser. Khim.*, **1966**, 930; *Chem. Abstr.*, **65**, 10451 (1966).
470. E. P. Stefl and M. F. Dull, *J. Am. Chem. Soc.*, **69**, 3037 (1947).
471. T. Urbański and W. Gurzynska, *Roczniki Chem.*, **25**, 213 (1951); *Chem. Abstr.*, **46**, 7994 (1952).
472. J. Kissel, *J. Russ. Phys.-Chem. Soc.*, **14**, 40 (1882)); *Ber. Referate*, **15**, 727 (1882).
473. J. U. Nef, *Chem. Ber.*, **29**, 1218 (1896).

474. L. W. Jones, *Ann. Chem.*, **20**, 1 (1898).
475. A. F. Holleman, *Rec. Trav. Chim.*, **15**, 356 (1896).
476. W. D. Emmons, *J. Am. Chem. Soc.*, **79**, 5739 (1957).
477. J. S. Splitter and M. Calvin, *J. Org. Chem.*, **23**, 651 (1958).
478. H. Wieland and Z. Kitasato, *Chem. Ber.*, **62B**, 1250 (1929).
478a. J. S. Belew, C. E. Grabiel, and L. B. Clapp, *J. Am. Chem. Soc.*, **77**, 1110 (1955).
479. K. Okon and G. Aluchna, *Bull. Acad. Polon. Sci., Ser. Sci., Chim., Geol. Geograph.*, **7**, 83 (1959); *Chem. Abstr.*, **54**, 17242 (1960).
480. H. Rheinboldt and M. Dewald, *Ann. Chem.*, **455**, 300 (1927).
481. L. C. Behr, *J. Am. Chem. Soc.*, **76**, 3672 (1954).
482. G. Heller and G. Spielmeyer, *Chem. Ber.*, **58**, 834 (1925).
483. K. Akashi, *Bull. Inst. Phys. Chem. Res. (Tokyo)*, **20**, 798 (1941); *Chem. Abstr.*, **43**, 7934 (1949).
484. A. Reissert and F. Lemmer, *Chem. Ber.*, **59**, 351 (1926).
485. N. Zinin, *Ann. Chem.*, **114**, 217 (1860).
486. F. R. Benson and W. L. Savell, *Chem. Rev.*, **46**, 1 (1950).
487. H. Goldstein and R. Stamm, *Helv. Chim. Acta*, **35**, 1470 (1952).
488. R. M. Milburn, *J. Am. Chem. Soc.*, **89**, 54 (1967).
489. E. W. Scott and J. F. Treon, *Ind. Eng. Chem. Anal. Ed.*, **12**, 189 (1940).
490. T. DeVries and R. W. Ivett, *Ind. Eng. Chem. Anal. Ed.*, **13**, 339 (1941).

Nitrones, Nitronates and Nitroxides
Edited by S. Patai and Z. Rappoport

CHAPTER 2

Nitrones and nitronic acid derivatives: their structure and their roles in synthesis

ELI BREUER
Department of Pharmaceutical Chemistry, The School of Pharmacy, The Hebrew University of Jerusalem, Jerusalem, Israel

I. INTRODUCTION

The nitrone function is represented by formula **1** (R = H, alkyl or aryl). If R^1 = OH, the compound is a nitronic acid. The present review is divided into two parts: the first dealing with nitrones, the second with nitronic acids.

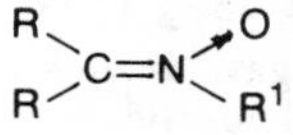

(1)

II. NITRONES

The subject of nitrones has been reviewed several times in the past[1-10]. The most recent general reviews concerning nitrones were published in the mid-sixties[2-4]. Specific aspects of the chemistry of nitrones, namely their rearrangements[5] and their participation in 1,3-dipolar cycloaddition reactions[6] were reviewed later. In view of the considerable expansion in the knowledge and understanding regarding the chemistry of nitrones during the past decade, the preparation of a new and up-to-date review seemed appropriate. The present review will not be concerned with the preparation of nitrones[7-9], and their uses in medicinal chemistry[10]. These subjects have been reviewed recently and adequately.

A. Structure of Nitrones

Nitrones are usually represented by formulae such as **2** or **3**, which imply that there is a positive charge on the nitrogen. However it has to be made clear that this positive charge is delocalized between the nitrogen and the α-carbon as represented by **3** ↔ **4**, resulting in a 1,3-dipolar structure.

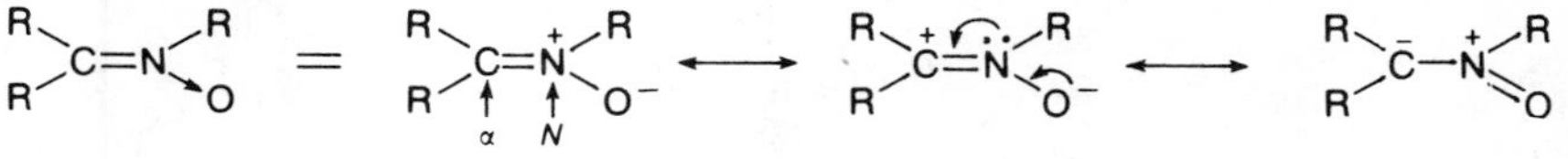

(2) (3) (4) (5)

The extent of this delocalization will naturally be influenced by the substituents in the α-position as well as on the nitrogen, which therefore will also have a marked influence upon the reactivity of the nitrone function. Another electronic structure that has to be considered is **5**. There is evidence that it contributes little in the ground state but certain phenomena can be best explained by invoking it (see below).

It is necessary to make a general remark regarding the various nomenclatures used in nitrones. There are two positions in the parent nitrone molecule: the α-position and the *N*-position (see formula **3**), the α-position is often referred to as the *C*-position. In addition nitrones are often called Schiff base *N*-oxides (e.g. *N*-benzylidene-*N*-methylamine *N*-oxide) or aldehyde *N*-alkyloximes or aldehyde *N*-alkyl nitrones. In the older literature nitrones were often called oxime *N*-alkyl ethers. The names of cyclic nitrones are usually derived from the name of the parent heterocycle.

1. *Theoretical calculations*

Ab initio molecular orbital calculations have been carried out for the parent molecule **6** and its eight isomers, showing that formamide is far more stable than the other isomers[11]. However, the calculated dipole moment of 4.99 D was quite far

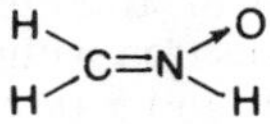

(6)

from the experimental value of 3.37–3.47 D (see Section II.A.2.c). Subsequent CNDO/2 and INDO calculations gave results closer to the experimental values[22]. CNDO/2 was also used to calculate the σ- and π-electron density of α-phenyl *N*-methyl nitrone (**7**)[12].

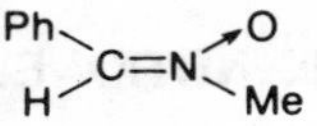

Ph(H)C=N(→O)Me

(7)

2. *Nonspectral physical methods*

a. X-ray studies. The structures of two nitrones have been determined by this method. Lipscomb and coworkers have determined the structures of *C*-*p*-chlorophenyl *N*-methyl nitrone (**8**) and compared its bond lengths to those of the isomeric *O*-methyloximes of *p*-chlorobenzaldehyde[13]. The bond angles of **8** are indicated in the formula. The N—O bond length in **8** was found to be 1.284 Å, considerably shorter than the N—O distance in the isomeric *syn* oxime(1.408 Å),

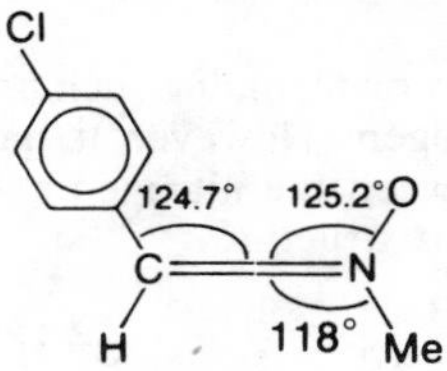

Cl
124.7° 125.2° O
C=N
H 118° Me

(8)

indicating the partial double-bond character of this bond in the nitrone. Moreover the C=N distance of 1.309 Å in the nitrone is longer than the corresponding bond in the *O*-methyloxime (1.260 Å). The data found for α,α,*N*-triphenyl nitrone (**9**) are

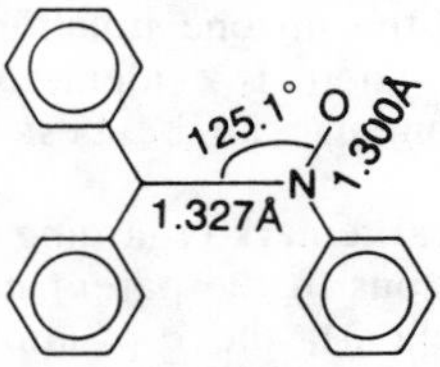

125.1° O
1.300Å
N
1.327Å

(9)

indicated in the formula[14]. These data lend support to the ideas regarding the structure of nitrones that were expressed in Section II.A.

b. Mesomorphism. Some nitrones are found to exhibit mesomorphism. This property can be exploited for the formation of liquid crystals[15–17].

c. Dipole moment and acid–base properties. Dipole moments have been determined for several series of nitrones and found to be 3.37–3.47 D; this confirms the dipolar structure indicated previously[12,18–21].

In a study of a series of *N*-phenyl nitrones having variously substituted aryl or heteroaryl groups in the α-position it was found that the *N*-oxide group is capable of acting either as an electron acceptor or as an electron donor[19], contributing to the effects of electron-donating and electron-withdrawing substituents. This can be illustrated for α-(*p*-nitrophenyl) (**10**) and α-(*p*-anisyl) (**11**) nitrones. Similar results have been found in a series of *N*-methyl nitrones[12].

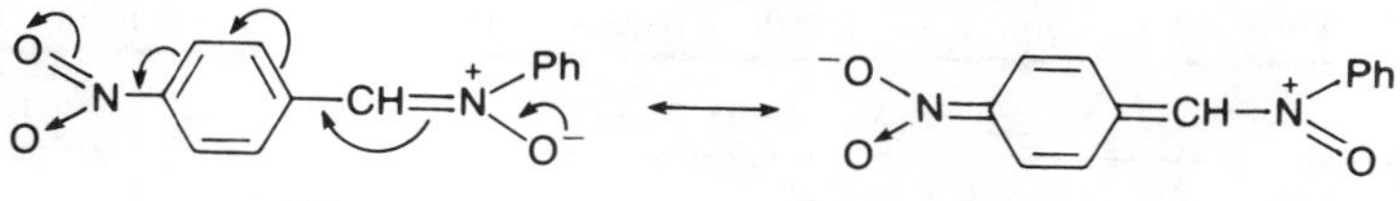

(**10**)

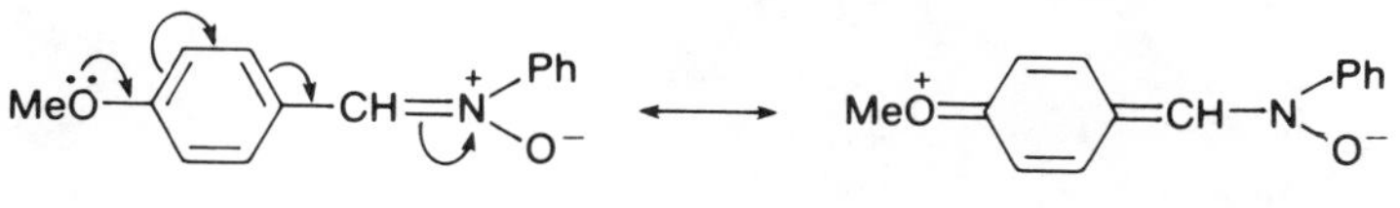

(**11**)

The ionization constants of a series of conjugate acids of substituted phenyl nitrones (e.g. **12**) and other aryl nitrones have been determined by potentiometric

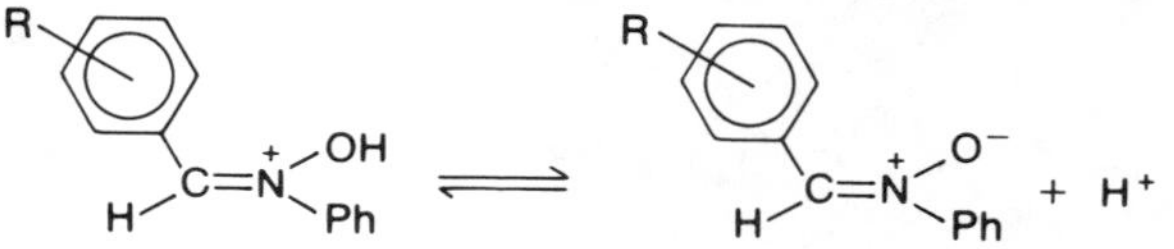

(**12**)

titration[23,24]. The values obtained could best be correlated with the σ^+ values of the substituent R. The basicity of nitrones towards a Lewis acid (BF_3) has also been studied. Complexation with boron trifluoride (equation 1) increased the dipole

$$\mathrm{RCH{=}\overset{+}{N}(R){-}O^- + BF_3 \rightleftharpoons RCH{=}\overset{+}{N}(R){-}O{-}\overset{-}{B}F_3} \qquad (1)$$

moment by approximately 5 D[25]. Calculations indicated that approximately 0.43 e was transferred from the oxygen to the boron.

The rate constants of the exchange of α-hydrogens in α,*N*-diaryl nitrones and the α and methyl hydrogens of α-aryl *N*-methyl nitrones with deuterium have also been correlated with the substituent constants of the groups in the aryl rings[26]. The electron-withdrawing effect of the *N*-oxide group has been found to have strong influence on the sensitivity of the methine hydrogen exchange to substituent effects[26].

The capability of nitrones to serve as electron donors is also manifested by their tendency to serve as ligands in nickel complexes[27,28] (see Section II.B.9.e).

3. Spectra of nitrones

a. Photoelectron spectra. The photoelectron spectra of a number of nitrones have been measured and various ionization potentials have been assigned[29–32]. The correlation of these with the mode of 1,3-cycloadditions of nitrones has also been discussed[30].

b. Ultraviolet spectra. A large number of papers report on the ultraviolet spectra of nitrones. Most of these are reported in the previous reviews[2,3]. More recent reports[33–37] deal with special aspects such as the influence of steric effects[33–35,37] and of solvent effects[34–36] upon the ultraviolet spectra of nitrones.

Spectral data of various types of nitrones are summarized in Table 1.

c. Infrared spectra. This subject has also been discussed in the previous reviews[2,3] and since then no papers have been especially devoted to this subject. It

TABLE 1. Ultraviolet spectral data of various types of nitrones in alcohol solvents

Type	λ_{max} (nm)	$\varepsilon_{max} \times 10^{-3}$	References
Aryl(H)C=N⁺(O⁻)Aryl	227–240 247–280 310–372	7–17 6–12 8–20	2, 36
Aryl(H)C=N⁺(O⁻)Alkyl	205 221 290	8 7 14–17	2
Aryl[a](H)C=N⁺(O⁻)Alkyl[a]	211 228 304	6 12 16	2
Alkyl(H)C=N⁺(O⁻)Alkyl	244	6	35
Alkyl[a](H)C=N⁺(O⁻)Alkyl[a]	230–235	8–9	2

[a]The two groups are connected to form a ring.

is generally accepted now that nitrones exhibit two characteristic bands resulting from N → O and C=N bond stretching vibrations. The N → O band appears in aromatic ketonitrones in the region of 1200–1300 cm^{-1}, while in aldonitrones it is seen between 1050–1170 cm^{-1}. The C=N stretching vibration appears in aromatic nitrones at 1550–1600 cm^{-1} and in aliphatic and alicyclic nitrones at 1570–1620 cm^{-1} [38,39]. In the related 4-substituted 3-imidazoline-3-oxides frequencies up to 1340 and 1630 cm^{-1} have been observed[40].

d. Nuclear magnetic resonance spectra. Only proton chemical shifts are available. Some typical chemical shifts of protons adjacent to the nitrone function are listed in Table 2.

In addition to these general data there are some additional features of interest in the NMR spectra of nitrones. The spectra of α-aryl aldonitrones and α,α-diaryl nitrones show signals corresponding to two hydrogens at a field lower than the rest of the aromatic hydrogens[35,36,43,44]. These low field signals have been assigned to the two *ortho* hydrogens of the phenyl group situated *syn* to the nitrone oxygen (see **13**) and apparently arise from its deshielding influence. The magnitude of this deshielding influence has been found comparable to that of the nitro group[35,36]. In

(13)

contrast, the deshielding effect of the nitrone group upon the *meta* protons of aryl groups attached to it is very weak[35,36]. From a study of a series of substituted

TABLE 2. Approximate chemical shifts of protons in the vicinity of the nitrone function[35,36,38,41,42]

Position	Chemical shift, δ (ppm)
Alkyl $CH{=}\overset{+}{N}(O^-)R$	6.4–6.7
Aryl $CH{=}\overset{+}{N}(O^-)R$	7–8
$CH_3{-}CH{=}\overset{+}{N}(O^-)R$	2
$RCH{=}\overset{+}{N}(O^-)CH_3$	3.4–4
$RCH{=}\overset{+}{N}(O^-)CH_2Aryl$	5

α,*N*-diaryl nitrones it appears that there are nonbonded interactions between the *ortho* substituent of the α-phenyl group and the *N*-oxygen, as there are between the *ortho* substituent of the *N*-phenyl group and the α-hydrogen. As a consequence of this the stable conformation of 2,2′-dimethyl-α-*N*-diphenyl nitrone would be as indicated by formula **14**, and the hydrogen at position 6 appears at the unusually low field of 9.5 ppm due to the deshielding effect of the oxygen[36]. In

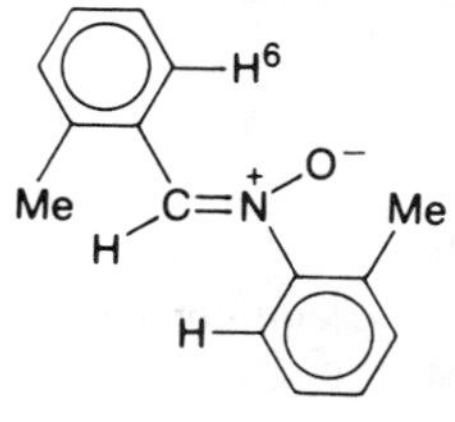

(**14**)

2,6-unsubstituted α-phenyl nitrones the phenyl group rotates freely and the two *ortho* hydrogens become magnetically equivalent. With more hindered 2,6-disubstituted aryl groups in the α-position of a nitrone it has been possible to obtain and isolate stable (*E*)-aldonitrones in addition to the predominant (*Z*) isomers (e.g. **15**)[45] (see Section II.A.4). The NMR spectra show the *N*-methyl of (*E*)-**15** at higher field (3.4 ppm) than that of the (*Z*) isomer (3.90 ppm), while the opposite trend is observed for the vinyl proton [(*Z*)-**15**: 7.6 ppm; (*E*)-**15**: 7.9 ppm][45].

In cyclic nitrones of various ring-size it has been found that there is a homoallylic coupling between two groups that lie in a transoid fashion across the nitrone function. The magnitude of the coupling constant varies somewhat with the ring-size (**16–18**)[46].

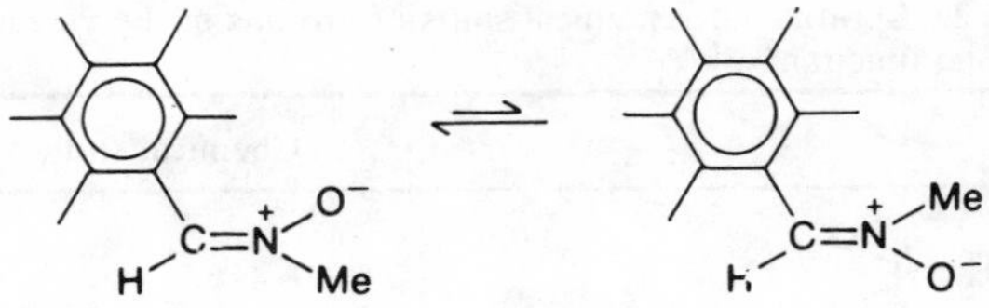

(Z-**15**) (E-**15**)

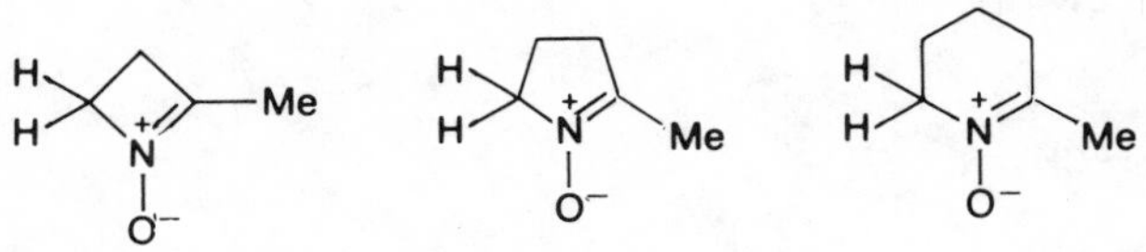

J_{ab} = 1.95 Hz (**16**) J_{ab} = 1.84 Hz (**17**) J_{ab} = 1.50 Hz (**18**)

NMR spectroscopy has also become a convenient tool to study geometrical and other isomerizations of nitrones since the spectra of the pairs of compounds involved are often found sufficiently different. For example the interconversion between the (*Z*) and (*E*) isomers of **19** has been followed by examining the

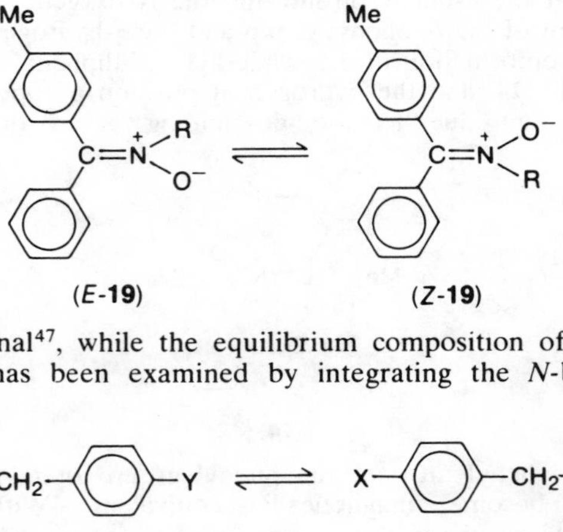

(*E*-**19**) (*Z*-**19**)

aromatic methyl signal[47], while the equilibrium composition of a series of α-aryl *N*-benzyl nitrones has been examined by integrating the *N*-benzyl CH_2 signal (equation 2)[39].

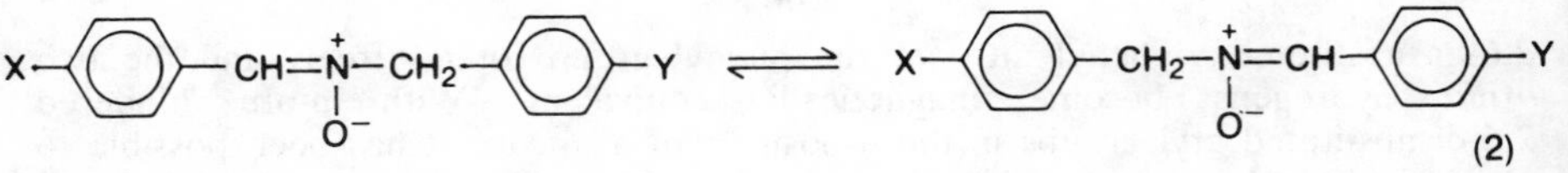

$$X-C_6H_4-CH=\overset{+}{N}(O^-)-CH_2-C_6H_4-Y \rightleftharpoons X-C_6H_4-CH_2-\overset{+}{N}(O^-)=CH-C_6H_4-Y \quad (2)$$

Specialized techniques such as the aromatic solvent-induced shift (ASIS) and shift reagents have also been used in connection with NMR spectra measurement of nitrones. The stereochemistry of the two geometric isomers of **20** has been assigned on the basis of ASIS[48]. The use of shift reagents has proved valuable in assigning the structures of the geometrically isomeric steroidal nitrones **21**[49,50], as well as the (*E*) configuration to α-phenyl-α,*N*-dimethyl nitrone (**22**)[51].

e. Mass spectra. The mass spectra of variously substituted and deuterated α,*N*-diaryl nitrones have been studied by several groups[52–54]. In all spectra appears

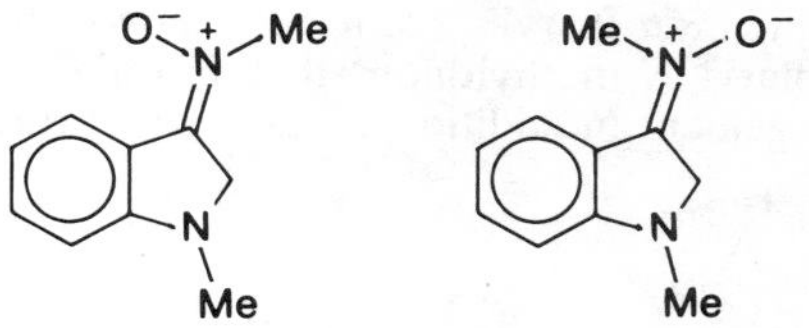

(*E*-**20**) (*Z*-**20**)

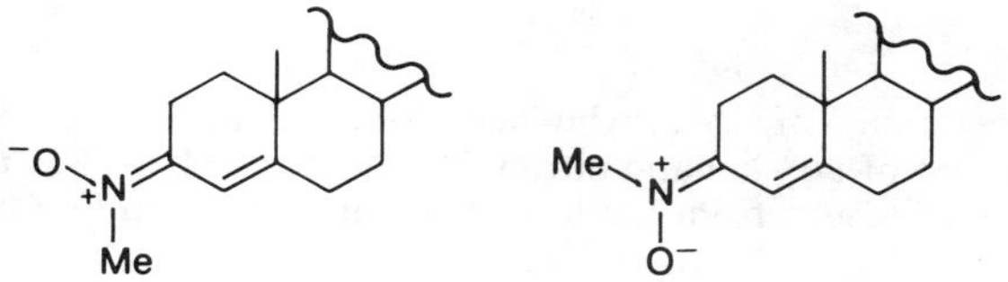

(*E*-**21**) (*Z*-**21**)

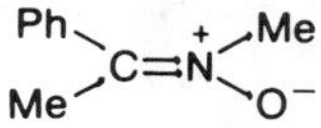

(*E*-**22**)

the M − 16 peak, which results from the loss of oxygen. This is a peak of some diagnostic value for *N*-oxides[55]. A summary of the mass spectral processes which α,*N*-diaryl nitrones undergo is presented in Scheme 1. The presence of the benzoyl cation requires oxygen migration from nitrogen to carbon and presumably the intermediacy of an oxaziridine, which may rearrange to an amide, which in turn may provide the fragment $PhC{\equiv}O^+$. In addition to this in all spectra the biphenylene radical cation and the 2-substituted benzisoxazolium cation are produced.

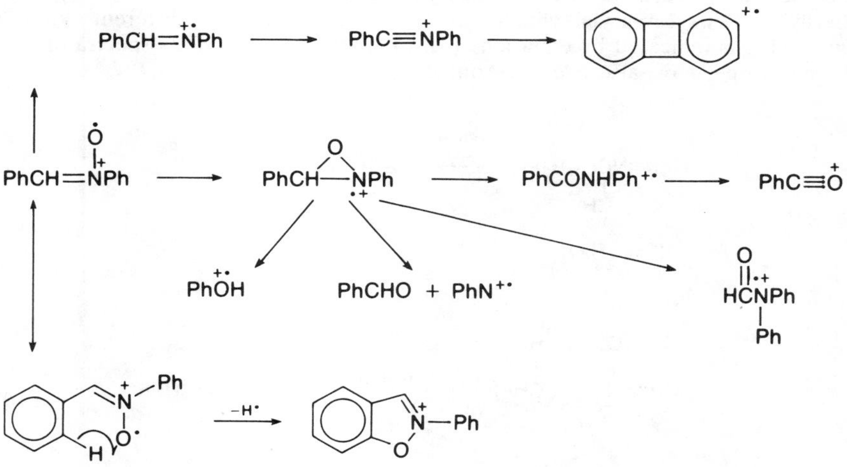

PhCH=N̈Ph (+•) → PhC≡N̈Ph (+) → biphenylene (+•)

PhCH=NPh (Ȯ, +) → PhCH—NPh (oxaziridine, •+) → PhCONHPh+• → PhC≡O (+)

PhOH (+•) PhCHO + PhN+• HCNPh (O, •+) Ph

−H•

SCHEME 1

N-methyl α-aryl[56] and α,α-diaryl[57] nitrones show analogous fragmentation patterns, as does α,α-diaryl *N*-methylthiomethyl nitrone **(23)** which undergoes a McLafferty-type rearrangement in addition to the normal fragmentation pattern[57].

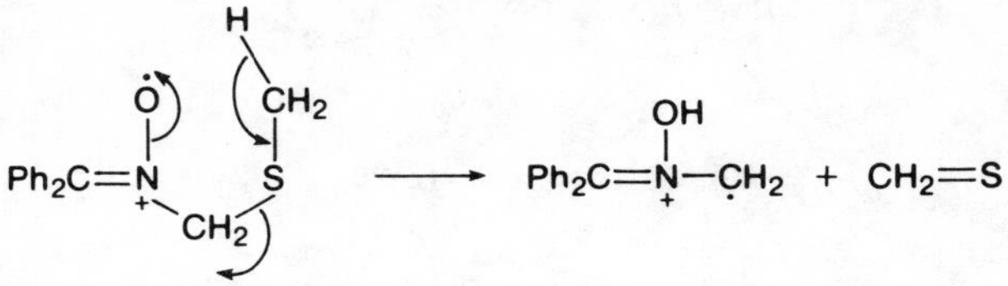

$$Ph_2C{=}\overset{+}{N}(\dot{O})CH_2SCH_2H \longrightarrow Ph_2C{=}\overset{+}{N}(OH){-}\dot{C}H_2 + CH_2{=}S$$

(23)

The mass spectrum of *N*-ethylidenecyclohexylamine *N*-oxide **(24)** is a representative of that of an aliphatic nitrone[58]. The peak M − 16 is not seen in this case. A peak M − 15 results from the loss of the methyl group by α-cleavage.

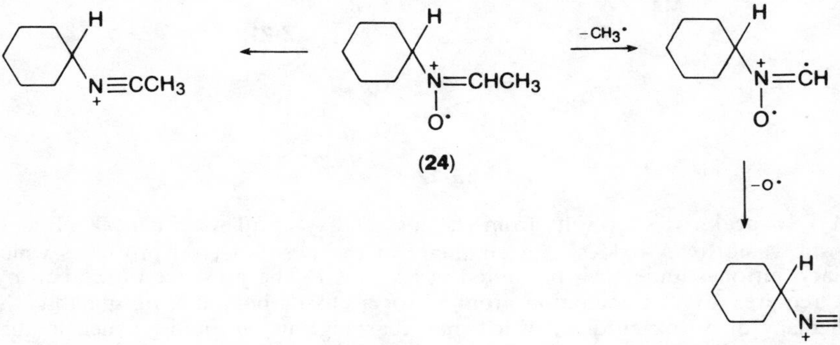

$$C_6H_{11}\overset{+}{N}{\equiv}CCH_3 \longleftarrow C_6H_{11}\overset{+}{N}(\dot{O}){=}CHCH_3 \xrightarrow{-CH_3^{\bullet}} C_6H_{11}\overset{+}{N}(\dot{O}){=}\dot{C}H \xrightarrow{-O^{\bullet}} C_6H_{11}\overset{+}{N}{\equiv}CH$$

(24)

The mass spectrum of the isomeric oxaziridine shows a similar pattern so interconversion with the nitrone cannot be ruled out. The mass spectra of *N*-alkylidene *N*-(1-cyanoalkyl) *N*-oxides have also been examined, showing similar patterns[58]. *N*-alkyl-*N*-arylalkylidene nitrones **(25)** undergo different kind of fragmentations indicated by equations (3), (4) and (5)[59]. The mass spectra of some α-aminonitrones have also been determined[60].

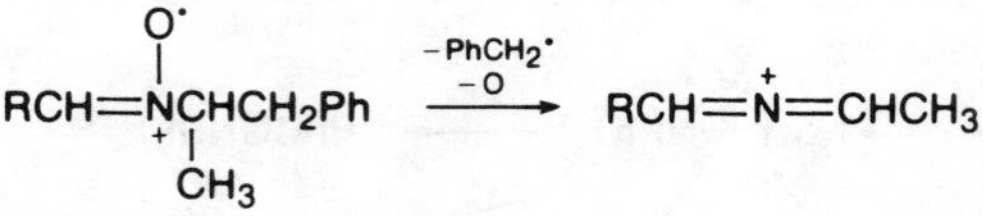

$$RCH{=}\overset{+}{N}(\dot{O})CH(CH_3)CH_2Ph \xrightarrow[-O]{-PhCH_2^{\bullet}} RCH{=}\overset{+}{N}{=}CHCH_3 \qquad (3)$$

(25) R≠H

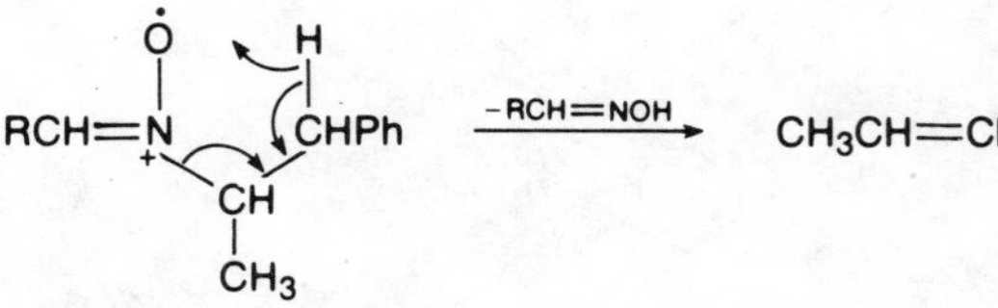

$$RCH{=}\overset{+}{N}(\dot{O})CH(CH_3)CH(H)Ph \xrightarrow{-RCH{=}NOH} CH_3CH{=}CHPh^{+\bullet} \qquad (4)$$

(25) R ≠ H

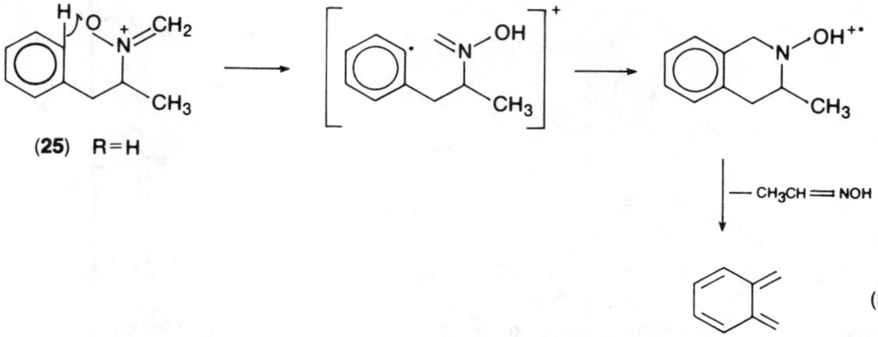

(**25**) R=H

$\xrightarrow{-\,CH_3CH{=}NOH}$ (5)

Alicyclic nitrones are somewhat different. 2,4,4-Trimethylpyrroline *N*-oxide gives a base peak of m/e = 41 which was assigned as $CH_3\overset{+\bullet}{C}N$[61], (equation 6), although

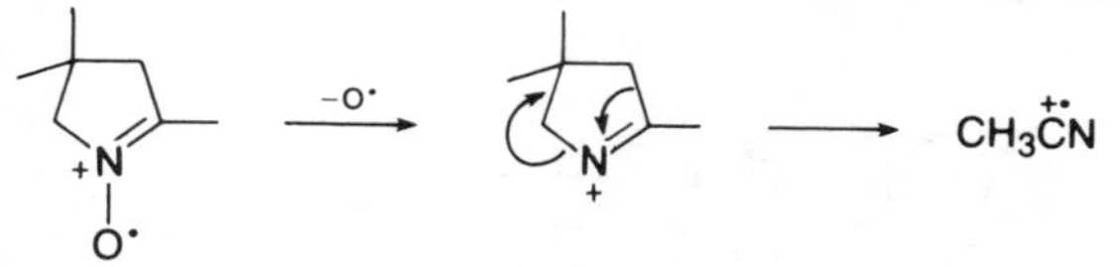

$CH_3\overset{+\bullet}{C}N$ (6)

the fragment m/e = 41 was observed in pyrroline *N*-oxides that have no methyl substituent at the 2-position, and therefore was assigned by others as $C_3H_5^+$ [55]. The isomeric 2,5,5-trimethylpyrroline *N*-oxide **(26)** shows a base peak believed to correspond to $[NOH]^{\bullet +}$, along with other peaks e.g. m/e = 83.

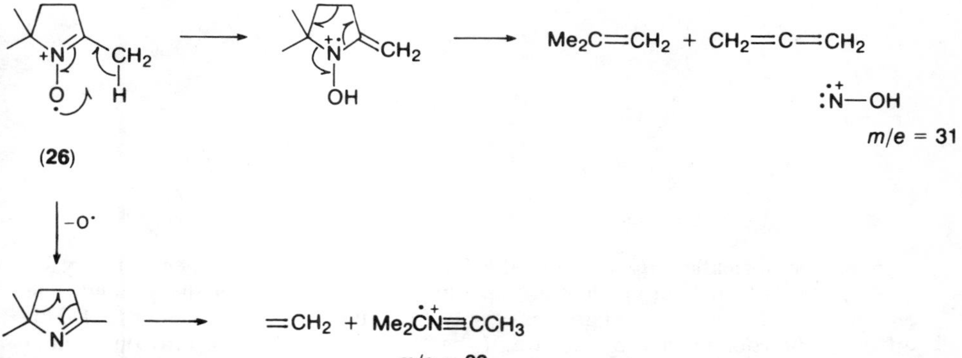

$Me_2C{=}CH_2 + CH_2{=}C{=}CH_2$

$:\overset{\bullet +}{N}{-}OH$

m/e = 31

(**26**)

$=CH_2 + Me_2\overset{\bullet +}{C}N{\equiv}CCH_3$

m/e = **83**

The negative-ion mass spectra of some α,*N*-diaryl nitrones have also been measured[62]. These spectra are found to be much simpler than the positive-ion spectra and the fragmentations are indicated in Scheme 2.

4. *Geometrical isomerism of nitrones*

The phenomenon of geometrical isomerism in nitrones has long been recognized and in many instances pairs (*E*) and (*Z*) isomers have been separated and configurations assigned[2,3,49,50]. However it is only recently that such pairs of isomers have been successfully isolated in acyclic aldonitrones, which usually exist entirely as the (*Z*) isomers **(27)**. It has been found that in derivatives of **27** with highly

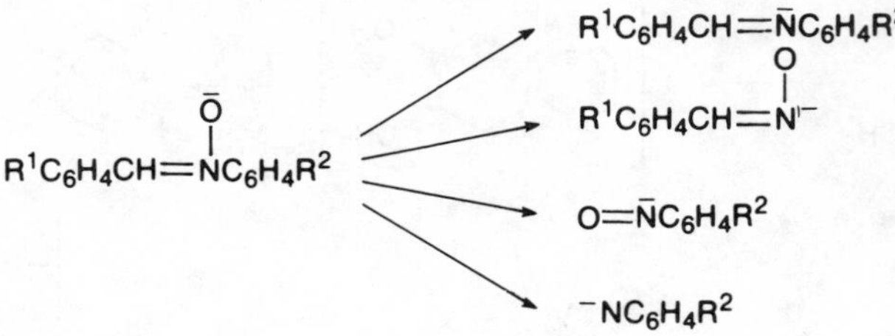

$$R^1C_6H_4CH{=}\overset{\bar{O}}{\overset{|}{N}}C_6H_4R^2 \longrightarrow \begin{cases} R^1C_6H_4CH{=}\bar{N}C_6H_4R^2 \\ R^1C_6H_4CH{=}\overset{O}{\overset{|}{N}}{}^{-} \\ O{=}\bar{N}C_6H_4R^2 \\ {}^{-}NC_6H_4R^2 \end{cases}$$

SCHEME 2

substituted aryl groups it is possible to isolate the stable (E) isomers (**28**)[45]. Apparently in these cases the increased steric requirements of the α-aryl group cause it to twist out of its normal coplanar conformation (**27**) and assume an orthogonal orientation (**29**).

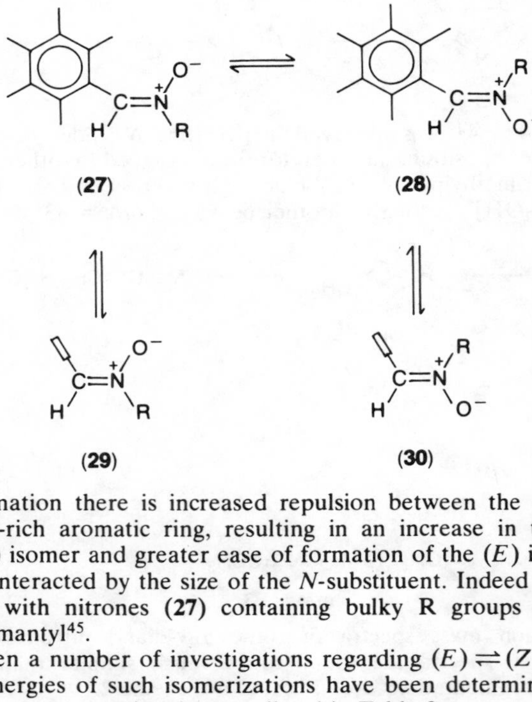

(**27**) ⇌ (**28**)

(**29**) (**30**)

In this conformation there is increased repulsion between the negative oxygen and the electron-rich aromatic ring, resulting in an increase in the ground-state energy of the (Z) isomer and greater ease of formation of the (E) isomer (**28**). This effect can be counteracted by the size of the N-substituent. Indeed only one isomer (Z) is observed with nitrones (**27**) containing bulky R groups such as t-butyl, neopentyl or adamantyl[45].

There have been a number of investigations regarding (E) ⇌ (Z) isomerizations. The activation energies of such isomerizations have been determined in a number of cases and some representative data are listed in Table 3.

The influence of the N-substituent upon the rate of isomerization has been studied in the α,α-diphenyl nitrone series (**31**)[63]. The approximately twofold rate of isomerization of **31b** as compared to **31a** is considered to be due to increased ground-state energies of the former caused by the stronger nonbounded interactions between the benzyl and the α-phenyl groups. However the fact that the benzhydryl derivative **31c** isomerizes 10 times faster than **31a** is taken as an indication that a new mechanism is operating in this case. It has been suggested that the

TABLE 3. Energy of activation of $(E) \rightleftharpoons (Z)$ isomerization of representative nitrones

Nitrone	E_A (kcal/mol)	References
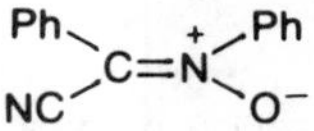 $Ph(NC)C{=}N^+(Ph)O^-$	24.6	64
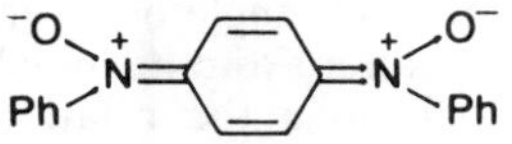 $^-O(Ph)N^+{=}C_6H_4{=}N^+(Ph)O^-$	12	65
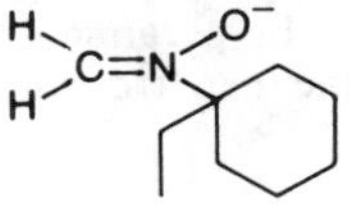 $H_2C{=}N^+(O^-)$-(1-ethylcyclohexyl)	23.2	66
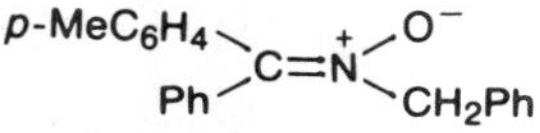 $p\text{-}MeC_6H_4(Ph)C{=}N^+(O^-)CH_2Ph$	33.6	63
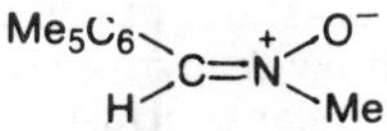 $Me_5C_6(H)C{=}N^+(O^-)Me$	33.1	45, 67
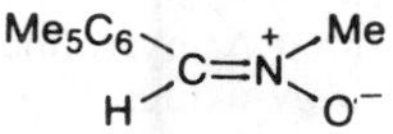 $Me_5C_6(H)C{=}N^+(Me)O^-$	34.6	45, 67
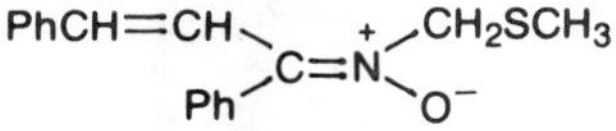 $PhCH{=}CH(Ph)C{=}N^+(CH_2SCH_3)O^-$	28.2	68

isomerization of *N*-benzhydryl nitrones proceeds, at least in part, via dissociation to iminoxy radicals followed by fast isomerization of the radical and recombination.

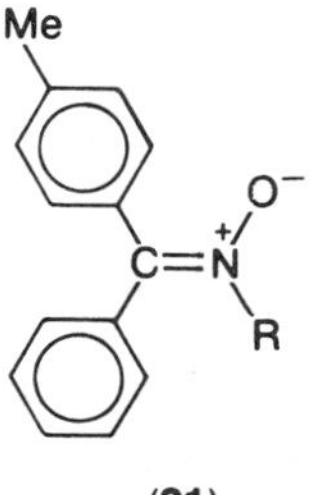

(a) R = Me
(b) R = CH_2Ph
(c) R = $CHPh_2$

(31)

The question of the mechanism of $(E) \rightleftharpoons (Z)$ isomerization has been approached theoretically. CNDO/2 and INDO calculations on the parent nitrone, **6**, indicate a high charge density on the oxygen and are consistent with a major contribution of structure **6a** in the ground state[22]. The twisted conformation, which is proposed to be the transition state for the rotation around the C=N bond, appears, however, to be well represented by **6b** since there is a high electron density on the carbon. However the calculated rotational barriers of 60 and 80 kcal/mol by these methods

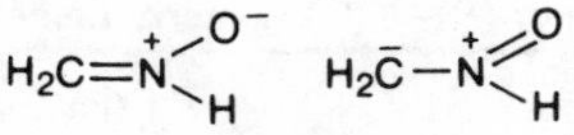

(**6a**) (**6b**)

are much higher than those observed experimentally[22]. Results from a more recent MINDO/3 study are consistent with a concerted rotation–pyramidalization pathway for the $(E) \rightleftharpoons (Z)$ isomerization[69]. According to this as the dihedral angle between the CH and the NH bonds increases the CH_2 undergoes deformation from its initial planar trigonal geometry and the carbon becomes pyramidal along the rotation coordinate. The driving force for this presumably arises from the negative charge density at the carbon (ϕυ) and the known preference of carbanions to be pyramidal. The MINDO/3 rotational barrier of 40.2 kcal/mol is closer to the experimental values (Table 3) than those obtained in the previous calculations.

5. *Tautomerism of nitrones*

a. Nitrone–hydroxyenamine tautomerism. This type of tautomerism is analogous to the keto–enol tautomerism of carbonyl compounds. The question of such tautomerism has been the subject of a number of papers. *N*-Hydroxy-Δ^2-pyrrolines (**33**) are not detectable by NMR in Δ^1-pyrroline *N*-oxides (**32**)[70]. Similar conclusions have been reached on the basis of deuterium exchange studies showing that only the hydrogens of the 2-methyl group undergo exchange in substituted Δ^1-pyrroline *N*-oxide derivatives (e.g. **34**)[61]. However in the case of indolenine *N*-oxides (e.g. **35**) the *N*-hydroxyindole tautomers (**36**) do exist, and their proportion can be increased using polar solvents[71,72].

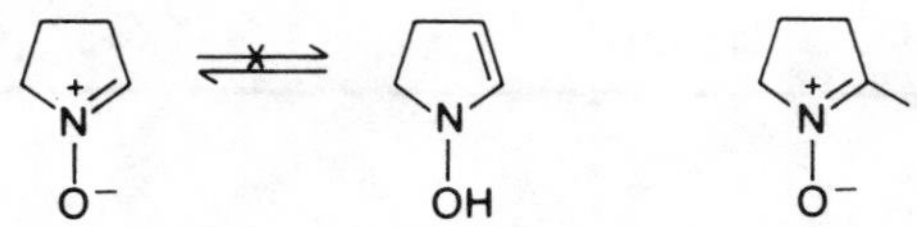

(**32**) (**33**) (**34**)

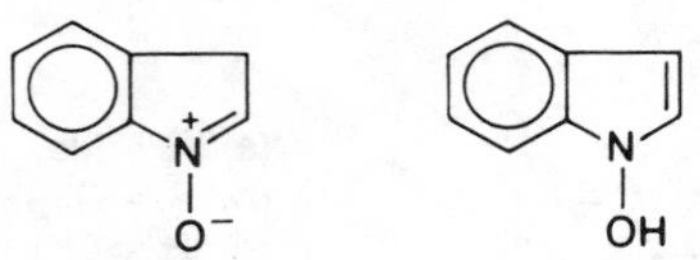

(**35**) (**36**)

This type of tautomerization is assumed to be involved also in the base-catalysed reaction of the flavanononitrone, which leads to the Schiff base as one of the products (equation 7)[73]. Interestingly the nitrone–hydroxyenamine equilibrium is found to be influenced by the heteroatom in the chromane system. Thus reaction of

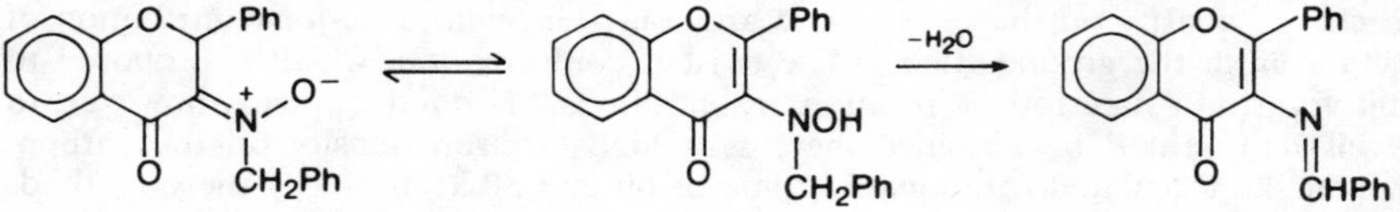

(7)

the oxime **37** (X = O) with diazomethane leads to nitrone **38**, whereas the thio analogue **37** (X = S) with the same reagent affords the hydroxylaminothiochromane **39**[74].

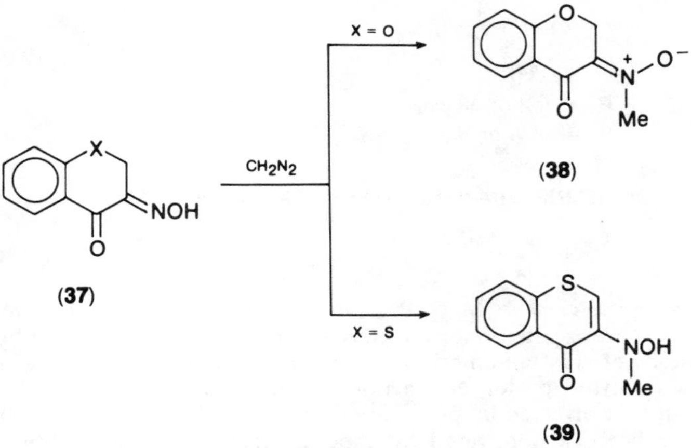

X = O
CH_2N_2
X = S

(**37**) (**38**) (**39**)

A keto group appropriately situated relative to the nitrone function will enhance tautomerism similarly to β-dicarbonyl compounds. Indeed the infrared spectrum of phenacylpyrroline *N*-oxide (**40**) shows bonds attributable to an intramolecular hydrogen bond which could result from either or both structures **40a** or **40b**[75].

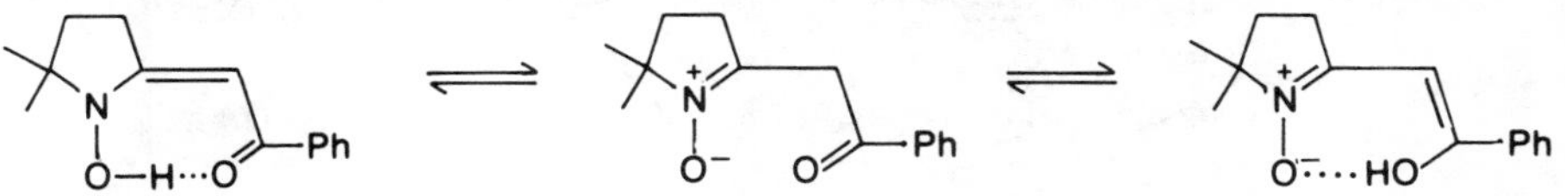

(**40b**) (**40**) (**40a**)

A band at 1700 cm^{-1} in the infrared spectrum of the cyclic hydroxamic acid **41** was interpreted as indicating the presence of the α-hydroxynitrone tautomer **41a**[76]. The

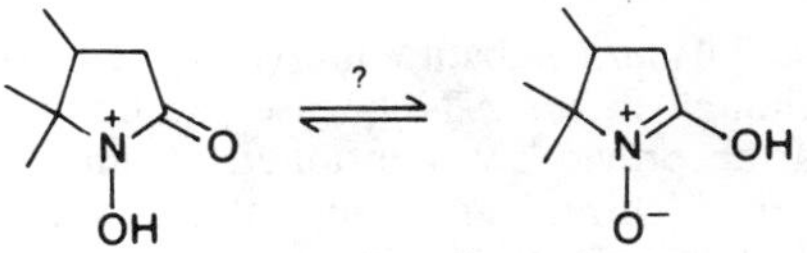

(**41**) (**41a**)

related question of α-aminonitrone/*N*-hydroxy-*N,N'*-disubstituted amidine tautomerism also should be considered in this section. A number of cases belonging to different categories have been examined by spectral methods and have been assigned the α-aminonitrone structure (see **42**, **44** and **45**) although acylation reactions of **42** afford acylation products derived from **43**[77].

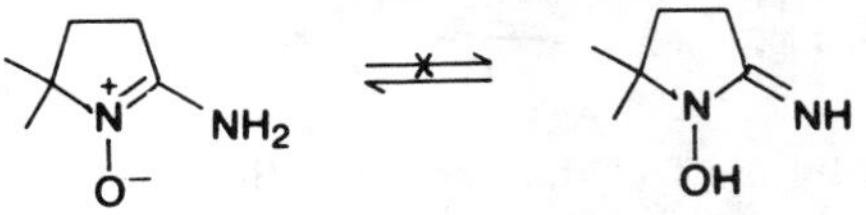

(Ref. 77)

(**42**) (**43**)

$$R^1NH—\overset{R^2}{C}=\overset{+}{N}R^1(O^-) \not\rightleftharpoons R^1N=\overset{R^2}{C}—NR^1(OH) \qquad \text{(Ref. 78)}$$

(**44**)

R^1 = aryl or alkyl
R^2 = alkyl or H

$$ArNHCH=\overset{+}{N}Ar(O^-) \not\rightarrow ArN=CH—NAr(OH) \qquad \text{(Ref. 79)}$$

(**45**)

b. Behrend rearrangement. Since this type of tautomerism was first reported by Behrend in 1891[80] it is often referred to by this name. It is represented by equation (8). The history of this tautomerism has been reviewed[2,3,5]. The reaction clearly involves base-catalysed proton abstraction leading to a delocalized carbanion (e.g. **47**), which can be converted by protonation to either of the two nitrones. However this has never been demonstrated by experimental means such as by deuterium incorporation. The influence of substituents in the aromatic ring upon the equilibrium in a series of α-aryl *N*-benzyl nitrones (**46**) has been studied by NMR[39].

$$R—C_6H_4—CH_2\overset{+}{N}(O^-)=CH—C_6H_5 \rightleftharpoons R—C_6H_4—CH=\overset{+}{N}(O^-)CH_2—C_6H_5 \qquad (8)$$

(**46a**) (**46b**)

$$R—C_6H_4—CH\cdots\overset{-}{N}(O)\cdots CH—C_6H_5$$

(**47**)

It has been found that all *para* substituents (NO_2, Cl, Me, OMe) favour nitrone type **46b** over **46a** although in the *p*-NO_2 case the equilibrium constant is not far from 1. This reasserts the previously mentioned assumption (Section II.A.2.c) that the nitrone function is electronically amphoteric and can interact both with electron-releasing and electron-withdrawing groups (see equations 9 and 10). This

$$CH_3O—C_6H_4—CH=\overset{+}{N}(O^-)CH_2Ph \longleftrightarrow CH_3\overset{+}{O}=C_6H_4=CH—N(O^-)CH_2Ph \qquad (9)$$

$$O_2N—C_6H_4—CH=\overset{+}{N}(O^-)CH_2Ph \longleftrightarrow (O)(^-O)N=C_6H_4=CH—\overset{+}{N}(=O)CH_2Ph \qquad (10)$$

conclusion is supported by the fact that *meta* substituents disfavour the substituted benzylidene isomers. In the *ortho*-substituted nitrones the substituted benzylidene is also disfavoured. This latter result can be rationalized in terms of steric interference of the substituent with conjugation. The question of the equilibrium in α,α-diphenyl *N*-benzyl nitrone (**48**), which was one of Behrend's first examples has also been reexamined[39]. It has been found that **49** predominates approximately by a factor of 2 over **48** presumably due to steric hindrance in the latter.

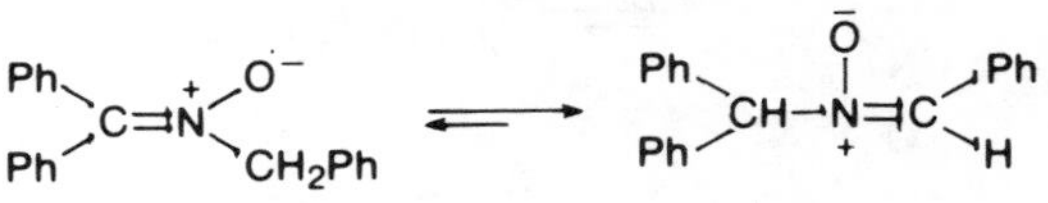

(**48**) (**49**)

In certain instances Behrend rearrangement is indicated by the structure of the end-product obtained. Thus the base-catalysed conversion of the flavanononitrone **50** to the oxazole **51** can be rationalized by assuming Behrend rearrangement followed

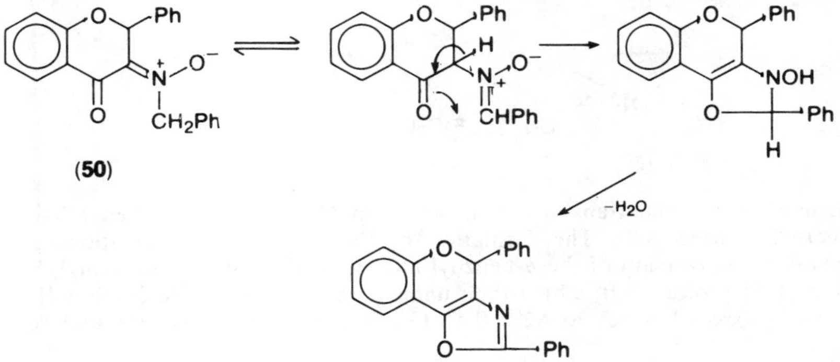

(**50**)

$-H_2O$

(**51**)

by cyclization and loss of a water molecule[73]. The base-catalysed reaction of α-phenyl *N*-methyl nitrone (**52**) with triethyl phosphonoacetate leads predominantly to 1-benzylaziridine-2-carboxylate (**55**) and less to its isomer **54** indicating the involvement of the unstable but reactive *N*-benzyl nitrone (**53**) in the reaction[81].

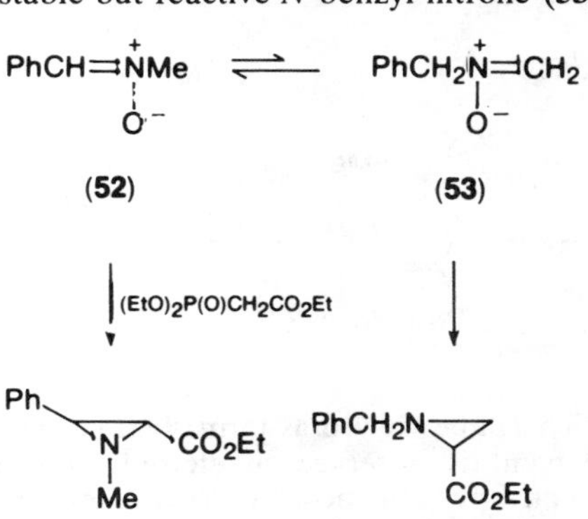

$PhCH{=}\overset{+}{N}Me(O^-)$ ⇌ $PhCH_2\overset{+}{N}{=}CH_2(O^-)$

(**52**) (**53**)

$(EtO)_2P(O)CH_2CO_2Et$

(**54**) (**55**)

Similarly the reaction of 2,4,4-trimethylpyrroline *N*-oxide (**56**) with triethyl phosphonoacetate gives only product **58** resulting from the less stable but more reactive aldonitrone (**57**), whereas the more reactive diethyl cyanomethylphosphonate is capable of reacting with the ketonitrone **56** to give product **59**[82].

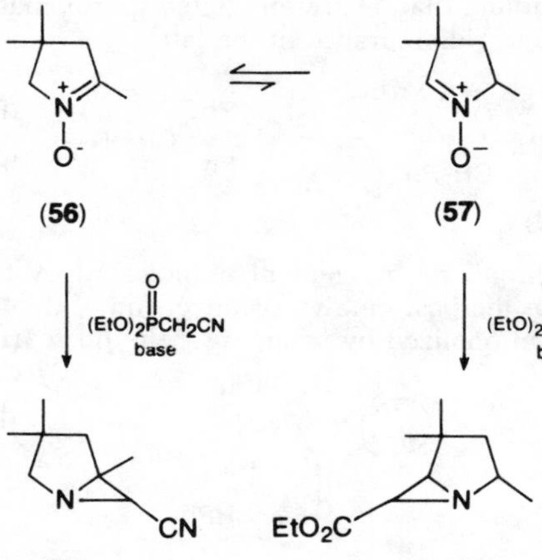

(**56**) (**57**)

(**59**) (**58**)

An unusual thermal transformation occurs in (*E*)-α-benzoyl-α-phenyl *N*-methylthiomethyl nitrone (**60**). The formation of the oxazole (**62**) is rationalized by assuming the conversion of the α-benzoyl nitrone to the *N*-(2-hydroxyvinyl) nitrone **61** by a [1,5] proton shift, which then undergoes cyclization (see Section II.A.5.c) followed by loss of water to **62**[68]. The (*Z*) isomer of 60 does not undergo this reaction[68].

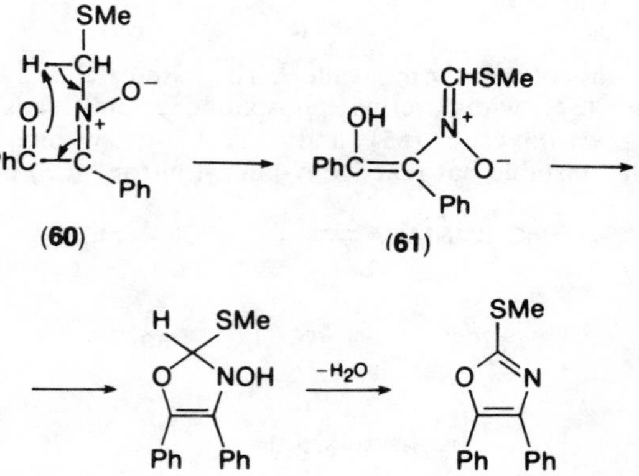

(**60**) (**61**)

(**62**)

An isomerization which can be viewed as formally if not mechanistically related to the Behrend rearrangement is observed in steroidal nitrones. It is found that hydroxynitrone **63** is converted by acid to its isomer **65**, presumably through intermediate **64**[83]. Both **63** and **65** furnish the same methoxynitrone **66** upon treatment with methanolic acid[83].

c. Ring–chain tautomerism. Early examples of this type of tautomerism have been reviewed[2]. More recently the preparations of a large series of α-aryl *N*-(2-hydroxyalkyl) nitrones (**67**)[84,85] and one example of α-aryl *N*-(3-hydroxyalkyl)

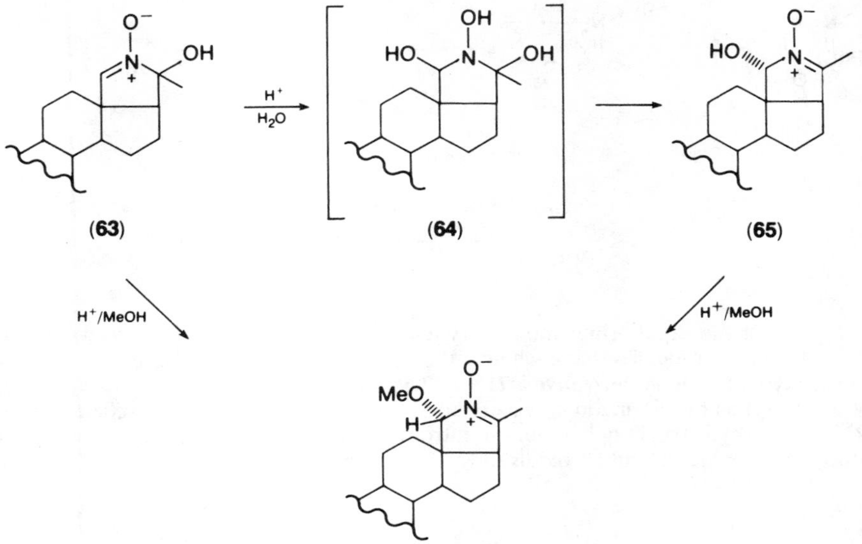

(**63**) (**64**) (**65**)

(**66**)

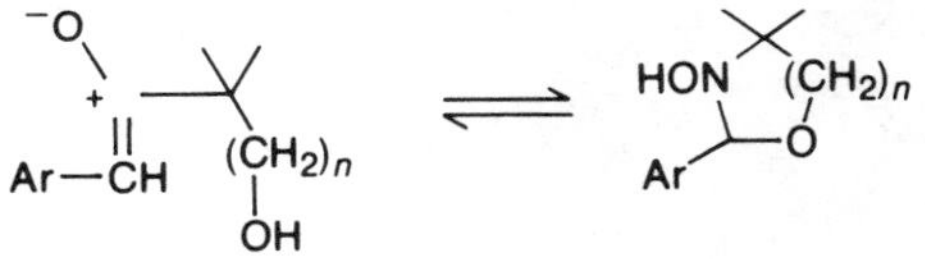

(**67**) $n = 1$
(**68**) $n = 2$
(**69**) $n = 3$

(**68**) and of α-aryl *N*-(4-hydroxyalkyl) (**69**) nitrone[85] have been reported. There is no spectral evidence found to indicate the presence of any cyclic tautomer in any of these cases. Acetylation of **67** gives, however, mixtures of the open-chain and cyclic acetates[84].

Oxidation of an *N*-(2-hydroxyalkyl) nitrone of type **67** with mercuric oxide leads to an oxazoline *N*-oxide indicating involvement of the cyclic tautomer in the reaction (equation 11)[84]. The *gem*-dimethyl nitrone **67** is oxidized by silver ion to the cyclic

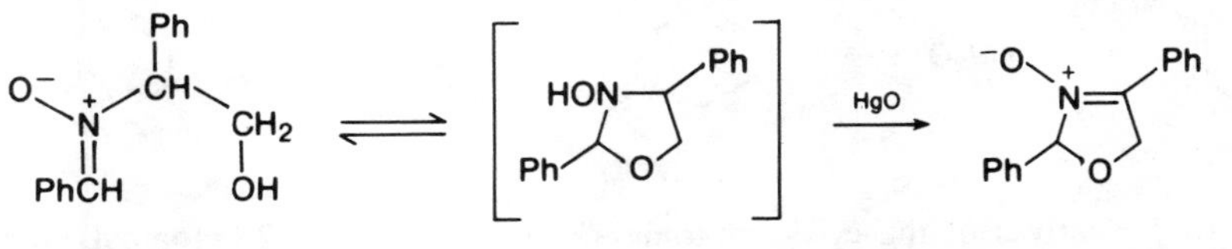

(11)

nitroxide radical **70**[85]. It has been suggested that in this case the cyclization is facilitated by the silver ion acting as a Lewis acid[85].

In contrast to the α-aryl nitrones mentioned, examination of a series of aliphatic aldonitrones by spectroscopic methods has revealed the presence of the cyclic

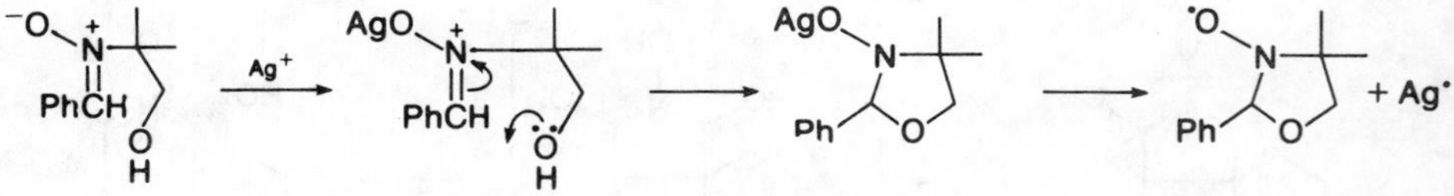

(67) (70)

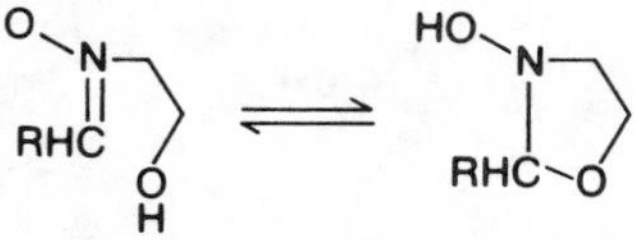

(12)

tautomers in the equilibrium mixture (equation 12). Acylation gives derivatives of the cyclic tautomer[86]. Ring–chain tautomerism has also been studied in 2-hydroxynaphthalene derivatives, **71**[87,88]. The equilibrium composition is influenced by R as well as by the medium. Acetylation gives derivatives of the cyclic tautomer **72**[89]. The *o*-hydroxyphenylaminoacyl nitrone **73** could be converted to the cyclic tautomer **74**, which could be oxidized to the new nitrone **75**[90,91]. Acetylation of both

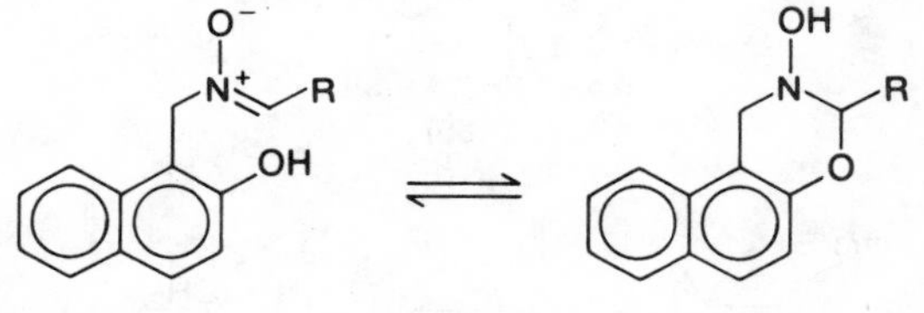

(71) (72)

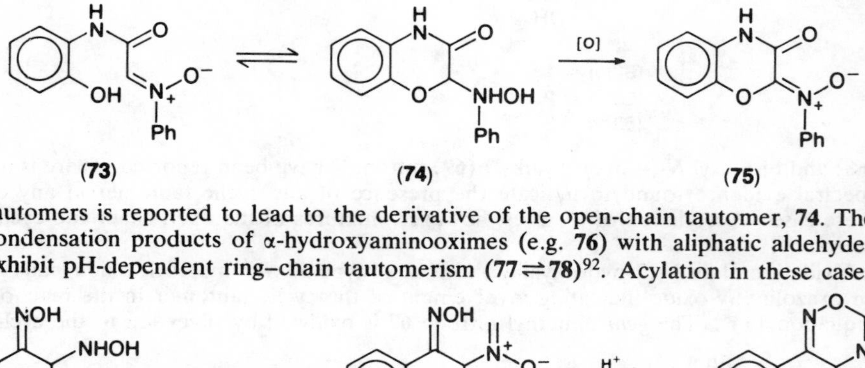

(73) (74) (75)

tautomers is reported to lead to the derivative of the open-chain tautomer, **74**. The condensation products of α-hydroxyaminooximes (e.g. **76**) with aliphatic aldehydes exhibit pH-dependent ring–chain tautomerism (**77** ⇌ **78**)[92]. Acylation in these cases

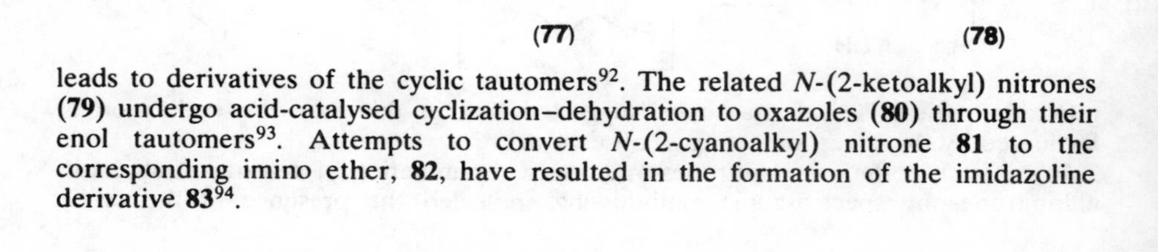

(77) (78)

leads to derivatives of the cyclic tautomers[92]. The related *N*-(2-ketoalkyl) nitrones **(79)** undergo acid-catalysed cyclization–dehydration to oxazoles **(80)** through their enol tautomers[93]. Attempts to convert *N*-(2-cyanoalkyl) nitrone **81** to the corresponding imino ether, **82**, have resulted in the formation of the imidazoline derivative **83**[94].

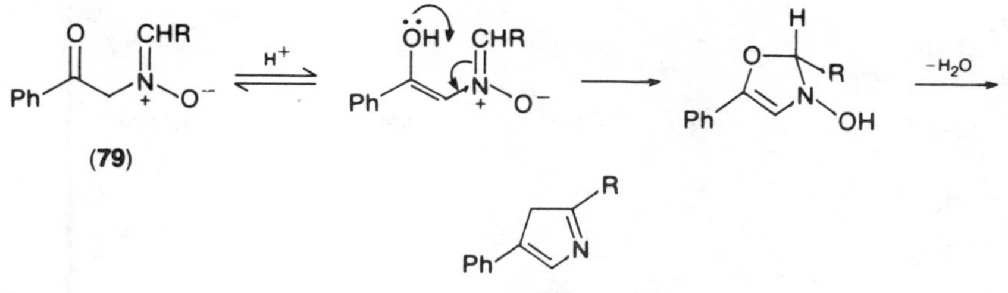

(79)

(80)

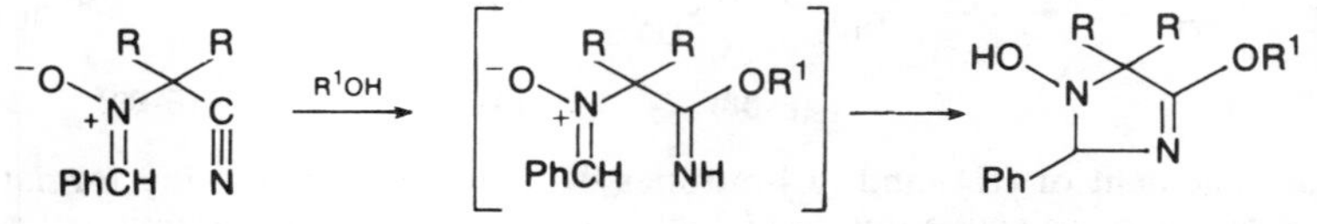

(81) (82) (83)

B. Reactions of Nitrones

1. Rearrangements of nitrones

Nitrone rearrangements have been reviewed extensively by Lamchen in 1968[5]; also some photochemical processes of nitrones can be found in a more general review concerning *N*-oxides[95]. Consequently this chapter will not be concerned with the historical background.

a. Formation of oxime O-*ethers – the Martynoff rearrangement.* This rearrangement, which is represented by the thermal conversion of α,α-diphenyl *N*-benzhydryl nitrone (**84**) to benzophenone oxime *O*-benzhydryl ether (**85**) is far

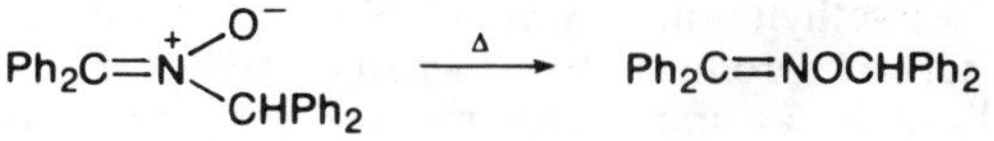

$$Ph_2C{=}\overset{+}{N}(O^-)CHPh_2 \xrightarrow{\Delta} Ph_2C{=}NOCHPh_2$$

(**84**) (**85**)

from being a general reaction. For example *N*-alkyl, *N*-aryl or *N*-benzyl nitrones do not undergo this reaction. In spite of its limited synthetic importance the mechanism of this reaction has been thoroughly studied by a variety of experimental techniques. The initially proposed concerted mechanism[96] involving a three-membered cyclic transition state (**86**) has been disproven. The mechanism which is supported by all

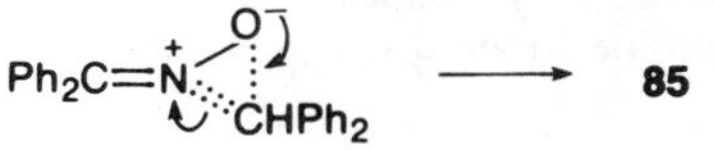

85

(**86**)

experimental results involves dissociation to iminoxy radicals (e.g. **88**) and benzhydryl radicals (**89**) which then recombine to form the oxime benzhydryl ether. ESR spectra indicate the presence of iminoxy radicals[97], while the presence of benzhydryl radicals has been shown by chemically induced nuclear polarization[98] and also be the isolation of tetraphenylethylene. By careful study of the steric course of

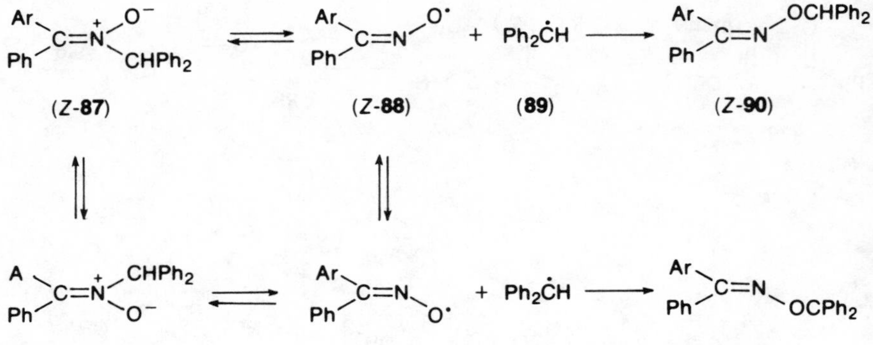

(*Z*-**87**) ⇌ (*Z*-**88**) + $Ph_2\dot{C}H$ (**89**) → (*Z*-**90**)

(*E*-**87**) ⇌ (*E*-**88**) + $Ph_2\dot{C}H$ (**89**) → (*E*-**90**)

the rearrangement of (*E*)- and (*Z*)-nitrones **87**, it was found that the formation of ethers is not stereoselective due to (*E*) ⇌ (*Z*) isomerization of the iminoxy radicals which is faster by several orders of magnitude than the isomerization of nitrones and of oximate anions under the reaction condition[99]. Other studies include a kinetic investigation of the effect of substituents in the aromatic ring. The lack of significant substituent effect upon the rate of the rearrangement argues against an ionic mechanism[100]. This rearrangement has also been studied using mixtures of α,α-diphenyl *N*-benzhydryl nitrone (**84**) and its tetradeuterio derivative **91**. On the

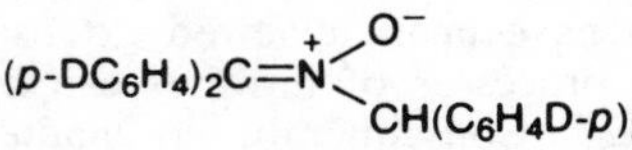

$(p\text{-}DC_6H_4)_2C{=}\overset{+}{N}(O^-)CH(C_6H_4D\text{-}p)_2$

(**91**)

basis of the crossover products that were observed, the extent of the intermolecularity of the reaction has been determined[101]. It was found that the intermolecular process is solvent-dependent. The thermal rearrangement of *N*-(diphenylmethylene)methylthiomethylamine *N*-oxide (**92**) yields, beside the oxime *O*-alkyl ether **93**, several products (see Scheme 3)[102]. The formation of these products is rationalized by assuming that several competing free-radical reactions take place simultaneously. Some free radicals are observed by ESR.

N-Methylthiomethyl nitrone **92** also undergoes rearrangement with acid catalysis to the same products. These results are rationalized by mechanisms that involve protonation of the nitrone oxygen followed by reaction with a nucleophile present[103]. The related methoxymethyl nitrone **94** rapidly isomerizes under the influence of acid catalysis to the corresponding oxime ether, which, however, decomposes slowly under the reaction conditions[104]. The mechanism suggested for this reaction consists of a protonation step, followed by dissociation to the oxime and methoxymethyl carbonium ion which recombine to yield the protonated oxime ether (**95**).

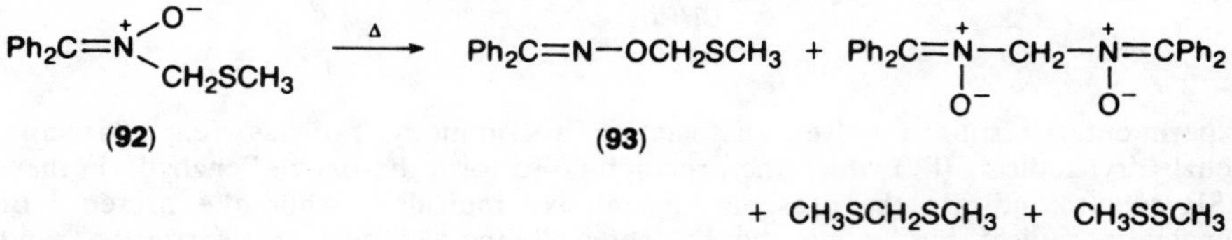

$Ph_2C{=}\overset{+}{N}(O^-)CH_2SCH_3$ (**92**) $\xrightarrow{\Delta}$ $Ph_2C{=}N{-}OCH_2SCH_3$ (**93**) + $Ph_2C{=}\overset{+}{N}(O^-){-}CH_2{-}\overset{+}{N}(O^-){=}CPh_2$ + $CH_3SCH_2SCH_3$ + CH_3SSCH_3

SCHEME 3

$$Ph_2C{=}\overset{+}{N}(O^-)CH_2OCH_3 + H^+ \rightleftharpoons Ph_2C{=}\overset{+}{N}(OH)CH_2OCH_3 \longrightarrow$$

(**94**)

$$Ph_2C{=}N{-}OH + \overset{+}{C}H_2OCH_3 \longrightarrow Ph_2C{=}N{-}\overset{+}{O}(H)CH_2OCH_3$$

(**95**)

The reverse reaction, namely the rearrangement of oxime *O*-ethers to nitrones has also been observed in the case of *O*-allyl ethers[105–107]. This reaction has been considered initially as a [2,3] sigmatropic rearrangement (**96–97**)[105]. Subsequently it

(**96**) (**97**)

has been suggested on the basis of ESR spectra[106] that, at least in part, it proceeds by a radical dissociation–recombination mechanism.

b. Formation of oxaziridines. This rearrangement is represented by equation (13). The stability of the three-membered ring products is varied, some are quite stable and others are formed as transient intermediates which undergo transformation to other products, mainly to one or both of the two possible amides (equation 14). Oxaziridines can sometimes undergo thermal rearrangement back to nitrones[2].

$$RCH{=}\overset{+}{N}(O^-){-}R \underset{\Delta}{\overset{h\nu}{\rightleftharpoons}} RCH\overset{O}{—}NR \qquad (13)$$

$$HCONR_2 \xleftarrow{\Delta} RCH\overset{O}{—}NR \xrightarrow{\Delta} RCONHR \qquad (14)$$

The earlier studies regarding this rearrangement can be found in the reviews[2–5]. Recent additions to the existing reports on stable oxaziridines are the *N*-monosubstituted compounds **99** that can be formed by irradiation of methylene nitrones, **98**[108]. These oxaziridines can be further rearranged to the corresponding formamides.

$$CH_2{=}\overset{+}{N}(O^-)R \xrightarrow{h\nu} CH_2\overset{O}{—}NR$$

(**98**) (**99**)

R = adamantyl, *t*-butyl

Studying the mechanism and stereochemistry of this rearrangement, Boyd and coworkers reported that irradiation of α-(*p*-nitrophenyl) *N*-alkyl nitrones (e.g. **100**) at room temperature gives rise to mixtures of *cis*- and *trans*-oxaziridines, with the latter predominating, but the composition of which depends upon the solvent[109]. These findings were disputed by Splitter and coworkers, who observed at −60°C the exclusive formation of *trans*-oxaziridines (**101**)[110]. They also found that under the conditions of partial irradiation a photostationary state consisting of the two

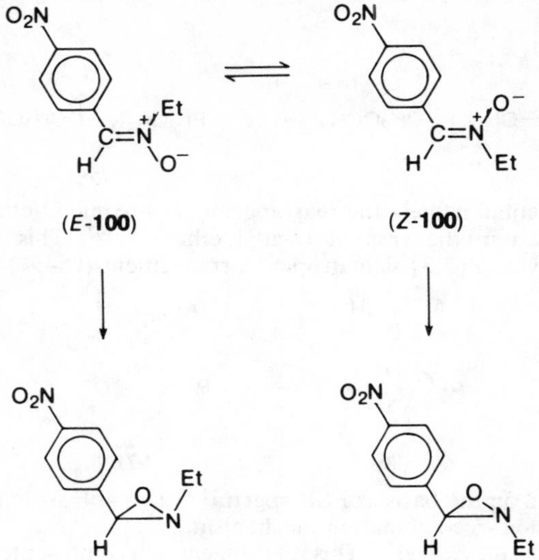

(*E*-**100**) (*Z*-**100**)

(*cis*-**101**) (*trans*-**101**)

geometrical isomers of nitrone **100** in the ratio 37% *E*/63% *Z* is attained. Comparing the ratio of products calculated from the quantum yields of the reaction of each nitrone isomer with the the experimental results it was found that the reaction of conversion of nitrones to oxaziridines is stereospecific[110].

In attempts to achieve asymmetric syntheses of oxaziridines nitrones were irradiated in a chiral solvent[111] as well as in an asymmetric liquid crystalline cholesteric medium[112]. While in the asymmetric solvent oxaziridines were obtained in optical yields of about 30%, the products from the experiments in cholesteric media showed negligible optical rotation.

The effect of intramolecular hydrogen bonding on the nitrone photoisomerization has also been studied[113]. The effect of different solvents upon the quantum yield of oxaziridine formation from *N*-phenyl salicylaldonitrone (**102**) with intramolecular

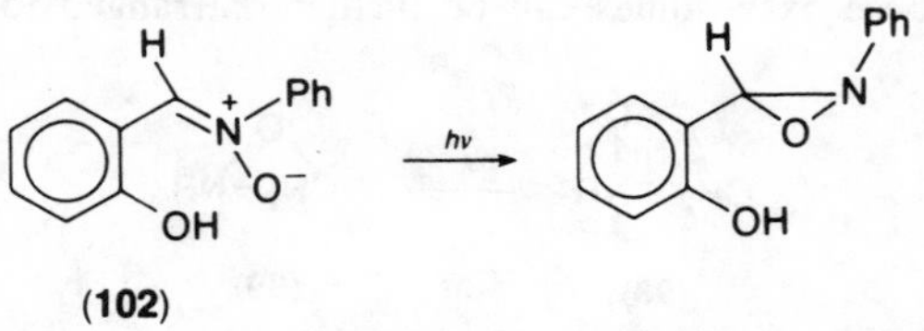

(**102**)

hydrogen bonds have been correlated. The quantum yields in the reaction of **102** increase with decreasing temperature. In contrast the quantum yield in the reaction of α,*N*-diphenyl nitrone is not temperature-dependent[113].

The photolysis of some interesting sulphur-containing nitrones has been shown to lead to methyleneoxaziridine radicals and via these to benzophenone and to sulphur–nitrogen heterocycles[114,115] (e.g. Scheme 4).

Oxaziridines derived from cyclic nitrones such as pyrroline *N*-oxides often undergo

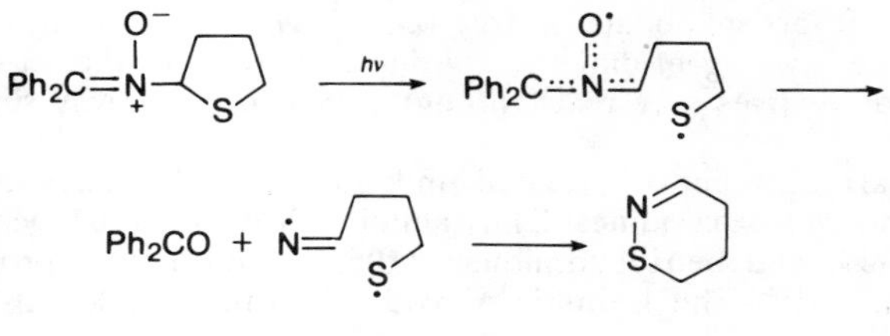

SCHEME 4

ring-contraction reactions that lead to 1-acylazetidines[5] (see equation 15). Some new examples of this reaction have been reported[116,117a,b]. Similarly to acyclic

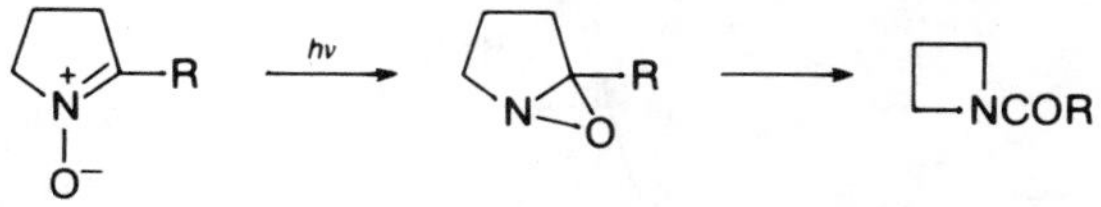

(15)

ketonitrones that upon irradiation yield imides[118,119] as a consequence of acyl migration in the intermediate oxaziridines, the irradiation of 3-oxo-Δ^1-pyrroline *N*-oxides yield β-lactams accompanied by 1,3-oxazine-6-one derivatives[120] (see Scheme 5). When the five-membered ring is condensed to a benzene ring only the ring-expansion reaction leading to benzoxazines is observed as has been demonstrated in several cases (equation 16)[121–123]. The first example of this series

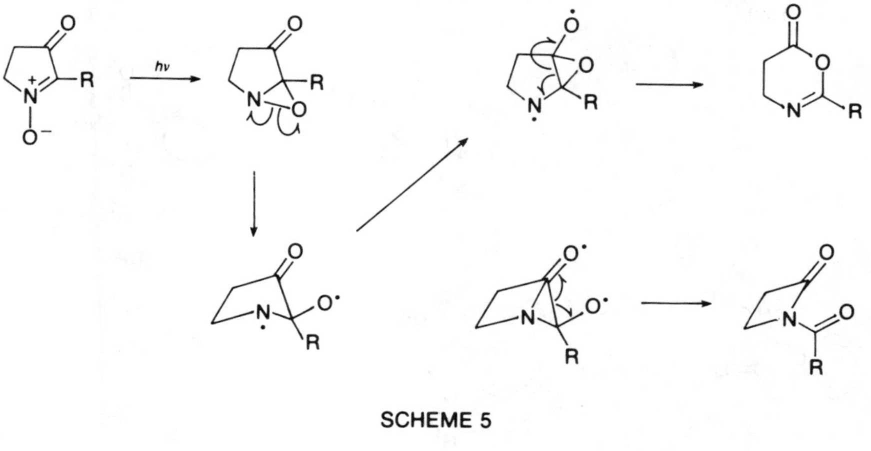

SCHEME 5

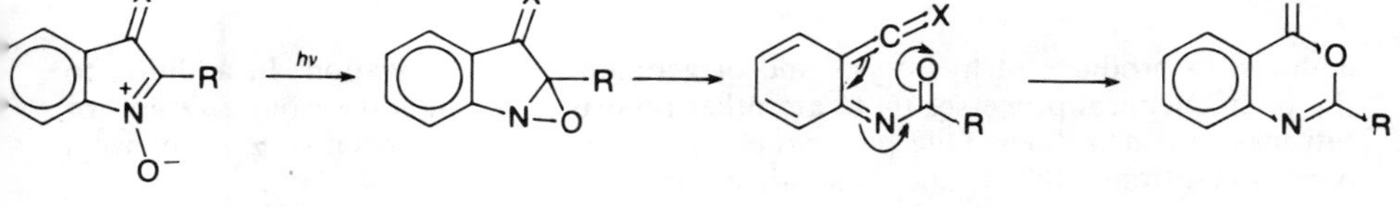

X = Me_2 (Ref. 121)
X = O (Ref. 122)
X = NPh (Ref. 123)

(16)

($X = Me_2$) cannot accommodate a free-radical-type mechanism because of the presence of the two *gem*-dimethyl groups. However the 3-keto and the 3-phenylimino derivatives may react through a biradical pathway similar to that in Scheme 5.

The photorearrangement of related indolinylidene *N*-oxides also results in ring-enlargements via oxaziridines. Thus amidine *N*-oxides (**103**) yield mixtures of benzoxazines (**104**) and benzopyrimidines (**105**) in addition to products resulting from deoxygenation[124]. The isomeric *N*-oxide derivatives (**106**) also yield **105** in

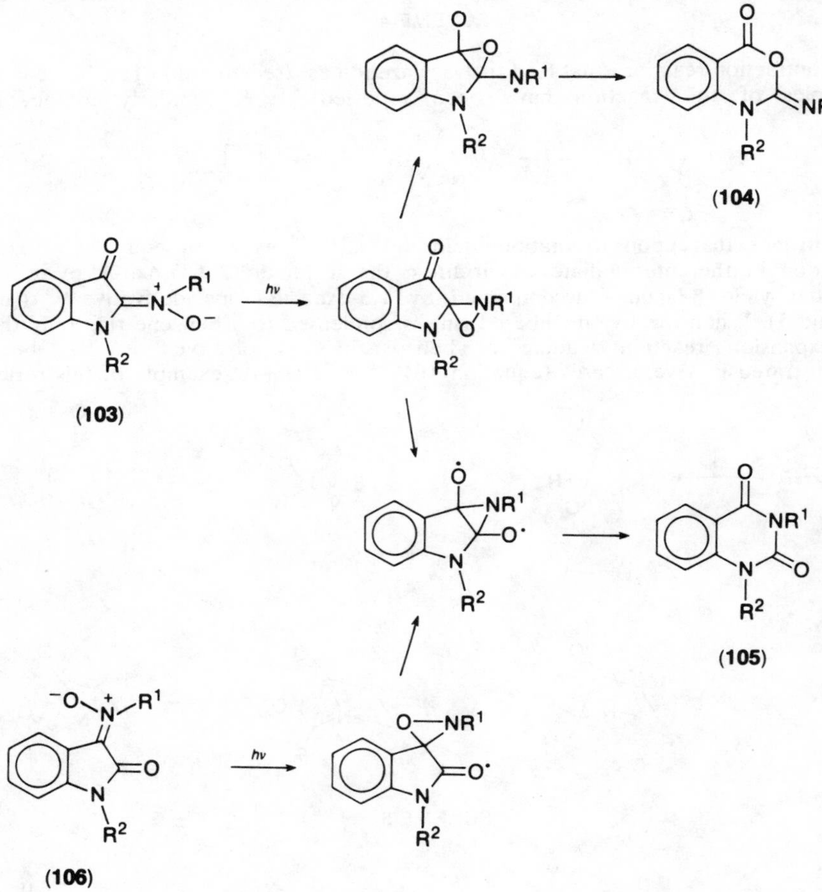

(**104**)

(**103**)

(**105**)

(**106**)

addition to products of hydrolysis and of geometrical isomerization. In addition to the photochemical processes there are other reports concerning the isomerizations of nitrones to oxaziridines. One paper reports the thermal conversion of α,α-dibenzoyl *N*-phenyl nitrone (**107**) to the oxaziridine **108**[125].

In another paper[126] it has been reported that the steroidal nitrone **109** is converted to the isomeric oxaziridine **110** by *p*-toluenesulphonyl chloride followed by base.

c. Formation of amides. The rearrangement of nitrones to amides has already been mentioned in the previous section, as amides are often the end-products of the

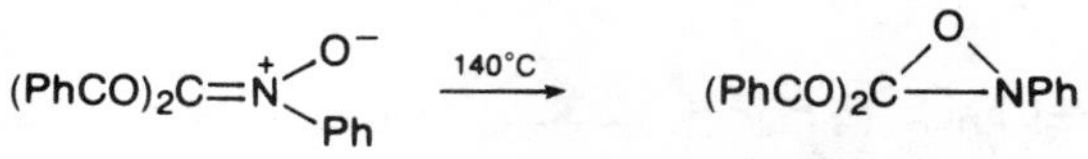

(107) (108)

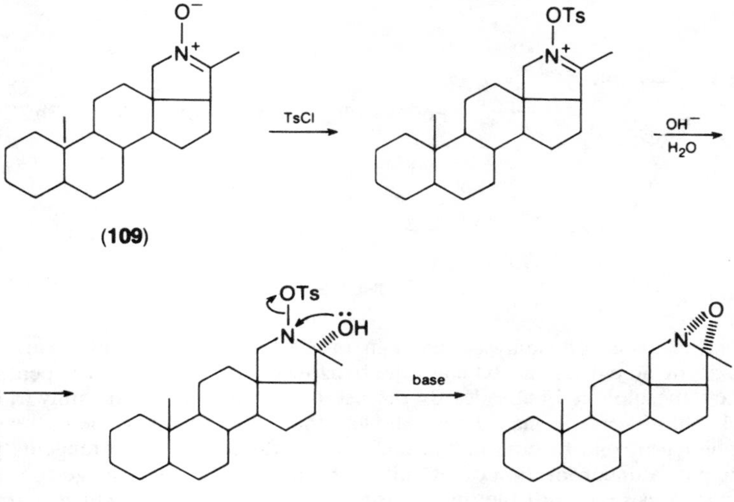

(109)

(110)

nitrone–oxaziridine rearrangement[25]. However in addition to this mode of amide formation there are several ways to effect the nitrone–amide rearrangement under conditions that clearly do not involve oxaziridines as intermediates. These include reactions which proceed under the catalytic effect of sodium alkoxides as well as acid anhydrides and acyl halides, catalysts normally used in the Beckmann rearrangement. Most of these examples can be found in the reviews[2–5,127]. Therefore it will suffice here to limit the discussion to the mechanism and to some recent new examples.

It seems reasonable to represent the mechanism of the base-catalysed rearrangement of nitrones to amides by the sequence of additions and eliminations depicted in Scheme 6.

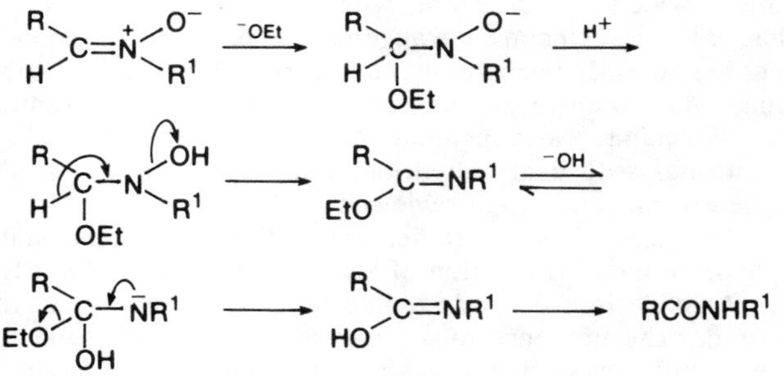

SCHEME 6

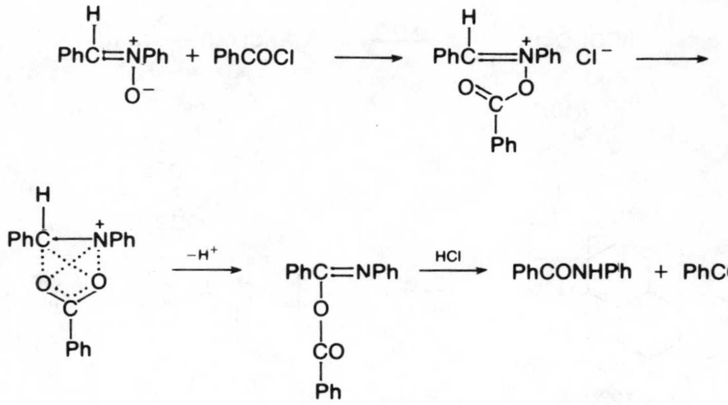

$$PhCH{=}\overset{+}{N}(O^-)Ph + PhCOCl \longrightarrow PhCH{=}\overset{+}{N}Ph(OCOPh)\ Cl^- \longrightarrow$$

$$[\text{cyclic transition state}] \xrightarrow{-H^+} PhC(OCOPh){=}NPh \xrightarrow{HCl} PhCONHPh + PhCOCl$$

SCHEME 7

The other types of catalysts that can effect the nitrone–amide rearrangement include acetic anhydride, acetyl chloride, benzoyl chloride, phosphorus pentachloride and *p*-toluenesulphonyl chloride. By the use of ^{18}O-labelled acetic anhydride[128] and benzoyl chloride[129] it has been shown that the rearrangement reaction of α,*N*-diphenyl nitrone to benzanilide under the influence of these reagents proceeds with the partial incorporation of ^{18}O into the amide. It has been suggested that these results are consistent with the initial formation of acyloxyimine which rearranges in part by a cyclic and in part by a 'sliding' mechanism as depicted in Scheme 7. A simple scrambling which would result from the dissociation–recombination of an acyloxyimino compound is ruled out on the basis of the ^{18}O incorporation data[129]. There is no incorporation of ^{18}O into the amide when *p*-toluenesulphonyl chloride-^{18}O or phosphorus oxychloride are used for the rearrangement. This is taken as an indication that in these cases the reaction proceeds entirely by the 'sliding' mechanism[129].

This rearrangement, which was initially only known for aldonitrones was more recently extended to ketonitrones and represents an analogy to the Beckmann rearrangement. In contrast to the latter, however, its outcome does not depend on the stereochemistry of the starting material since (*Z*)- and (*E*)-nitrones (**21**) were found to interconvert rapidly under the reaction conditions and they both give the same lactame **111**[50].

A mechanism which is consistent with the experimental results and which resembles that of the Beckmann rearrangement is presented in equation (17)[50]. This rearrangement has recently been applied to a series of bicyclo[4.2.0]octanones (**112**), which rearrange via the corresponding nitrones **113** to perhydroindolone, **114**. The conventional Beckmann rearrangement leads to the isoindole skeleton[130]. Other reactions of nitrones with acid anhydrides and acyl halides, that do not result in Beckmann type rearrangement are reviewed in Section II.B.7.a.

d. Oxygen migration. This is a rather unusual reaction somewhat reminiscent of the acetic-anhydride-induced reaction of aromatic *N*-oxides. It was found[131] that the oxygen of the *N*-oxide in a benzodiazepine derivative migrates to the position α to the nitrogen under the influence of a Lewis acid in a nitrile solvent (equation 18). The presence of nitrile was claimed essential to the success of the rearrangement, however, it was not specified in what solvents the experiments failed.

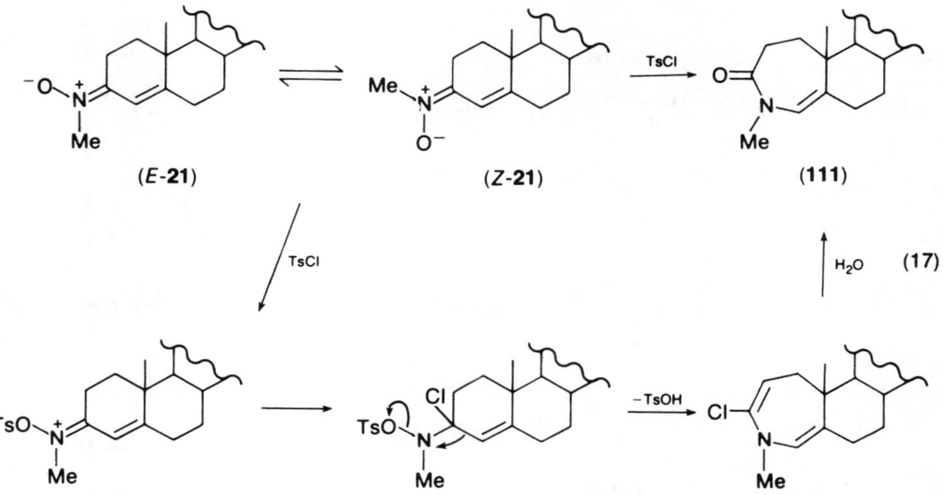

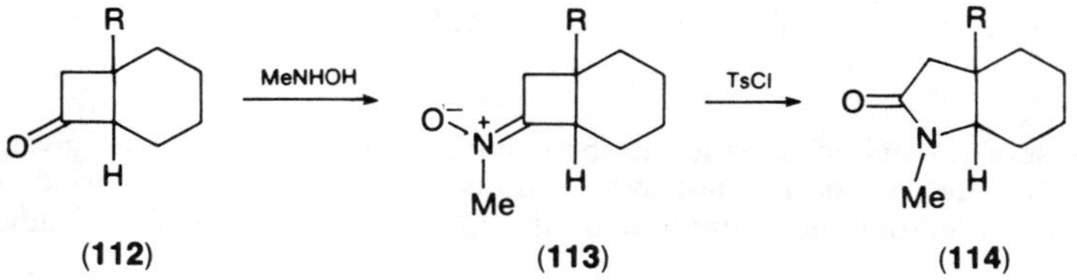

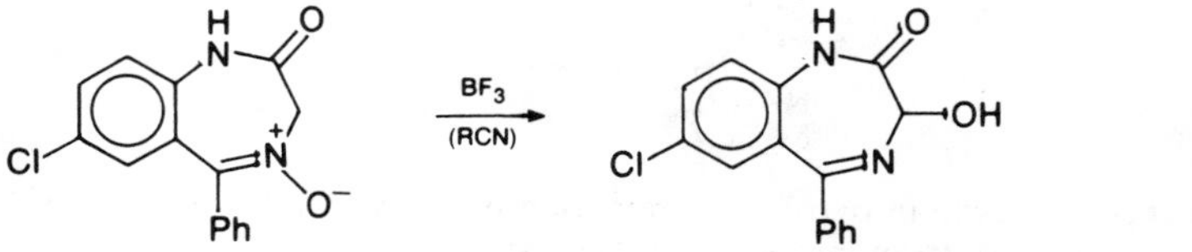

(E-21) ⇌ (Z-21) —TsCl→ (111)

TsCl; −TsOH; H_2O (17)

(112) —MeNHOH→ (113) —TsCl→ (114)

BF_3 (RCN) (18)

2. Formation of olefins by elimination of nitrones

The possibility of this reaction, which resembles the Cope elimination, was first pointed out by Kim and Weintraub in order to explain the formation of oxime and nitroolefin products in an attempted preparation of nitrones[132] (see Scheme 8). It was suggested that the nitrone is an intermediate in the reaction. This behaviour was found to be general for aldehydes possessing electron-withdrawing groups. The first examples of authentic elimination reactions of nitrones were reported by Boyd[133,134]. It was shown that *N*-alkyl-*N*-fluorenylidene *N*-oxides undergo elimination to olefins under mild conditions (equation 19). The proportion of 1-butene (58%) and 2-butene from the *N*-2-butyl-*N*-fluorenylidene nitrone and other results indicate that the direction of elimination is influenced by statistical and steric factors. There have been kinetic studies of eliminations from *N*-fluorenylidene-*N*-alkyl *N*-oxides[135] as well as from various α,α-diaryl *N*-alkyl nitrones[135–137]. It was concluded on the basis of activation entropy values that were found to be close to zero, that the reaction is concerted and proceeds through a cyclic transition state[135–137]. Examination of the

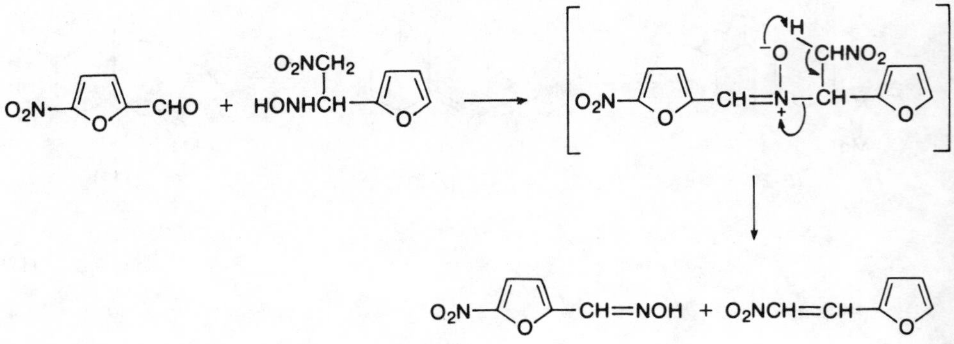

$$O_2N\text{-furyl-}CHO + HONHCH(CH_2NO_2)\text{-furyl} \longrightarrow [O_2N\text{-furyl-}CH{=}\overset{+}{N}(O^-)\text{-}CH(CHNO_2)\text{-furyl}] \longrightarrow O_2N\text{-furyl-}CH{=}NOH + O_2NCH{=}CH\text{-furyl}$$

SCHEME 8

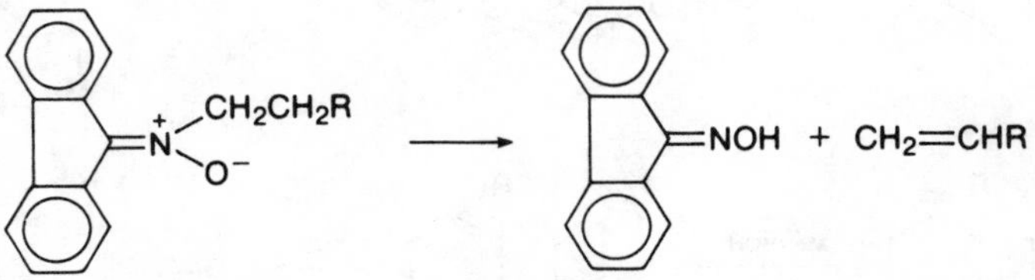

$$\text{Fluorenylidene}{=}\overset{+}{N}(O^-)CH_2CH_2R \longrightarrow \text{Fluorenylidene}{=}NOH + CH_2{=}CHR \qquad (19)$$

effect of substituents in substituted benzophenones upon the rate gives further evidence that the reaction is not synchronous, but that the N—C bond breakage precedes considerably the formation of the O—H[135] (equation 20). Radicals that

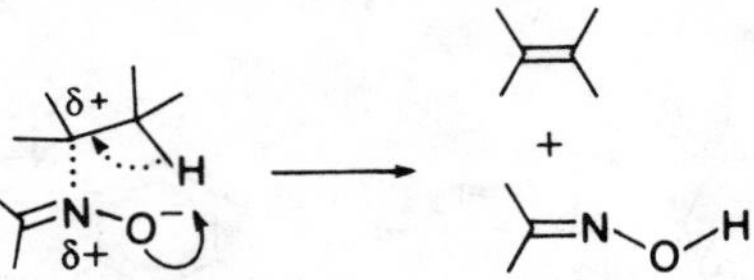

$$\text{(transition state, } \delta+ \text{ on C and N, } N\text{—}O^- \cdots H) \longrightarrow \text{alkene} + {>}C{=}N\text{—}O\text{—}H \qquad (20)$$

have been observed in pyrolytic elimination of α-aryl *N-t*-butyl nitrones[138] have been suggested to arise from minor side-reactions[135].

In a series of α,α-diphenyl *N*-alkylthioalkyl nitrones olefin-forming eliminations were sometimes accompanied by N to O migration. However the reaction could be used for thiacycloalkene formation (equation 21)[137].

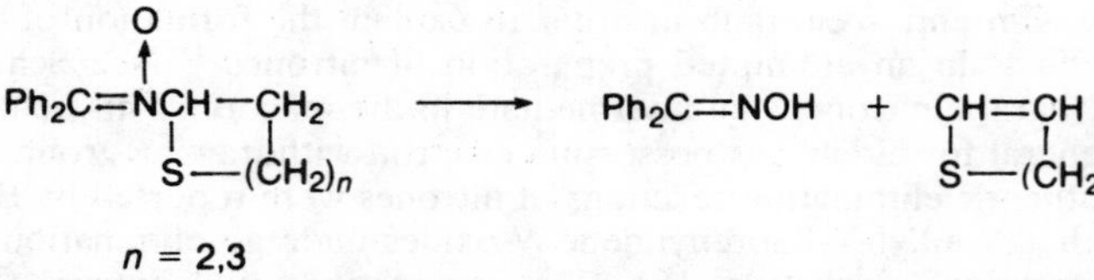

$$Ph_2C{=}\overset{O\uparrow}{N}CH\text{—}CH_2,\ \text{ring } S\text{—}(CH_2)_n \longrightarrow Ph_2C{=}NOH + CH{=}CH,\ \text{ring } S\text{—}(CH_2)_n \qquad (21)$$

$n = 2,3$

In conclusion this elimination reaction seems to be fairly general and might find uses for special cases of olefin synthesis.

3. *Oxidation of nitrones*

The influence of a variety of oxidizing agents upon nitrones has been examined.

a. Lead tetraacetate (LTA). This reagent reacts with several types of nitrones.

Similarly to its reaction with aromatic *N*-oxides, arylaldonitrones yield *N*-acetoxy-*N*-acylamines, that can also be viewed as *(N)O*-acylhydroxamic acids **(115)**[139–142]. A similar result has been reported for the cyclic aldonitrone **(116)** which gives *N*-acetoxy-2-pyrrolidone **(117)** in good yield[76].

$$\mathrm{ArCH{=}\overset{+}{N}(O^-)Ar \xrightarrow{LTA} ArCON(OAc)Ar}$$

(115)

(116) **(117)**

Various mechanisms can be considered for this reaction; however, it seems reasonable to assume that the reaction leads in the first step to a pentocoordinated lead derivative **(118)** which subsequently rearranges to **119**. The latter collapses to the α-acetoxynitrone **(120)** with the loss of acetic acid and lead (II) acetate. Intramolecular transfer of the acetyl group in **120** leads to the end-product **(115)**. In

$$\mathrm{ArCH{=}\overset{+}{N}(Ar)O^- + Pb(OCOCH_3)_4 \longrightarrow}$$

(118)

(119)

$$\mathrm{+\ Pb(OAc)_2 + AcOH}$$

(120)

$$\mathrm{ArCON(Ar)OCOCH_3}$$

(115)

the reaction of α-phenyl *N*-*t*-butyl nitrone with LTA free radicals are observed; however, it has not been shown that hydroxamic acetates are formed therefore there is no conclusive evidence for a free-radical mechanism in this reaction.

In a study using labelled starting material it has been shown that the reaction of α-phenyl *N*-benzyl nitrone **(121)** with LTA gives a scrambled product **(122)**[143]. Consequently it has been assumed that a symmetrical intermediate **(123)** is involved in the reaction.

Ketonitrones react with LTA giving a different product since there is no possibility

$$\underset{(\mathbf{121})}{Ph^{14}CH{=}\overset{+}{N}(O^-)CH_2Ph} \xrightarrow{LTA} Ph^{14}CH(OAc)\overset{+}{N}(O^-){=}CHPh \xrightarrow{LTA} \underset{(\mathbf{123})}{Ph^{14}CH(OAc){-}N(OPb(OAc)_3){-}CH(OAc)Ph}$$

$$Ph^{14}CH(OAc){-}\overset{+}{N}(O^-){=}C(OAc)Ph \qquad Ph^{14}C(OAc){=}\overset{+}{N}(O^-){-}CH(OAc)Ph$$

$$\downarrow \qquad\qquad \downarrow$$

$$\underset{(\mathbf{122})}{Ph^{14}CH(OAc){-}N(OAc){-}COPh \;+\; Ph^{14}CON(OAc){-}CH(OAc)Ph}$$

of eliminating acetic acid from the intermediate. α,α,*N*-triphenyl nitrone gives with LTA benzophenone diacylal and nitrosobenzene[144] presumably via the sequence shown in equation (22).

$$Ph_2C{=}\overset{+}{N}(O^-)Ph + Pb(OCOCH_3)_4 \longrightarrow Ph_2C{=}\overset{+}{N}Ph{-}O{-}Pb(OCOCH_3)_3 \cdots OC(O)CH_3 \longrightarrow \tag{22}$$

$$\longrightarrow Ph_2C(OCOCH_3){-}NPh{-}O{-}Pb(OCOCH_3)_2 \cdots OC(O)CH_3 \longrightarrow Ph_2C(OAc)_2 + PhN{=}O + Pb(OAc)_2$$

b. Iron (III) salts. Ferric chloride has been shown to oxidize aldonitrones of the pyrroline *N*-oxide series to hydroxamic acids (e.g. equation 23)[145]. The kinetics and

$$\text{5,5-dimethyl-1-pyrroline } N\text{-oxide } (\overset{+}{N}{-}O^-) \xrightarrow{FeCl_3} \text{5,5-dimethyl-1-hydroxy-2-pyrrolidinone } (N{-}OH,\ C{=}O) \tag{23}$$

the mechanism of this reaction have been studied[146,147]. In contrast to this, ketonitrones can be oxidized either to nitrosoketones or to products devoid of nitrogen, depending upon the amount of oxidizing agent used. For example using two moles of ferric chloride the hydroxynitrone **124** may be isolated; however, excess reagent leads to the keto acid (**125**).

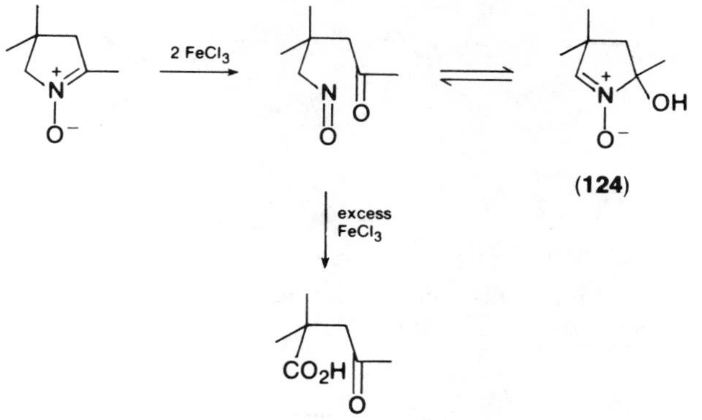

(124)

(125)

The oxidation of some 2-amino-Δ^1-pyrroline *N*-oxides with potassium ferricyanide has also been studied[77]. It has been shown that aminonitrone **42** yields upon treatment with alkaline ferricyanide (or neutral permanganate) the dimeric azopyrroline *N*-oxide (**126**) presumably via the radical intermediate **127**.

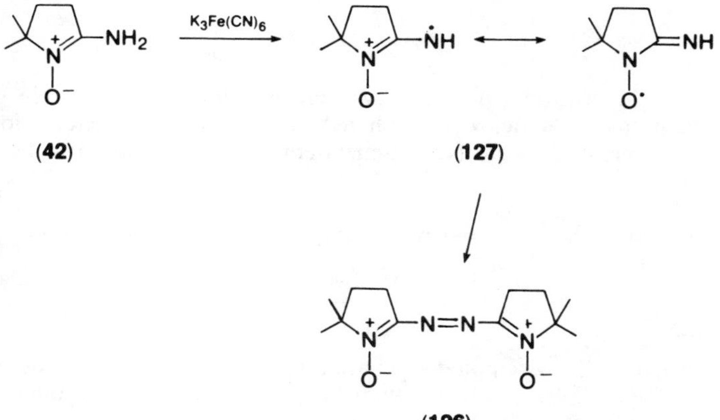

(42) (127)

(126)

Oxidation of the related hydrazonitrone **126a** (which is obtained by hydrogenation of **126**) can be effected by air, giving the blue radical anion **128** which can be kept indefinitely in the absence of air[77] (see also Section II.B.4a).

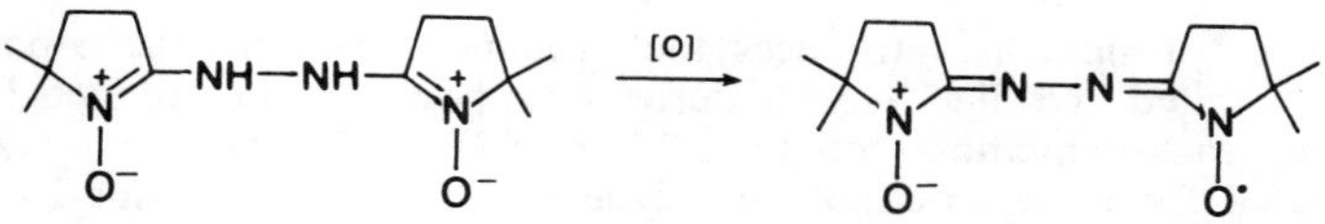

(126a) (128)

In contrast to this, ferricyanide oxidation of the monomethyl nitrone **129** leads to laevulinonitrile presumably via the sequence shown in equation (24)[77].

c. Periodate. Sodium and tetraethylammonium periodate have been used for the oxidation of pyrroline *N*-oxides[148,149]. It has been shown that variously substituted

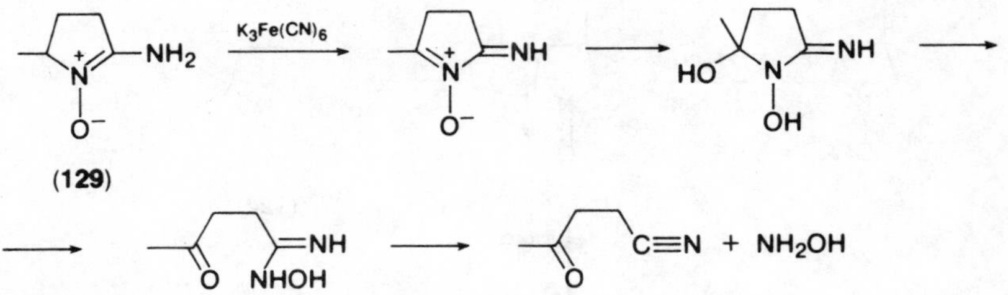

(24)

pyrroline *N*-oxide aldonitrones can be oxidized by periodate to give nitrosopentanoic acids, isolated as the dimers (equation 25).

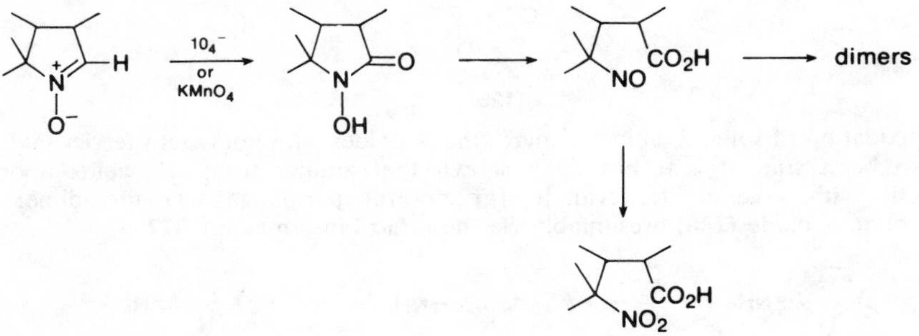

(25)

In contrast a 2-substituted pyrroline *N*-oxide was found to be stable under the reaction conditions[149]. α-Benzoyl *N*-phenyl nitrone **(130)** reacted slowly with periodate; however, its hydrate **(131)** underwent a facile oxidation to benzoic acid and nitrosobenzene[149].

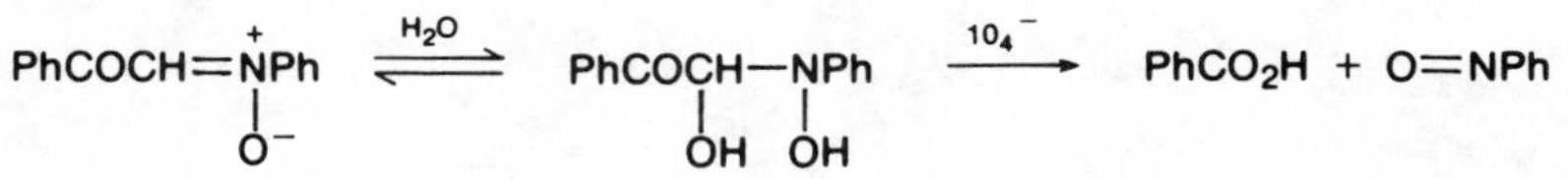

d. Halogenation. An attempted preparation of an α-bromonitrone by direct bromination of α,N-diphenyl nitrone with bromine resulted in bromination of the *N*-phenyl ring[150] (equation 26).

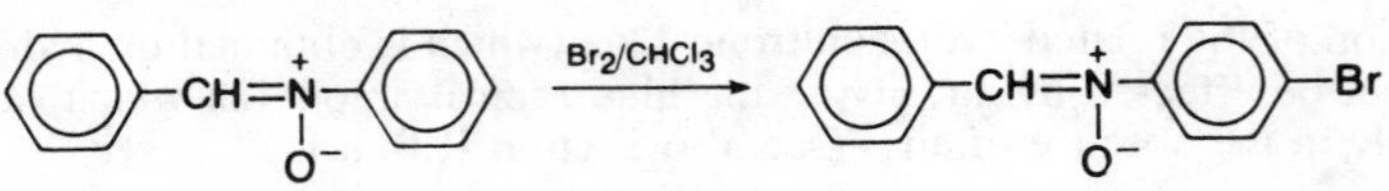

(26)

Hypobromite presumably introduces the bromine atom into the α-position of variously substituted pyrroline *N*-oxide derivatives; however, the products hydrolyse rapidly in the reaction mixture (equations 27 and 28)[151]. The cyclic hydroxamic acid can be isolated from the oxidation of 5,5-dimethylpyrroline *N*-oxide-2-carboxylic acid if excess hypobromite is avoided[151] (equation 28).

Chlorination of some aliphatic open-chain nitrones was achieved by *N*-chlorosuccinimide. This reagent introduced the chlorine atom into the β-position[152] (e.g. equation 29).

2-Substituted 5,5-dimethylpyrroline *N*-oxides **(132)** undergo bromination at the 3-position by *N*-bromosuccinimide. In case of the 2-cyano derivative **(132**, R = CN)

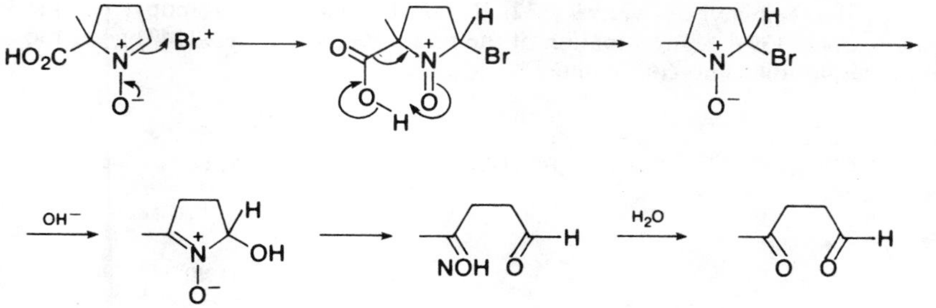

(27)

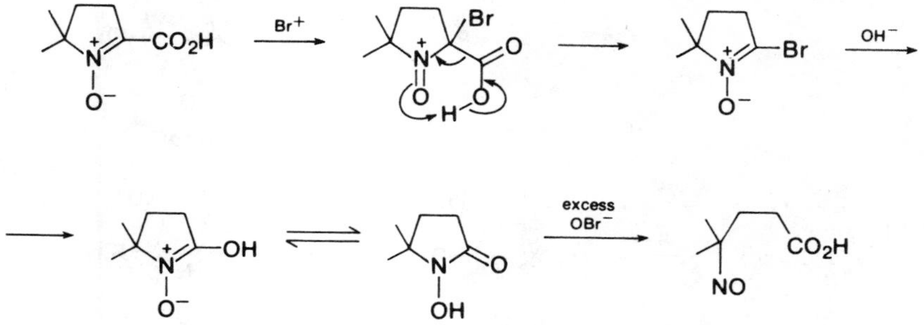

(28)

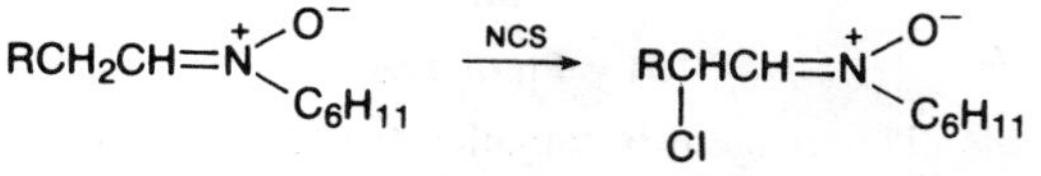

$$RCH_2CH{=}\overset{+}{N}(O^-)C_6H_{11} \xrightarrow{NCS} RCH(Cl)CH{=}\overset{+}{N}(O^-)C_6H_{11} \qquad (29)$$

R = H, Me

the 3-bromo compound (**133**) is obtained[153]. The latter can undergo various transformations indicated by the formulae. The reaction of the 2-*t*-butyl-5,5-dimethylpyrroline *N*-oxide (**132**, R = *t*-Bu) with NBS leads to the 4-bromo derivative (**138**) presumably through the intermediate **137**[154]. The same

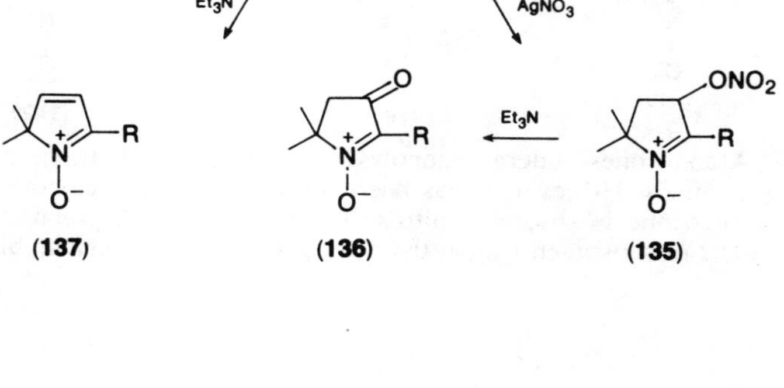

(**132**) (**133**) (**134**) (**137**) (**136**) (**135**)

reaction with the 2-phenyl derivative (**132**, R = Ph) gives the dibromopyrrolenine *N*-oxide derivative (**139**)[159]. Bromination of the related 3-oxonitrones (**140**) gives the 4-bromo or the 4,4-dibromo compounds[159].

132 (R = *t*-Bu) $\xrightarrow{\text{NBS}}$ (**138**)

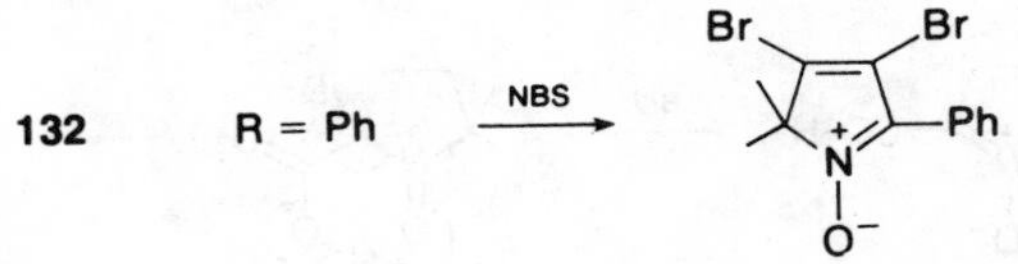

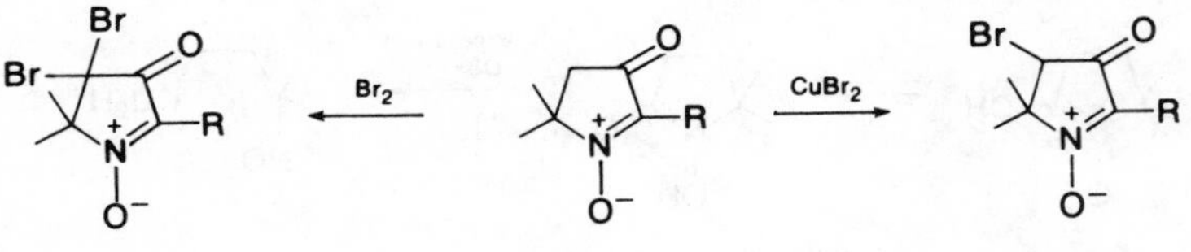

132 R = Ph $\xrightarrow{\text{NBS}}$ (**139**)

$\xleftarrow{Br_2}$ (**140**) $\xrightarrow{CuBr_2}$

R = Ph or *t*-Bu

e. Selenium dioxide. This reagent is capable of oxidizing in variable yield a methyl or methylene group attached to the α-position of a nitrone function (e.g. equation 30)[155]. 2,4,4-Trimethylpyrroline *N*-oxide had previously been reported to yield upon

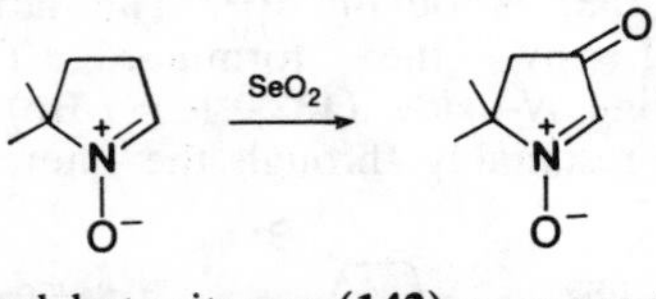

$\xrightarrow{SeO_2}$ (30)

oxidation the six-membered ketonitrone (**142**) as a result of rearrangement of the initially formed aldehyde **141**[3]. More recently it has been reported that the aldehyde **141** can be obtained in satisfactory yield if the oxidation is conducted at room temperature[156].

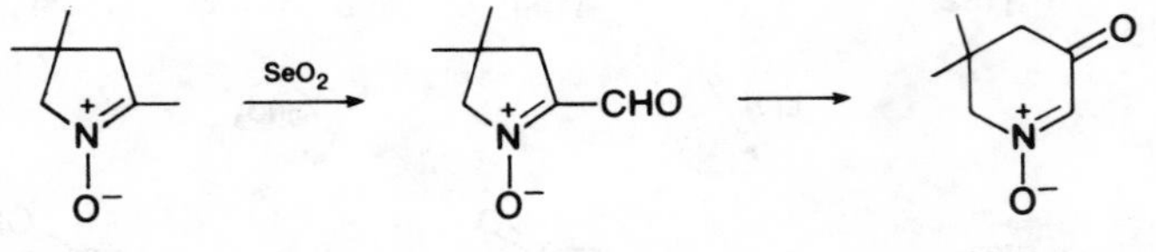

$\xrightarrow{SeO_2}$ (**141**) $\longrightarrow$ (**142**)

f. Ozone. Aldonitrones undergo ozonolysis rapidly at −78°C to yield aldehydes and nitro compounds. However, it has been shown that if the reaction is stopped after 1 mole of ozone is absorbed, nitroso compounds can be isolated (equation 31)[157]. The effect of substituents upon the rate ozonization has been studied, and on

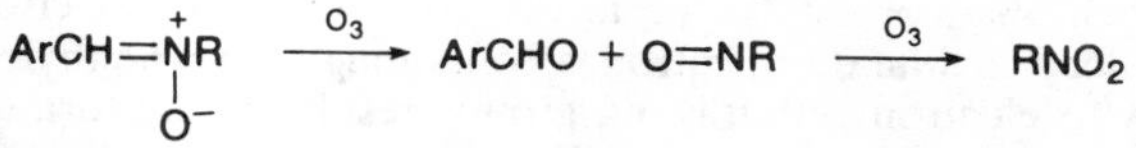

$$ArCH{=}\overset{+}{N}(O^-)R \xrightarrow{O_3} ArCHO + O{=}NR \xrightarrow{O_3} RNO_2 \qquad (31)$$

the basis of the results the conclusion has been reached that the reaction proceeds via an electrophilic attack of the ozone molecule upon the α-position of the nitrone (equation 32)[158].

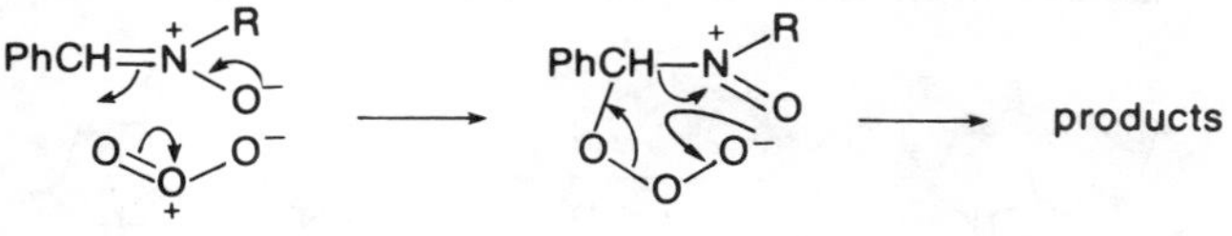

$$PhCH{=}\overset{+}{N}(R)O^- + O{=}\overset{+}{O}{-}O^- \longrightarrow [PhCH{-}\overset{+}{N}(R){=}O,\ O{-}O{-}O^-] \longrightarrow \text{products} \qquad (32)$$

g. Photooxidation. Ultraviolet irradiation of some α,*N*-diaryl nitrones of α-phenyl *N-t*-butyl nitrone in solution was shown to produce aroyl nitroxide radicals (equation 33)[159].

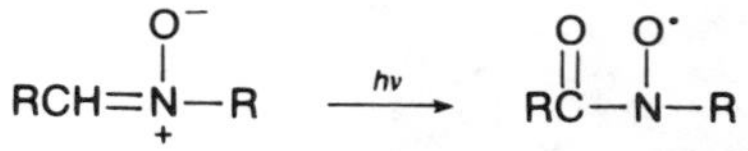

$$RCH{=}\overset{+}{N}(O^-){-}R \xrightarrow{h\nu} RC({=}O){-}N(O^\bullet){-}R \qquad (33)$$

Various mechanisms could be suggested to rationalize this observation. The source of oxygen was assumed to be either the nitrone or the isomeric oxaziridine that could be formed under the reaction conditions.

Singlet oxygen reacts with 2,4,4-trimethylpyrroline *N*-oxide to give the explosive hydroperoxide **143**[160]. The aldonitrone 4,5,5-trimethylpyrroline *N*-oxide is recovered unchanged from such a reaction.

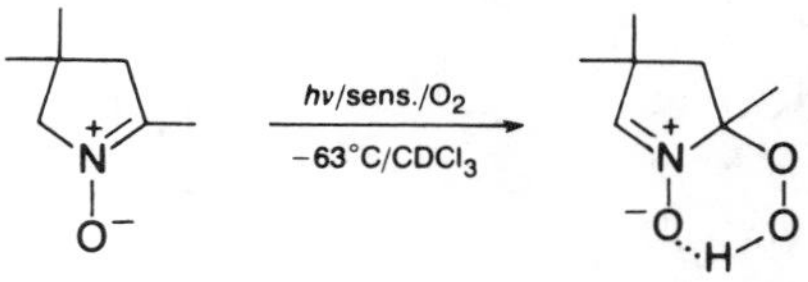

$$\xrightarrow[-63^\circ C/CDCl_3]{h\nu/sens./O_2}$$

(**143**)

h. Miscellaneous. Methylene groups attached to the α-position of the nitrone function that are adjacent to additional activating groups exhibit particular sensitivity to oxidation.

The activated nitrones (**144**) undergo easy oxidation by nickel peroxide or by lead dioxide to the stable radicals vinylaminyl oxides or vinyl nitroxides **145**[161]. The

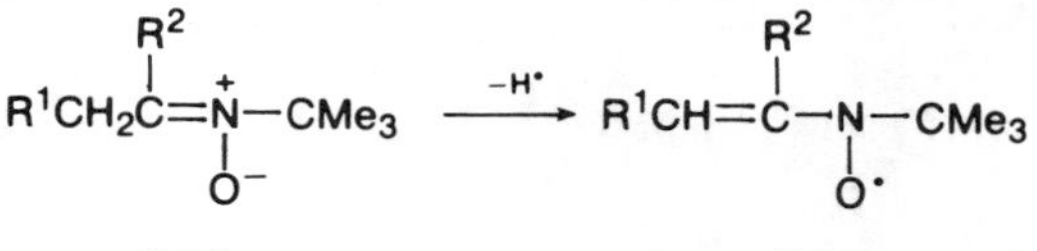

$$R^1CH_2C(R^2){=}\overset{+}{N}(O^-){-}CMe_3 \xrightarrow{-H^\bullet} R^1CH{=}C(R^2){-}N(O^\bullet){-}CMe_3$$

(**144**) (**145**)

R^1	R^2
$PhSO_2$	H
$PhSO_2$	Ph
MeOCO	MeOCO

stability of the radicals is probably due to the presence of the electron-withdrawing groups R′[161]. Indeed, oxidation of nitrones containing active methylene groups that are not adjacent to electron-withdrawing groups results in the formation of dimers probably via the reactive free radicals. This is demonstrated by the reactions of various pyrroline *N*-oxide derivatives.

Attempts to prepare β-dinitrone **147** by the oxidation of the hydroxylaminonitrone **146** resulted in the formation of the dimeric products **148** and **149**[162,163]. Similarly the

[O]Cu^{2+} or MnO_2 or chloranil or Hg^{2+}

(**146**) (**147**)

[O] $-H^{\bullet}$ dimerization

[O] $-2\,H^{\bullet}$

(**149**) (or geometrical isomer) (**148**)

[O]Cu^{2+}

(**150**) (**151**)

$-H^{\bullet}$

(**153**) (**152**)

α-benzyl nitrone **151** could not be isolated. The oxidation of the hydroxylamine **150** led to the dimer **153** presumably via **151** and **152**[164].

The γ-ketonitrone **154** is a stable compound but it can also be easily oxidized to the dimer **155**[75].

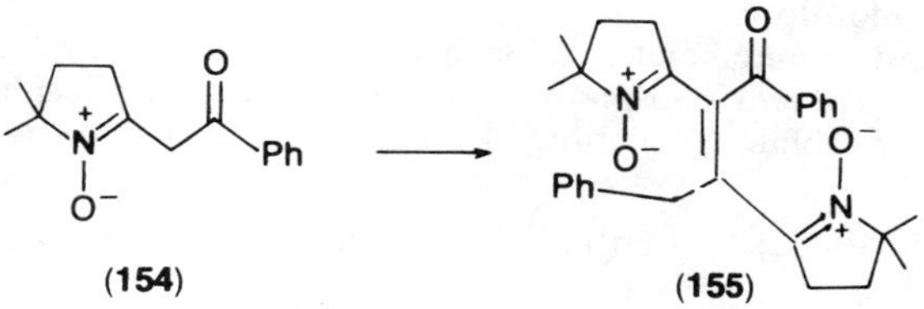

(**154**) (**155**)

Chromate oxidation of the six-membered ketonitrone **156** also gives a dimeric product **157**[165].

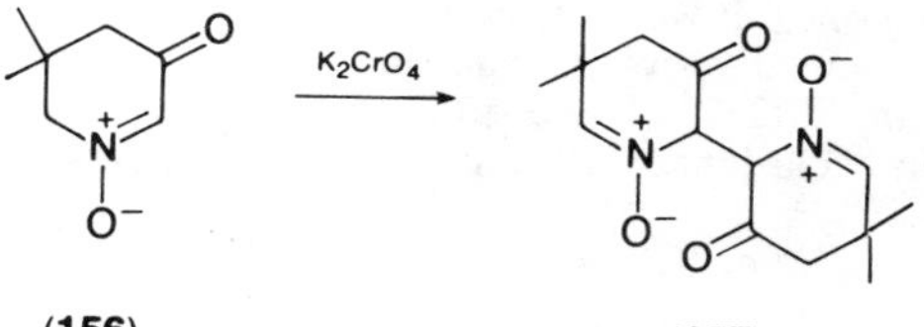

K_2CrO_4

(**156**) (**157**)

4. Reduction of nitrones

Nitrones can be reduced to the corresponding radical anions by the addition of one electron. They can be deoxygenated to imines as well as reduced to hydroxylamines or to amines by a variety of reducing agents. The earlier knowledge of these aspects is summarized in the previous reviews[2,3].

a. Reduction to radical anions. The addition of one electron to a nitrone converting it to a radical anion has been achieved using sodium metal as the reducing agent. α,α,*N*-Triphenyl nitrone has been converted to **158**[166] and **126** to **128**[77] by this method (see also Section II.B.3.b).

$Ph_2C{=}\overset{+}{N}(O^-)Ph \longrightarrow Ph_2\dot{C}{-}NPh(O^-\ Na^+)$

(**158**)

Na

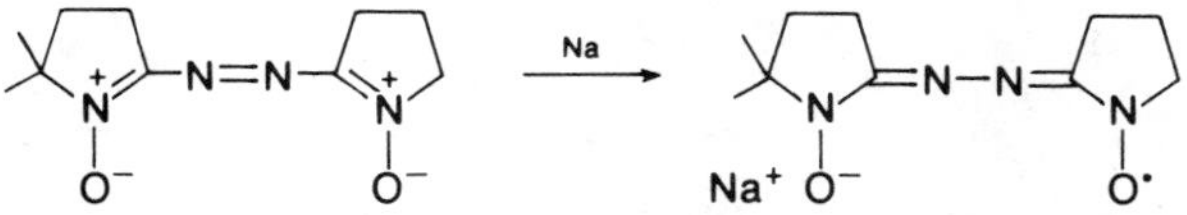

(**126**) (**128**)

b. Deoxygenation. A summary of typical deoxygenation procedures of aromatic *N*-oxides that are also applicable to nitrones is available in a recent book[166]. Only more recent methods will be discussed in this section.

Deoxygenation of a nitrone is reported to be induced by iron pentacarbonyl[167]. Some nitrones derived from cinnamaldehyde are successfully deoxygenated by treatment with sodium borohydride at room temperature[168]. At higher temperatures further reduction is observed. Hydrogenation over a freshly prepared W4 Raney nickel has been found a reliable method for the deoxygenation of Δ^1-pyrroline *N*-oxides[75]. Recently hexachlorodisilane has been shown to act as a mild reagent, capable of selective deoxygenation of nitrones and *N*-oxides in good yields at room temperature in chloroform[169].

One of the most common methods for the deoxygenation of nitrones and *N*-oxides is by the use of trivalent phosphorus compounds such as phosphorus trichloride, phosphines and phosphites (equation 34). The reaction of nitrones with phosphites

$$\mathrm{RCH{=}\overset{+}{N}(O^-)R} + \mathrm{R_3P} \longrightarrow \mathrm{RCH{=}NR} + \mathrm{R_3PO} \qquad (34)$$

has been studied in some detail. Using α,N-diaryl nitrones substituted in both rings[170], as well as α-aryl *N*-alkyl nitrones substituted in the aromatic ring[171], it was found that the rate is enhanced by electron-withdrawing substituents, and retarded by electron-donating ones. It was suggested that the deoxygenation reaction proceeds by nucleophilic attack of the phosphorus upon the nitrogen[171]. However, a mechanism involving attack of the phosphorus upon the oxygen seems more reasonable and is in accordance with the substituent effects observed (equation 35).

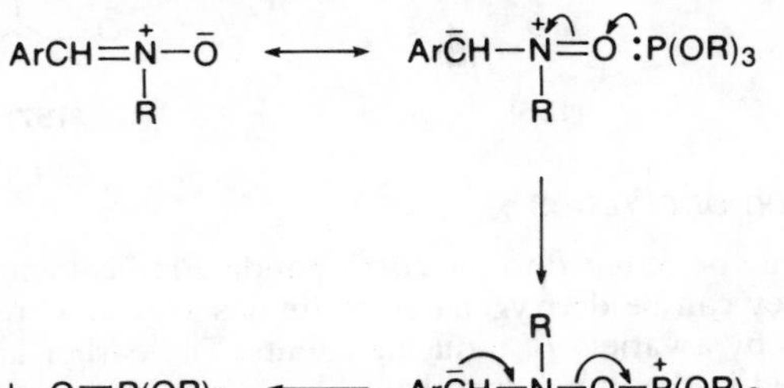

$$\mathrm{ArCH{=}\overset{+}{N}(R){-}\bar{O}} \longleftrightarrow \mathrm{Ar\bar{C}H{-}\overset{+}{N}(R){=}O}\ \mathrm{:P(OR)_3}$$

$$\downarrow$$

$$\mathrm{ArCH{=}NR} + \mathrm{O{=}P(OR)_3} \longleftarrow \mathrm{Ar\bar{C}H{-}N(R){-}O{-}\overset{+}{P}(OR)_3} \qquad (35)$$

(MeO)$_2$P(O), N, H, H — **(161)** $\xleftarrow{\text{TMP, AcOH}}$ O$^-$, N$^+$, H, H — **(159)** $\xrightarrow{\text{TMP, MeOH}}$

O, OMe, (MeO)$_2$P, N, H, H

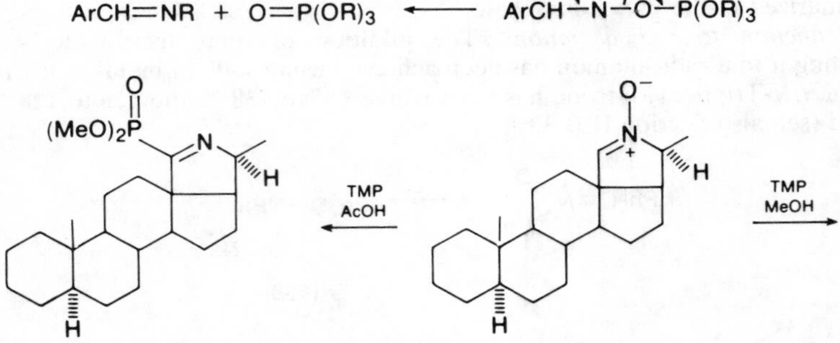

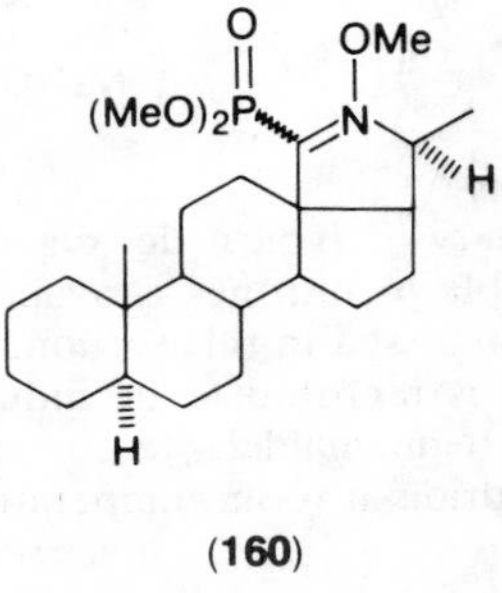

(160)

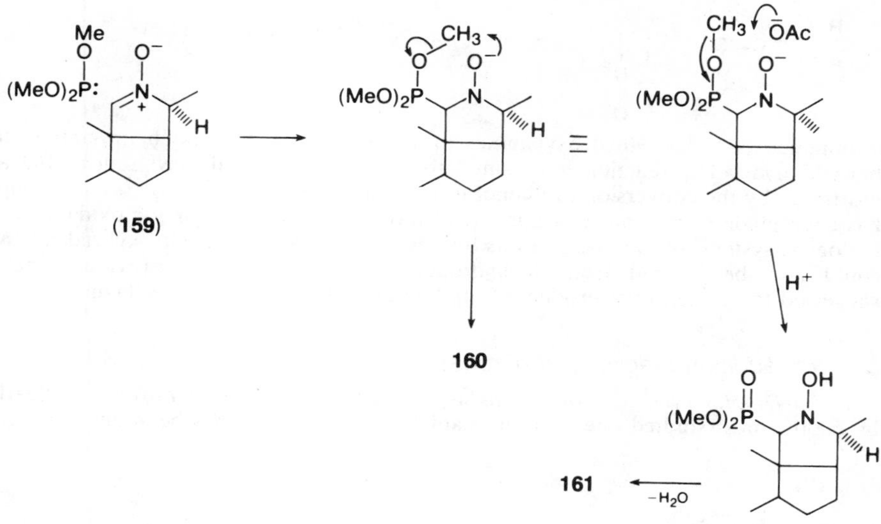

(159)

160

161

SCHEME 9

Recently it has been reported that the reaction of a steroidal cyclic nitrone with trimethyl phosphite may take a variety of courses, but can be controlled to give, selectively, different products[172]. Thus, reaction of *N*-demethyl5α-Δ^{18}-conenine *N*-oxide (**159**) with trimethyl phosphite (TMP) in refluxing methanol leads to a 18-epimeric mixture of *N*-methoxyphosphonates **160**. On the other hand reaction of **159** with TMP in acetic acid gives iminophosphonate **161**[172]. The formation of these two products is rationalized by Scheme 9.

Finally simple deoxygenation of **159** can be achieved by using TMP in the presence of triethylamine.

c. Reduction to hydroxylamines and to amines. Complex metal hydrides have been reported to reduce nitrones to *N,N*-disubstituted hydroxylamines[2,3]. These include lithium aluminium hydride and sodium or potassium borohydride. Some recent examples of partial and total reduction of dinitrones by such reagents have been reported[165,173, 174].

Trichlorosilane has been reported to effect reduction of *N*-aryl and *N*-alkylketo- and aldo-nitrones to the corresponding *N,N*-dialkylhydroxylamines at low temperatures and in high yields[169]. The polarographic reduction of nitrones to secondary amines via the corresponding imines has also been reported[175].

d. Nitrones as oxidizing agents. It is pertinent to this section to mention reports concerning the ability of nitrones to act as oxygen donors. There are numerous reactions of nitrones with ketenes and ketene imines in which the major product consists of the elements of the heterocumulene with the addition of the oxygen, which on mechanistic grounds can be assumed as originating from the nitrone. There is also the imine by-product isolated in these reactions. These reactions can be viewed as nitrone-induced oxidative rearrangements and are discussed in Sections II.B.6.a and b. A simple case of oxidation by a nitrone is supplied by the reaction of thioketenes with pyrroline *N*-oxides which yields α-thiollactones (equation 36)[176].

Recently it has been reported that flavin *N*(5)-oxide **162** oxidizes under illumination at ambient temperature a variety of substrates, such as phenols to

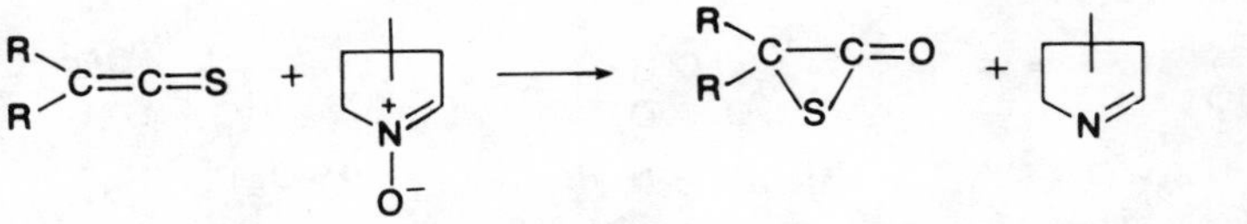

(36)

p-quinones, *N,N*-dialkylhydroxylamines to nitrones and *N,N*-dialkylbenzylamines to benzaldehyde. The reaction is assumed to proceed via the nitroxyl radical **163** as illustrated by the conversion of phenol to hydroquinone (Scheme 10)[177]. An attempt has been made to rationalize a number of flavin-dependent enzymatic oxidations in biological systems on the basis of this finding. It is proposed that nitroxyl radical **163** could also be formed from 4*a*-hydroperoxyflavin, which has previously been suggested to be the initial product of binding of molecular oxygen to flavin.

5. Cycloaddition reactions of nitrones

a. 1,3-Dipolar cycloadditions. This aspect of the chemistry of nitrones is clearly by far the most studied one. The mechanism of the reaction has been given a great

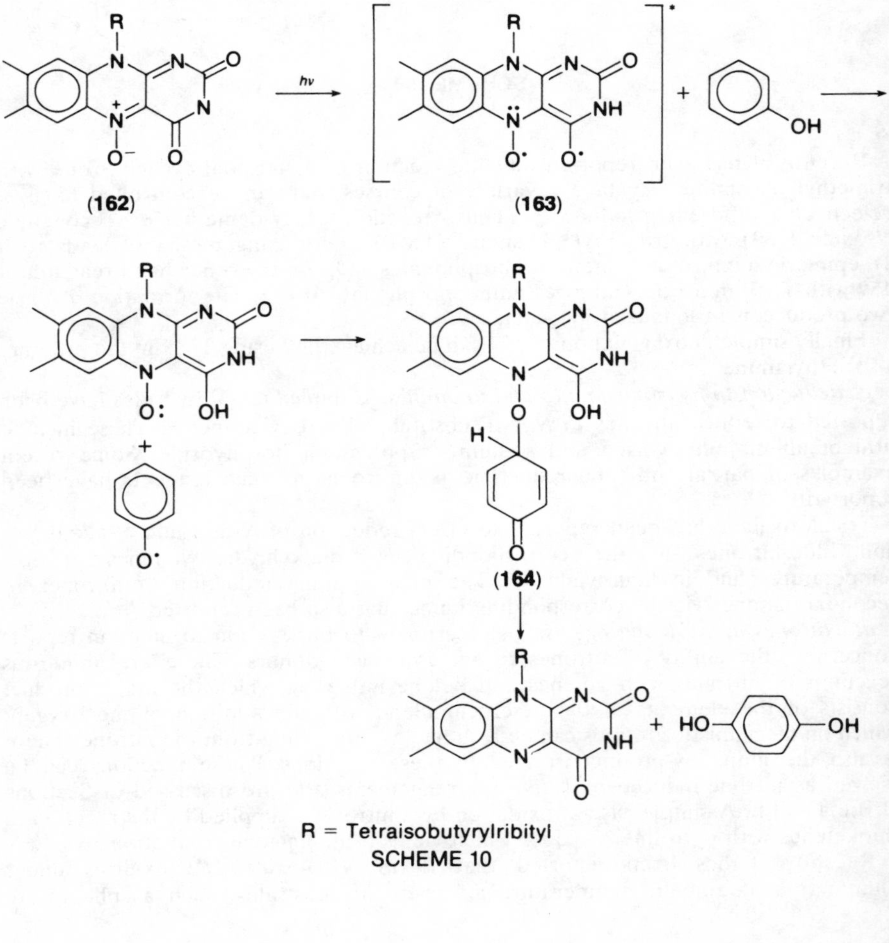

(**162**) (**163**) (**164**)

R = Tetraisobutyrylribityl

SCHEME 10

deal of attention within the context of the general 1,3-dipolar cycloaddition which encompasses reactions of a variety of 1,3-dipoles[178]. There also exist a number of recent reviews dealing with the synthetic aspects of this reaction[179–181]. The emphasis in this section will be placed on more recent papers not included in the previous reviews.

Huisgen, who was the first to formulate 1,3-dipolar cycloadditions in modern, generalized terms, advocates a concerted mechanism which assumes that the two sigma bonds are formed simultaneously[182] (equation 37a). The diradical mechanism of

(37a)

Firestone depicts the formation of the ring in two steps, the first leading to a diradical, the second involving the cyclization (equation 37b)[183] An attempt to

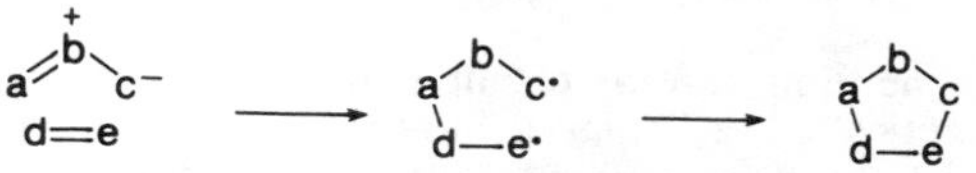

(37b)

reconcile the conflicting points between the two mechanisms has been made by Harcourt, who proposes a concerted diradical mechanism for these reactions[184]. The apparent lack of influence of solvent polarity in most 1,3-dipolar cycloadditons rules out mechanism involving ions or zwitterions[182,185], although this type of mechanism (equation 38) needs to be considered for reactions of highly polarized alkene dipolarophiles[186].

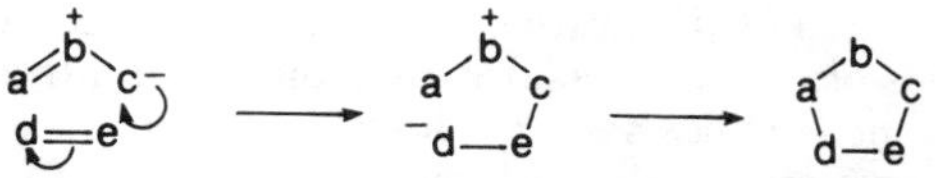

(38)

Most cycloaddition reactions between a nitrone and a monosubstituted ethylene lead to 5-substituted isoxazolidines **165** rather than the 4-substituted heterocycles **166**, regardless of whether the R^2 group is electron-withdrawing or

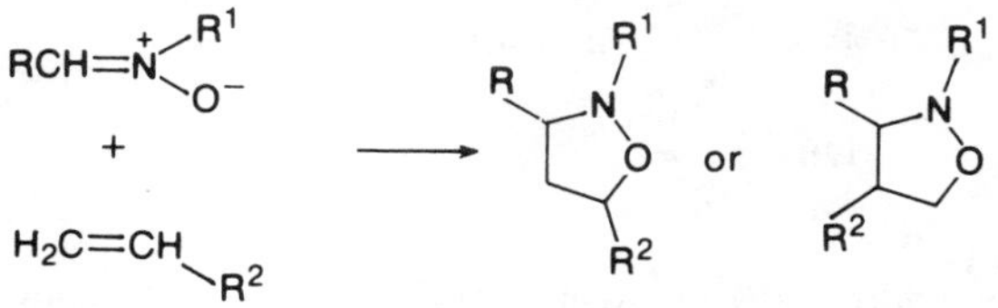

(**165**) (**166**)

electron-releasing. Only cycloadditions of nitrones with very electron-deficient alkenes lead predominantly to 4-substituted isoxazolidines. Initial attempts to explain this behaviour invoked both electronic and steric effects[187–189]; however, subsequently this aspect has been successfully treated by application of the perturbation theory. According to this the course of the reaction will be determined by the selection of the favoured frontier orbital interaction[190]. More recently it has also been shown that the amount of 4-substituted isoxazolidine also depends upon the nature of the nitrone[30]. The more electron-rich the nitrone the higher the relative proportion of the 4-substituted product. This can also be predicted by experiment, namely by determining the ionization potential of the nitrone by means of photoelectron spectroscopy. The lower the ionization potential of a nitrone the larger the tendency for a 4-substituted isoxazolidine formation, provided that the second condition is fulfilled, namely that the dipolarophile is sufficiently electron-deficient. These results are rationalized in terms of frontier orbital interactions[30]. The unusual nitrone *N*-cycloheptatrienylidenemethylamine oxide **(167)**, reacts with the strongly electron-deficient phenyl vinyl sulphone to give only the unstable 4-substituted isoxazolidine **168** which rapidly rearranges by a 1,7-sigmatropic shift to the final stable product **169**[191].

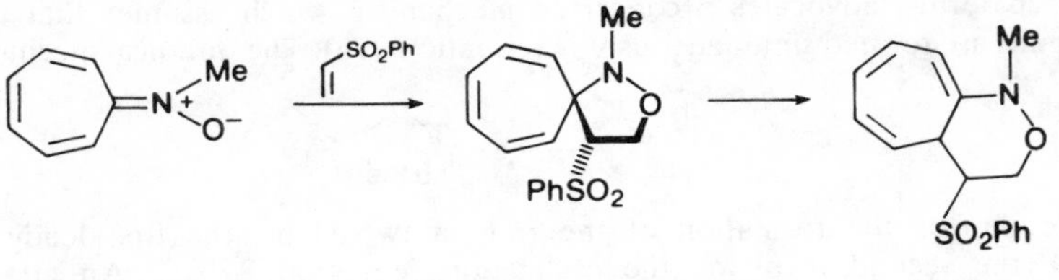

(167) **(168)** **(169)**

It is necessary to mention several recent papers with mechanistic or theoretical objectives. The kinetics of α,*N*-diaryl nitrones with *N*-phenylmaleimide[192] and dibenzoylethylene[193] have been studied. The stereochemistry of the reaction was studied using 3,4-dichlorocyclobutene[194], indene[195], 3,4-dihydronaphthalene[195], bicycloheptadiene[196] and 2-methylene[2,2,1′]bicycloheptane[197] as dipolarophiles. The influence of secondary orbital interactions upon the mode of cycloadditions of nitrones with unsaturated esters has been studied by the group of Carrie and Hamelin[198]. These interactions are of two types, one between the nitrogen of the nitrone and the ester carbonyl group of the dipolarophile and the second type, between the α-phenyl group of the nitrone and the alkyl oxygen of the ester group.

Synthetic applications of the 1,3-dipolar cycloadditions of nitrones can be found in the previously mentioned reviews[179–181]; however, there are significant recent contributions that deserve mentioning. A convenient method for cycloaddition with *in situ* prepared *N*-methyl nitrone **(170)** has been published[199] (equation 39). **170** can

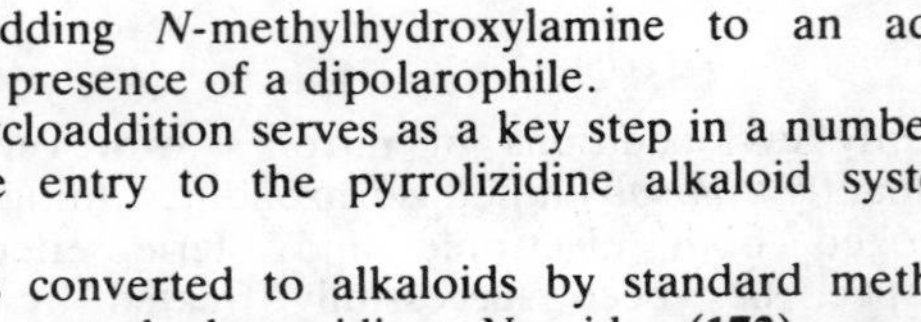

$$CH_2{=}\overset{+}{N}(O^-)Me + CH_2{=}CHR \longrightarrow \text{5-R-2-methylisoxazolidine} \quad (39)$$

(170) 53%

be prepared by adding *N*-methylhydroxylamine to an aqueous solution of formaldehyde in the presence of a dipolarophile.

The 1,3-dipolar cycloaddition serves as a key step in a number of natural product syntheses. A simple entry to the pyrrolizidine alkaloid system is presented in Scheme 11[200].

Compound **171** is converted to alkaloids by standard methods. An analogous synthesis based on tetrahydropyridine *N*-oxide **(172)** provides entry to the quinolizidine group[201]. Compound **172** has also been reacted with styrene and with propene providing syntheses of sedridine **(173)**, sedamine **(174)** and allosedamine **(175)** (Scheme 12)[202].

The synthesis of cocaine has also been reported recently by Tufariello and coworkers[203–205] (Scheme 13). The key intermediate in this synthesis is the unsaturated nitrone ester **177**, which was expected to cyclize intramolecularly as indicated. The efforts to synthesize **177** were initially met with severe difficulties[203].

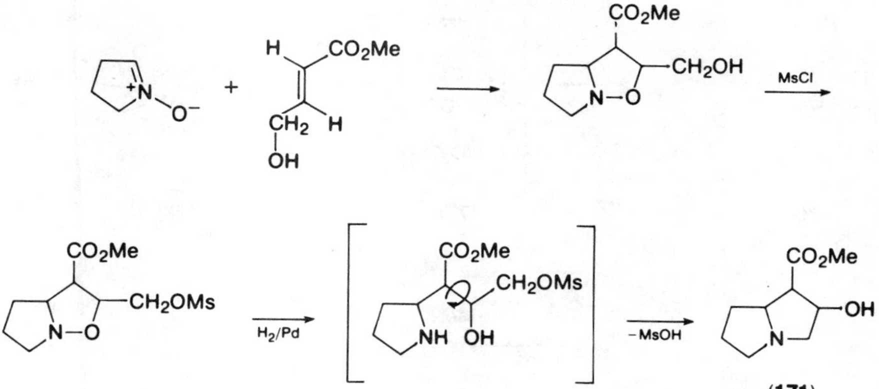

H_2/Pd

$-MsOH$

MsCl

CO_2Me CH_2OH CH_2OMs

(**171**)

SCHEME 11

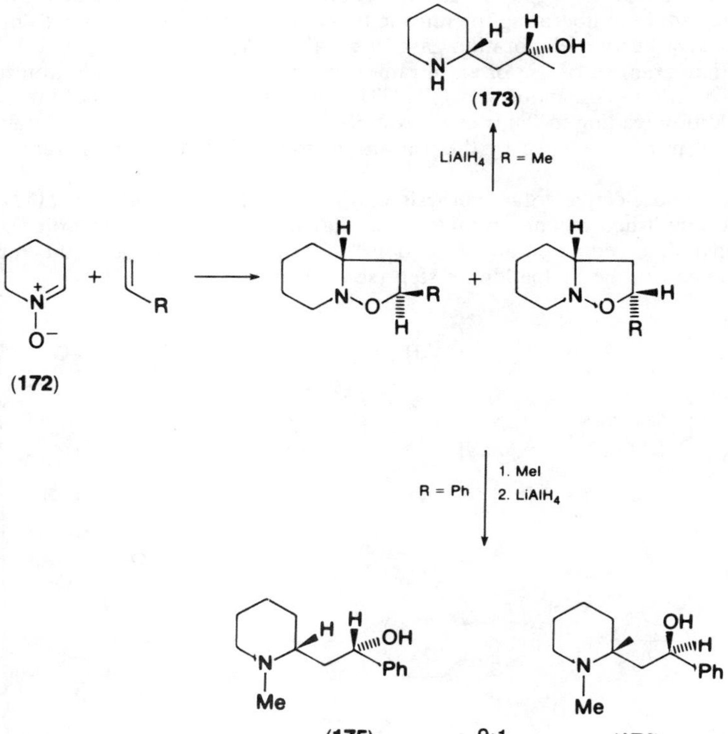

(**173**)

$LiAlH_4$ | R = Me

(**172**)

R = Ph | 1. MeI 2. $LiAlH_4$

(**175**) 9:1 (**174**)

SCHEME 12

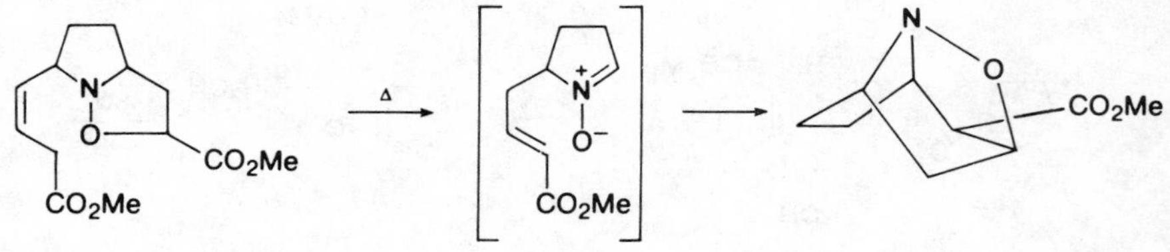

(176) (177) (178)

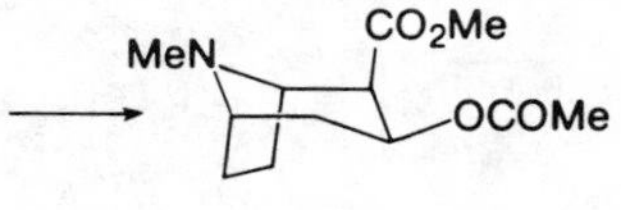

(179)

SCHEME 13

However, when the nitrone function was protected as the 1,3-cycloadduct (**176**) the side-chain could be elaborated in good yield. Finally heating **176** caused retrocycloaddition liberating the nitrone function which *in situ* cyclized to the bridged product **178**. Further elaboration gave *dl*-cocaine (**179**).

Another example of use of an intramolecular 1,3-dipolar cycloaddition reaction is a new synthesis of biotin (**181**)[206]. The key step again is the intramolecular cycloaddition leading to the tricyclic compound **180**. This can be carried out either by the use of nitrile oxide, or by the use of a nitrone 1,3-dipole. The latter is indicated in Scheme 14.

An enantioselective total synthesis of the alkaloid (+)-luciduline (**182**) has also been accomplished recently by intramolecular nitrone–olefin cycloaddition followed by methylation, reduction and oxidation[207]. Fortunately only the 'right' regioisomer was obtained in the cycloaddition step (see Scheme 15).

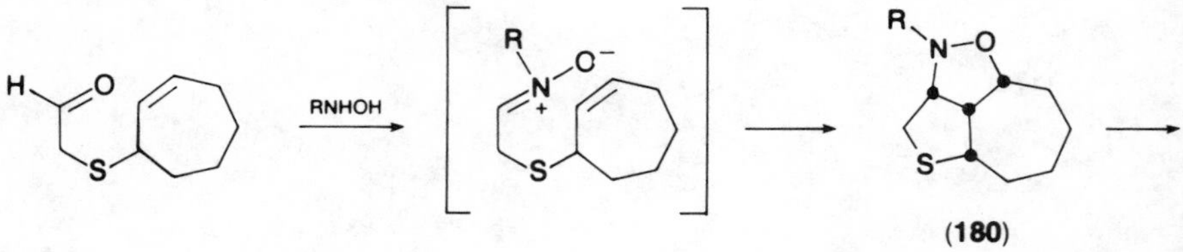

(180)

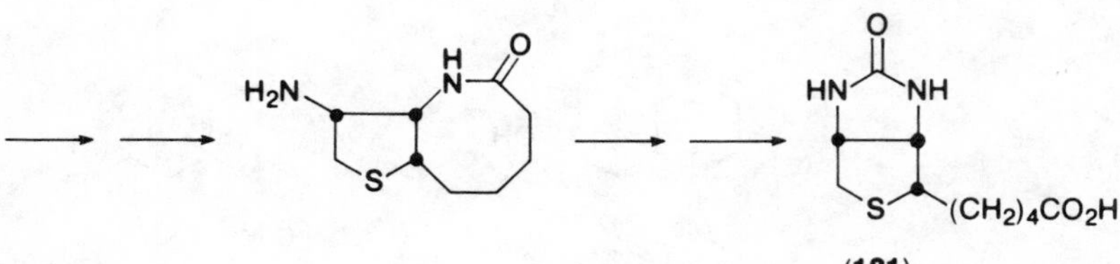

(181)

SCHEME 14

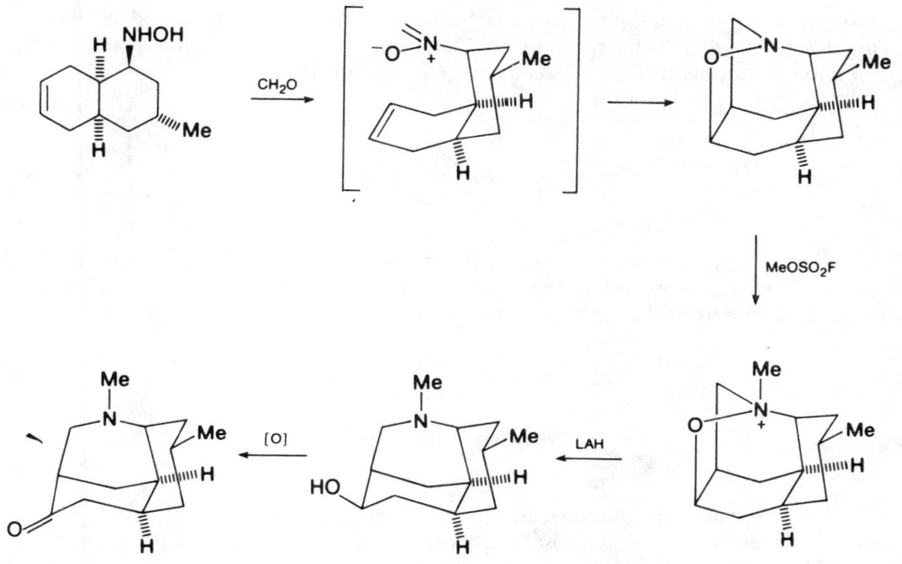

SCHEME 15

The stereospecificity of the intramolecular nitrone–olefin cycloaddition has recently been exploited for the synthesis of the sesquiterpene bisabalol (**183**)[208]. Intramolecular cycloaddition of the nitrone derived from farnesol is followed by several steps that have to be taken in order to remove the nitrogen from the molecule (Scheme 16).

An elegant way to utilize intramolecular 1,3-dipolar cycloaddition for the

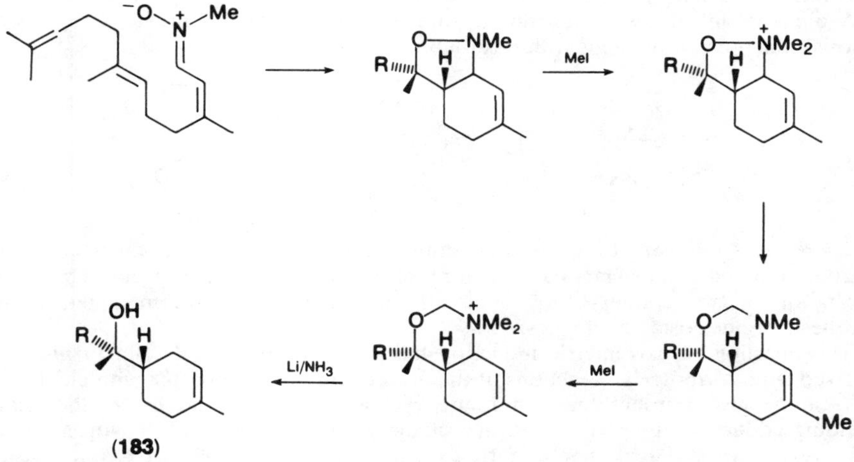

SCHEME 16

construction of two heterocyclic rings has been devised by Black and coworkers. They have shown that the reaction of *C*-acyl nitrones with allylamine results in the formation of tetrahydropyrroloisoxazoles (equation 40)[209].

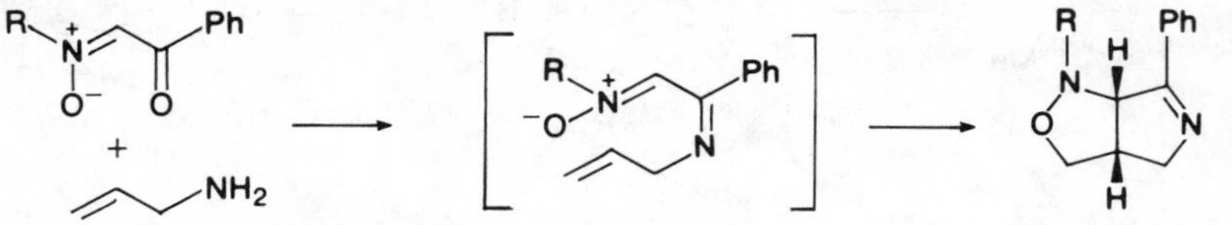

(40)

In a related intramolecular cycloaddition, vinyl nitrone **184** undergoes thermal cyclization to the isoxazoline **185**[68]. This reaction is the first reported case of a nitrone participating in a 1,5-dipolar cyclization[210].

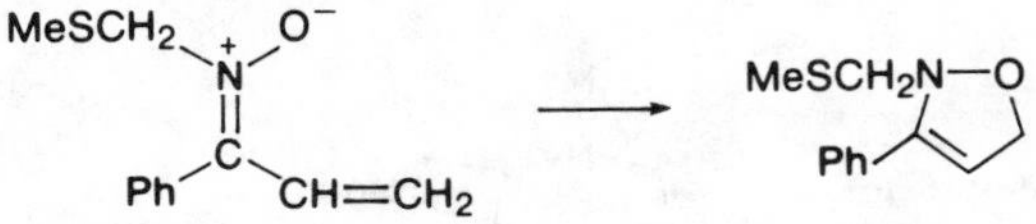

(**184**) (**185**)

Unusual products are obtained in the reactions of perfluoro-alkenes and -alkynes with diaryl nitrones. Although isoxazolines, which are the primary products of the cycloaddition of nitrones with alkynes, have been shown to undergo a variety of rearrangements[179], only recently has it been reported that the reaction of α-aryl *N*-phenyl nitrone with hexafluoro-2-butyne leads to indole **186**[211].

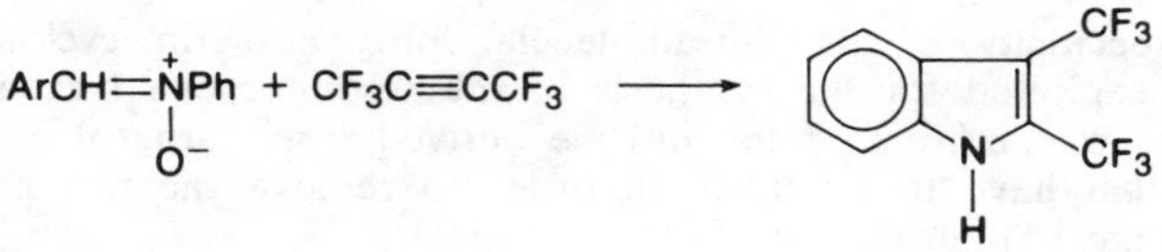

$$\mathrm{ArCH{=}\overset{+}{N}(O^-)Ph + CF_3C{\equiv}CCF_3 \longrightarrow}$$

(**186**)

Another unusual observation is the formation of β-lactam **187** in the reaction of *C,N*-diphenyl nitrone with hexafluoropropene[212], in contrast to previous reports of normal reactions of nitrones with hexafluoropropene[213].

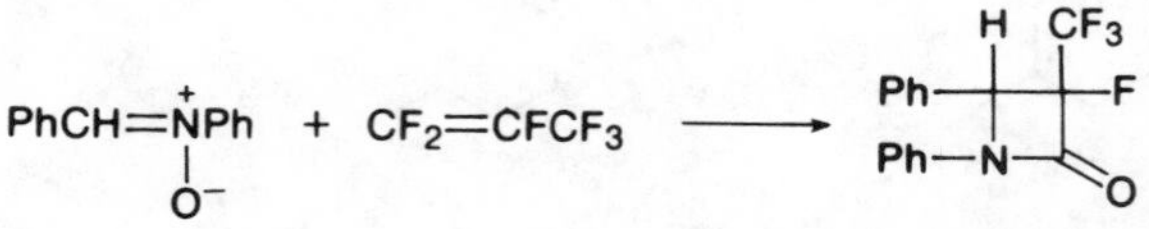

$$\mathrm{PhCH{=}\overset{+}{N}(O^-)Ph + CF_2{=}CFCF_3 \longrightarrow}$$

(**187**)

There are a number of reports concerning cycloadditions of chiral nitrones, Some nitrones derived from sugars have been reported to undergo such reactions with the formation of two epimers[214,215]. Some of the products are intermediates in the synthesis of nucleoside analogues.

The question of asymmetric induction has been studied by Vasella in nitrones derived from D-ribose[216]. Reactions of the protected ribose oxime **188** with aldehydes or acetone give nitrones that enter into cycloadditions with methyl methacrylate yielding adducts of type **189**. Cleavage of the glycoside bond and dehydrogenation give isoxazolines **190** in enantiomeric excess (e.e.) of 67–95%. An example using actaldehyde is depicted in Scheme 17.

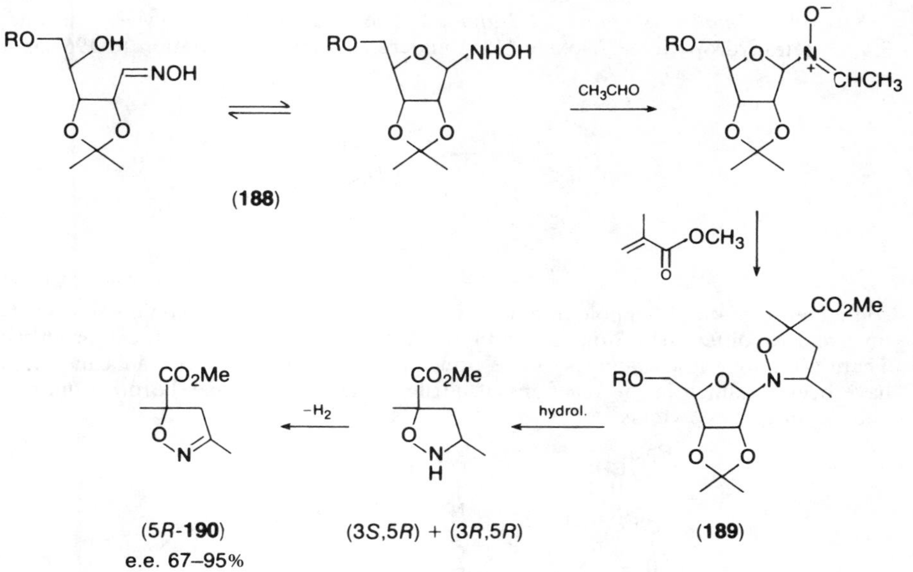

(188)

(5R-190)
e.e. 67–95%

(3S,5R) + (3R,5R)

(189)

SCHEME 17

The presence of a chiral group linked to the nitrogen has been found to be very effective in inducing asymmetry. For example *C*-phenyl *N*-(*S*)-1-phenylethyl nitrone (**191** gives dipolar cycloaddition products in some cases in enantiomeric excess of 100%[217]. Unfortunately reactions of this nitrone with many dipolarophiles give mixtures of three or all four of the possible regio- and stereo-isomers. Another nitrone that gives products of high optical purity is a derivative of L(−)-menthyl glyoxalate, **192**[217].

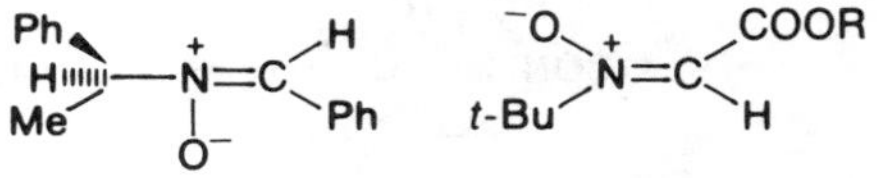

(191) (192)

R = (−)-menthyl

A different use of the 1,3-dipolar cycloaddition is its application to the polymerization of appropriately designed olefinic nitrones. One such compound is α-(*p*-maleimidophenyl) *N*-phenyl nitrone (**193**) which polymerizes readily by intermolecular cycloaddition to give a polymer of the structure **194**[218].

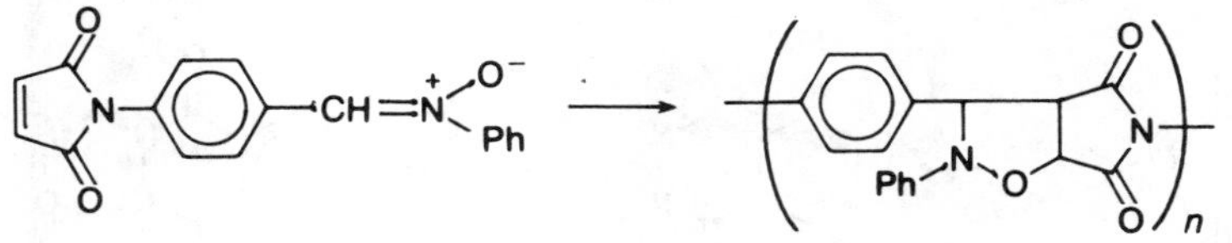

(193) (194)

b. Cycloaddition to other 1,3-dipoles. The six-membered cyclic nitrone, 3,4,5,6-tetrahydropyridine *N*-oxide (**195**) undergoes rapid dimerization to **196**[2] as a

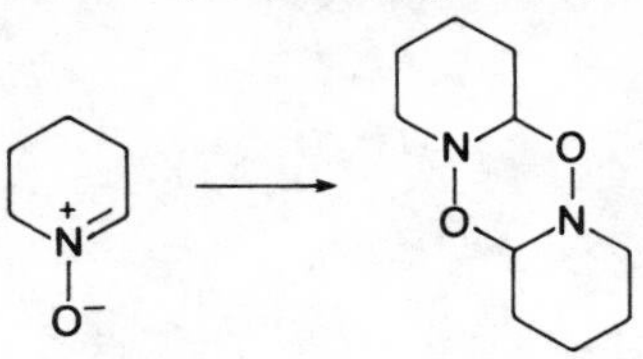

(**195**) (**196**)

consequence of dipole–dipole interaction. This type of interaction could conceivably involve two different 1,3-dipoles with the formation of unsymmetrical products. There are only a few examples of this principle in the literature. Dioxadiazine (**179**) have been obtained in the reactions of nitrile oxides with nitrone–boron trifluoride adducts in 20–45% yields[219].

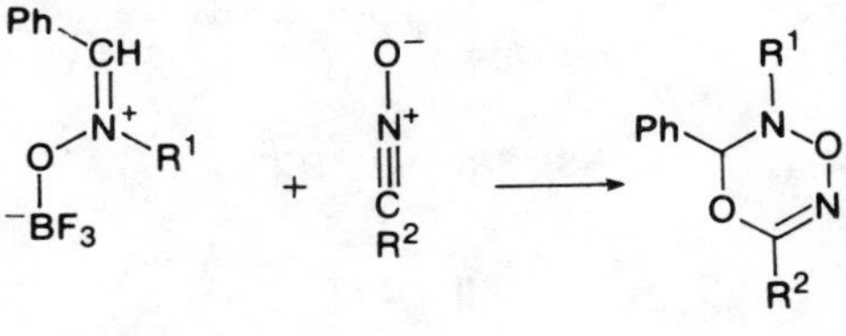

(**197**)

Some aziridines and diaziridines behave as masked 1,3-dipoles in reactions toward certain dipolarophiles. Such are 1-aroylaziridines, which undergo cycloaddition, when heated with nitrones, to oxadiazines (equation 41)[220]. Similar behaviour is shown by azaspiropentanes (equation 42)[221] and diaziridines (equation 43[222].

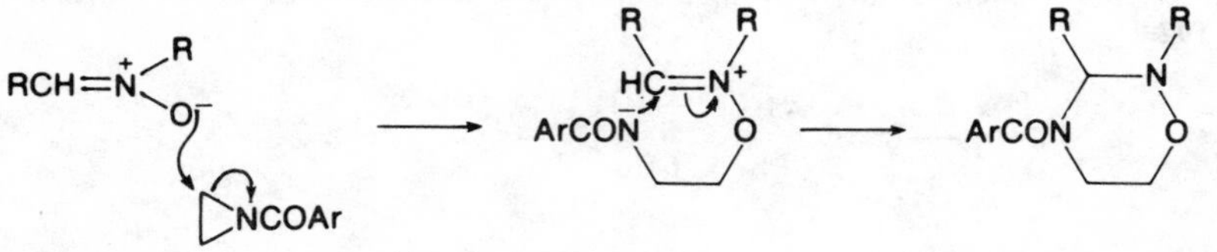

(41)

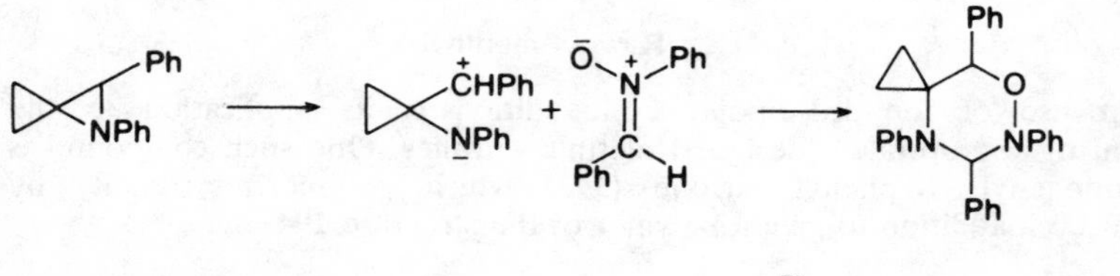

(42)

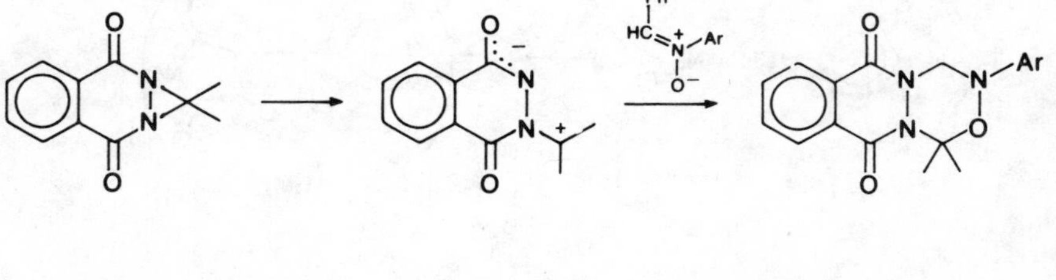

(43)

Related to this section is the reaction of nitrones with an aziridinium compound which is defined as a '1,3-polar' moiety since it behaves as a β-aminocarbonium ion. Its reaction with a nitrone is shown in equation (44)[223].

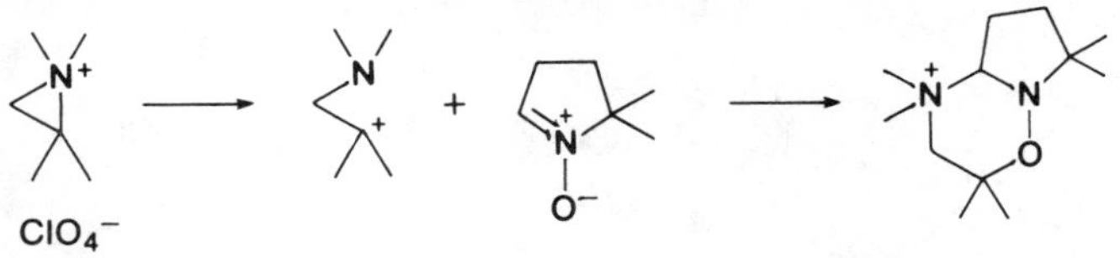

ClO_4^- (44)

c. *Reactions of C-(1-chloroalkyl)nitrones ('α-chloronitrones')*. This class of compounds, represented by formula **198** was introduced by Eschenmoser and named by him 'α-chloronitrones'[152]. This name is incorrect and contrary to *Chemical Abstracts* nomenclature, since the clorine is not located at the α-position of the nitrone function. The name α-chloronitrone is also misleading in view of the existence of true α-chloronitrones[224,225]. The usefulness of this class of compounds

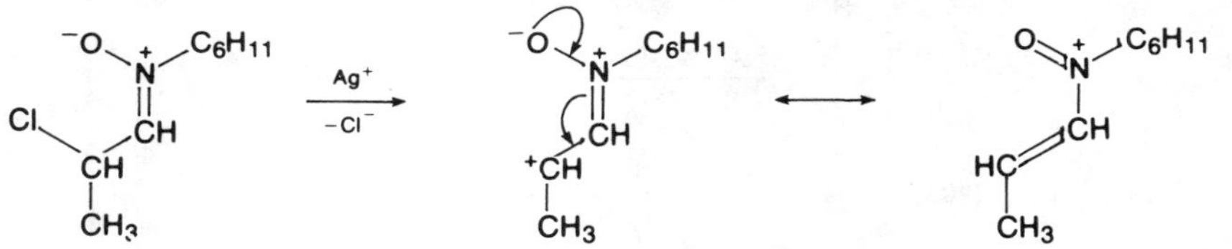

(198) **(199)**

stems from the fact that ionization of the chloride (brought about by silver ion) leads to the development of a positive charge in the α-position of the masked aldehyde function (*Umpolung*), which can be stabilized by the nitrone function to form the heterodienic *N*-alkyl-*N*-vinylnitrosonium cation (**199**). This is an electron-deficient diene and is capable of undergoing Diels–Alder-type cycloaddition to unactivated olefins[152] and acetylenes[226]. Double bonds that are capable of undergoing such cycloaddition may have up to four alkyl substituents or may be of enol ethers[227–229]. The adducts obtained from the cycloaddition of *N*-alkyl-*N*-vinylnitrosonium cations (e.g. **199** → **200**) can be used in a number of ways. For example stabilization of the

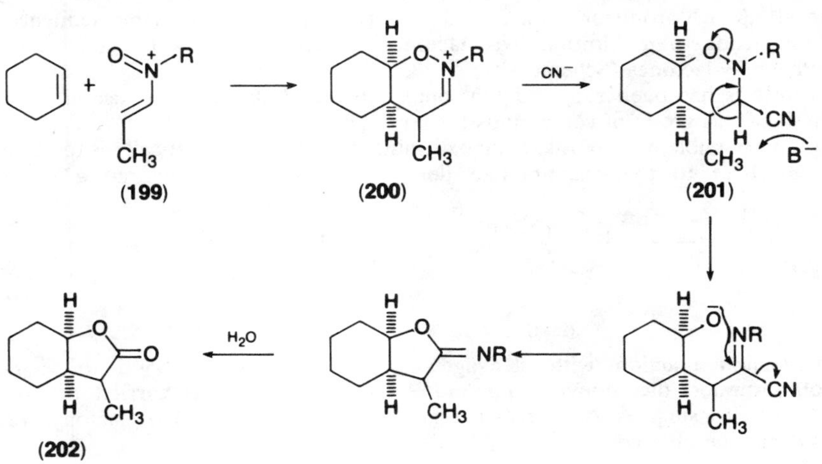

(199) **(200)** **(201)**

(202)

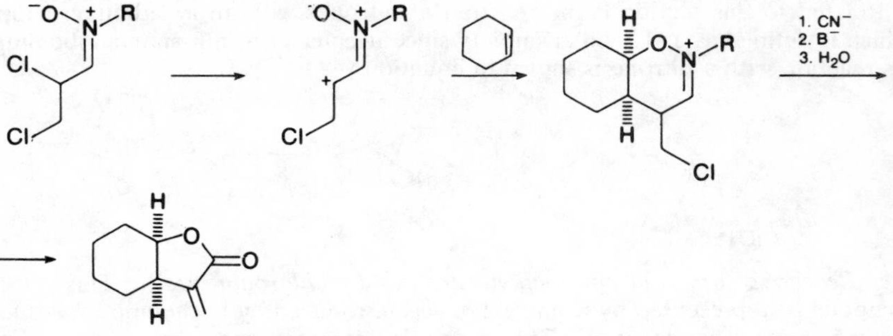

SCHEME 18a

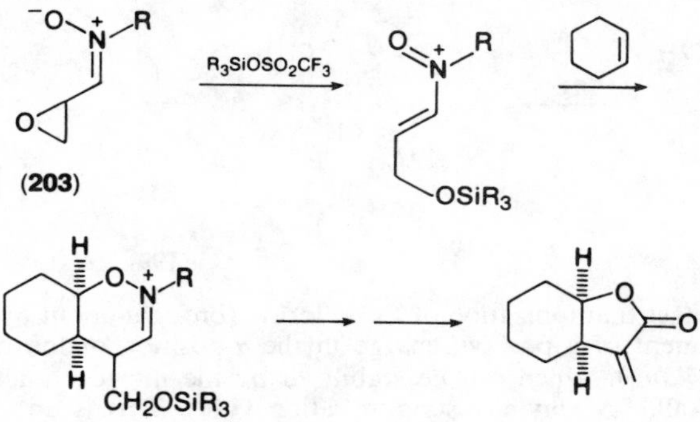

(203)

SCHEME 18b

oxazinium cation **200** by cyanide ion leads to a stable molecule **201** that can easily be converted to a γ-lactone **(202)**[230]. If *C*-(1,2-dichloroethyl) nitrones('α,β-dichloronitrones') are used as starting materials, the same sequence of reactions leads to α-chloromethyl-γ-lactones, which can easily be converted to α-methylene-γ-lactones (Scheme 18a)[231].

Recently is has been reported that epoxynitrones **(203)** are also useful for the synthesis of this group of compounds (Scheme 18b)[232].

Another possible way to utilize the oxazinium salts **(200)** synthetically is to convert them by base to the enamine-like derivatives **204** which undergo a thermal

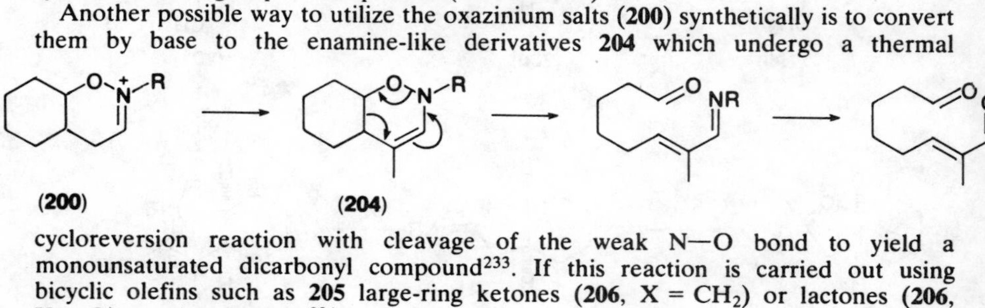

(200) (204)

cycloreversion reaction with cleavage of the weak N—O bond to yield a monounsaturated dicarbonyl compound[233]. If this reaction is carried out using bicyclic olefins such as **205** large-ring ketones (**206**, X = CH_2) or lactones (**206**, X = O) may be obtained[234].

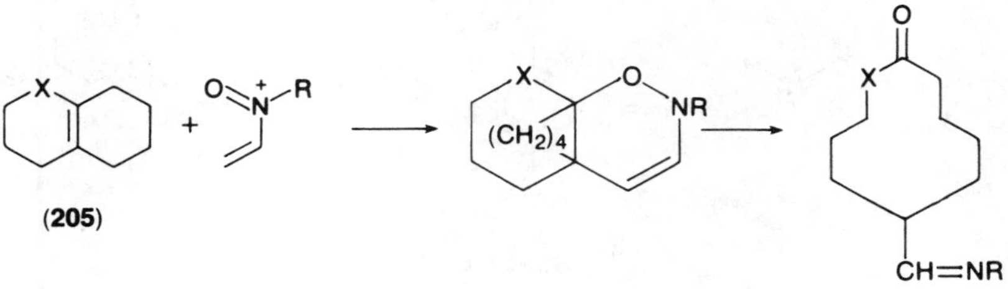

(205)

(206)

Cycloaddition of the chloronitrone-derived heterodienes with acetylenes provides a route to α,β-unsaturated ketones[226] (Scheme 19).

The cations obtained by the ionization of chloronitrones may also react as electrophilic reagents in Friedel–Crafts-like substitution reactions toward olefins and activated aromatic compounds[235]. If the reactions of *N*-alkyl-*N*-vinylnitrosonium cations with olefins are carried out in polar solvents such as liquid sulphur dioxide the main products are β,γ-unsaturated aldehydes (Scheme 20)[235]. Similarly reactions with activated aromatic systems yield α-arylaldehydes (Scheme 21)[235]. In these reactions the *N*-alkyl-*N*-alkenylnitrosonium ions serve as masked α-acylcarbonium ions (*Umpolung*).

N-Alkyl-*N*-alkenylnitrosonium ions derived from chloronitrones have also been found to cleave ethers[229,234]. A particularly elegant application of such a nitrosonium ion was its use by Eschenmoser and coworkers for selective hydrolysis of the amide function in cobyrinic acid hexamethyl ester monoamide to the

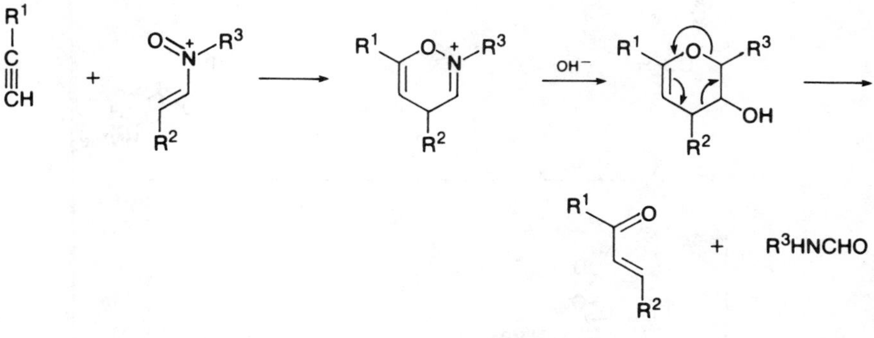

R^3HNCHO

SCHEME 19

60%
+
30% cycloadduct

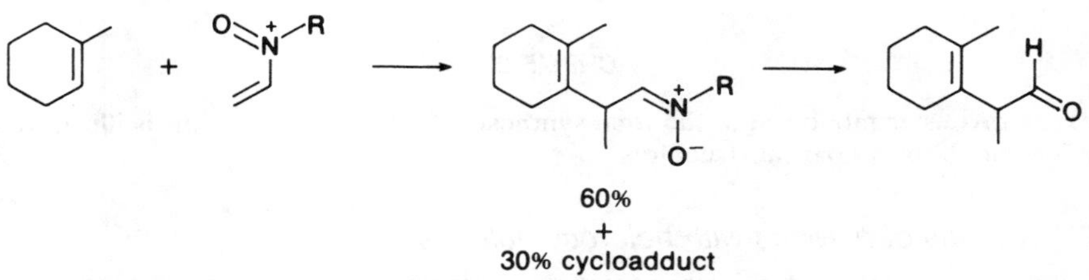

SCHEME 20

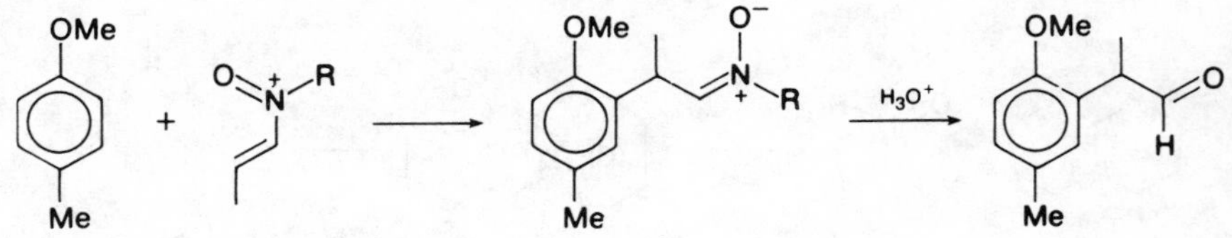

SCHEME 21

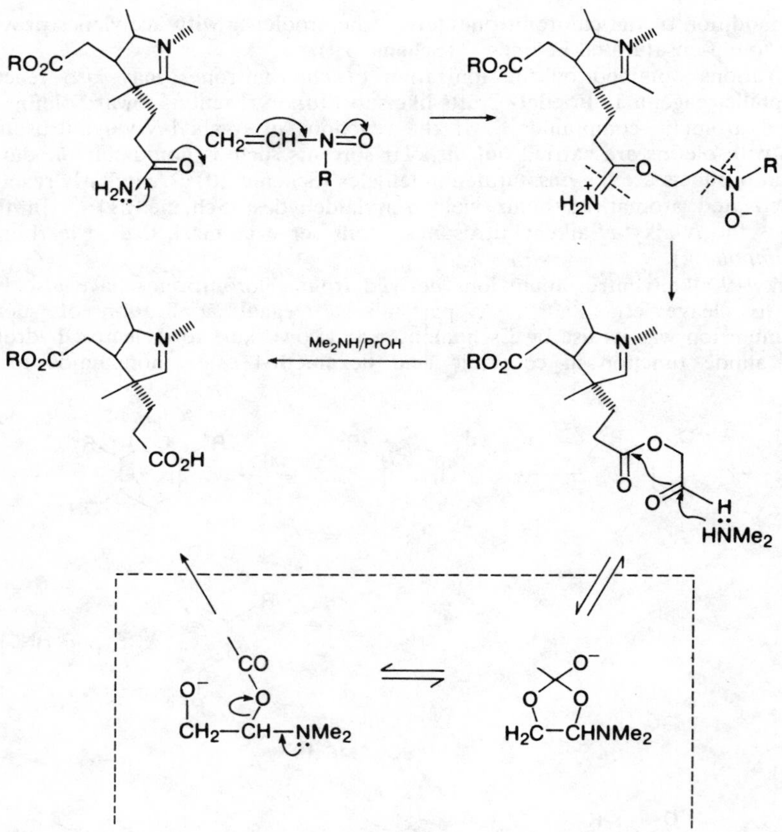

SCHEME 22

hexamethyl ester monoacid in the total synthesis of vitamin B_{12}[236]. This is illustrated in Scheme 22 using partial structures.

6. Reactions of nitrones with heterocumulenes

With the exception of a recent report describing an unusual mode of reaction of nitrones with allenes leading to a rearranged product depicted in Scheme 23[237], in

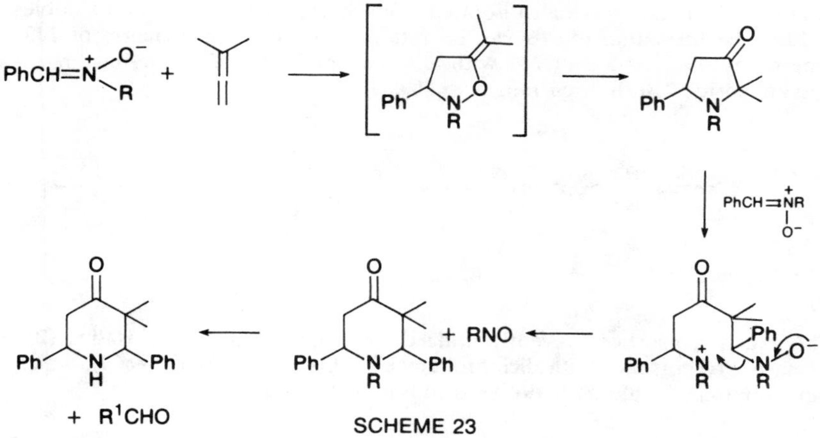

SCHEME 23

general the reactions of nitrones with allenes lead to cycloaddition products in analogy to simple olefins. In contrast, reactions of many heterocumulenes with nitrones take a course entirely different from that of 1,3-dipolar cycloaddition. This difference in behaviour can be attributed in part to the polar nature of the double bonds (or at least one double bond) in heterocumulenes. There are also similarities that can be noted between many of the reactions discussed in this section and the one dealing with electrophilic reagents (Section II.B.7.a) due to the same reasons.

a. Ketenes. The reaction of nitrones with ketenes was first studied by Staudinger and Miescher[238]. They, and later others[239,240], found that *N*-aryl nitrones react with ketenes to produce *o*-imino-α,α-disubstituted-arylacetic acids (**207**). The mechanism suggested differs from cycloaddition, and involves attack of the nitrone oxygen upon the carbonyl carbon of the ketene, followed by sigmatropic rearrangement and proton transfer (Scheme 24).

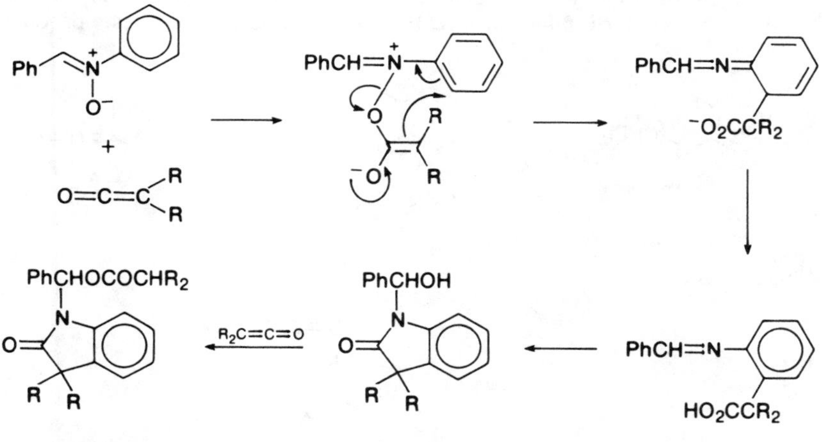

(**209**) (**208**) (**207**)

SCHEME 24

Other products that were also isolated from these reactions are the oxindoles **208** and **209**. The formation of **209** can be rationalized either by reaction of **208** with ketene or by the reaction of **207** with a second molecule of ketene leading to the mixed anhydride **210** that rearranges to **209**.

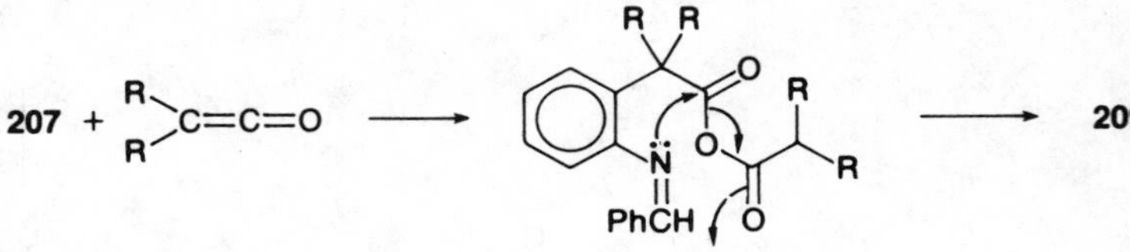

207 + $R_2C{=}C{=}O$ ⟶ (**210**) ⟶ **209**

This reaction has been recently utilized for a new synthesis of isatins (**211**) by reacting *N*-aryl nitrones with dichloroketene[241]. Oxindole **209** (R = Cl), which was obtained initially, could be hydrolysed to isatins in good yields.

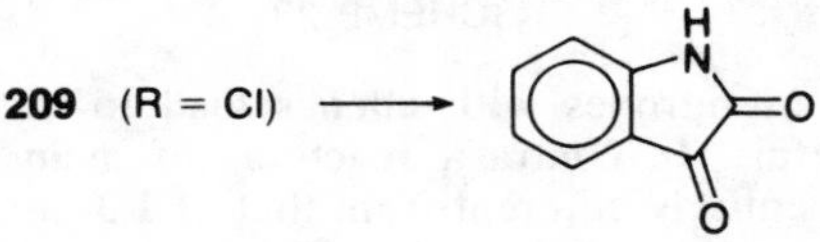

209 (R = Cl) ⟶ (**211**)

In contrast the reaction of the *N*-2,6-disubstituted-aryl nitrone gives lactone **212**, presumably through a mechanism which is similar to that described previously (Scheme 25)[242].

Further work by Taylor and coworkers has concerned reactions of nitrones derived from 9-fluorenone. *N*-Alkyl nitrones of this series have given with dimethylketene spiranic products **213** and **214**[243].

It was assumed that the β-lactam, **213**, resulted from deoxygenation of the nitrone followed by cycloaddition of the fluorenone imine to the ketene. The reaction of the same nitrones with diphenylketene give the analogous oxazolidinones (**214**, R^1 = Ph) but a different kind of β-lactam (**215**)[244]. The formation of these products is

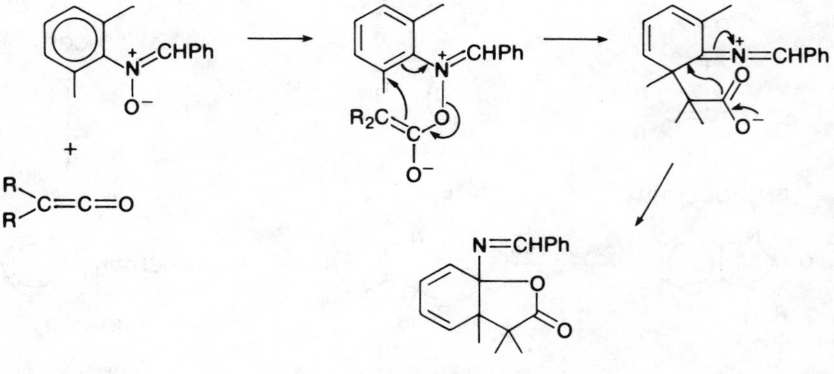

(**212**)

SCHEME 25

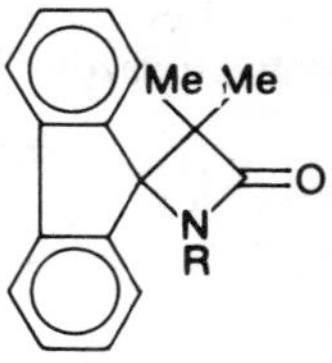

(213) 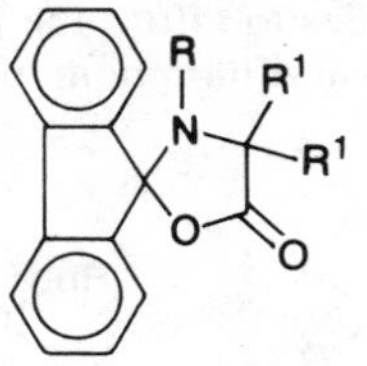 (214)

R^1 = Me

rationalized by assuming the formation of the zwitterionic intermediate **216**, which may undergo two types of sigmatropic shifts leading to the two products (Scheme 26).

The more electron-deficient α-cyano-*t*-butylketene reacts with *N*-alkyl-*N*-

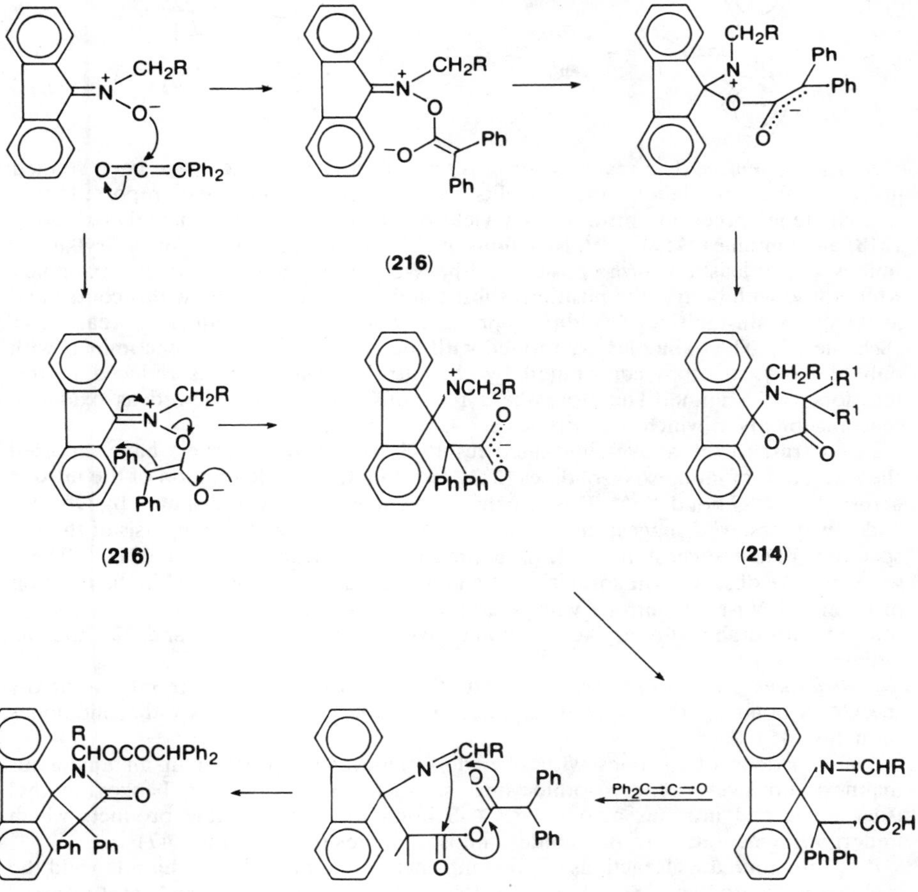

(216)

(216) (214)

(215)

SCHEME 26

fluorenylidene *N*-oxides to afford the product of 1,3-cycloaddition, **217**[245]. The same ketene reacts with *N*-arylfluorenone nitrone to give oxazolidinones (equation 45)[245].

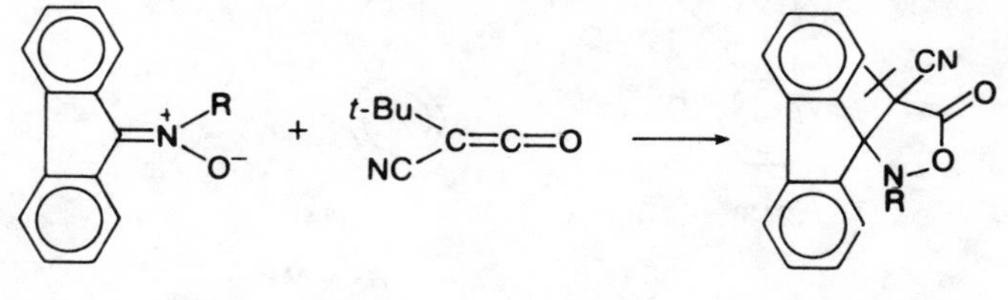

(217)

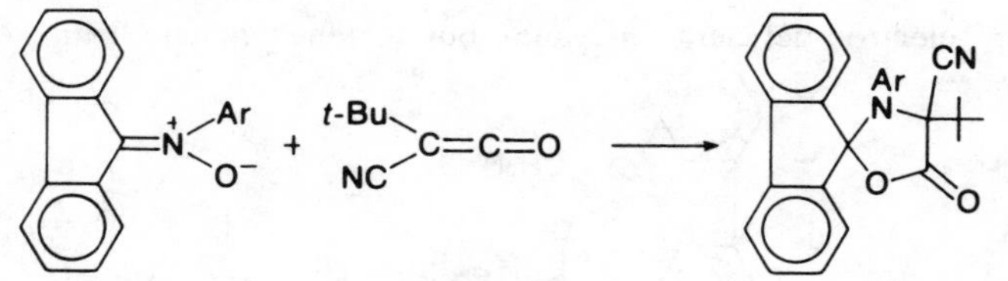

(45)

b. Ketene imines. The reactions of nitrones with ketene imines have been studied in a number of laboratories. Del'tsova and coworkers have reported that *N*-arylketene imines and nitrones may yield two types of products, namely oxindoles **(218)** and amidines **(219)**[246,247]. Reactions of *C,N*-diphenyl nitrones with *N*-arylketene imines with at least one *ortho* position of the *N*-aryl group unoccupied give oxindoles, while those with both *ortho* positions substituted are unable to follow this course and lead to an unstable cycloaddition product **220** which subsequently rearranges (Scheme 27). The oxindoles are formed with the elements of the ketene imines, with only the oxygen atom contributed by the nitrone. The latter is reduced in this reaction to the imine. Therefore these reactions can also be viewed as oxidative rearrangements in which the nitrones act as oxidizing agents.

In contrast to the above-mentioned results Barker and coworkers have reported the formation of iminoisoxazolidines **(221)** resulting from cycloaddition of the nitrone across the C=C bond[248,249]. This structure assignment has been disputed by Ohshiro and coworkers, who suggest the imidazolidinone structure **222** on the basis of the ^{13}C spectrum. They suggest that **222** is a product of rearrangement of **221**[250]. These workers have observed the formation of an imidazolidinone of type **222** in the reaction of *C*-phenyl *N-t*-butyl nitrone with *N*-aryldimethylketene imine. This nitrone, when reacted with diphenyl-*N*-arylketene imine gives the oxindoles **223** and **224** and the aldimine[250].

c. Isocyanates and isothiocyanates. The functional groups add the nitrone across the C=N bond. In the reactions of isocyanates the products are oxadiazolidinones (equation 46)[209].

The reactions of nitrones with phenyl isothiocyanate proceed in an analogous manner; however, other isothiocyanates, such as *p*-tolyl- or benzoyl-methyl isocyanates add nitrones across the C=S bond to yield unstable products which undergo fragmentation to thioamides and isocyanates (e.g. equation 47)[251].

d. Carbodiimides. Reactions of alkyl nitrones with diphenylcarbodiimide yield the cycloaddition product **225**[252]. However, the analogous products from *N*-aryl nitrones are apparently not as stable and therefore only secondary products are isolated (Scheme 28)[253].

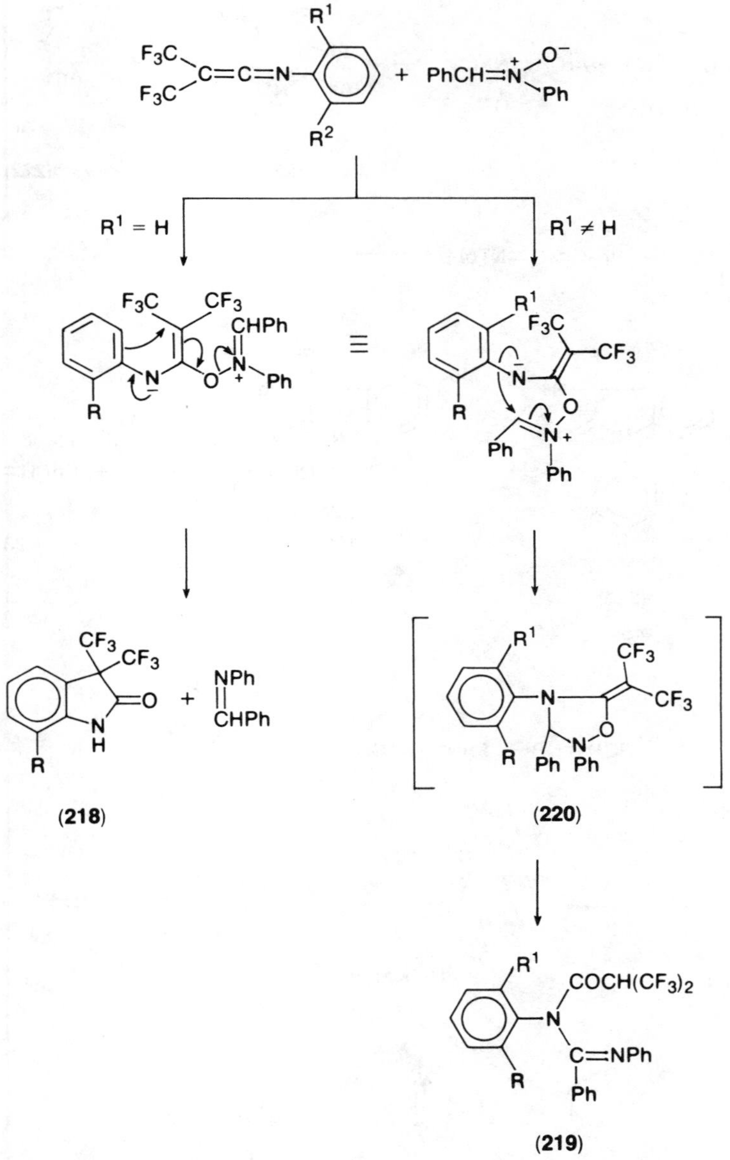

(218)

(220)

(219)

SCHEME 27

The fragmentation leading to amidine with loss of isocyanate is reminiscent of what was observed in the case of isothiocyanate. The amidine formation is sometimes accompanied by the migration of a group from the carbon to nitrogen[253].

e. Sulphines. There is only one paper concerned with the reaction of this

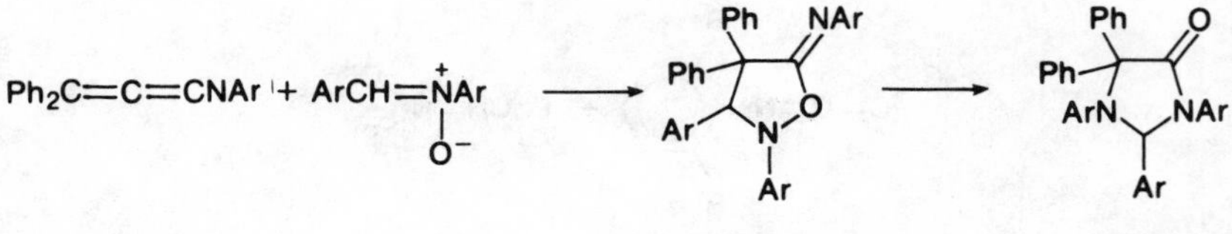

$Ph_2C{=}C{=}CNAr + ArCH{=}\overset{+}{N}(O^-)Ar \longrightarrow$ (221) $\longrightarrow$ (222)

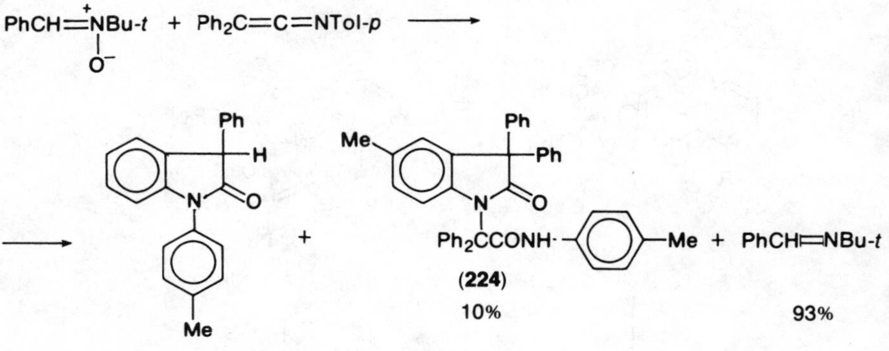

$PhCH{=}\overset{+}{N}(O^-)Bu\text{-}t + Ph_2C{=}C{=}NTol\text{-}p \longrightarrow$ (223) 45% + (224) 10% + $PhCH{=}NBu\text{-}t$ 93%

$RCH{=}\overset{+}{N}(O^-)R + PhN{=}C{=}O \longrightarrow$ (46)

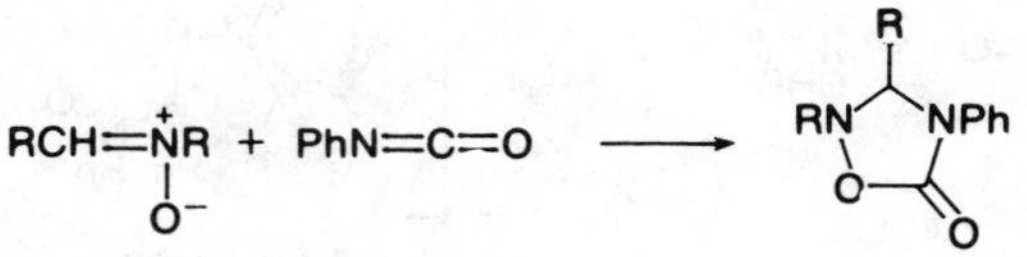

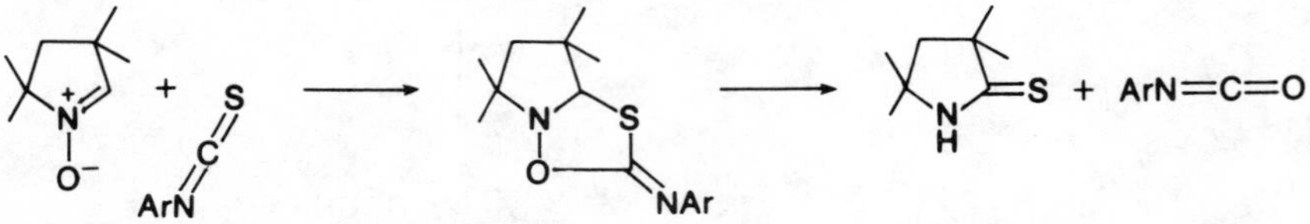

(47)

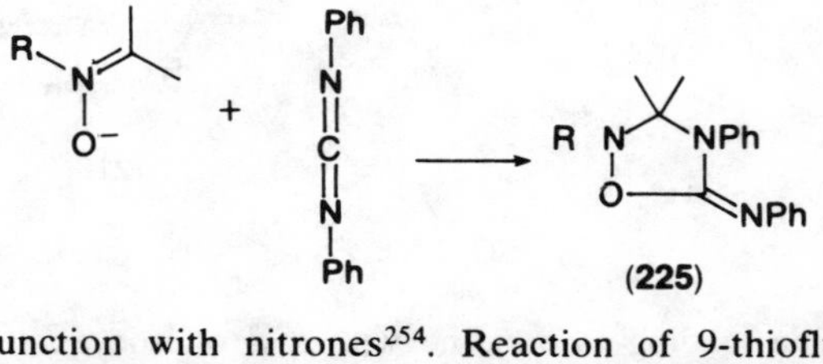

(225)

heterocumulene function with nitrones[254]. Reaction of 9-thiofluorenone S-oxide (**226**) with C-phenyl N-methyl nitrone gives in high yield product **227** formed from two moles of thiofluorenone and one mole of nitrone with the loss of sulphur dioxide (Scheme 29).

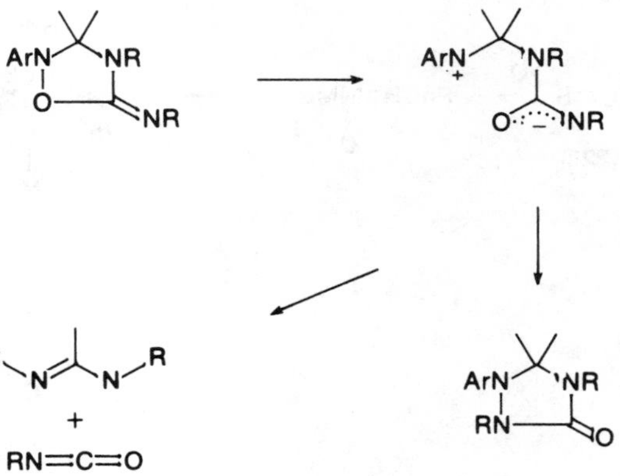

SCHEME 28

Sulphines derived from thiobenzophenone and its substituted derivatives were found to be not sufficiently reactive towards this nitrone. Bis(phenylthio)sulphine (**228**) was found to react slowly with *C*-phenyl *N*-methyl nitrone to give in high yield the enamine **229**[254].

f. N-*Sulphinyl compounds*. The reaction of *N*-sulphinylbenzenesulphonamide (**230**) with two types of nitrones has been studied. The reaction between **230** and *C,N*-diaryl nitrones yields *N*-benzenesulphonyl-*N′,N′*-diarylformamidine (**231**) (Scheme 30). The formation of this product is postulated to proceed via the fragmentation of the cycloadduct **232** involving carbon to nitrogen migration of an aryl group similarly to what was observed with the products of carbodiimides[255]. The reaction of an *N*-alkyl nitrone also yields an amidine (**233**) which in this case is

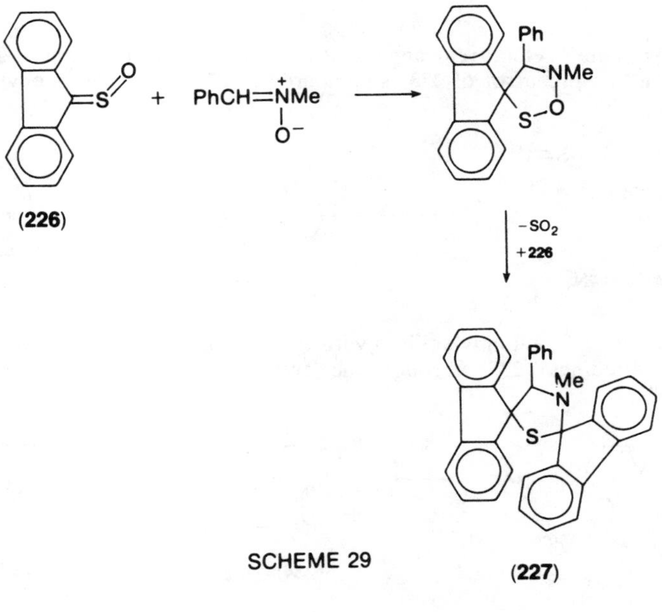

(**226**)

$-SO_2$
+**226**

SCHEME 29

(**227**)

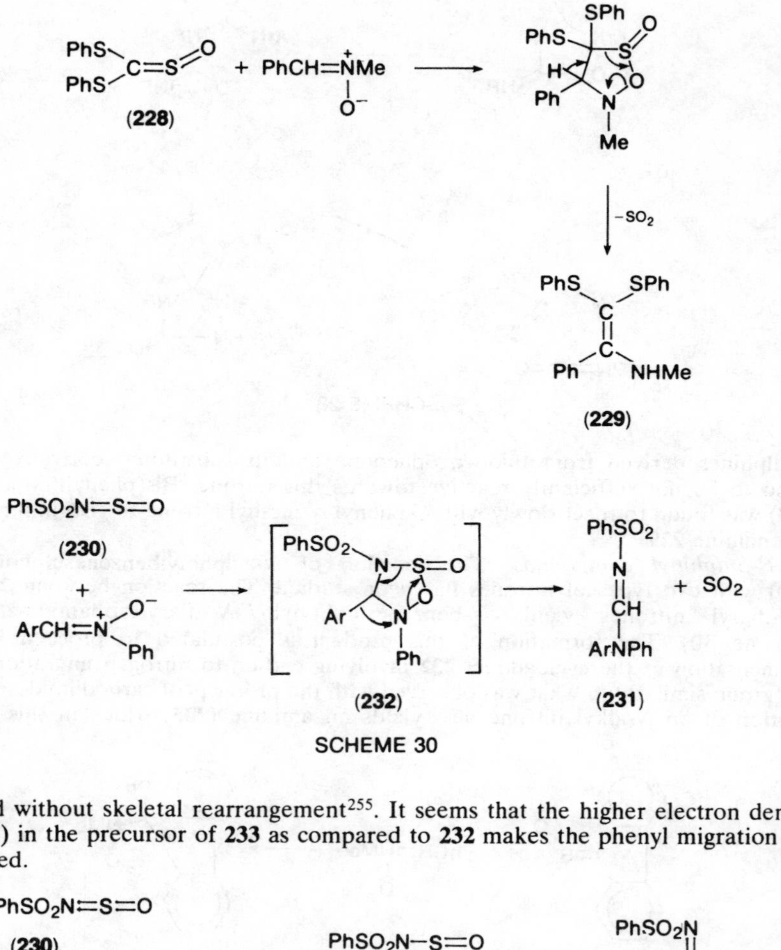

(228)

(229)

(230)

(232)

(231)

SCHEME 30

formed without skeletal rearrangement[255]. It seems that the higher electron density of N(2) in the precursor of **233** as compared to **232** makes the phenyl migration less favoured.

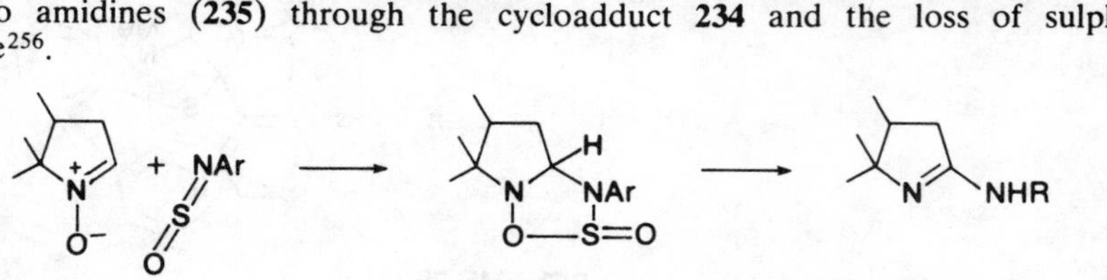

(230)

(233)

Reactions of *N*-sulphinylanilines with cyclic aliphatic nitrones have been shown to lead to amidines (**235**) through the cycloadduct **234** and the loss of sulphur dioxide[256].

(234)

(235)

g. Sulphenes. The reaction of *N*-aryl nitrones with sulphene has been shown by Truce and coworkers to lead to benzoxathiazepines **237** via rearrangement of the initial cycloadduct **236**[257–259]. The rearrangement is essentially a migration from

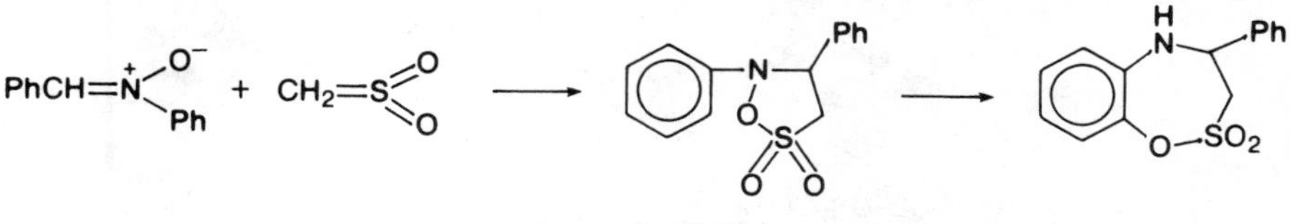

(**236**) (**237**)

nitrogen to the *ortho* position of the aromatic ring, to which there are many precedents in the literature. Several mechanisms have been considered for this rearrangement[258].

α-Ketosulphenes have been reported to react with *N*-aryl nitrones similarly to sulphene[260]. Thus the reaction of benzoylsulphene with a *C,N*-diaryl nitrone yields two benzoxathiazepines **238** and **239**, the first of which has been shown to lose a benzaldehyde in a fragmentation involving an unusual rearrangement leading to **239**, for which the pathway depicted in Scheme 31 has been proposed[260].

Cyclic nitrones of the pyrroline *N*-oxide series react with sulphene or with phenylsulphene in benzene to yield the corresponding iminosulphonic acids[261,262] (equation 48).

In contrast the reactions of α-ketosulphenes with these nitrones in dioxane lead to ketoenamines (**240**) which may react with a second mole of ketosulphene[262] (Scheme 32). Similar behaviour is shown by cyclic ketosulphenes derived from indane and tetralin.

The reaction of α-methyl nitrone (**241**) with a ketosulphene yields a product with a rearranged skeleton **242**, which may add a second mole of ketosulphene to yield **243**[262]. The reaction of ketosulphenes can be directed towards the formation of iminosulphonic acids by performing the reactions in methylene chloride or acetonitrile instead of dioxane[262].

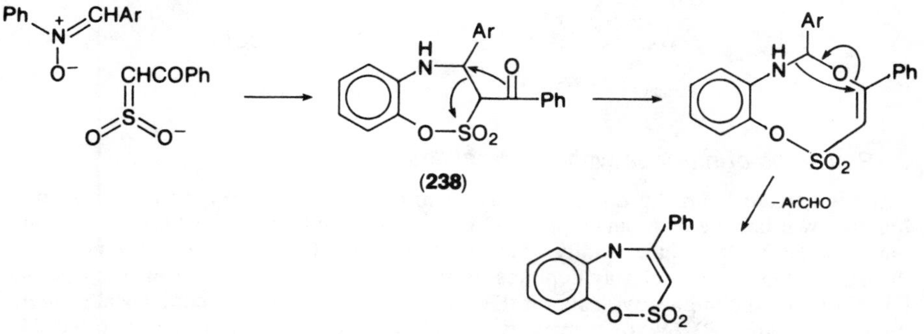

(**238**)

−ArCHO

(**239**)

SCHEME 31

PhH

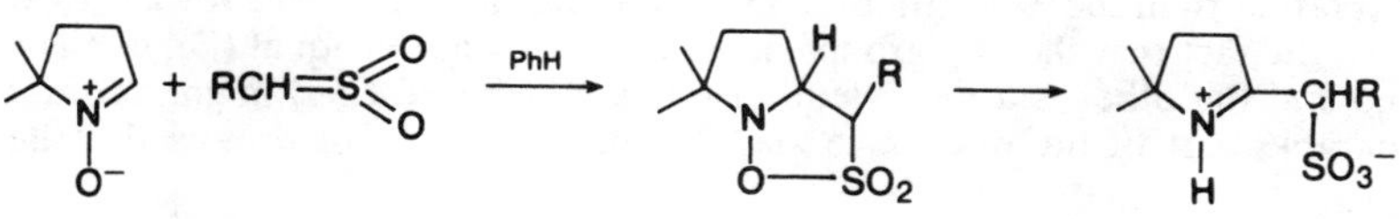

(48)

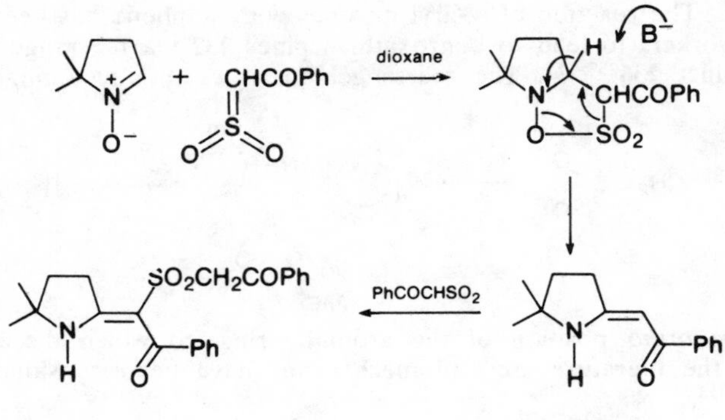

dioxane

PhCOCHSO$_2$

(240)

SCHEME 32

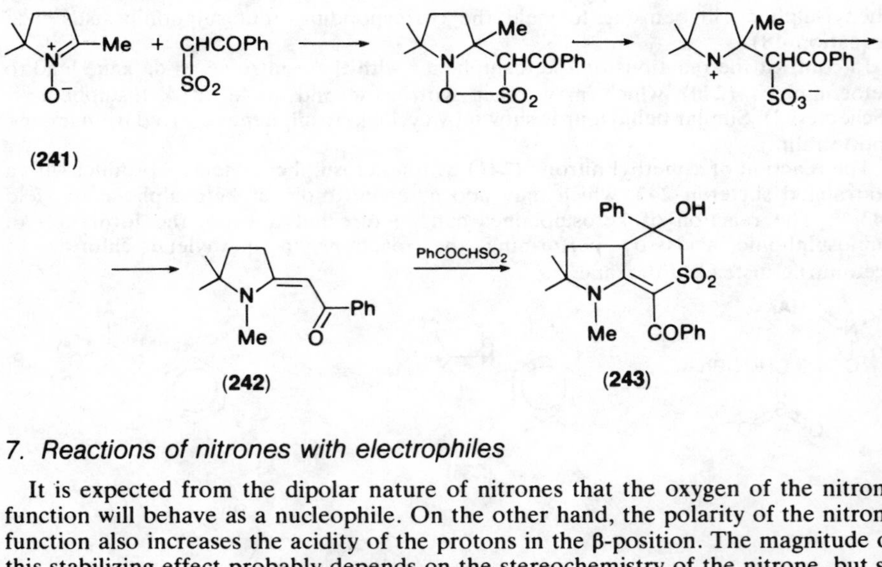

(241)

PhCOCHSO$_2$

(242)

(243)

7. *Reactions of nitrones with electrophiles*

It is expected from the dipolar nature of nitrones that the oxygen of the nitrone function will behave as a nucleophile. On the other hand, the polarity of the nitrone function also increases the acidity of the protons in the β-position. The magnitude of this stabilizing effect probably depends on the stereochemistry of the nitrone, but so far there are no comparative data on the acidities of β-carbons in geometrical isomers such as (*E*)- and (*Z*)-**244**, or the relative stabilities of the two types of carbonions **245** and **246**.

The tendency of 2-methylpyrroline *N*-oxide derivatives to undergo deuterium exchange[61] (equation 49) and aldol-type condensation[2,3] (equation 50) involving the methyl group rather than the 3-position seems to indicate the preference for **245** over **246**. However, the fact that the two groups being compared are unequal (i.e. one is a methyl group and the other is a methylene in a five-membered ring) might account for a considerable part of the difference and therefore one cannot draw a definite conclusion.

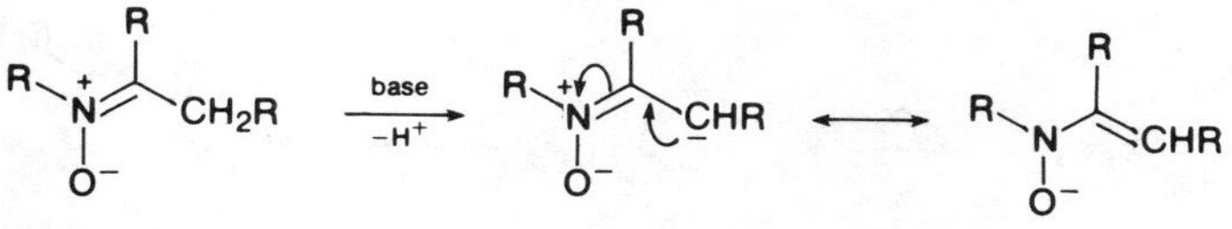

(*E*-**244**) (**245**)

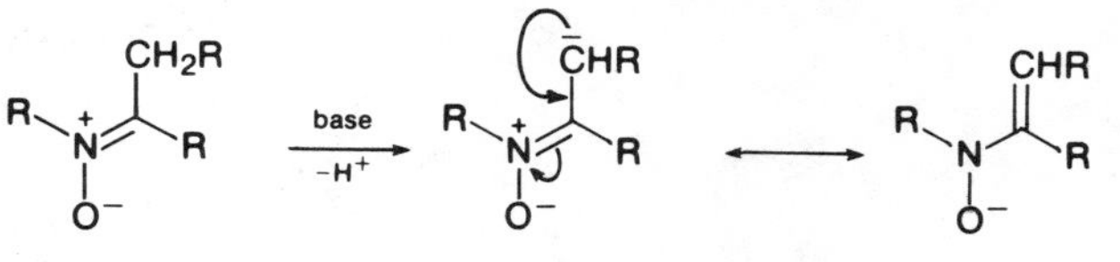

(*Z*-**244**) (**246**)

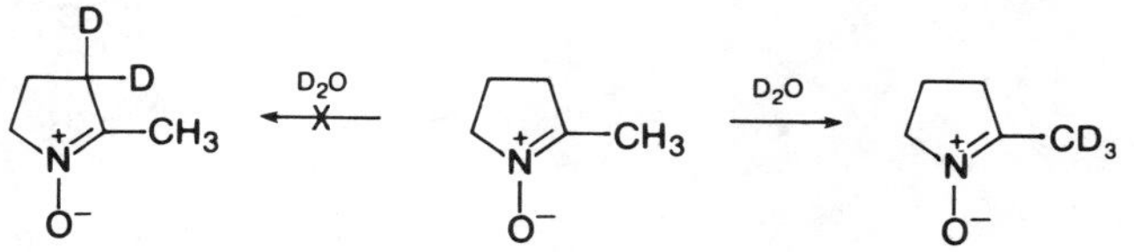

(49)

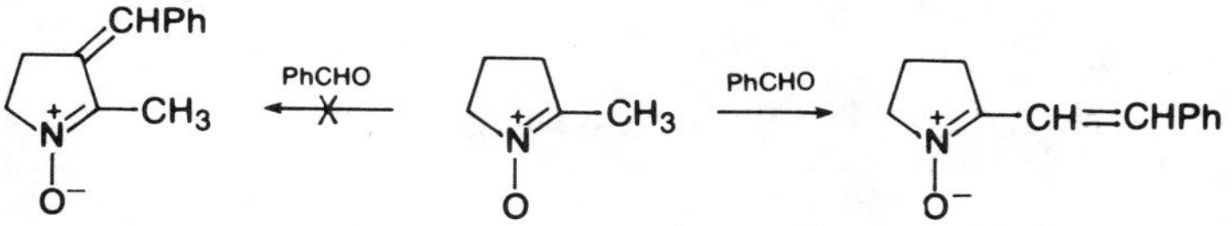

(50)

Another relevant aspect is the question of *O*- versus *C*-alkylation, which has received a great deal of attention in carbonyl compounds, but has not been studied systematically in nitrones.

However, nitrones clearly are ambident nucleophiles and they may react with electrophilic reagents either as oxygen or as carbon nucleophiles and accordingly this section will be divided into two parts.

a. Nitrones as oxygen nucleophiles. In addition to the reactions that will be reviewed here it is appropriate to point out that the acid-catalysed Beckmann rearrangement of nitrones (Section II.B.1.c) as well as many of the reactions of nitrones with heterocumulenes (Section II.B.6) may be classified as belonging to this section, as their first step is a nucleophilic attack of the nitrone oxygen upon the electrophilic site of the reagent. The present discussion deals almost exclusively with reactions of nitrones with acid derivatives.

The reaction of 2,4,4-trimethylpyrroline *N*-oxide with acetic anhydride[156] or benzoyl chloride[263] yields the diacyl derivatives of type **247**. It is reasonable to assume the formation of **247** as taking place by the mechanism indicated by Scheme 33[61].

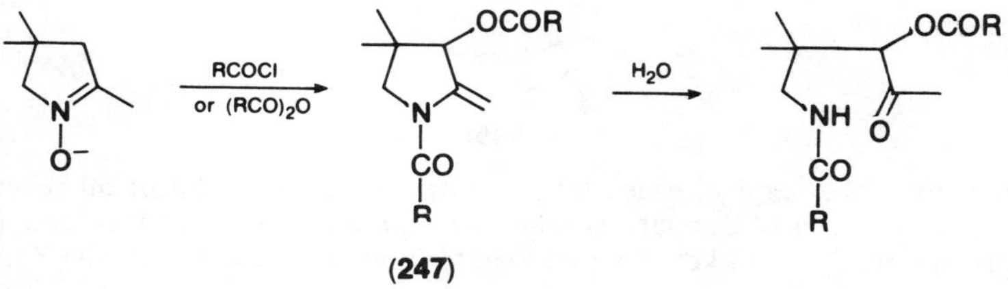

(**247**)

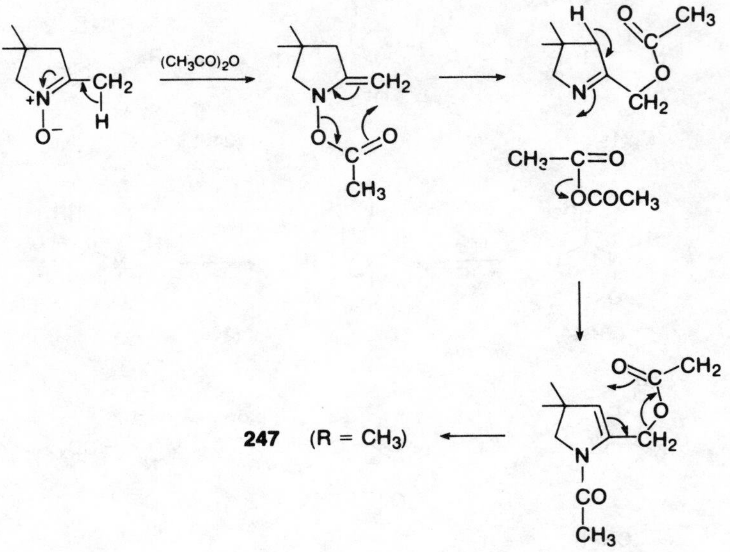

SCHEME 33

The steroidal nitrone **109** (cf. Section II.B.1.b) exhibits a different kind of behaviour. Its treatment with benzoyl chloride followed by water gives the α-hydroxy *N*-benzoyloxy compound (**248**) which can be rearranged thermally to the imine **249**[126].

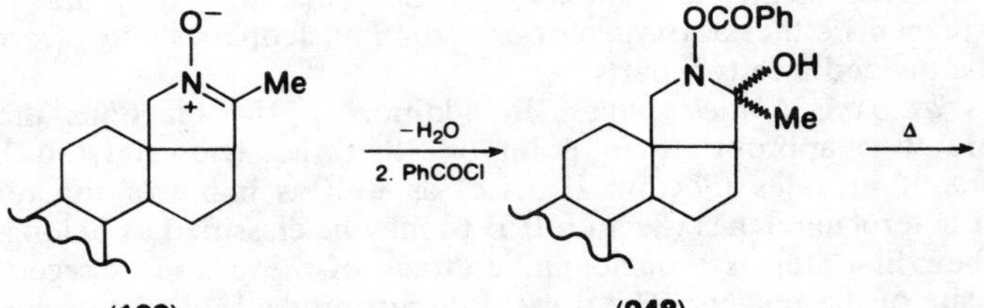

(**109**) (**248**)

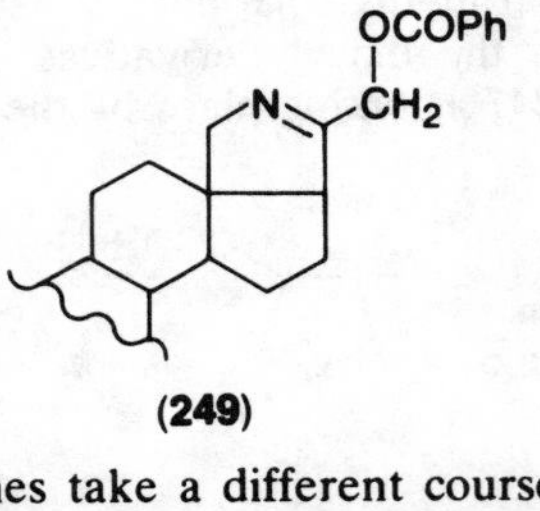

(**249**)

The reactions of *N*-aryl nitrones take a different course. Almost all reactions of *N*-aryl nitrones with acid derivatives lead to *ortho* substitution in the aromatic ring resembling the reactions of ketenes that lead to indoles (Section II.B.6.a). *C,N*-diaryl nitrones react with phosgene or with thionyl chloride to give in high yield, and in most cases exclusively, the *ortho*-chlorinated imine **250**[264,265]. A cyclic mechanism has

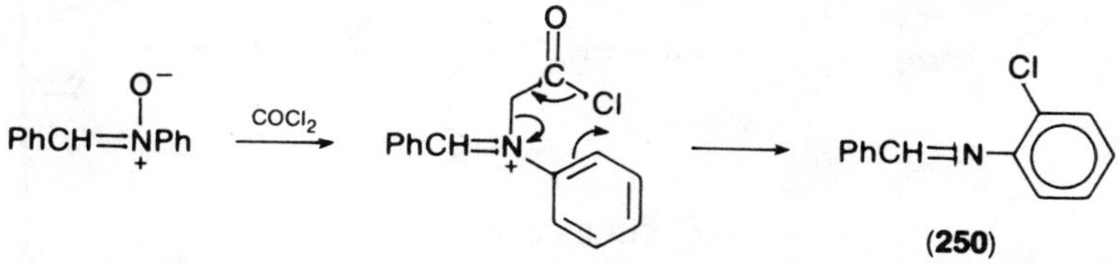

PhCH=N(O⁻)Ph →($COCl_2$) PhCH=N⁺(OCOCl)Ph → PhCH=N–C₆H₄Cl-o

(250)

been suggested for the reaction. The related reaction of *N*-aryl nitrones with oxalyl chlc ide leads to the introduction of the chlorooxalyl group into the *ortho* position of the *N*-aryl group (equation 51)[266].

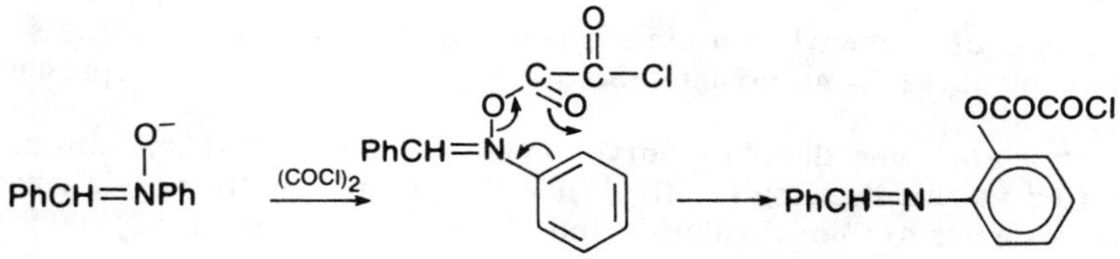

PhCH=N(O⁻)Ph →($(COCl)_2$) PhCH=N⁺(OCOCOCl)Ph → PhCH=N–C₆H₄(OCOCOCl)-o (51)

Reaction of *C,N*-diphenyl nitrone with formic acid results also in a product of *ortho* substitution and reduction, **251**[267]. Its formation is rationalized in Scheme 34.

Seeking evidence for the existence of *N*-aroyloxy-*N*-benzylideneammonium chloride (**252**), the assumed intermediate in the acyl-halide-catalysed Beckmann rearrangement of nitrones, Heine and coworkers treated α-phenyl *N*-alkyl nitrones with aroyl chlorides in wet solvents. The products, *N*-alkyl-*O*-aroyloxyhydroxylamines (**253**) which were isolated in good yields can be considered to arise from the postulated intermediates (**252**) through hydrolysis[268].

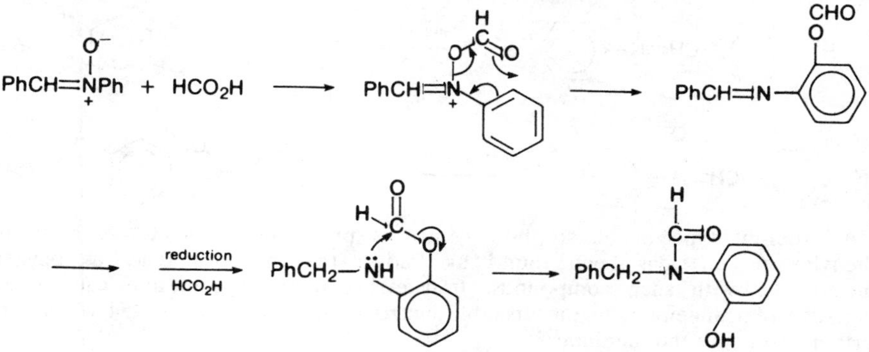

PhCH=N⁺(O⁻)Ph + HCO_2H → PhCH=N⁺(OCHO)Ph → PhCH=N–C₆H₄(OCHO)-o →(reduction, HCO_2H) PhCH₂–NH–C₆H₄(OCHO)-o → PhCH₂–N(CHO)–C₆H₄OH-o

(251)

SCHEME 34

PhCH=N⁺(O⁻)R + ArCOCl → [PhCH=N⁺R(OCOAr)] Cl⁻ →(H_2O) ArCOONHR

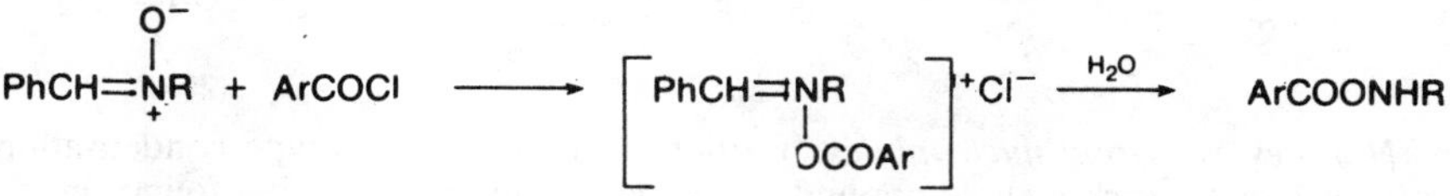

(252) **(253)**

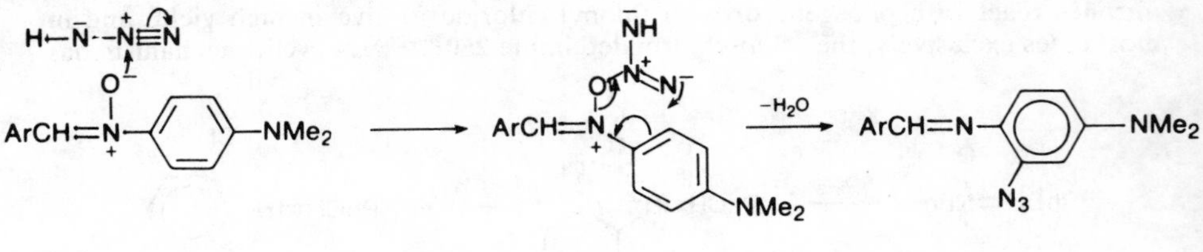

SCHEME 35

In contrast, treatment of *N*-aryl nitrones with aroyl chlorides resulted in *ortho* substitution and the *N,O*-diacyl derivative of the corresponding *o*-aminophenol was isolated[268].

The reaction of α-phenyl *N-p*-dimethylaminophenyl nitrone that leads to the introduction of an azido group into the *ortho* position[269] may be represented by Scheme 35.

The reactions of some thiophosphoryl compounds with *C,N*-diaryl nitrones have been reported to give puzzling results. While thiophosphoryl trichloride gave *ortho* chlorination, similar to thionyl chloride and phosgene[270] (equation 52), the reaction

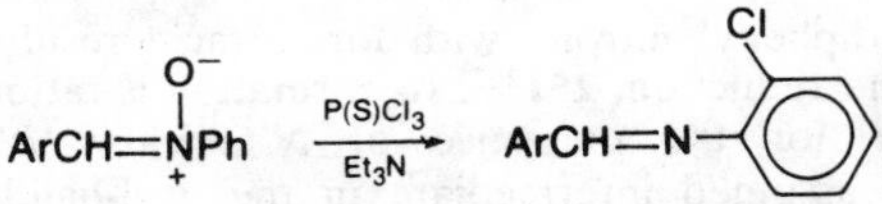

$$\mathrm{ArCH{=}\overset{+}{N}(O^-)Ph} \xrightarrow[\mathrm{Et_3N}]{\mathrm{P(S)Cl_3}} \mathrm{ArCH{=}N{-}C_6H_4{-}Cl}\ (ortho) \tag{52}$$

of phenylphosphonothioic dichloride with *C,N*-diaryl nitrones gave benzothiazoles[270] (equation 53). In contrast reactions of methyl diphenylphosphonothioate gave the corresponding benzoxazoles (equation 54)[270].

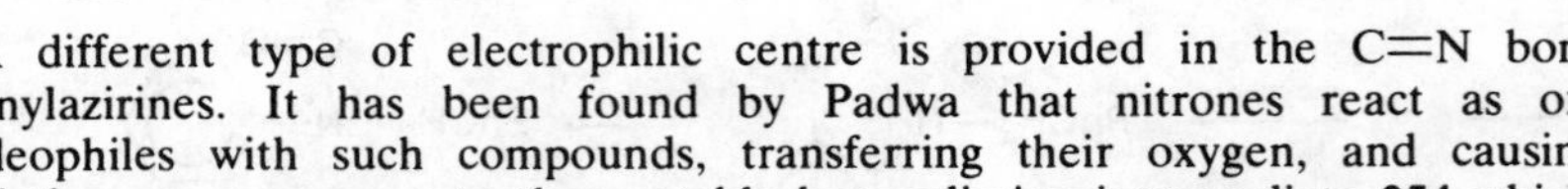

$$\mathrm{R^2{-}C_6H_4{-}CH{=}\overset{+}{N}(O^-){-}C_6H_4{-}R^1} \xrightarrow{\mathrm{PhP(S)Cl_2}} \text{2-(}R^2\text{-phenyl)benzothiazole bearing } R^1 \tag{53}$$

$$\mathrm{R^2{-}C_6H_4{-}CH{=}\overset{+}{N}(O^-){-}C_6H_4{-}R^1} \xrightarrow{\mathrm{Ph_2P(S)OMe}} \text{2-(}R^2\text{-phenyl)benzoxazole bearing } R^1 \tag{54}$$

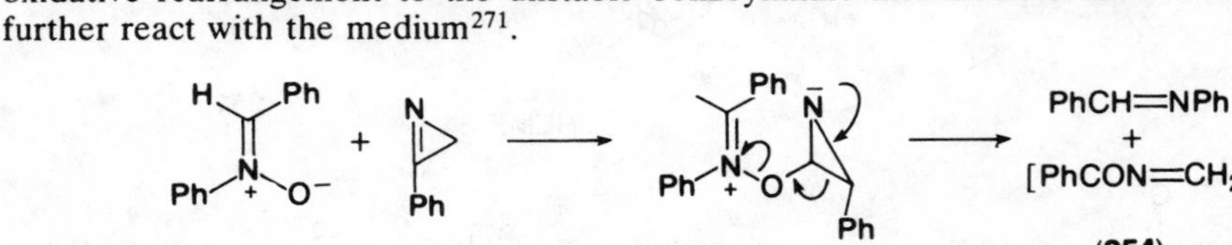

A different type of electrophilic centre is provided in the C=N bond of phenylazirines. It has been found by Padwa that nitrones react as oxygen nucleophiles with such compounds, transferring their oxygen, and causing an oxidative rearrangement to the unstable benzoylimine intermediate **254** which will further react with the medium[271].

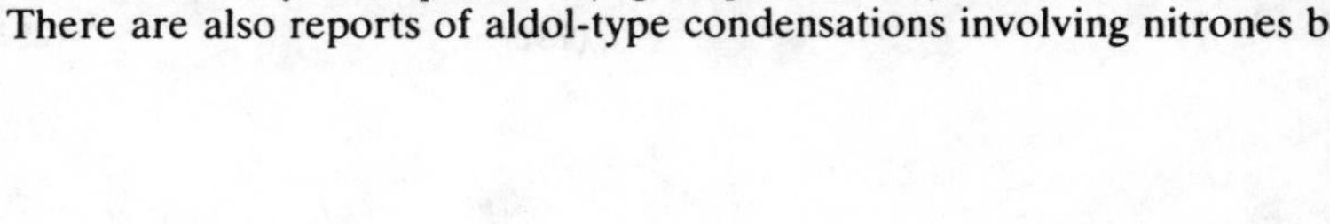

PhCH=NPh + [PhCON=CH$_2$]

(254)

b. Nitrones as carbon nucleophiles. Earlier examples of aldol-type condensations of nitrones with carbonyl compounds (e.g. equation 50) can be found in the reviews[2,3]. There are also reports of aldol-type condensations involving nitrones both as the nucleophile and the electrophile. However older structure assignments should be viewed with caution. Since the publication of the review of Hamer and Macaluso[2] the structures of the *N*-phenyl nitrones derived from acetone and from butyraldehyde have been reexamined by X-ray crystallography as well as by mass and NMR spectra, and have been revised. The dimers are isoxazolidines as shown by formulae **255**[272] and **256**[273,274].

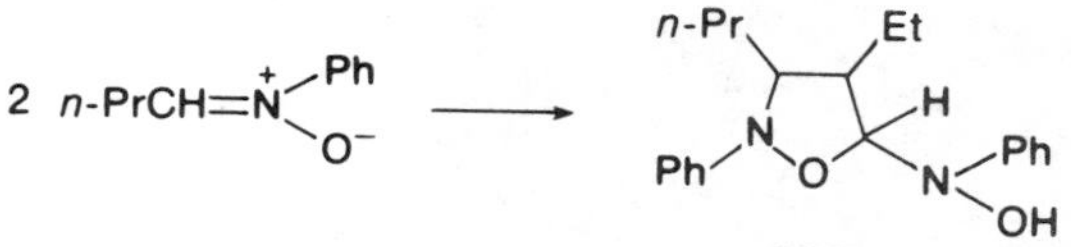

(**256**)

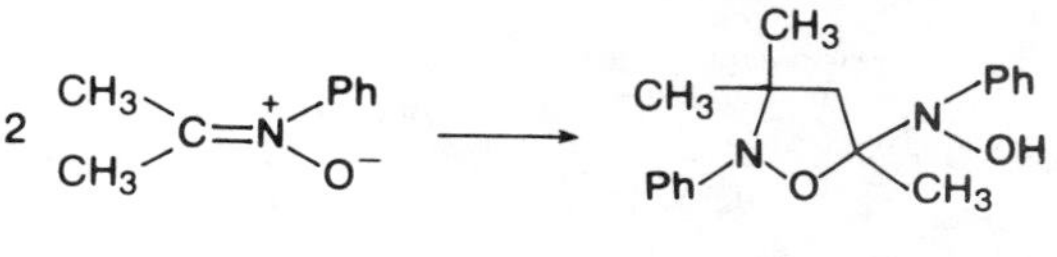

(**255**)

The formation of these dimers can be rationalized by assuming an aldol-type dimerization followed by intramolecular nucleophilic attack of the oxide anion on the α-position of the nitrone (Scheme 36). Recently it has been reported that such dimerization may be accompanied by the loss of hydroxylamine leading to an isoxazoline (equation 55)[275].

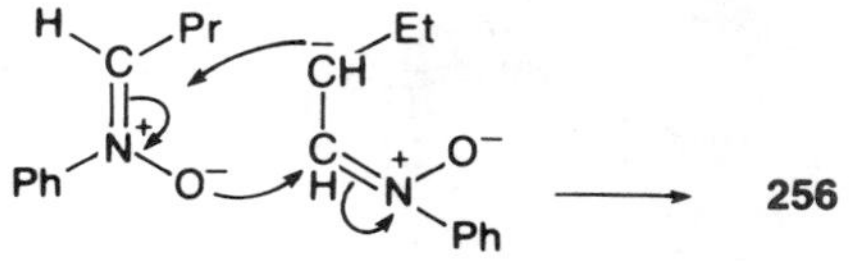

SCHEME 36

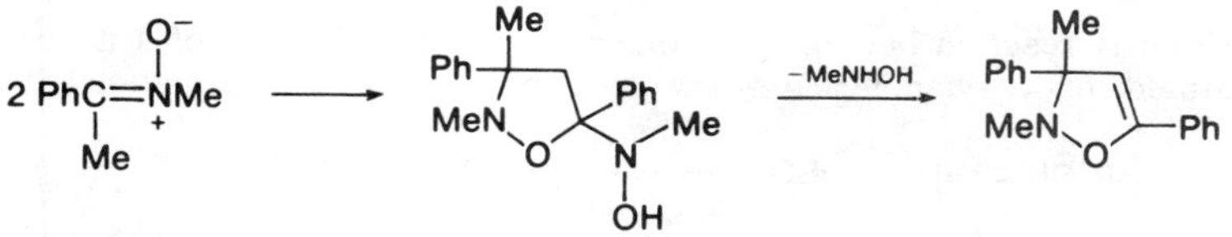

(55)

Derivatives of pyrroline *N*-oxide have been reported to undergo reactions with 3,5-dicarbomethoxypyridinium tosylate to yield 1,4- and 1,2-dihydropyridines[276]. An aldonitrone of this series reacted through the 3-carbanion (equation 56) while a 2-methylpyrroline *N*-oxide formed the anion on the methyl carbon affording products in which the two rings are connected by a methylene group (equation 57)[276]. Equation (56) represents a rare example indicating the formation of a carbanion of type **246**, *anti* to a nitrone oxygen (see introduction to Section II.B.7).

A different kind of aldol-like condensation, involving a carbanion obtained by deprotonation of the *N*-alkyl group was found in the process of elucidating the mode of base-induced formation of desoxybenzoin from *N*-benzyl α-phenyl nitrone[277]. It was found that under base catalysis this nitrone dimerizes to form the hydroxylaminonitrone **257** which subsequently undergoes a series of reactions that produce, among other products, desoxybenzoin[277].

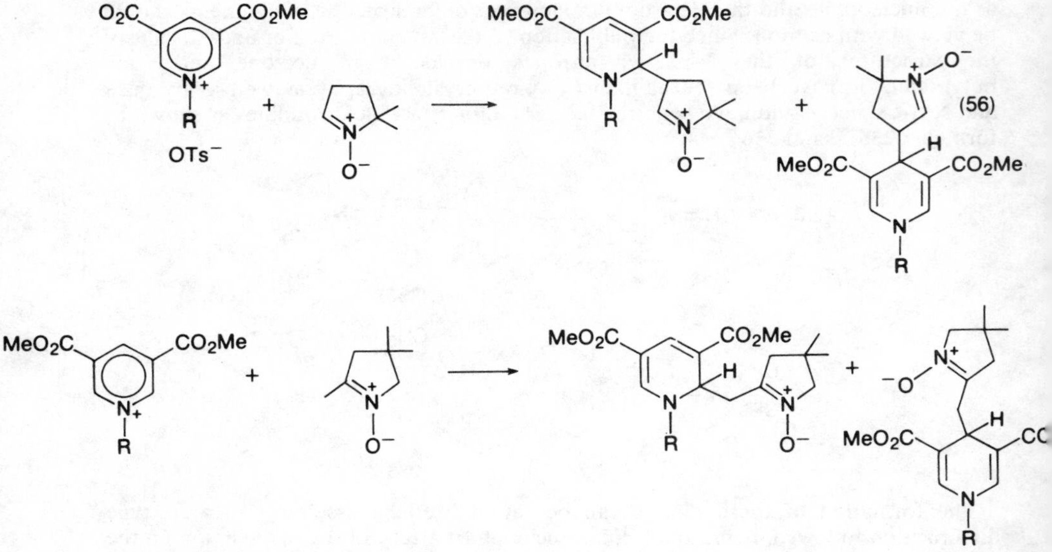

(56)

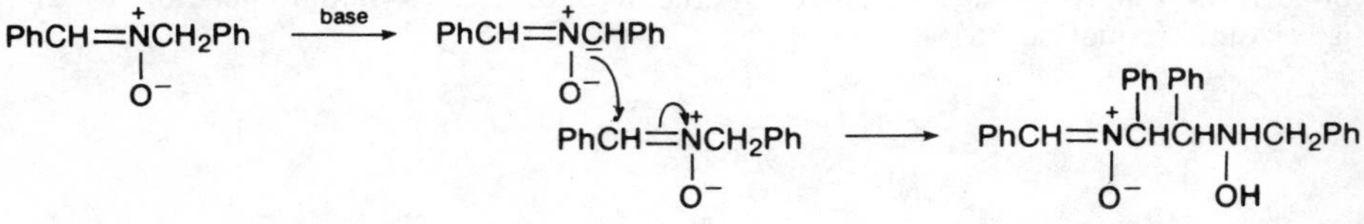

8. *Reactions of nitrones with nucleophiles*

The simplest reaction belonging to this class is the hydrolysis of a nitrone to an *N*-substituted hydroxylamine and the carbonyl compound (equation 58). This

$$R^1CH{=}\overset{+}{N}(O^-)R^2 + H_2O \longrightarrow R^1CHO + HN(OH)R^2 \tag{58}$$

reaction constitutes the final step of the Kröhnke aldehyde synthesis[278], and is usually carried out by acid catalysis. A number of more recent papers report acid-catalysed hydrolysis of nitrones[279–285] in addition to those cited earlier[2,3]. Nitrones also undergo cleavage by hydrazine or its derivatives[2,280,286] or by hydroxylamine[2,287,288] to furnish directly the hydrazone or oxime of the carbonyl compound and the *N*-substituted hydroxylamine (e.g. equation 59).

$$R_2C{=}\overset{+}{N}(O^-)R^1 + NH_2OH \longrightarrow R_2C{=}NOH + HONHR^1 \tag{59}$$

The reaction of aryl nitrones with arylnitroso compounds leading to azoxyarenes has been shown to involve hydrolysis to the *N*-arylhydroxylamine in the first step[289].

Methanol can add to suitably activated nitrones such as 2-cyanopyrroline *N*-oxides with the formation of the corresponding hydroxylamine, which upon heating loses hydrogen cyanide to furnish the α-methoxynitrone (equation 60)[290]. This reaction

(60)

may serve as a general method to α-alkoxynitrones. Similarly suitably activated nitrones add azide ion to give tetrazoles, presumably via α-azidoimines[291] (equation 61) in addition to other competing reactions.

$$\mathrm{RCH{=}\overset{+}{N}Ar(O^-)} \xrightarrow{N_3^-} \left[\mathrm{RCH(N_3){-}NAr(OH)} \longrightarrow \mathrm{RC(N_3){=}NAr}\right] \longrightarrow \text{tetrazole} \quad (61)$$

C,C-Diaryl nitrones may be converted to the corresponding thiobenzophenones and *N*-alkylhydroxamic acids in good yields by treatment with thio acids[292]. This reaction involves nucleophilic attack of the thiocarboxylate anion upon the nitrone, followed by decomposition for which a cyclic mechanism has been suggested (equation 62)[292]. In contrast reactions of *N*-(1)cyanoalkyl)-*N*-alkylideneamine

$$\mathrm{Ar_2C{=}\overset{+}{N}(Me)O^-} + \mathrm{HSC(O)Me} \longrightarrow \mathrm{Ar_2C(SC(O)Me){-}N(Me){-}OH} \longrightarrow \mathrm{Ar_2C{=}S} + \mathrm{MeCON(Me)OH} \quad (62)$$

N-oxides with thiols yield imidazoles. Nucleophilic attack of the thiolate anion upon the cyano group initiates the reaction, followed by cyclization and loss of water (equation 63)[293].

(63)

Cyanide has long been known to add to nitrones providing the corresponding cyanoimines (equation 64)[2,294,295]. Recently it has been shown that the reaction of

$$\mathrm{RCH{=}\overset{+}{N}R^1(O^-)} + \mathrm{HCN} \longrightarrow \mathrm{RCH(CN){-}NR^1(OH)} \xrightarrow{-H_2O} \mathrm{RC(CN){=}NR^1} \quad (64)$$

C-aryl *N*-alkyl nitrones with cyanide can form imidazoles directly via the corresponding cyanoimines[296]. The mechanism in Scheme 37 has been suggested for this process[296].

The ability of carbanions to add to the nitrone function has also been recognized.

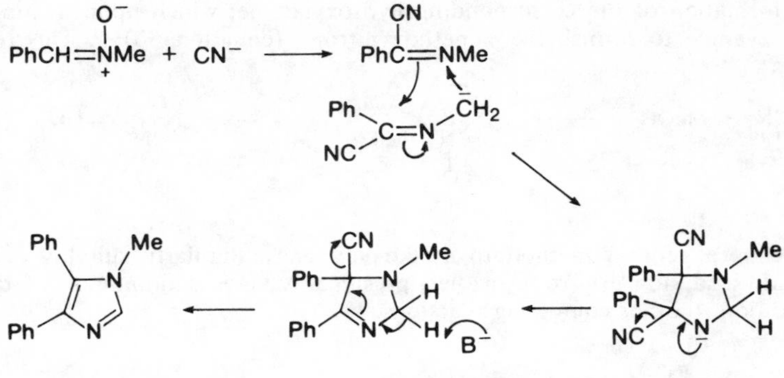

SCHEME 37

Early examples that include nitromethane and 2-methylpyrroline *N*-oxide have been reviewed[2].

More recently reactions of active methylene compounds and of some ylids have been studied. The reactions of diethyl malonate with Δ^1-pyrroline *N*-oxides proceeds beyond the initially obtained product and with the loss of ethanol isoxazolidines are formed (equation 65)[297]. The reaction has been further studied by Stamm reacting a

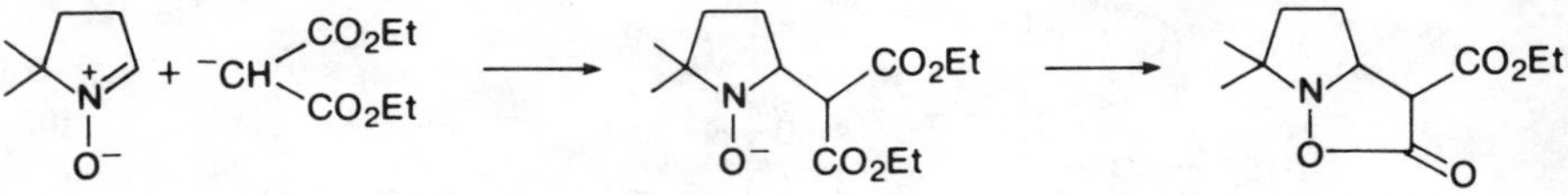

(65)

series of *C*-aryl *N*-alkyl nitrones with carbanions derived from diethyl malonate, malonamide, α-ethoxycarbonyl-γ-butyrolactone, ethyl phenylacetate and ethyl cyanoacetate to yield predominantly the 3,4-*trans*-isoxazolidinones[298–300] (e.g. equation 66). The yields have been found to vary with the steric requirements of the *N*-alkyl group of the nitrone and with the nature of the carbanion.

(66)

X = CO_2Et, CN, Ph, etc.

Although simple enamines add to nitrones in a 1,3-dipolar fashion[179] it was found recently by Zbaida and Breuer that the enaminonitrile **258** can add as a nucleophile to a nitrone function. The hydroxylamine product **259** is capable of adding to a second mole of enaminonitrile if an excess of the latter is present in the reaction mixture (Scheme 38)[301].

The reaction of nitrones with phosphorus ylids has been studied in some depth. Huisgen and Wulff have reported that the reaction of triphenylphosphonium ylids with nitrones leads to stable oxazaphospholidines (equation 67)[302]. Later it was shown by Black and Davis that reaction of ethoxycarbonylmethylenephosphorane with 5,5-dimethyl-Δ^1-pyrroline *N*-oxide leads, with the extrusion of triphenylphosphine oxide, to a mixture of aziridine (**260**) and the enamino ester **261** (equation 68)[303].

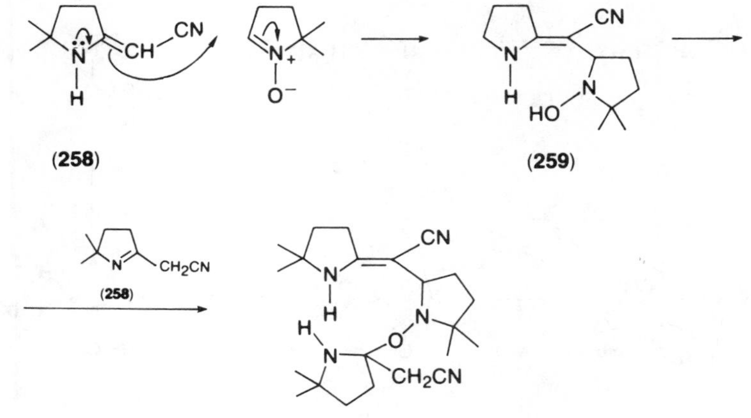

(258) (259)

(258)

SCHEME 38

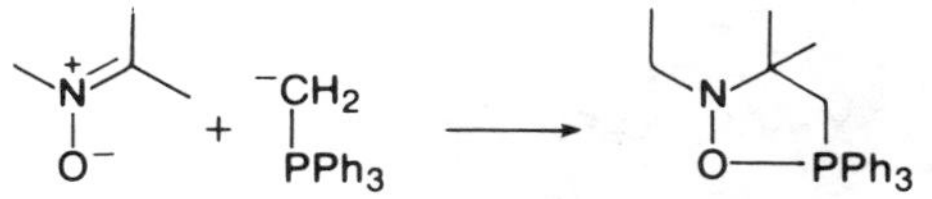

(67)

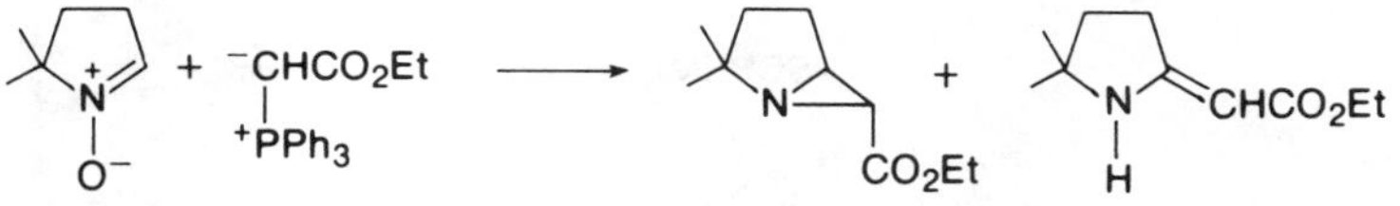

(68)

(260) (261)

The formation of aziridines from nitrones by the action of phosphono ylids was first reported by Breuer and coworkers, who found that the reaction of triethyl phosphonoacetate and *C*-phenyl *N*-methyl nitrone (**52**) gives a mixture of two aziridines **54** and **55** due to tautomerism in nitrone **52** under the reaction conditions[81] (see Section II.A.5.b).

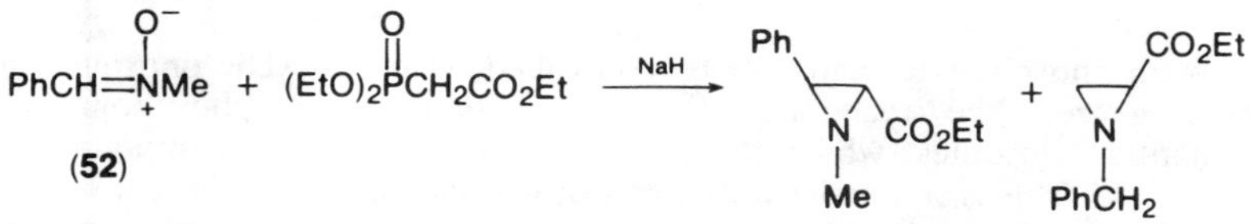

(52) (54) (55)

The reaction of phosphonates with nitrones of the pyrroline *N*-oxide series has been studied by Black and Davis[303] and by Breuer and coworkers[304,305]. It has been found that the reaction leads to *trans* aziridines (**260**) and enamines (**261**)[303–305], and that in certain cases it can be directed to lead predominantly, or even exclusively, to yield one of the products by appropriate choice of solvent and counterion[304,305]. Protonation inhibits aziridine formation and then base-catalysed elimination leads to enamine (Scheme 39).

Similarly the reactions of 3,4-dihydroisoquinoline *N*-oxide with phosphonates lead to the aziridine **262** (X = CO_2Et) or enamines **263** (where X may be either CN or CO_2Et)[306]. Subsequently it has been shown that the course of the reactions of

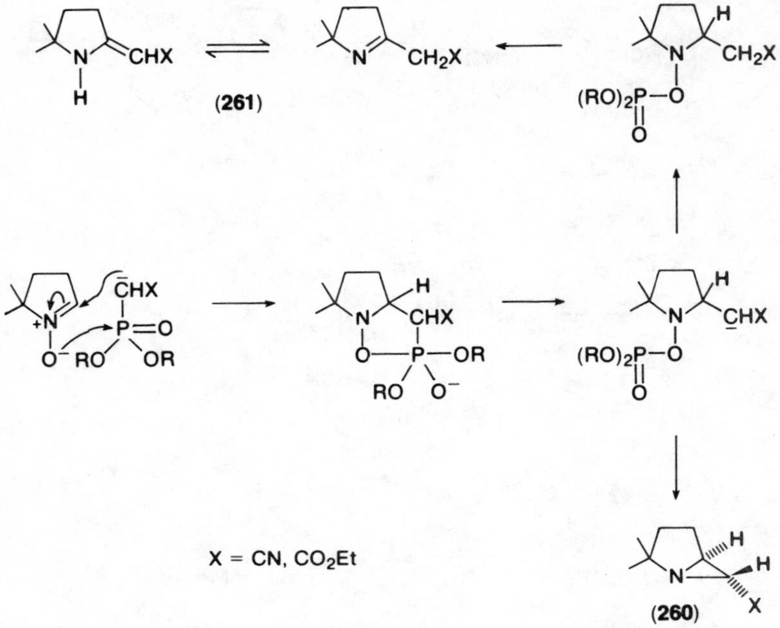

X = CN, CO_2Et

SCHEME 39

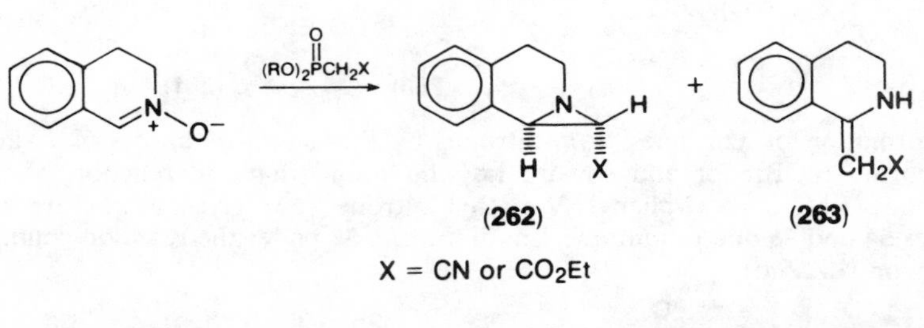

X = CN or CO_2Et

nitrones with phosphonates can also be controlled by using cyclic phosphonates of varying ring-size. The reactions of five-membered cyclic phosphonates give predominantly aziridines while those of the six-membered compounds yield enamines[307–309]. This is rationalized in terms of varying basicities of the intermediate **264** as a function of the size of the dioxaphosphacycloalkane ring and their differing tendencies to undergo protonation (Scheme 40). The existence of intermediate **264** ($n = 2$) has been established by means of ^{31}P-NMR spectroscopy[310]. The reactions of some nitrones with phosphinoxy ylids have also been shown to give aziridines and enamines[301].

9. Reactions of nitrones with organometallic compounds

a. Organo-lithium and -magnesium compounds. The Grignard reaction of nitrones has been known for over 60 years[2,3]. It is one of the reactions upon which the recognition of the similarity between the carbonyl and nitrone functions was

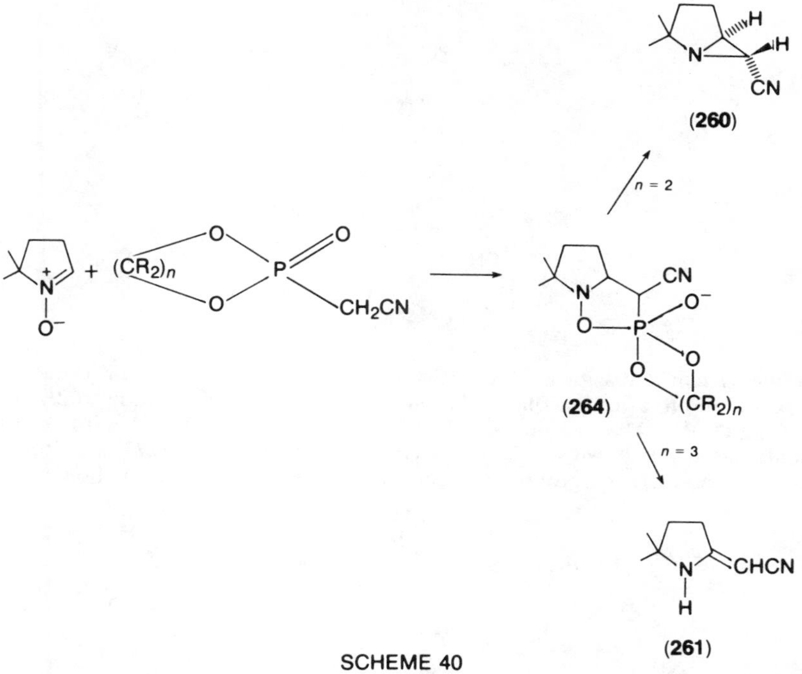

(260)

$n = 2$

(264)

$n = 3$

(261)

SCHEME 40

based. The increased interest in this subject in recent years is stimulated by the usefulness of the products, *N,N*-disubstituted hydroxylamines. Oxidation of compounds of this type yields stable nitroxide radicals used as spin labels. In some cases the tendency for oxidation of the *N,N*-dialkylhydroxylamines is so great that their isolation is difficult and the Grignard reactions yield the nitroxide radicals directly. Benzhydryl-*t*-butylnitroxide radicals have been synthesized through the addition of phenylmagnesium bromide or phenyllithium to α-phenyl *N*-*t*-butyl nitrone and oxidation of the hydroxylamine (equation 69)[311]. Radicals derived from pyrrolidine (**265**, X = CH_2)[312] or oxazolidine (**265**, X = 0)[313] have been prepared by a similar reaction sequence.

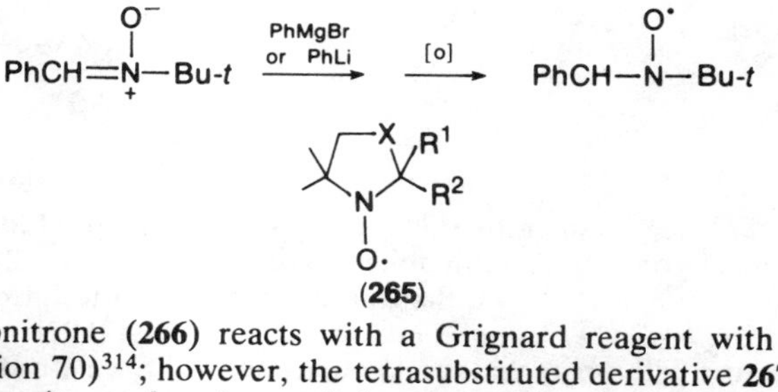

$$PhCH{=}\overset{+}{N}(O^-){-}Bu\text{-}t \xrightarrow{PhMgBr \text{ or } PhLi} \xrightarrow{[o]} PhCH{-}N(O^\bullet){-}Bu\text{-}t \quad (69)$$

(265)

The α-cyanonitrone (**266**) reacts with a Grignard reagent with the loss of the cyanide (equation 70)[314]; however, the tetrasubstituted derivative **267** is too hindered to react on the nitrone function and the α-acetylnitrone **268** is obtained from the reaction of the cyanide group[314].

Reactions of β-ketonitrones with Grignard reagents have been studied in the

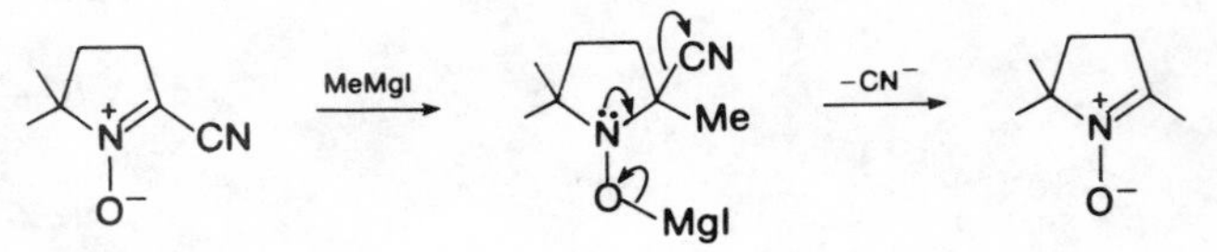

(70)

(266)

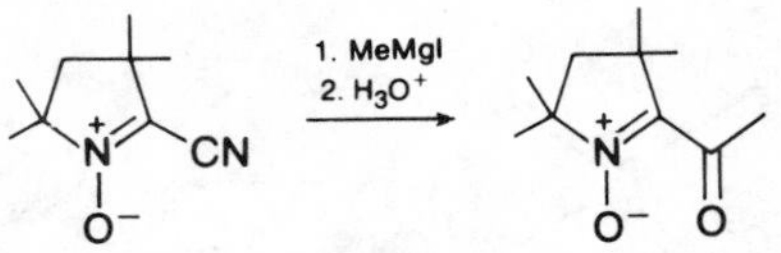

(267) (268)

pyrroline system and with isatogens. 3-Oxo-1-pyrroline *N*-oxides **(269)** react at the carbonyl group to affo:d alcohols **270** that easily undergo dehydration to 2*H*-pyrrole 1-oxides **(271)**[315]. The reaction of phenylisatogens **(272)** with Grignard or lithium reagents has been shown to yield two types of products, one **(273)** resulting from addition to the carbonyl group and a second **(274)** to the nitrone function[316].

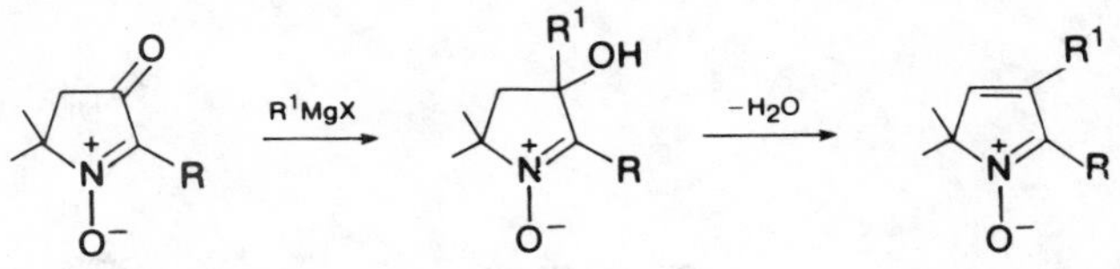

(269) (270) (271)

R = Ph, *t*-Bu

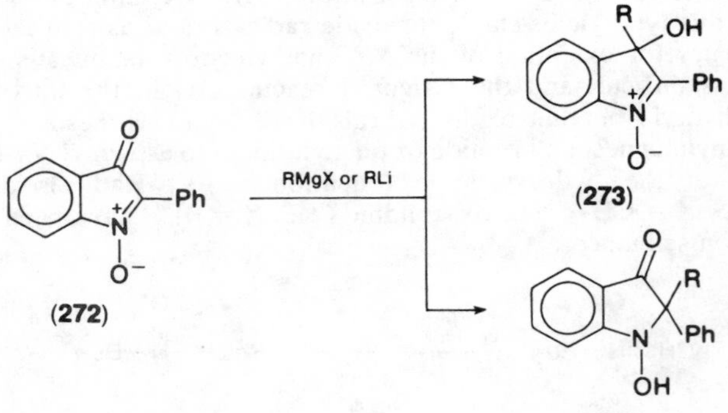

(272) (273) (274)

Oxidation of **274** leads to nitroxide radicals[316]. Related studies concerning Grignard reactions of 2-alkylquinoline *N*-oxides[317] and 2,3-disubstituted benzimidazole 1-oxide[318] have shown that these can also lead to nitroxide radicals.

b. Organo-zinc and -copper compounds. C-Aryl *N*-alkyl nitrones undergo the Reformatzky reaction with organozinc compounds derived from α-bromo esters, such as α-bromo-acetate, -propionate, -butyrate and -isobutyrate[319,320]. The reactions lead to the formation of isoxazolidinones, **275** (Scheme 41) resembling the reaction of nitrones with diethyl malonate (Section II.B.8).

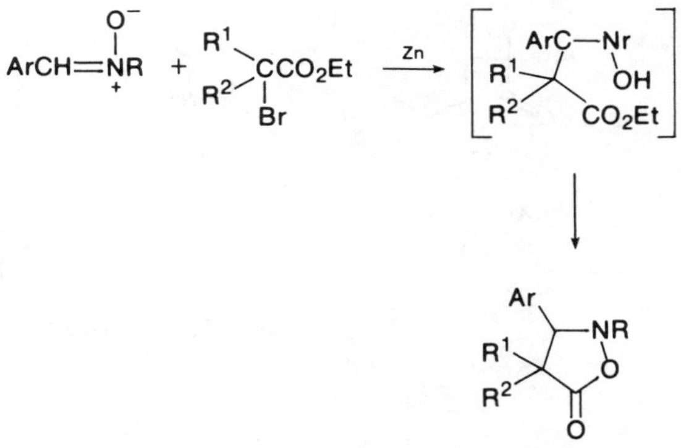

ArCH=N(O⁻)R + R¹R²C(Br)CO₂Et —Zn→ [ArC(R¹)(R²)(CO₂Et)—N(R)OH] → (275)

SCHEME 41

The reaction of a copper acetylide with nitrones was shown to give β-lactams[321,322]. Tracer studies indicate that the carbonyl oxygen originates from the nitrone, while the hydrogen in the 3-position originates from the solvent[322]. A mechanism which is consistent with these results is presented in Scheme 42.

Reactions of copper acetylides with 5- or 6-membered cyclic nitrones result in bicyclic β-lactams **276** and **277**[322].

c. Organogermanium compounds. Germanium hydrides can react with nitrones via three different pathways: (*i*) polar addition of Ge—H to the nitrone to yield Ge—C-bonded products (**278**), (*ii*) a dark homolytic reaction to give a Ge—O-bonded product (**279**), and (*iii*) a photocatalysed homolytic reaction to give a Ge—C-bonded product[323–325].

d. Organo-lead, -tin and -mercury compounds. The photolysis of these organometallic compounds yields radicals (see Section II.B.10).

e. Transition metal derivatives. This aspect of the chemistry of nitrones is almost completely unexplored. Iron pentacarbonyl has been reported to cause

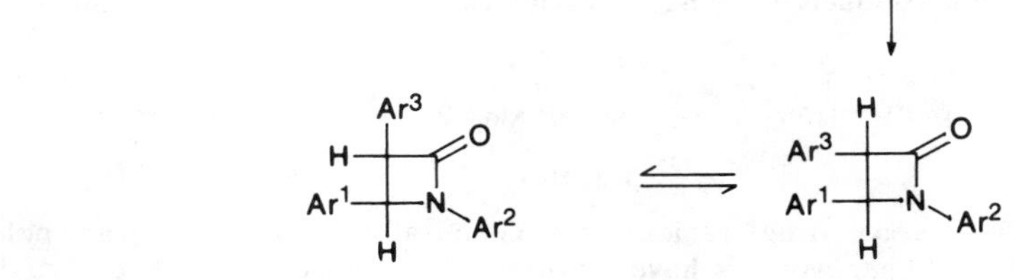

SCHEME 42

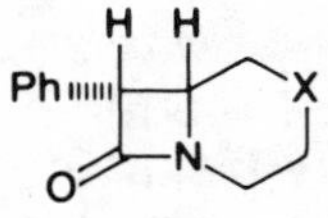

(276)

X = O, CH_2

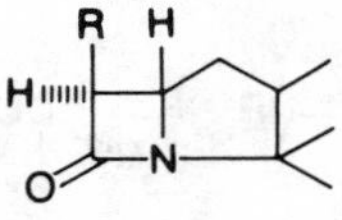

(277)

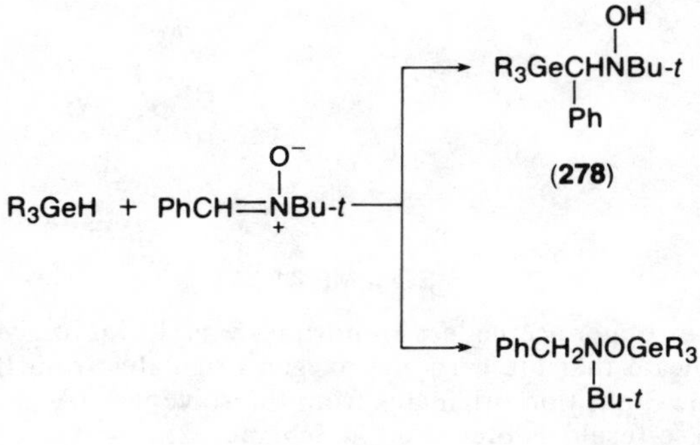

$$R_3GeH + PhCH{=}\overset{+}{N}(O^-)Bu\text{-}t \longrightarrow R_3GeCH(Ph)N(OH)Bu\text{-}t\ (278) \quad + \quad PhCH_2N(Bu\text{-}t)OGeR_3\ (279)$$

deoxygenation of a nitrone[167]. Nitrones can serve as ligands in nickel complexes[27,28]. For example, N-methyl C-(2)pyridyl) nitrone forms complex **280** with nickel.

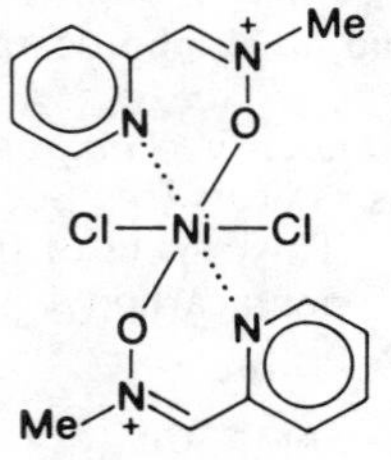

(280)

10. Reactions of nitrones with free radicals

It has been shown that free radicals generated thermally from azobisisobutyronitrile (AIBN) add in the 1,3-manner to α,N-diphenyl nitrone (equation 71)[326]. Similar adducts have been obtained from related nitrones, and the reaction

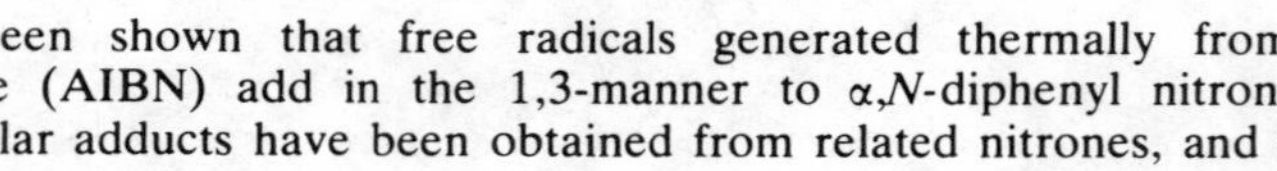

$$PhCH{=}\overset{+}{N}(O^-)Ph + \underset{\text{(AIBN)}}{Me_2C(CN)N{=}NC(CN)Me_2} \xrightarrow{\Delta} PhCH(CMe_2CN)N(Ph)OC(CN)Me_2 \qquad (71)$$

shown to involve free radicals by chemically induced nuclear polarization (CIDNP)[327]. Other workers have reported, in contrast, that α-,N-diphenyl nitrone reacts with radicals to give nitroxides[328].

The formation of nitroxide radicals is a general reaction with the more hindered α-phenyl *N-t*-butyl nitrone (PBN). Reaction of this nitrone with free radicals generated from AIBN gives the crystalline stable free radical **281**[329].

$$\underset{\text{PBN}}{\text{PhCH}{=}\overset{+}{\text{N}}(\text{O}^-)\text{Bu-}t} + \text{AIBN} \longrightarrow \underset{\text{Me}_2\text{CCN}}{\text{PhCH}}{-}\text{N}(\text{O}^\bullet)\text{Bu-}t$$

(**281**)

m.p. 121–122°C, red needles

5,5-Dimethylpyrroline 1-oxide can also trap free radicals with the formation of nitroxides, but these products found to be relatively unstable and to disproportionate readily (equation 72)[329]. Earlier the arylation of pyrroline *N*-oxides by arenediazonium salts was postulated to proceed by a free-radical mechanism[165].

$$2\ \text{(5,5-dimethylpyrroline 1-oxide)} + 2\,\text{R}^\bullet \longrightarrow 2\ \text{(2-R-2-H-5,5-dimethylpyrrolidin-1-yloxyl, N–O}^\bullet) \longrightarrow \text{(2-R-5,5-dimethylpyrroline 1-oxide, N}^+\text{–O}^-) + \text{(2-R-2-H-5,5-dimethyl-1-hydroxypyrrolidine, N–OH)} \quad (72)$$

Subsequently it was rapidly recognized that nitrones can be useful traps for short-lived reactive free radicals, since their reactions can produce more stable radicals detectable by ESR and whose hyperfine coupling parameters permit identification of the initial radical trapped. This technique has been named spin trapping[330]. Radicals generated by the decomposition of organo-lead, -tin and -mercury compounds were thus trapped by PBN and the structures of the resulting radicals of type **281** were confirmed by alternate synthesis[142]. Later the complicating influence of the presence of traces of air in the reaction mixture was recognized, especially since the nitrones used were not selective and they reacted with alkyl, alkoxy and alkylmetaloxy radicals giving a variety of products[331]. PBN was reported to add alkoxy and alkylperoxy radicals obtained from photolysis of di-*t*-butyl ketone in $^{17}O_2$-saturated solvents[332] and also from alkyl halides by reaction with potassium superoxide[333]. The trapping of hydroxy and hydroperoxy radicals with PBN and 5,5-dimethyl pyrroline 1-oxide was also reported[334], but disputed recently[335].

Less reactive radicals such as succinimidyl (**282**) have been successfully trapped by *N-t*-butyl nitrone (equation 73)[336,337]. Recently special nitrones have been designed

$$\text{(succinimidyl N}^\bullet\text{)} + \text{CH}_2{=}\overset{+}{\text{N}}(\text{O}^-)\text{Bu-}t \longrightarrow \text{(succinimido)NCH}_2{-}\text{N}(\text{O}^\bullet)\text{Bu-}t \quad (73)$$

(**282**)

to serve as radical traps. An interesting bifunctional trap is α-(3,5-di-*t*-butyl-4-hydroxyphenyl) *N-t*-butyl nitrone (**283**)[338]. This nitrone differentiates between alkyl and alkoxy radicals. Alkyl radicals add to the nitrone function (equation 74) to give nitroxy radicals **284**, and alkoxy radicals abstract the phenol proton leading to the phenoxy radical **285** (equation 75). The ESR spectra of the two radicals are markedly different. α-(4-Pyridyl 1-oxide) *N-t*-butyl nitrone (**286**) is particularly suitable for trapping hydroxyl radicals in aqueous solutions[335].

The most recent additions to this arsenal of nitrones, with a nonpolar hydrocarbon

$$\text{HO}-\text{C}_6\text{H}_2(\text{Bu-}t)_2-\text{CH}{=}\overset{+}{\text{N}}(\text{O}^-)\text{Bu-}t + \text{R}^\bullet \longrightarrow \text{HO}-\text{C}_6\text{H}_2(\text{Bu-}t)_2-\text{CH(R)}-\text{N(O}^\bullet)\text{Bu-}t \quad (74)$$

(283) (284)

$$\mathbf{283} + \text{RO}^\bullet \longrightarrow {}^\bullet\text{O}-\text{C}_6\text{H}_2(\text{Bu-}t)_2-\text{CH}{=}\overset{+}{\text{N}}(\text{O}^-)\text{Bu-}t + \text{ROH} \quad (75)$$

(285)

chain that are particularly suitable to serve as spin traps in complex biphasic systems containing regions of different viscosity and polarity, are the two general structures **287** and **288**[339]. The methods developed for these two nitrones permit the synthesis of nitrones with a variety of R, R^1 and R^2 side-chains.

$$\text{O}{\leftarrow}\text{N}(\text{C}_5\text{H}_4)-\text{CH}{=}\overset{+}{\text{N}}(\text{O}^-)\text{Bu-}t$$

(286)

$$\text{PhCH}{=}\overset{+}{\text{N}}(\text{O}^-)-\text{C(Me)}<\text{(CH}_2\text{O)}_2\text{C}R^1R^2$$

(287)

$$\text{R}-\overset{+}{\text{N}}(\text{C}_5\text{H}_4)-\text{CH}{=}\overset{+}{\text{N}}(\text{O}^-)\text{Bu-}t$$

(288)

An unusual type of reaction of a nitrone with a free radical is shown by the perchloronitrone **289**, which loses a chlorine atom under the influence of radicals to yield a new species **290**[225].

$$\text{Cl}_2\text{C}{=}\overset{+}{\text{N}}(\text{O}^-)\text{CCl}_3 + \text{Cl}^\bullet \longrightarrow \text{Cl}_2\text{C}{=}\overset{+}{\text{N}}(\text{O}^-)-\dot{\text{C}}\text{Cl}_2 + \text{Cl}_2$$

(289) (290)

III. NITRONIC ACID DERIVATIVES

A. Nitroalkane–Nitronic Acid Tautomerism

Nitronic acids (**292**) are tautomers of aliphatic nitro compounds (**291**)[340], sharing a common anion, the structure of which can be represented by formulae **293–295**.

The tautomeric equilbrium usually favours the nitro compounds **291** rather than the nitronic acids **292** since the former are normally weaker acids. The mechanism of the tautomerization **291** $\rightleftharpoons$ **292** and data concerning the acidity constants of aliphatic nitro compounds and nitronic acids were presented by Nielsen[340].

In recent years the thermodynamic and kinetic acidities of nitroalkanes have been

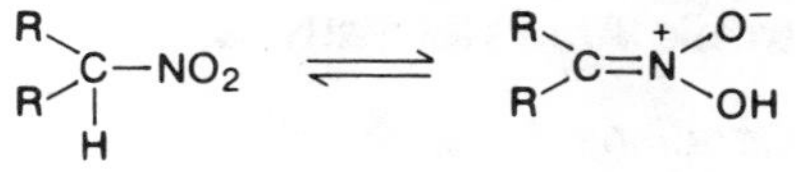

(291) ⇌ (292)

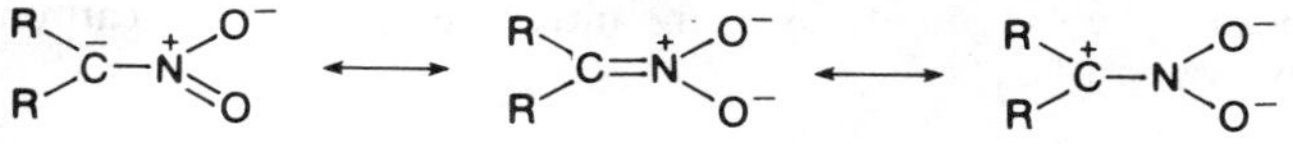

(293) ⟷ (294) ⟷ (295)

more intensively studied[341-350] and some anomalies have become apparent. In measuring rates of deprotonation and equilibrium acidity constants of nitroalkanes and nitrocycloalkanes it became apparent that an increase in equilibrium acidity is not necessarily paralleled by the same trend in rates of deprotonation. For example, 2-nitropropane is more acidic (by 2.5 pK_a units) than nitromethane, yet it undergoes deprotonation 89 times slower than the latter[347]. Also nitrocyclobutane is less acidic than nitrocyclopentane by nearly 2 pK units, but deprotonates at a four times faster rate. Two series of arylnitroalkanes, namely 1-aryl-2- nitropropanes[344] and 1-arylnitroethanes[343,344] bearing various substituents on the aromatic ring have also been studied. One surprising result was that the substituent effect in one series paralleled that in the other in spite of the difference in distance between the aromatic ring and the nitro group between the two series. This is explained by indirect resonance effects relayed by induction since in the 1-aryl-2-nitropropanes the transmission of direct resonance effects is not possible. It has also been found[344,346], that substituent effects are greater on the rate of deprotonation of the nitroalkane, than on the equilibrium acidities. All the results can be rationalized by visualizing the mode of deprotonation, which leads first to the pyramidal carbanion, **293**. The rate of formation of this species will be the kinetic acidity, and it is easy to see that in an arylnitroalkane substituents in the aromatic ring will influence it. This carbanion has to undergo reorganization to the planar nitronate, **294**, the stability of which will determine the equilibrium acidity of the nitro compound.

Consider the case of nitromethane and 2-nitropropane. The tertiary hydrogen of the latter would be expected to be less acidic than the hydrogens of nitromethane, and this is reflected by the lower ionization rate or kinetic acidity of 2-nitropropane. The higher equilibrium acidity of the latter clearly results from the increased stability of $Me_2C{=}NO_2^-$ (as compared to $CH_2{=}NO_2^-$) and its relatively low rate of protonation on carbon. The situation is analogous in arylnitroalkanes. For example, a substituent such as m-NO_2 increases the rate of deprotonation of $ArCHMeNO_2$, but it also increases the rate of protonation of $ArCMe{=}NO_2^-$[344].

Recently it has been suggested[350], that since the nitroalkane anomalies are observed always in protic solvents, they can be explained by assuming that stabilization of the nitronate anion **294** by hydrogen bonding of the solvent is the major factor determining equilibrium acidity, while hydrogen bonding has little or no stabilizing influence upon the transition state leading to **293**. Therefore, substituents will affect more the stability of **293** than that of **294**. Indeed a linear Brønsted plot was found for a series of substituted arylnitromethanes, whose equilibrium acidities and rates of deprotonation were measured in dimethyl sulphoxide.

The stereochemistry of protonation of nitronates was also discussed previously[340]. Later papers conclude that this is a kinetically controlled process in which steric hindrance may control the direction of approach and give preponderantly one isomer[351-353].

B. Structure of Nitronic Acid Derivatives

1. Theoretical considerations

Hückel molecular orbital calculations have been carried out for a series of delocalized bicyclic anions (**296**)[354]. These calculations indicate that the negative charge resides on the oxygen atoms of the nitronate and that the carbon framework is slighly positive.

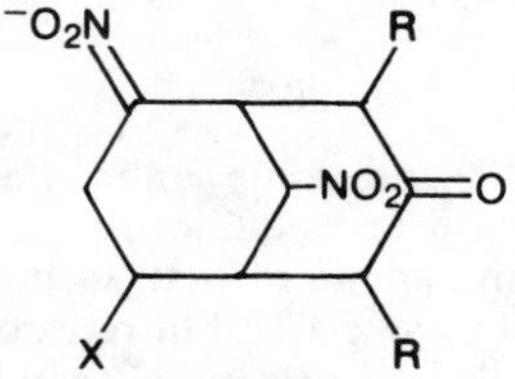

(**296**)
X = NO_2, CN

2. X-ray studies

The molecular dimensions of the following nitronates have been determined by X-ray crystallography: dipotassium nitroacetate (**297**)[355], dirubidium tetranitroethanediide (**298**)[356], bis(propane-2-nitronato)copper (**299**)[357], and 2-nitro-1,3-indanedione hydrate (**300**)[358]. The results obtained in these works are consistent with

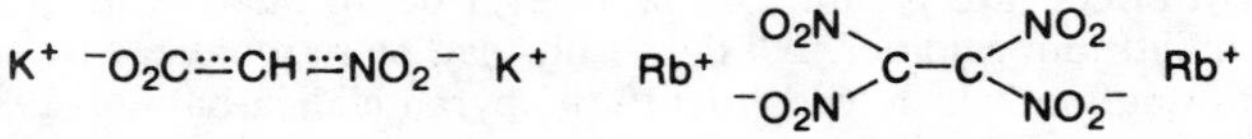

$K^+ \; ^-O_2C\text{⋯}CH\text{⋯}NO_2^- \; K^+$

(**297**)

(**298**)

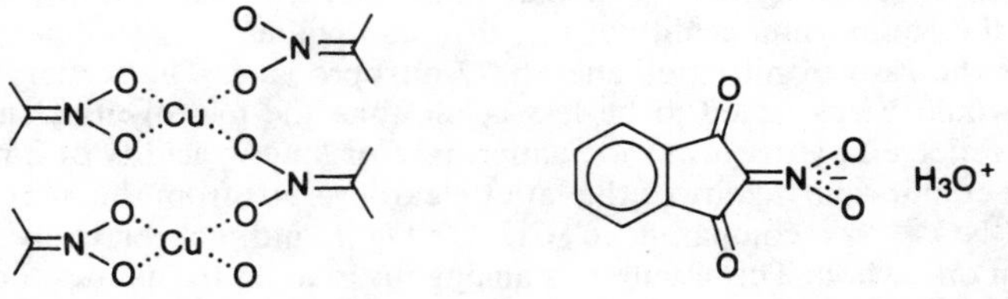

(**299**)

(**300**)

the ideas previously accepted for the structure of nitronate anions. It is found that the nitronate anion is planar. The C—N bond length is between 1.35–1.38 Å, which is shorter than a single (1.45 Å) and longer than a double (1.23Å) bond indicating partial double-bond character. The N—O bond lengths found (1.25–1.27 Å) are somewhat longer than those in nitro groups, indicating considerable contribution of resonance structures in which both oxygens are equally negatively charged.

3. Dipole moments

This method is useful for the determination of the structures of geometrically isomeric nitronate esters. Grée and Carrié separated the isomeric pair **301**[359] and

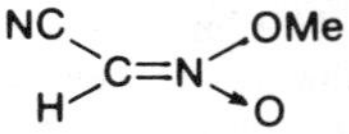

(*Z*-**301**) μ = 2.25 D

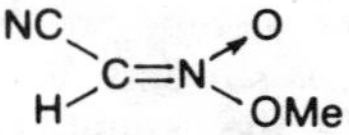

(*E*-**301**) μ = 4.71 D

determined their dipole moments. The experimental values (indicated) corresponded well with those calculated by two different methods for various rotational conformers around the N—OMe bond.

4. Spectra

Fundamental spectral properties of nitronic acids, nitronates (salts and esters) are available in the previous review[340]. Apart from some additional ^{1}H-NMR data of (*E*) and (*Z*) nitronic esters[359], almost all recent work has been devoted to the study of the effect of solvent upon various spectral properties of nitronate salts. It is now apparent[343,360] that the ultraviolet absorption maximum can be shifted to higher wavelengths by as much as 150 nm upon changing from a hydroxylic solvent (e.g. MeOH) to a dipolar aprotic solvent such as hexamethylphosphoric triamide. This change in the spectrum reflects the transfer of the negative charge from the oxygens of the nitronate anion, where it can no longer be stabilized in the absence of a hydrogen-bonding solvent, to the carbon of the nitronate, resulting in lesser stabilization, and accordingly the $\pi \rightarrow \pi^*$ transition occurs at lower energy (higher wavelength). This phenomenon is illustrated by some examples. The potassium salt of 1-nitroindene (**302**) has an absorption maximum of 338 nm in ethanol and 418 nm in HMPA[360], whereas the sodium salt of 1-*p*-nitrophenylnitroethane (**303**) absorbs at 394 nm in methanol but at 557 nm in HMPA[343].

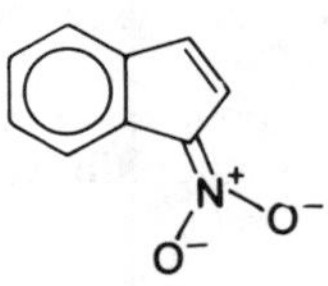

(**302a**) 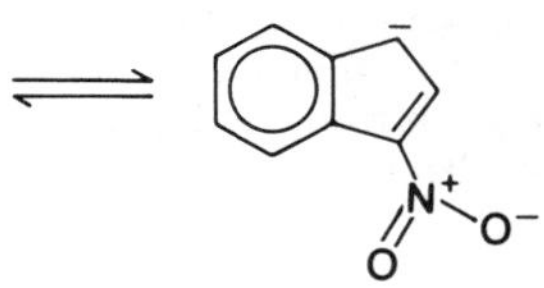 (**302b**) 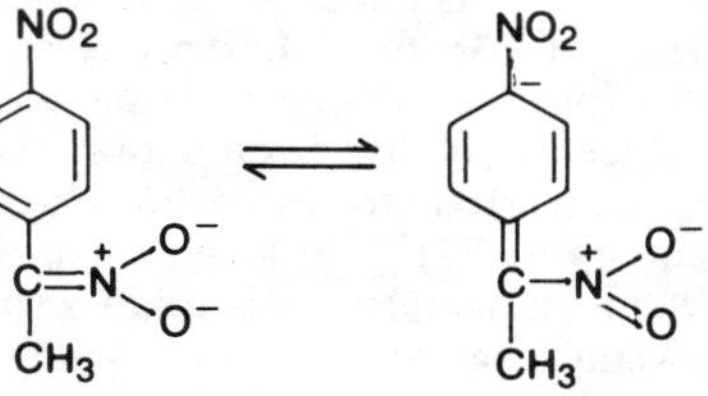 (**303**)

Earlier infrared studies also indicate that the charge of nitronates is concentrated on the oxygen atoms[361,362], but investigation of the infrared spectra of a series of lithium, sodium, and potassium nitronates derived from 2-nitropropane has shown that the C=N stretching bond shifts from 1603 cm^{-1} (for the K^+ salt) to 1637 (for Li^+), showing that the latter has more C=N double-bond character due to tighter coordination between the lithium and the oxygen[343].

Similar conclusions can be reached from examining NMR spectra. Although the NMR data found for some sodium alkylnitronates in water are consistent with a structure in which the negative charge resides on the oxygens, studies of solvent effects upon the ^{1}H-NMR spectrum of 1-nitroindene indicate[360] that there is a downfield shift of H(2) and an upfield shift of H(3) upon passing from trifluoroethanol to HMPA. In addition, there is a change in the magnitude of the coupling constant of the olefinic protons $J_{2,3}$. The latter has a value of 5.6 Hz (as in indene) in the hydroxylic solvents, but the value decreases to 5.0 Hz in HMPA. All these are consistent with the assumption that in the aprotic solvent there is increased electron density at C(3) and a lower π-bond order between C(2)–C(3) (see structure **302b**).

Some ^{13}C-NMR results confirm these conclusions, as it was found that the chemical shift of the nitronate carbon is also highly solvent-dependent[363]. In a series of sodium nitronates the nitronate carbon appears at a lower field by 10 ppm in methanol than in dimethyl sulphoxide.

C. Reactions of Nitronic Acid Derivatives

1. Oxidation of nitronates

Oxidation of nitronates is one of the most popular alternatives for the usual hydrolytic conditions of the Nef reaction (Section III.C.5). Permanganate oxidation has been shown previously to give carbonyl products[340]. More recently this reaction has been employed in prostaglandin synthesis[364,365]. Recent kinetic results from studies of this reaction are consistent with a mechanism which consists of rate-determining attack of permanganate upon the C=N double bond leading to species **304** which is transformed to products by the sequence shown in Scheme 43[366–369]. Other oxidizing agents that provide high-yield alternatives to the Nef reaction are ozone[370–371] and singlet oxygen[372].

Nitronate salts are also oxidized by persulphate. This reagent can also affect the conversion of nitronate salts to carbonyl compounds[373,374], but it can also be directed to yield vicinal dinitro compounds if carried out in the presence of an organic solvent which extracts the product formed (equation 76)[373,374]. In the absence of such a solvent, various other secondary oxidation products, such as nitro olefins (**305**) or 3,4,5-trisubstituted isoxazoles (**306**) can be formed[374,375]. This persulphate oxidation reaction is catalysed by silver ions[374].

Recently, it has been shown that nitronates derived from arylnitromethanes can be converted to 3,4,5-triarylisoxazoles by silver ion in dimethyl sulphoxide (equation 77)[376]. *t*-Butyl hydroperoxide oxidation of nitronates catalysed by vanadium acetylacetonate has also been shown to give Nef products[377].

Many examples of direct halogenation of nitronates have been known for some time[378]. Recently, it has been shown that nitronates can also be fluorinated directly

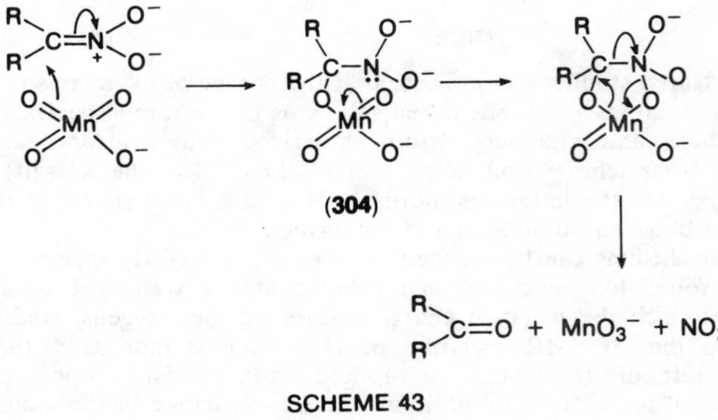

(**304**)

$R_2C{=}O + MnO_3^- + NO_2^-$

SCHEME 43

$$RCHNO_2^- + S_2O_8^{2-} \longrightarrow RCH(NO_2)-CHR(NO_2) + 2\,SO_4^{2-} \quad (76)$$

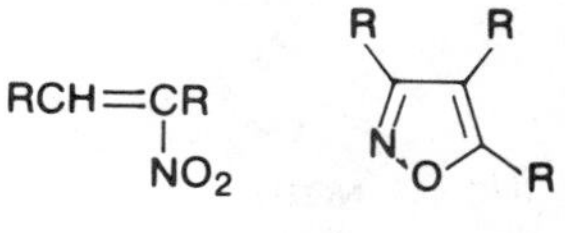

$RCH{=}CR(NO_2)$ (305)

(306)

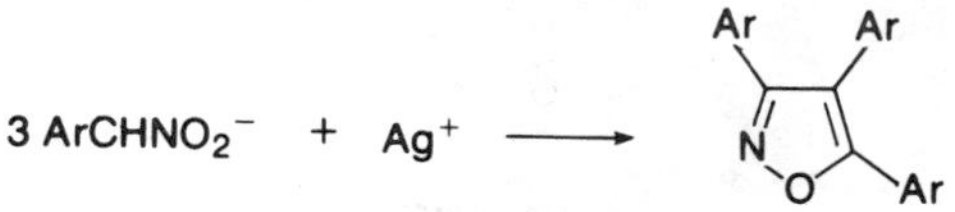

$$3\,ArCHNO_2^- + Ag^+ \longrightarrow \text{(3,4,5-triarylisoxazole)} \quad (77)$$

by treatment with fluorine in aqueous solution to give α-fluoronitro compounds[379–382].

2. Reduction of nitronates

The treatment of nitronate salts with reducing agents under hydrolytic conditions provides an additional alternative to the Nef reaction. These reactions presumably proceed via the formation of the oxime, which undergoes hydrolysis to the carbonyl compound (equation 78). Reducing agents that have been employed include

$$RCH{=}\overset{+}{N}(O^-)O^- \xrightarrow[H^+]{\text{red.}} RCH{=}NOH \xrightarrow{H_2O} RCHO \quad (78)$$

titanium trichloride[383,384] vanadium II chloride[385], chromium II chloride[386] and ascorbic acid[387].

Similarly to other *N*-oxide-type compounds, nitronates can also be deoxygenated by appropriate oxygen acceptors. Thus, silyl nitronates can be converted to *O*-silyloximes by trimethyl phosphite (equation 79)[388].

$$R_2C{=}\overset{+}{N}(O^-)OSiR_3 + (MeO)_3P \longrightarrow R_2C{=}N{-}OSiR_3 + (MeO)_3PO \quad (79)$$

A fragmentation of nitronate esters that results in the formation of carbonyl compounds from the alcohol moiety of the nitronate may be viewed as an intramolecular oxidation by a nitronate (equation 80)[340]. This fragmentation has

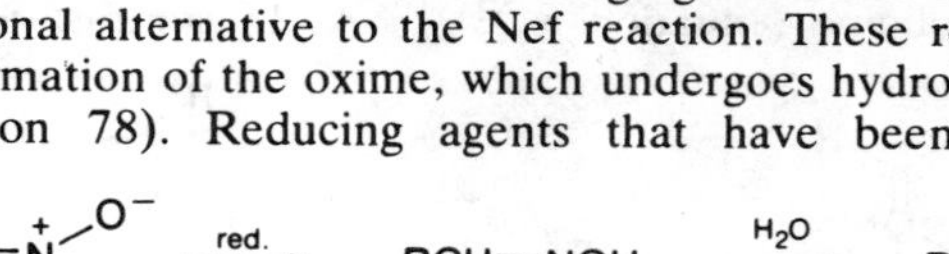

$$\text{(nitronate ester)} \longrightarrow >C{=}O + >{=}NOH \quad (80)$$

recently been developed into a method for the oxidation of alcohols via nitronates. These can be synthesized in good yield from the alcohol and 2,6-di-*t*-butyl-4-nitrophenol and subsequently decomposed to the carbonyl compounds (Scheme 44)[389].

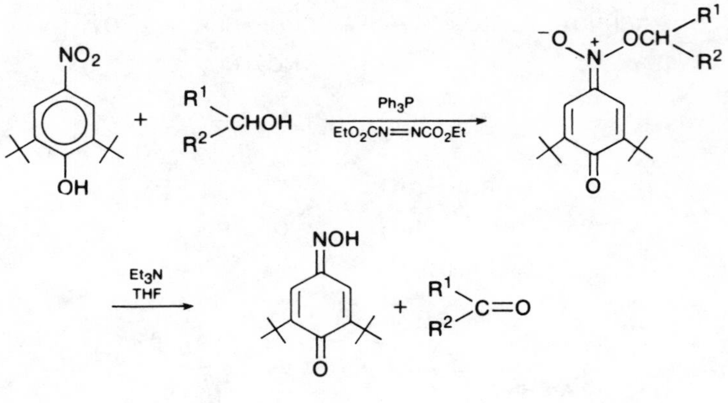

SCHEME 44

3. *1,3-Dipolar cycloaddition reactions of nitronates*

1,3-Dipolar cycloadditions have been discussed in Section II.B.5.a and in reviews of this subject[178,182–184,340].

The regiospecificity of the nitronate cycloaddition has been reconfirmed. In all known cases cycloaddition of a nitronate to a monosubstituted olefin leads to a 5-substituted isoxazolidine (equation 81)[34,390–394]. In contrast to nitrones (Section II.B.5.a) no variation in regioselectivity has been observed in reactions of nitronic esters. Disubstituted olefins such as methyl crotonate or crotononitrile give products with the electron-withdrawing group in position 5[393], and trisubstituted olefins react with formation of 5,5-disubstituted products[392b]; 1,3- and 1,4-dienes react with nitronic esters with the participation of each double bond separately[395]. Boron trifluoride has been shown to catalyse the 1,3-dipolar cycloaddition of a nitronic ester[400].

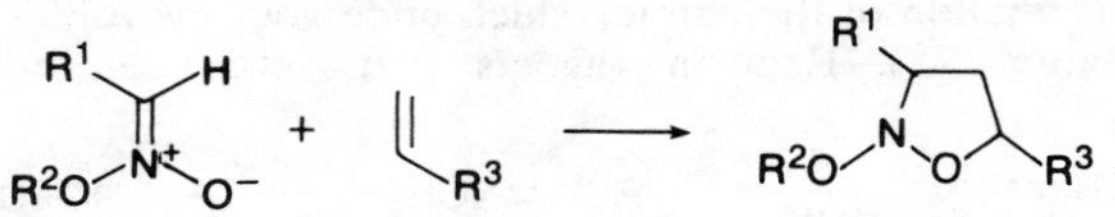

(81)

The influence of the structure of the nitronic ester upon its reactivity in 1,3-dipolar cycloaddition and the stereoselectivity of the reactions have been extensively studied by Carrié and his group[396–399]. They have found that (*Z*) nitronic esters (**307**) are more reactive in these reactions than their (*E*) isomers (**308**)[396]. (*Z*) Nitronic esters react with olefins with the exclusive formation of

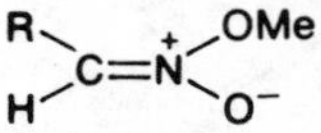

(*Z*-**307**) 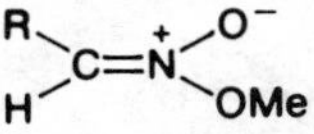 (*E*-**308**)

isoxazolidines with the OMe and R groups *cis*-related (equation 82), while the reaction of the (*E*) isomer (**308**) yields only the stable *trans* invertomer (equation 83)[396,397]. These results were discussed in terms of secondary orbital interactions of the two reactants. It was concluded that the governing factor is the interaction between the orbitals of the central atom of the 1,3-dipole and those of the

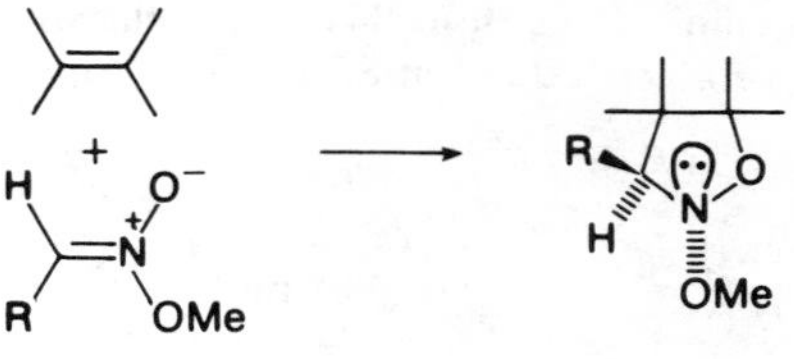

(307) (82)

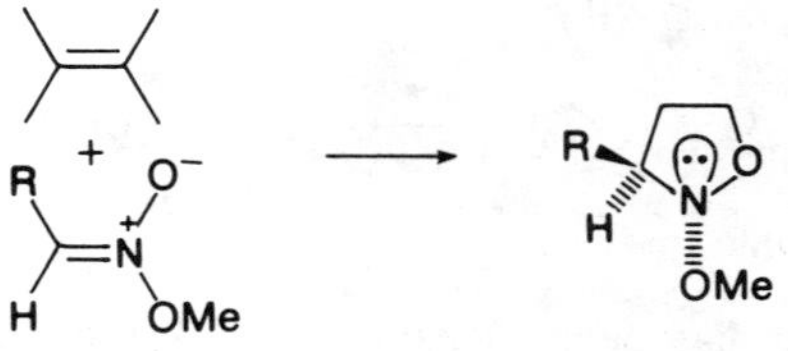

(308) (83)

substituent on the dipolarophile[398]. The mode of approach of nitronic esters to dipolarophiles has also been determined. Using maleic anhydride and maleic imides as models for 1,2-disubstituted dipolarophiles, it was found that the reaction proceeds by the *endo* approach illustrated in Scheme 45[399], yielding, stereoselectively, the corresponding *N*-alkoxyisoxazolidines **309** and **310**, which exist as stable invertomers on the nitrogen. In contrast to this, the mode of approach in the cycloaddition reactions of nitronates with monosubstituted olefins varies with the substituent of the dipolarophile and with the reaction conditions; both *endo* and *exo* approaches have been observed[397].

The products of nitronate ester cycloadditions may undergo various further transformations. Acid catalysis may induce the elimination of alcohol with the

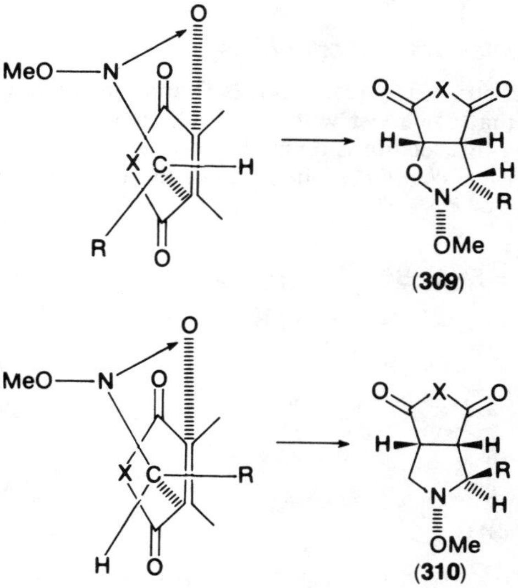

(309)

(310)

SCHEME 45

formation of an isoxazoline (equation 84)[391] or cleavage of the ring and the formation of an acyclic product (equation 85)[391]. This type of ring-opening can also

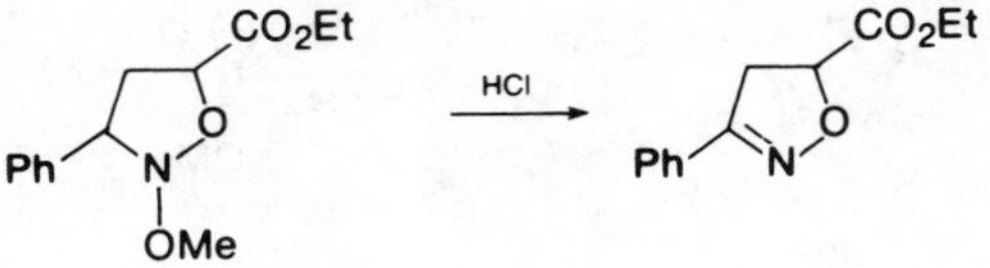

$$\text{5-(CO}_2\text{Et)-3-Ph-2-OMe-isoxazolidine} \xrightarrow{\text{HCl}} \text{5-(CO}_2\text{Et)-3-Ph-isoxazoline} \qquad (84)$$

$$\text{5-(CO}_2\text{Et)-3-Ph-2-OMe-isoxazolidine} \xrightarrow{\text{HCl}} PhC(=NOMe)CH_2CHOHCO_2Et \qquad (85)$$

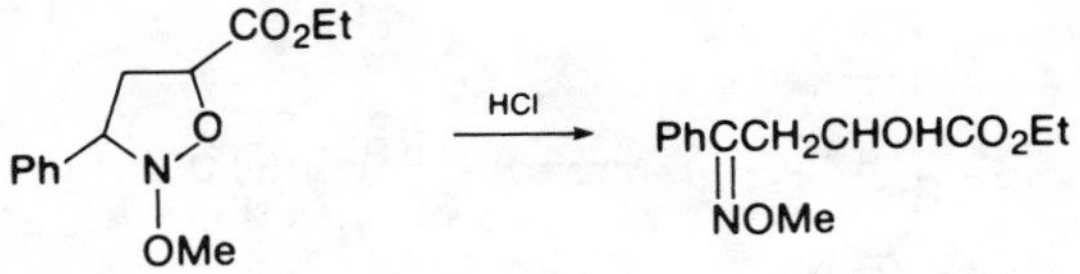

be caused thermally, as has been elegantly used for the synthesis of γ-amino-α-hydroxybutyric acid, which is a constituent of some aminoglycoside antibiotics (see Scheme 46)[392a]. Ethyl nitronate reacts with methyl acrylate to give the isoxazolidine **311**. Thermolysis of **311** gives the oxime ether, which can be reduced catalytically to the desired product in good yield[392a].

The cycloaddition of acetylenes to nitrones leads to 4-isoxazolines, which rearrange easily to acylaziridines[401]. In contrast nitronate esters with alkenes give directly *N*-alkoxy-2-acylaziridines also, presumably, through 4-isoxazolines[402]. Grée and Carrié have shown that the reaction is stereoselective and from each geometric isomer of the nitronic ester a different pair of products is formed[402]. Thus a (*Z*) nitronic ester gives two aziridines **312** and **313** with R and OMe *cis*-related, while the (*E*) isomer gives the two other stable invertomers **314** and **315**. Various mechanistic possibilities for the isoxazoline–acylaziridine rearrangement are discussed[402].

4. *Reactions of nitronates with electrophiles*

It follows from their structures (**293** and **294**) that nitronates may behave as ambident nucleophiles that can react with electrophiles either at the oxygen or at the carbon. Accordingly, this section is divided into two parts.

a. Nitronates as oxygen nucleophiles. In general, the carbon atom of a nitronate

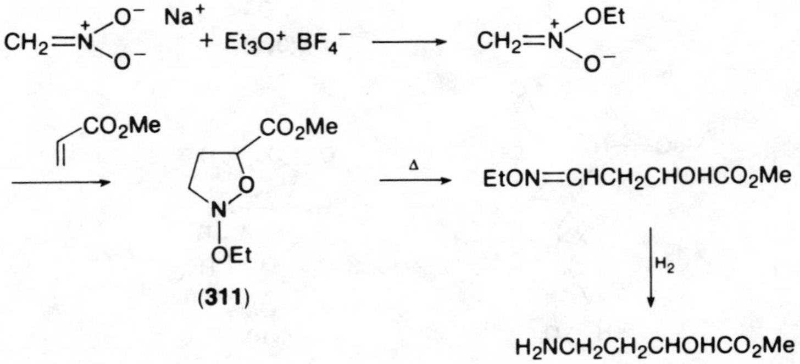

$$CH_2{=}\overset{+}{N}(O^-)O^-\ Na^+ + Et_3O^+\ BF_4^- \longrightarrow CH_2{=}\overset{+}{N}(OEt)O^-$$

$$\xrightarrow{CH_2=CHCO_2Me} \text{5-(CO}_2\text{Me)-2-OEt-isoxazolidine } (\mathbf{311}) \xrightarrow{\Delta} EtON{=}CHCH_2CHOHCO_2Me \xrightarrow{H_2} H_2NCH_2CH_2CHOHCO_2Me$$

SCHEME 46

(Z) HC≡CY → [isoxazoline intermediate] → (312) + (313)

(E) HC≡CY → [isoxazoline intermediate] → (314) + (315)

anion is a poor nucleophile, therefore alkylation or acylation of the oxygen is the reaction that is most often observed[403]. Quenching of nitronate anions, generated from nitroalkanes with lithium diisopropylamide, with a trialkylsilyl chloride results in silyl nitronates that are useful synthetic intermediates[393,394,404,405] (equation 86).

$$R^1R^2CHNO_2 + LiN(i\text{-}Pr)_2 \longrightarrow R^1R^2C{=}NO_2^-\ Li^+ + R_3SiCl \longrightarrow R^1R^2C{=}\overset{+}{N}(O^-)OSiR_3 \quad (86)$$

The O-alkylation of nitronates by methyl bromoacetate is the first step in a novel synthesis of β-amino-α-hydroxy acids of some importance in antibiotics (Scheme 47)[406]. The primary product of this reaction (**317**) undergoes fragmentation into an oxime and methyl glyoxylate (cf. Section III.C.2); the latter enters the Henry reaction (Section III.C.4.b) with a second mole of nitronate **316**, resulting in an α-hydroxy-β-nitro acid, which can further be reduced to the end-product[406].

The reactions of nitronates with acylating agents result in end-products that are different from the O-acylnitronates that are the postulated primary products.

Nitronates derived from primary nitroparaffins are converted by chlorodiethoxyphosphine to nitriles via the O-acylated product **318** as illustrated in Scheme 48[407].

More recently it has been proposed that the conversion of primary nitroalkanes to carboxylic acids[408,409] by acetic anhydride–sodium acetate proceeds through the

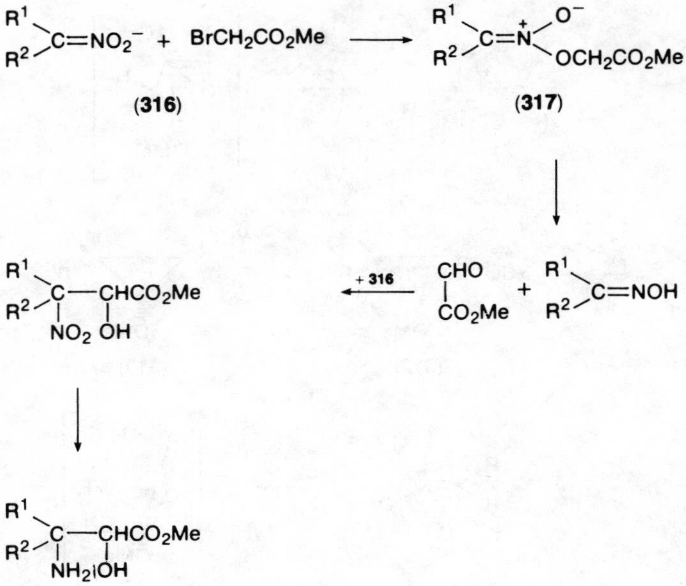

$R^1R^2C{=}NO_2^- + BrCH_2CO_2Me \longrightarrow R^1R^2C{=}\overset{+}{N}(O^-)OCH_2CO_2Me$

(**316**) (**317**)

$\longrightarrow CHO{-}CO_2Me + R^1R^2C{=}NOH$

$\xrightarrow{+\,316} R^1R^2C(NO_2){-}CH(OH)CO_2Me \longrightarrow R^1R^2C(NH_2){-}CH(OH)CO_2Me$

SCHEME 47

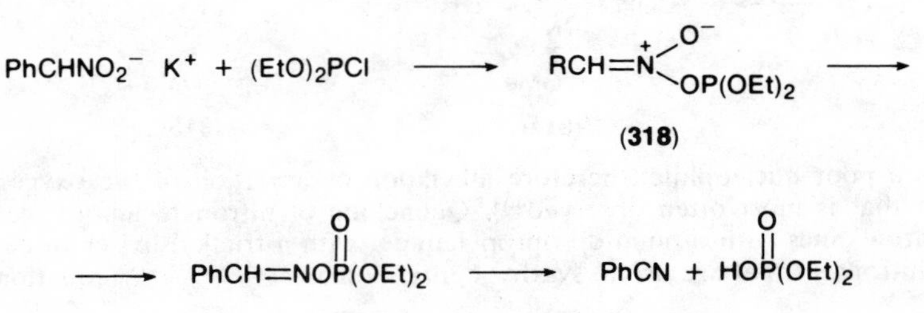

$PhCHNO_2^-\ K^+ + (EtO)_2PCl \longrightarrow RCH{=}\overset{+}{N}(O^-)OP(OEt)_2 \longrightarrow$

(**318**)

$\longrightarrow PhCH{=}NOP(O)(OEt)_2 \longrightarrow PhCN + HOP(O)(OEt)_2$

SCHEME 48

$RCH{=}\overset{+}{N}(O^-)O^- + Ac_2O \longrightarrow RCH{=}\overset{+}{N}(O^-)OCOCH_3 \longrightarrow$

(**319**)

$\longrightarrow R{-}C(H){-}N(OCOCH_3)$ (oxaziridine, C–N–O ring) $\longrightarrow RCONHOCOCH_3 \xrightarrow{H_2O} RCOOH$

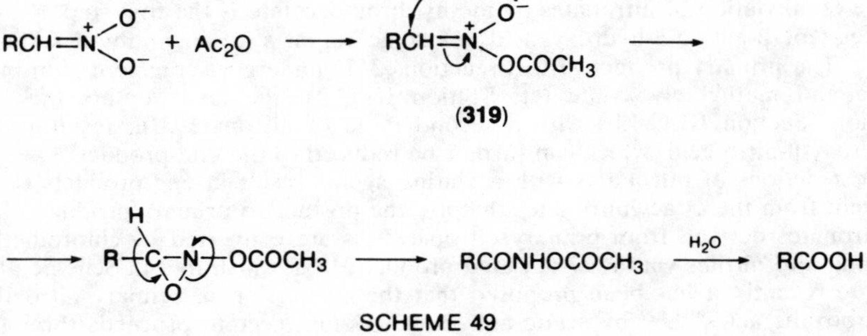

SCHEME 49

intermediacy of a mixed carboxylic nitronic anhydride (**319**) (Scheme 49)[410]. Mixed anhydrides of type **319** may also lose carboxylic acid to give nitrile oxides which can be trapped by dipolarophiles[410,411].

b. Nitronates as carbon nucleophiles. This aspect of the reactivity of nitronates is probably the most interesting one, since the importance and synthetic utility of any synthon will be related to its ability to serve as partner in carbon–carbon bond-formation reactions. Although nitronates are rather weak carbon nucleophiles, there are examples indicating that they can be alkylated and acylated on the carbon as well as they can add to a carbonyl group or to a polar carbon–carbon double bond (Michael reaction). Recently, it has been discovered that, by the addition of two equivalents or butyllithium to a nitroalkane, 'doubly deprotonated' species such as **320** can be produced (equation 87)[412].

$$RCH_2NO_2 \xrightarrow{BuLi} (R)(H)C{=}\overset{+}{N}(OLi)O^- \xrightarrow{BuLi} (R)(Li)C{=}\overset{+}{N}(OLi)O^- \quad \textbf{(320)} \qquad (87)$$

(i) *C*-Alkylation of nitronates. Alkylations of nitronates that proceed by radical chain mechanisms[413] are reviewed in a separate chapter in this volume[414]. The earlier literature contains only intramolecular *C*-alkylations of nitronates as indicated in equations (88)[415] and (89)[416].

$$\text{3-(3-tosyloxypropyl)nitrocyclohexane (}NO_2\text{, OTs)} \xrightarrow{base} \text{bicyclic nitro compound (}NO_2\text{)} \qquad (88)$$

$$I(CH_2)_3NO_2 \xrightarrow{base} \text{cyclopropyl-}NO_2 \qquad (89)$$

Since in dipolar aprotic solvents the negative charge resides upon the carbon of the nitronate anion (Section III.B.4) carbon alkylations are promoted by the use of solvents such as *N,N*-dimethylacetamide and yields up 88%, depending upon the alkyl halide, can be obtained (equation 90)[417–419]. This reaction has been developed into a synthesis of amino acids.

$$O_2NCH_2CO_2Me \xrightarrow[CH_3CONMe_2]{RX/NaOEt} RCH(NO_2)CO_2Me \qquad (90)$$

Katritzky and coworkers have reported recently that nitronate anions can be carbon-alkylated by 1-substituted 2,4,6-triphenylpyridinium cations, also in dipolar solvents[420]. However, the mechanism of this reaction has not yet been clarified.

Nitroalkanes that cannot be alkylated by alkyl halides via nitronates have been shown to undergo smooth alkylation using the 'doubly deprotonated' (cf. **320**) derivatives[421].

(ii) *C*-Acylation of nitronates. Although most acylating agents attack nitronate anions on the oxygen, some give *C*-acyl products. Such reagents are methyl methoxymagnesium carbonate (equation 91)[422], aroyl cyanides (equation 92)[423–425] acylimidazoles (equation 93)[426] and isatoic anhydride (**321**) (equation 94)[427].

Dilithiated derivatives are considerably more reactive than nitronates toward

$$RCHNO_2^- + CH_3OCO_2MgOCH_3 \longrightarrow RCH(CO_2CH_3)NO_2 \quad (91)$$

$$RCHNO_2^- + R^1COCN \longrightarrow R^1COCH(R)NO_2 \quad (92)$$

$$CH_2{=}NO_2^- + RCON(\text{imidazolyl}) \longrightarrow RCOCH_2NO_2 \quad (93)$$

$$CH_2{=}NO_2^- + \textbf{(321)} \longrightarrow \text{2-}H_2NC_6H_4COCH_2NO_2 \quad (94)$$

(321)

acylating agents as, in contrast to the monoanions, they can be acylated by anhydrides and esters to give cleanly *C*-acylnitro compounds[388,412].

(iii) *C*-Hydroxylalkylation of nitronates. This reaction, the addition of nitronates to the carbonyl group of aldehydes and ketones, is known as the Henry or the nitroaldol reaction[403]. The formation of nitro alcohols is often accompanied by spontaneous loss of water with the formation of nitro olefins (equation 95). In this

$$R^1CHNO_2 + R^2R^3CO \longrightarrow R^1CH(NO_2){-}C(OH)R^2R^3 \longrightarrow R^1(O_2N)C{=}CR^2R^3 \quad (95)$$

case too, the use of dianions presents an improvement since the product formed from the dianion is stable and the reaction is not reversible[388,412].

Another improvement in the nitroaldol condensation has been reported by Seebach. It was found that silyl nitronates derived from primary nitroalkanes add to the carbonyl group of aldehydes in the presence of a catalytic amount of tetrabutylammonium fluoride (equation 96)[405].

$$R^1CHO + R^2CH{=}\overset{+}{N}(O^-)OSiR^3_3 \xrightarrow{Bu_4N^+F^+} R^1CH(OSiR^3_3)CH(NO_2)R^2 \quad (96)$$

(iv) Addition of nitronates to polar C=C bonds. The Michael addition of nitroalkanes to α,β-unsaturated aldehydes, ketones, esters and nitriles, as well as to unsaturated sulphones and nitro olefins is a very efficient reaction. An extensive survey of it can be found in Houben–Weyl[403]. Some recent applications have been used in the synthesis of natural products. The Michael addition of nitromethane to the cyclopentenone **322** has been used by two groups to introduce carbon-13 of the prostaglandin molecule (Scheme 50)[364,428].

The Michael addition of a nitronate may sometimes be followed by the elimination of nitrite. In such cases the net result is the Michael addition of a vinyl

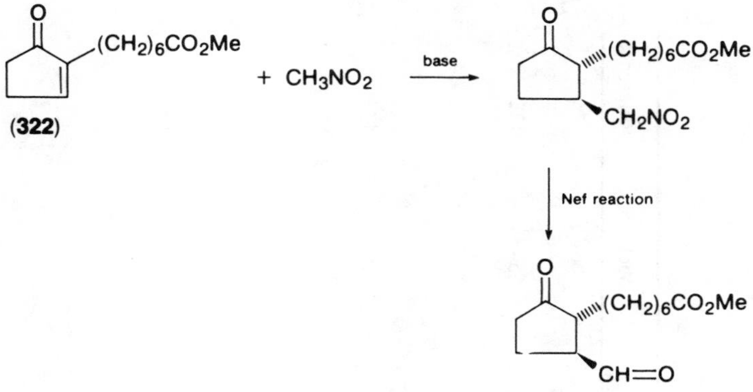

SCHEME 50

carbanion. This is illustrated in the reaction of β-nitropropionate with cyclopentenone (Scheme 51)[429].

Another application of the Michael reaction is the addition of nitronates derived from 2-nitro alcohols to acrolein. These reactions yield, nitropolydeoxy sugars, which can be converted to aminopolydeoxy sugars; some of these are components of antibiotics (e.g. Scheme 52)[430].

5. *Reactions of nitronates with nucleophiles*

One of the most important reactions of nitroalkanes is the Nef reaction[431]. This reaction is the hydrolysis of a nitronic acid derived from a primary or secondary nitroalkane to yield an aldehyde or a ketone (equation 97).

The yields in this reaction are not always high, and therefore, some effort has been directed towards improvements. Some variants, using oxidizing or reducing conditions, have already been presented in Sections III.C.1 and III.C.2, respectively. In addition to these, there exist several recent reports concerning the Nef reactions using solvolytic or neutral conditions or solid catalysts. These are presented in Table 4.

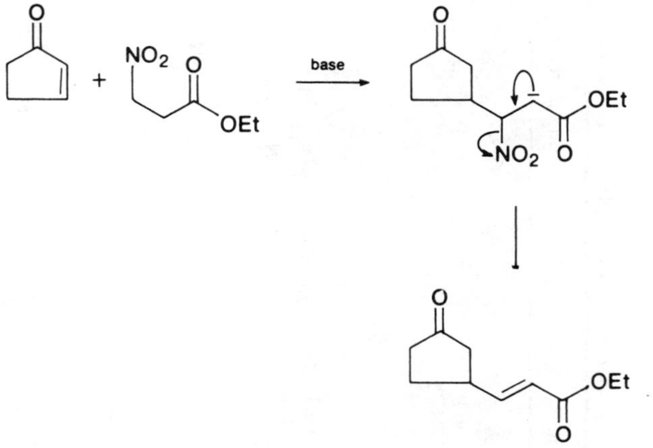

SCHEME 51

TABLE 4. Some recent examples of the Nef reaction

Reagent	Starting material	Product	Yield (%)	References
1. Me_2NH 2. HCl/H_2O	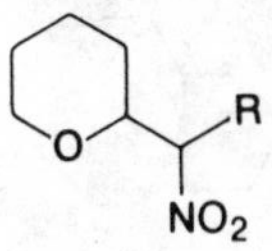 O, R, NO_2	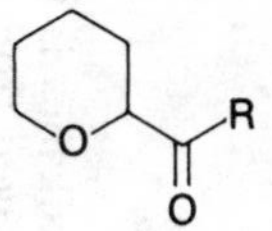 O, R, O	55–65	432
1. NaOH/EtOH 2. HCl/H_2O	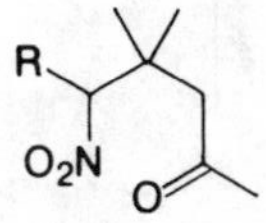 R, O_2N, O	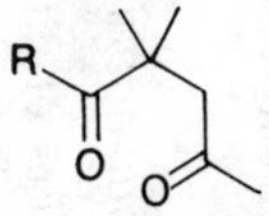 R, O, O	70	433
1. NaOMe/MeOH 2. H_2SO_4/MeOH	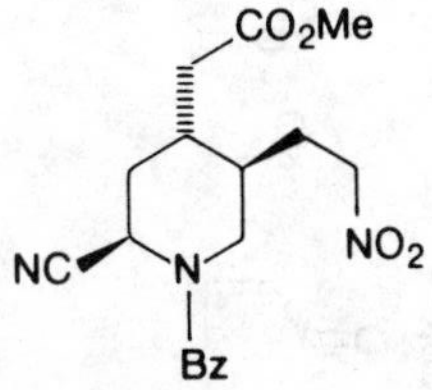 CO_2Me, NC, N, Bz, NO_2	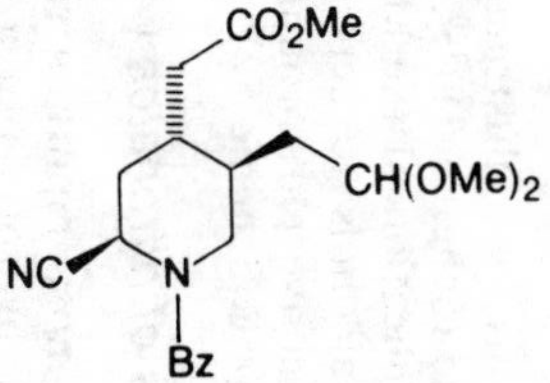 CO_2Me, $CH(OMe)_2$, NC, N, Bz	78	434

1. NaOMe 2. H_2SO_4/H_2O	O, $(CH_2)_3CO_2Me$, NO_2	O, $(CH_2)_3CO_2Me$, CHO	56	428
1. NaOEt/H_2O/THF 2. H_2SO_4	NO_2, O	CHO, O	45	435
$NaNO_2$/*n*-PrONO/DMSO	R_2CHNO_2	$R_2C{=}O$	70–85	408, 409
NaOMe/SiO_2	R_2CHNO_2	$R_2C{=}O$	80–95	436, 437
SiO_2	$RC(=O)-C(R)=CH-C(=NO_2^-)CH_3$	$RC(=O)-C(R)=CHC(=O)CH_3$	35–65	373, 438

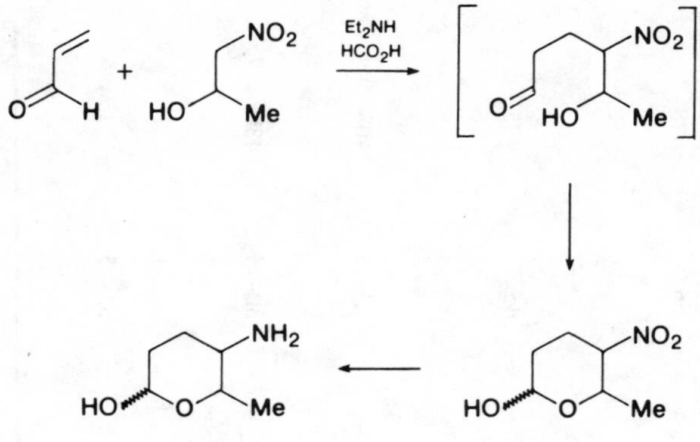

SCHEME 52

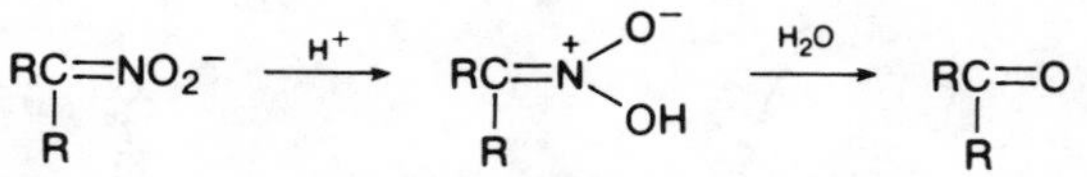

$$RC(R)=NO_2^- \xrightarrow{H^+} RC(R)=\overset{+}{N}(O^-)OH \xrightarrow{H_2O} RC(R)=O \qquad (97)$$

Finally, attention is drawn to two recently published review articles concerning the synthetic utility of aliphatic nitro compounds[388,439].

IV. ACKNOWLEDGEMENTS

This chapter was written while the author was Visiting Scientist in the Department of Chemical Research, Searle Research and Development, Division of G. D. Searle and Co., Chicago, Illinois. The author wishes to express thanks to Searle for hospitality, library and secretarial services, and to Drs. D. St.C. Black, W. H. Rastetter and D. A. Wilson for communicating results prior to publication.

V. REFERENCES

1. L. I. Smith, *Chem. Rev.*, **23**, 193 (1938).
2. J. Hamer and A. Macaluso, *Chem. Rev.*, **64**, 473 (1964).
3. G. R. Delpierre and M. Lamchen, *Quart. Rev.*, **19**, 329 (1965).
4. P. A. S. Smith, *Open-chain Nitrogen Compounds*, Vol. 2, W. A. Benjamin, New York, 1966, pp. 29–63.
5. M. Lamchen in *Mechanisms of Molecular Migrations*, Vol. 1 (Ed. B. S. Thyagarajan), Wiley–Interscience, New York, 1968, pp. 1–60.
6. D. St. C. Black, R. I. Crozier and V. C. Davis, *Synthesis*, 205 (1975).
7. W Rundel in *Houben—Weyl: Methoden der Organischen Chemie*, Vol. 10/4 (Ed. E. Müller), Georg Thieme Verlag, Stuttgart, 1968, pp. 311–448.
8. H. Stamm in *Methodicum Chimicum*, Vol. 6 (Ed. F. Zymalkowski) Georg Thieme Verlag, Stuttgart, 1975, pp. 333–350.
9. S. R. Sandler and W. Karo, *Organic Functional Group Preparations*, Vol. III, Academic Press, New York, 1972, pp. 301–317.
10. W. Kliegel, *Pharmazie*, **32**, 643 (1977); **33**, 331 (1978).
11. M. A. Robb and I. G. Csizmadia, *J. Chem. Phys.*, **50**, 1819 (1969).
12. V. I. Minkin, E. A. Medyantseva, I. M. Andreeva and G. V. Gorshkova, *Zh. Org. Khim*, **9**, 148 (1973); *Chem. Abstr.*, **79**, 4494t (1973).
13. K. Folting, W. N. Lipscomb and B. Jerslev, *Acta Cryst.*, **17**, 1263 (1964).
14. J. N. Brown and L. M. Trefonas, *Acta Cryst.*, **B29**, 237 (1973).

15. W. R. Young, *Mol. Cryst. Liquid Cryst.*, **10**, 237, (1970).
16. W. R. Young, I. Haller and A. Aviram, *Mol. Cryst. Liquid Cryst.*, **13**, 357 (1971).
17. N. A. Akmanova, N. R. Khairullin and Yu V. Svetkin, *Zh. Obshch. Khim.*, **46**, 704 (1976); *Chem. Abstr.*, **84**, 163766c (1976).
18. I. Yu Kokoreva, L. A. Neiman, Ya. K. Sirkin and S. T. Kirillova, *Dokl. Akad. Nauk SSSR*, **156**, 412 (1964); *Chem. Abstr.*, **61**, 5045h (1964).
19. E. A. Medyantseva, I. M. Andreeva and V. I. Minkin, *Zh. Org. Khim.*, **8**, 146 (1972); *Chem. Abstr.*, **76**, 112556m (1972).
20. V. Baliah and V. Chandrasekharan, *Indian J. Chem.*, **8**, 1096 (1970).
21. E. A. Medyantseva, I. M. Andreeva and V. A. Bren, *Izv. Sev.-Kavk. Nauchn. Tsentra. Vyssh. Shk., Ser. Estesty. Nauk*, **6**, 63 (1978); *Chem. Abstr.*, **89**, 196805w (1978).
22. W. B. Jennings, D. R. Boyd and L. C. Waring, *J. Chem. Soc., Perkin II*, 610 (1976).
23. V. A. Bren, E. A. Medyantseva and V. I. Minkin, *Reakts. Sposobn. Org. Soedin.*, **5**, 988 (1968); *Chem. Abstr.*, **71**, 29944n (1969).
24. V. A. Bren, E. A. Medyantseva, I. M. Andreeva and V. I. Minkin, *Zh. Org. Khim.*, **9**, 767 (1973), *Chem. Abstr.*, **79**, 4793q (1973).
25. V. I. Minkin, E. A. Medyantseva, I. M. Andreeva and V. S. Yureva, *Zh. Org. Khim.*, **10**, 2597 (1974); *Chem. Abstr.*, **82**, 85923u (1975).
26. N. N. Zatsepina, I. F. Tupitsyn, A. I. Belyashova, E. A. Medyantseva, I. M. Andreeva and V. I. Minkin, *Reakts. Sposobn. Org. Soedin*, **12**, 223 (1975); *Chem. Abstr.*, **85**, 32286u (1976).
27. J. R. Hutchison, G. N. LaMar and W. D. Horrocks, Jr., *Inorg. Chem.*, **8**, 126 (1969).
28. R. W. Kluiber and W. D. Horrocks, Jr., *Inorg. Chim. Acta*, **4**, 183 (1970).
29. K. N. Houk, P. Caramella, L. L. Munchausen, Y.-M. Chang, A. Battaglia, J. Sims and D. C. Kaufman, *J. Electron Spectry.*, **10**, 441 (1977).
30. K. N. Houk, A. Bimanand, D. Mukherjee, J. Sims, Y.-M. Chang, D. C. Kaufman and L. N. Domelsmith, *Heterocycles*, **7**, 293 (1977).
31. J. Bastide, J. P. Maier and T. Kubota, *J. Electron Spectry.*, **9**, 307 (1976).
32. D. Mukherjee, L. N. Domelsmith and K. N. Houk, *J. Amer. Chem. Soc.*, **100**, 1954 (1978).
33. P. Brocklehurst, *Tetrahedron*, **18**, 299 (1962).
34. T. Kubota, M. Yamakawa and Y. Mori, *Bull. Soc. Chem. Japan*, **36**, 1552 (1963).
35. K. Koyano and H. Suzuki, *Tetrahedron Letters*, 1859 (1968).
36. K. Koyano and H. Suzuki, *Bull. Soc. Chem. Japan*, **42**, 3306 (1969).
37. P. Grammaticakis, *Compt. Rend., Ser. C*, **268**, 78 (1969).
38. M. Masui and C. Yijima, *J. Chem. Soc.* (*C*), 2022 (1967.
39. P. A. S. Smith and S. E. Gloyer, *J. Org. Chem.*, **40**, 2504 (1975).
40. M. M. Mitasov, I. A. Grigorev, G. I. Shchukin, I. K. Korobeinicheva, and L. B. Volodarskii, *Izv. Sib. Otd. Akad. Nauk SSSR, Ser. Khim. Nauk.*, 112 (1978); *Chem. Abstr.*, **88**, 200214b (1978).
41. R. Bonnett and D. E. McGreer, *Can. J. Chem.*, **40**, 177 (1962).
42. J. E. Baldwin, A. K. Qureshi and B. Sklarz, *Chem. Commun.*, 373 (1868); *J. Chem. Soc.* (*C*), 1073 (1969).
43. D. A. Kerr and D. A. Wilson, *J. Chem. Soc.* (*C*), 1718 (1970).
44. E. J. Grubbs, R. J. Milligan and M. H. Goodrow, *J. Org. Chem.*, **36**, 1780 (1971).
45. J. Bjorgo, D. R. Boyd, D. C. Neill and W. B. Jennings, *J. Chem. Soc., Perkin I*, 254 (1977).
46. D. St. C. Black, R. F. C. Brown, B. T. Dunstan and S. Sternhell, *Tetrahedron Letters*, 4283 (1974).
47. T. S. Dobashi, M. H. Goodrow and E. J. Grubbs, *J. Org. Chem.*, **38**, 4440 (1973).
48. H. G. Aurich and U. Grigo, *Chem. Ber.*, **109**, 200 (1976).
49. P. M. Weintraub and P. L. Tiernan, *J. Org. Chem.*, **39**, 1061 (1974).
50. D. H. R. Barton, M. J. Day, R. H. Hesse and M. M. Pechet, *J. Chem. Soc., Perkin I*, 1764 (1975).
51. Y. Yoshimura, Y. Mori and K. Tori, *Chem. Letters*, 181 (1972).
52. T. H. Kinstle and J. G. Stam, *Chem. Commun.*, 185 (1968).
53. B. S. Larsen, G. Schroll, S.-O. Lawesson, J. H. Bowie and R. G. Cooks, *Chem. Ind.* (*London*), 321 (1968); *Tetrahedron*, **24**, 5193 (1968).

54. L. A. Neiman, V. I. Maimind, M. M. Shemyakin, V. A. Puchkov, V. N. Bochkarev, Yu. S. Nekrasov and N. S. Vulfson, *Zh. Obsch. Khim.*, **37**, 1600 (1967); *Chem. Abstr.*, **68**, 48863q (1968).
55. R. Grigg and B. G. Odell, *J. Chem. Soc.* (*B*), 218 (1966).
56. H. Stamm, J. Hoenicke and H. Steudle, *Arch. Pharm.*, **305**, 619 (1972).
57. W. M. Leyshon and D. A. Wilson, *Org. Mass. Spectrom.*, **7**, 251 (1973).
58. M. Masui and C. Yijima. *Chem. Pharm. Bull.*, **17**, 1517 (1969)
59. R. T. Coutts, G. R. Jones, A. Benderly and A. L. C. Mak, *Can. J. Pharm. Sci.*, **13**, 61 (1978).
60. E. Baranowska, I. Panfil and C. Belzecki, *Bull. Acad. Pol. Sci. Ser., Sci. Chim.*, **25**, 93 (1977); *Chem. Abstr.*, **87**, 52293p (1977).
61. J. F. Elsworth and M. Lamchen. *J. S. Afr. Chem. Inst.*, **23**, 61 (1970).
62. J. H. Bowie, S.-O. Lawesson, B. S. Larsen, G. E. Lewis and G. Schroll, *Australian J. Chem.*, **21**, 2031 (1968).
63. T. S. Dobashi, M. H. Goodrow and E. J. Grubbs, *J. Org. Chem.*, **38**, 4440 (1973).
64. K. Koyano and I. Tanaka, *J. Phys. Chem.*, **69**, 2545 (1965).
65. R. W. Layer and C. J. Carman, *Tetrahedron Letters*, 1285 (1968).
66. L. W. Boyle, M. J. Peagram and G. H. Witham, *J. Chem. Soc.* (*B*), 1728 (1971).
67. J. Bjorgo, D. R. Boyd and D. C. Neill, *J. Chem. Soc., Chem. Commun.*, 478 (1974).
68. N. S. Ooi and D. A. Wilson, private communication.
69. W. B. Jenning and S. D. Worley, *Tetrahedron Letters*, 1435 (1977).
70. R. Bonnett and D. E. McGreer, *Can. J. Chem.*, **40**, 177 (1962).
71. M. Mousseron-Canet and J.-P. Boca, *Compt. Rend.*, **260**, 2851 (1965); *Bull. Soc. Chim. Fr.*, 1296 (1967).
72. R. M. Acheson, P. G. Hunt, D. M. Littlewood, B. A. Murrer and H. E. Rosenberg, *J. Chem. Soc., Perkin I*, 1117 (1978).
73. M. Michalska and I. Orlich, *Bull. Acad. Pol. Sci., Ser. Sci. Chim.*, **23**, 655 (1975); *Chem. Abstr.*, **84**, 121698b (1976).
74. B. Eistert and G. Holzer, *Chem. Ber.*, 109, 3462 (1976).
75. D. St. C. Black, V. M. Clark, B. G. Odell and Lord Todd, *J. Chem. Soc., Perkin I*, 1944 (1976).
76. N. J. A. Gutteridge and F. J. McGillan, *J. Chem. Soc.* (*C*). 641 (1970).
77. A. R. Forrester and R. H. Thomson, *J. Chem. Soc.*, 5632 (1963).
78. H. G. Aurich, *Angew. Chem.* (*Internat. Ed.*), **6**, 810 (1967).
79. E. C. Taylor and R. E. Buntrack, *J. Org. Chem.*, **36**, 634 (1971).
80. R. Behrend and E. Konig, *Ann. Chem.*, **263**, 355 (1891).
81. E. Breuer and I. Ronen-Braunstein, *J. Chem. Soc., Chem. Commun.*, 949 (1974).
82. E. Breuer and S. Zbaida, *J. Org. Chem.*, **42**, 1904 (1977).
83. H. Dadoun, J. P. Alazard, J. Parello and X. Lusinchi, *Tetrahedron*, **34**, 2639 (1978).
84. W. Kliegel and H. Becker, *Chem. Ber.*, **110**, 2067 (1977).
85. E. G. Janzen and R. C. Zawalski, *J. Org. Chem.*, **43**, 1900 (1978).
86. W. Kliegel and H. Becker, *Chem. Ber.*, **110**, 2090 (1977).
87. H. Mohrle and M. Lappenberg, *Chem. Ber.*, **109**, 1106 (1976).
88. H. Mohrle, M. Lappenberg and D. Wendisch, *Monatsh.*, **108**, 273 (1977).
89. H. Mohrle and M. Lappenberg, *Arch. Pharm.*, **311**, 806 (1978).
90. Yu. V. Svetkin, N. A. Akmanova and G. P. Plotnikova, *Zh. Org. Khim.*, **10**, 561 (1974); *Chem. Abstr.*, **80**, 133367y (1974).
91. N. A. Akmanova, V. S. Anisimova and Yu. V. Svetkin, *Org. Khim.*, 93 (1976); *Chem. Abstr.*, **88**, 152519f (1978).
92. L. B. Volodarskii, Yu. G. Putsykin and V. I. Mamatyuk, *Zh. Org. Khim.*, **5**, 355 (1969); *Chem. Abstr.*, **70**, 115137n (1969).
93. T. K. Sevastyanova and L. B. Volodarskii, *Zh. Org. Khim.*, **7**, 1974 (1971); *Chem. Abstr.*, **76**, 3740w (1972).
94. G. Zinner and U. Kruger, *Chemiker Z.*, **101**, 154 (1977).
95. G. S. Spence, E. C. Taylor and O. Buchardt, *Chem. Rev.*, **70**, 231 (1970).
96. A. C. Cope and A. C. Haven, *J. Amer. Chem. Soc.*, **84**, 1197 (1962).
97. J. S. Vincent and E. J. Grubbs, *J. Amer. Chem. Soc.*, **91**, 2022 (1969).

98. D. G. Morris, *J. Chem. Soc., Chem. Commun.*, 221 (1971).
99. T. S. Dobashi, D. R. Parker and E. J. Grubbs, *J. Amer. Chem. Soc.*, **99**, 5382 (1977).
100. J. A. Villarreal, T. S. Dobashi and E. J. Grubbs, *J. Org. Chem.*, **43**, 1890 (1978).
101. J. A. Villarreal and E. J. Grubbs, *J. Org. Chem.*, **43**, 1896 (1978).
102. I. W. Jones, D. A. Kerr and D. A. Wilson, *J. Chem. Soc.* (*C*), 2595 (1971).
103. I. W. Jones, D. A. Kerr and D. A. Wilson, *J. Chem. Soc.*, (*C*), 2591 (1971).
104. H. Morgan and D. A. Wilson, *J. Chem. Soc., Perkin I*, 2803 (1972).
105. S. Ranganathan, D. Ranganathan, R. S. Sidhu and A. K. Mehrotra, *Tetrahedron Letters*, 3577 (1973).
106. A. Eckersley and N. A. J. Rogers, *Tetrahedron Letters*, 1661 (1974).
107. T. Kusumi, K. Yoneda and H. Kakisawa, *Synthesis*, 221 (1979).
108. M. L. Druelinger, R. W. Shelton and S. R. Lammert, *J. Heterocycl. Chem.*, **13**, 1001 (1976).
109. D. R. Boyd, W. B. Jennings, R. Spratt and D. M. Jerina, *Chem. Commun.*, 745 (1970).
110. J. S. Splitter, T.-M. Su, H. Ono and M. Calvin, *J. Amer. Chem. Soc.*, **93**, 4075 (1971).
111. D. R. Boyd and D. C. Neill, *J. Chem. Soc., Chem. Commun.*, 51 (1977).
112. C. Eskenazi, J. F. Nicoud and H. B. Kagan, *J. Org. Chem.*, **44**, 995 (1979).
113. H. Harada, Y. Mori and I. Tanaka, *Mol. Photochem.*, **2**, 153 (1970).
114. J. C. Evans, E. D. Owen and D. A. Wilson, *J. Chem. Soc., Perkin II*, 557 (1974).
115. W. M. Leyshon and D. A. Wilson, *J. Chem. Soc., Perkin I*, 1925 (1975).
116. D. St. C. Black and K. G. Watson, *Australian J. Chem.*, **26**, 2491 (1973).
117. (a) W. Sliwa, *Roczn. Chem.*, **50**, 667 (1976).
(b) D. St. C. Black, N. A. Blackman and A. B. Boscacci, *Tetrahedron Letters*, 175 (1978); *Australian J. Chem.*, **32**, 1775 (1979).
118. T. Sasaki and M. Takahashi, *Bull. Chem. Soc. Japan*, **41**, 1967 (1968).
119. A. Padwa, *J. Amer. Chem. Soc.*, **87**, 4365 (1965).
120. D. St. C. Black and A. Boscacci, *Australian J. Chem.*, **30**, 1109 (1977).
121. D. Dopp, *Tetrahedron Letters*, 3215 (1972).
122. D. R. Eckroth and R. H. Squire, *J. Org. Chem.*, **36**, 224 (1971).
123. M. Colonna and M. Poloni, *Ann. Chem. Roma*, **63**, 287 (1973).
124. H. G. Aurich and U. Grigo, *Chem. Ber.*, **109**, 200 (1976).
125. H. O. Larson, K. Y. W. Ing and D. L. Adams, *J. Heterocycl. Chem.*, **7**, 1227 (1970).
126. J. P. Alazard, B. Khemis and X. Lusinchi, *Tetrahedron*, **31**, 1427 (1975).
127. G. Zinner, *Chemiker Z.*, **102**, 58 (1978).
128. S. Tamagaki, S. Kozuka and S. Oae, *Tetrahedron*, **26**, 1795 (1970).
129. S. Tamagaki and S. Oae, *Bull. Chem. Soc. Japan*, **44**, 2851 (1971).
130. P. W. Jeffs and G. Molina, *J. Chem. Soc., Chem. Commun.*, 3 (1973).
131. L. H. Schlager, *Tetrahedron Letters*, 4519 (1970).
132. H. K. Kim and P. M. Weintraub, *J. Org. Chem.*, **35**, 4282 (1970).
133. D. R. Boyd, *Tetrahedron Letters*, 3467 (1973).
134. D. R. Boyd and D. C. Neill, *J. Chem. Soc., Perkin I*, 1309 (1977).
135. D. R. Boyd, D. C. Neill and M. E. Stubbs, *J. Chem. Soc., Perkin II*, 30 (1978).
136. M. H. Goodrow, J. A. Villarreal and E. J. Grubbs, *J. Org. Chem.*, **39**, 3447 (1974).
137. W. M. Leyshon and D. A. Wilson, *J. Chem. Soc., Perkin I*, 1920 (1975).
138. K. Sommermeyer, W. Seifert and W. Wilker, *Tetrahedron Letters*, 1821 (1974).
139. L. A. Neiman, S. V. Zhukova, L. B. Senyavina and M. M. Shemyakin, *Zh. Obsch. Khim.*, **38**, 1480 (1968).
140. S. Tamagaki and S. Oae, *Bull. Chem. Soc. Japan*, **43**, 1573 (1970).
141. N. Singh and K. Krishan, *Tetrahedron Letters*, 2711 (1973).
142. E. G. Janzen and B. J. Blackburn, *J. Amer. Chem. Soc.*, **91**, 4481 (1969).
143. L. A. Neiman, S. V. Zhukova and V. A. Tyurikov, *Tetrahedron Letters*, 1889 (1973).
144. L. A. Neiman and S. V. Zhukova, *Tetrahedron Letters*, 499 (1973).
145. J. F. Elsworth and M. Lamchen, *J. Chem. Soc.* (*C*), 1477 (1966).
146. J. F. Elsworth and M. Lamchen, *J. S. Afr. Chem. Inst.*, **24**, 196 (1971).
147. J. F. Elsworth and M. Lamchen, *J. S. Afr. Chem. Inst.*, **25**, 1 (1972).
148. V. M. Clark, B. Sklarz and A. Todd, *J. Chem. Soc.*, 2123 (1959).
149. A. K. Qureshi and B. Sklarz, *J. Chem. Soc.* (*C*), 412 (1966).

150. J. K. Sutherland and D. A. Widdowson, *J. Chem. Soc.*, 3495 (1964).
151. G. W. Alderson, D. St. C. Black, V. M. Clark and Lord Todd, *J. Chem. Soc., Perkin I*, 1955 (1976).
152. U. M. Kempe, T. K. Das Gupta, K. Blatt, P. Gygax, D. Felix and A. Eschenmoser, *Helv. Chim. Acta*, **55**, 2187 (1972).
153. D. St. C. Black, N. A. Blackman and R. F. C. Brown, *Australian J. Chem.*, **32**, 1785 (1979).
154. D. St. C. Black and N. A. Blackman, *Australian J. Chem.*, **32**, 1795 (1979).
155. D. St. C. Black and A. B. Boscacci, *Australian J. Chem.*, **29**, 2511 (1976).
156. D. H. R. Barton, N. J. G. Gutteridge, R. H. Hesse and M. M. Pechet, *J. Org. Chem.*, **34**, 1473 (1969).
157. A. H. Riebel, R. E. Erickson, C. J. Abshire and P. S. Bailey, *J. Amer. Chem. Soc.*, **82**, 1801 (1960).
158. R. F. Erickson and T. M. Myszkiewicz, *J. Org. Chem.*, **30**, 4326 (1965).
159. A. L. Bluhm and J. Weinstein, *J. Amer. Chem. Soc.*, **92**, 1444 (1970).
160. T.-Y. Ching and C. S. Foote, *Tetrahedron Letters*, 3771 (1971).
161. H. G. Aurich, K. Hahn and K. Stork, *Angew. Chem.* (*Internat. Ed*), **14**, 551 (1975).
162. R. F. C. Brown, V. M. Clark, I. O. Sutherland and A. Todd, *J. Chem. Soc.*, 2109 (1959).
163. D. St. C. Black, V. M. Clark, B. G. Odell, I. O. Sutherland and Lord Todd, *J. Chem. Soc., Perkin I*, 1942 (1976).
164. D. St. C. Black and A. B. Boscacci, *Australian J. Chem.*, **30**, 1353 (1977).
165. R. F. C. Brown, L. Subrahmanyan and C. P. Whittle, *Australian J. Chem.*, **20**, 339 (1967).
166. A. R. Katritzky and J. M. Lagowski, *Chemistry of the Heterocyclic* N-*Oxides*, Academic Press, New York–London, 1971, pp. 166–231.
167. H. Alper and J. T. Edward, *Can. J. Chem.*, **48**, 1543 (1970).
168. K. Krishan and N. Singh, *Indian J. Chem.*, **12**, 222 (1974).
169. A. G. Hortmann, J. Koo and C.-C. Yu, *J. Org. Chem.*, **43**, 2289 (1978).
170. S. Hashimoto, I. Furakawa and S. Fujumoto, *Nippon Kagaku Kaishi*, 391 (1972); *Chem. Abstr.*, **76**, 112403j (1972).
171. S. Hashimoto, I. Furukawa and T. Katami, *Nippon Kagaku Kaishi*, 511 (1974); *Chem. Abstr.*, **80**, 145023w (1974).
172. P. Milliet and X. Lusinchi, *Tetrahedron*, **35**, 43 (1979).
173. J. E. Baldwin and N. H. Rogers, *Chem. Commun.*, 524 (1965).
174. M. Lamchen and T. W. Mittag, *J. Chem. Soc.* (*C*), 2300 (1966).
175. P. Zuman and O. Exner, *Coll. Czech. Chem. Commun.*, **30**, 1832 (1965).
176. E. Schaumann and U. Behrens, *Angew. Chem.* (*Internat. Ed.*), **16**, 722 (1977).
177. W. H. Rastetter, T. R. Gadek, J. P. Tane and J. W. Frost, *J. Amer. Chem. Soc.*, **101**, 2228 (1979).
178. G. Bianchi, C. DeMicheli and R. Gandolfi in *Supplement A: The Chemistry of Double-bonded Functional Groups* (Ed. S. Patai), John Wiley and Sons, London, 1977, pp. 369–532.
179. D. St. C. Black, R. F. Crozier and V. C. Davis, *Synthesis*, 205 (1975).
180. A. Padwa, *Angew. Chem.* (*Internat. Ed.*), **15**, 123 (1976).
181. W. Oppolzer, *Angew. Chem.* (*Internat. Ed.*), **16**, 10 (1977).
182. R. Huisgen, *J. Org. Chem.*, **41**, 403 (1976).
183. R. A. Firestone, *Tetrahedron*, **33**, 3009 (1977).
184. R. D. Harcourt, *Tetrahedron*, **34**, 3125 (1978).
185. Y.-M. Chang, J. Sims and K. N. Houk, *Tetrahedron Letters*, 4445 (1975).
186. D. St. C. Black, R. F. Crozier and I. D. Rae, *Australian J. Chem.*, **31**, 2239 (1978).
187. R. Huisgen, *Angew. Chem.* (*Internat. Ed.*), **2**, 565, 633 (1963).
188. R. Huisgen, *J. Org. Chem.*, **33**, 2291 (1968).
189. R. Huisgen, R. Sustmann and K. Bunge, *Chem. Ber.*, **105**, 1324 (1972).
190. K. N. Houk, *Accounts Chem. Res.*, **8**, 361 (1975) and references cited therein.
191. D. Mukherjee, L. N. Domelsmith and K. N. Houk, *J. Amer. Chem. Soc.*, **100**, 1954 (1978).

192. Ya. D. Samuilov, S. E. Soloveva, T. F. Gurutskaya and A. I. Konovalov, *Zh. Org. Khim.*, **14**, 1693 (1978); *Chem. Abstr.*, **89**, 214586g (1978).
193. K. Tada, T. Yamada and F. Toda, *Bull. Chem. Soc. Japan*, **51**, 1839 (1978).
194. C. DeMicheli, A. Gamba-Invernizzi and R. Gandolfi, *Tetrahedron Letters*, 2493 (1975).
195. G. Bianchi, C. DeMicheli and R. Gandolfi, *J. Chem. Soc., Perkin I*, 1518 (1976).
196. H. Taniguchi, T. Ikeda, Y. Yoshida and E. Imoto, *Bull. Chem. Soc. Japan*, **50**, 2694 (1977).
197. H. Taniguchi, T. Ikeda and E. Imoto, *Bull. Chem. Soc. Japan*, **51**, 1859 (1978).
198. M. Joucla and J. Hamelin, *J. Chem. Res.*, 276(S), 3535(M) (1978) and references cited therein.
199. E. J. Fornefeld and A. J. Pike, *J. Org. Chem.*, **44**, 835 (1979).
200. J. J. Tufariello and J. P. Tette, *J. Org. Chem.*, **40**, 3866 (1975).
201. J. J. Tufariello and J. J. Tegeler, *Tetrahedron Letters*, 4037 (1976).
202. J. J. Tufariello and Sk. Asrof Ali, *Tetrahedron Letters*, 4647 (1978).
203. J. J. Tufariello, J. J. Tegeler, S. C. Wong and Sk. Asrof Ali, *Tetrahedron Letters*, 1733 (1978).
204. J. J. Tufariello and G. B. Mullen, *J. Amer. Chem. Soc.*, **100**, 3638 (1978).
205. J. J. Tufariello, G. B. Mullen, J. J. Tegeler, E. J. Trybulski, S. C. Wong and Sk. Asrof Ali, *J. Amer. Chem. Soc.*, **101**, 2435 (1979).
206. P. N. Confalone, E. D. Loller, G. Pizzolato and M. R. Uskokovic, *J. Amer. Chem. Soc.*, **100**, 6291 (1978).
207. W. Oppolzer and M. Petrzilka, *Helv. Chim. Acta*, **61**, 2755 (1978).
208. M. A. Schwartz and G. E. Swanson, *J. Org. Chem.*, **44**, 953 (1979).
209. D. St. C. Black, R. F. Crozier and I. D. Rae, *Australian J. Chem.*, **31**, 2013 (1978).
210. E. C. Taylor and I. J. Turchi, *Chem. Rev.*, **79**, 181 (1979).
211. Y. Kobayashi, I. Kumadaki and T. Yoshida, *Heterocycles*, **8**, 387 (1977).
212. K. Tada and F. Toda, *Tetrahedron Letters*, 563 (1978).
213. I. L. Knunyants, E. G. Bykhovskaya, V. N. Frosin, I. V. Galakhov and L. I. Regulim, *Zh. Vses. Khim. Obsch.*, **17**, 356 (1972); *Chem. Abstr.*, **77**, 101432n (1972).
214. J. M. J. Tronchet and E. Mihaly, *Helv. Chim. Acta*, **55**, 1266 (1972).
215. H. Paulsen and M. Budzis, *Chem. Ber.*, **107**, 2009 (1974).
216. A. Vasella, *Helv. Chim. Acta*, **60**, 426, 1273, (1977).
217. C. Belzecki and I. Panfil, *J. Org. Chem.*, **44**, 1212 (1979).
218. G. Manecke and J. Klawitter, *Makromol. Chem.*, **142**, 253 (1971).
219. W. Kliegel, *Chemiker Z.*, **100**, 236 (1976).
220. M. A. Calcagno, H. W. Heine, C. Kruse and W. A. Kofke, *J. Org. Chem.*, **39**, 162 (1974).
221. O. Tsuge and H. Watanabe, *Heterocycles*, **7**, 907 (1977).
222. H. W. Heine and L. Heitz, *J. Org. Chem.*, **39**, 3192 (1974).
223. N. J. Leonard, D. A. Durand and F. Uchimaru, *J. Org. Chem.*, **32**, 3607 (1967).
224. V. Astley and H. Sutcliffe, *Tetrahedron Letters*, 2707 (1971).
225. V. Astley and H. Sutcliffe, *Chem. Commun.*, 1303 (1971).
226. S. Shatzmiller and A. Eschenmoser, *Helv. Chim. Acta*, **56**, 2975 (1973).
227. A. Ruttimann and D. Ginsburg, *Helv. Chim. Acta*, **58**, 2237 (1975).
228. S. Shatzmiller and R. Neidlein, *Tetrahedron Letters*, 4151 (1976).
229. S. Levinger and S. Shatzmiller, *Tetrahedron*, **34**, 563 (1978).
230. T. K. Das Gupta, D. Felix, U. M. Kempe and A. Eschenmoser, *Helv. Chim. Acta*, **55**, 2198 (1972).
231. M. Petrzilka, D. Felix and A. Eschenmoser, *Helv. Chim. Acta*, **56**, 2950 (1973).
232. M. Riediker and W. Graf, *Helv. Chim. Acta*, **62**, 205 (1979).
233. P. Gygax, T. K. Das Gupta and A. Eschenmoser, *Helv. Chim. Acta*, **55**, 2205 (1972).
234. E. Shalom, J.-L. Zevon and S. Shatzmiller, *J. Org. Chem.*, **42**, 4213 (1977).
235. S. Shatzmiller, P. Gygax, D. Hall and A. Eschenmoser, *Helv. Chim. Acta*, **56**, 2961 (1973).
236. A. Eschenmoser, cited by R. B. Woodward, *Pure Appl. Chem.*, **33**, 171 (1973).
237. G. Cum. G. Sindona and N. Uccella, *J. Chem. Soc., Perkin I*, 719 (1976).
238. H. Staudinger and K. Miescher, *Helv. Chim. Acta*, **2**, 554 (1919).

239. C. H. Hassall and A. E. Lippman, *J. Chem. Soc.*, 1059 (1953).
240. R. N. Pratt, D. P. Stokes, G. A. Taylor and P. C. Brookes, *J. Chem. Soc. (C)*, 2086 (1968).
241. A. D. Baker, D. Wong, S. Lo, M. Bloch, G. Horozoglu, N. L. Goldman and R. Engel, *Tetrahedron Letters*, 215 (1978).
242. D. P. Stokes and G. A. Taylor, *J. Chem. Soc. (C)*, 2334 (1971).
243. R. N. Pratt, D. P. Stokes and G. A. Taylor, *J. Chem. Soc., Perkin I*, 498 (1975).
244. A. F. Gettins, D. P. Stokes, G. A. Taylor and C. B. Judge, *J. Chem. Soc., Perkin I*, 1849 (1977).
245. A. M. Abu-Gharbia, M. M. Joullie and I. Miura, *Heterocycles*, **9**, 457 (1978).
246. D. P. Del'tsova, N. P. Gambaryan and I. L. Knunyants, *Dokl. Akad. Nauk SSSR*, **206**, 620 (1972); *Chem. Abstr.*, **78**, 15959d (1973).
247. D. P. Del'tsova, Z. V. Safronova, N. P. Gambaryan, M. Y. Antipin and Yu. T. Struchkov, *Izv. Akad. Nauk SSSR, Ser. Khim.*, 1881 (1978); *Chem. Abstr.*, **89**, 215280q (1978).
248. M. W. Barker and J. H. Gardner, *J. Heterocycl. Chem.*, **5**, 881 (1968).
249. M. W. Barker and C. J. Wierengo, *J. Heterocycl. Chem.*, **12**, 175 (1975).
250. N. Murai, M. Komatsu, Y. Ohshiro and T. Agawa, *J. Org. Chem.*, **42**, 448 (1977).
251. D. St. C. Black and K. G. Watson, *Australian J. Chem.*, **26**, 2473 (1973).
252. M. Komatsu, Y. Ohshiro and T. Agawa, *J. Org. Chem.*, **37**, 3192 (1972).
253. M. Radau and K. Hartke, *Arch. Pharm.*, **305**, 737 (1972).
254. B. F. Bonini, G. Maccagnani, G. Mazzanti, P. Pedrini and B. Zwanenburg, *Gazz. Chim. Ital.*, **107**, 283 (1977).
255. B. P. Stark and M. H. G. Ratcliffe, *J. Chem. Soc.*, 2640 (1964).
256. O. Tsuge, M. Tashiro and S. Mataka, *Tetrahedron Letters*, 3877 (1968).
257. W. E. Truce, J. R. Norell, R. W. Campbell, D. G. Brady and J. W. Fieldhouse, *Chem. Ind. (London)*, 1870 (1965).
258. W. E. Truce, J. W. Fieldhouse, D. J. Vrencur, J. R. Norell, R. W. Campbell and D. G. Brady, *J. Org. Chem.*, **34**, 3097 (1969).
259. F. Eloy and A. Van Overstraeten, *Bull. Soc. Chim. Belges*, **76**, 63 (1967).
260. O. Tsuge and M. Noguchi, *J. Org. Chem.*, **41**, 2438 (1976).
261. D. St. C. Black and V. C. Davis, *Tetrahedron Letters*, 1993 (1975); *Australian J. Chem.*, **30**, 1573 (1977).
262. O. Tsuge and M. Noguchi, *Chem. Letters*, 749 (1977).
263. N. J. A. Gutteridge and J. R. M. Dales, *J. Chem. Soc. (C)*, 122 (1971).
264. D. Liotta, A. D. Baker, F. Weinstein, D. Felsen, R. Engel and N. L. Goldman, *J. Org. Chem.*, **38**, 3445 (1973).
265. D. Liotta, A. D. Baker, S. Goldstein, N. L. Goldman, F. Weinstein-Lanse, D. Felsen-Reingold and R. Engel, *J. Org. Chem.*, **39**, 2718 (1974).
266. D. Liotta, A. D. Baker, N. L. Goldman and R. Engel, *J. Org. Chem.*, **39**, 1975 (1974).
267. K. Suzuki, T. Sukamoto and M. Sekiya, *Chem. Pharm. Bull. Tokyo*, **23**, 1611 (1975).
268. R. H. Heistand, II, M. A. Stahl and H. W. Heine, *J. Org. Chem.*, **43**, 3613 (1978).
269. D. Moderhack, *Synthesis*, 299 (1973).
270. R. Nagase, T. Kawashima, M. Yoshifuji and N. Inamoto, *Heterocycles*, **8**, 243 (1977).
271. A. Padwa and K. Crosby, *J. Org. Chem.*, **39**, 2651 (1974).
272. R. Foster, J. Iball and R. Nash, *Chem. Commun.*, 1414 (1968).
273. A. D. Baker, J. E. Baldwin, D. P. Kelly and J. DeBernardis, *Chem. Commun.*, 344 (1969).
274. W. Kliegel, *Tetrahedron Letters*, 2627 (1969).
275. H. Stamm and H. Steudle, *Arch. Pharm.*, **309**, 1014 (1976).
276. R. M. Wilson and A. J. Eberle, *J. Org. Chem.*, **39**, 2804 (1974).
277. J. H. Hall and M. R. Gisler, *J. Org. Chem.*, **42**, 1133 (1977).
278. F. Kröhnke, *Angew. Chem. (Internat. Ed.)*, **2**, 380 (1963).
279. E. Buehler and G. B. Brown, *J. Org. Chem.*, **32**, 265 (1967).
280. E. Bellasio, F. Parravicini, T. LaNoce and E. Testa, *Ann. Chim. (Rome)*, **58**, 407 (1968).
281. E. F. Schoenewaldt, R. B. Kinnel and P. Davis, *J. Org. Chem.*, **33**, 4270 (1968).
282. D. St. C. Black, R. F. C. Brown and A. M. Wade, *Australian J. Chem.*, **25**, 2155 (1972).

283. C. W. Eaton, G. H. Denny, Jr., M. A. Ryder, M. G. Ly and R. D. Babson, *J. Med. Chem.*, **16**, 289 (1973).
284. H. Maehr and M. Leach, *J. Org. Chem.*, **39**, 1166 (1974).
285. B. Liberek and Z. Palacz, *Rocz. Chem.*, **45**, 1173 (1971).
286. K. J. Ross-Petersen and H. Hjeds, *Dansk. Fidsskr. Farm.*, **43**, 188 (1969).
287. L. B. Volodarskii and T. K. Sevastyanova, *Zh. Org. Khim.*, **7**, 1687 (1971).
288. T. Polonski and A. Chimiak, *J. Org. Chem.*, **41**, 2092 (1976).
289. H. M. Rosenberg and M. P. Serve, *J. Org. Chem.*, **37**, 1443 (1972).
290. S. Zbaida and E. Breuer, unpublished results.
291. D. Moderhack, *J. Heterocycl. Chem.*, **14**, 757 (1977).
292. K. Kimura, H. Niwa and S. Motoki, *Bull. Chem. Soc. Japan*, **50**, 2751 (1977).
293. M. Masui, K. Suda, M. Yamauchi and C. Yijima, *J. Chem. Soc., Perkin I*, 1955 (1972).
294. M. Masui and C. Yijima, *J. Chem. Soc.* (*C*), 2022 (1967).
295. N. Singh and S. Mohan, *Chem. Commun.*, 868 (1969).
296. N. G. Clark and E. Cawkill, *Tetrahedron Letters*, 2717 (1975).
297. L. S. Kaminsky and M. Lamchen, *J. Chem. Soc.* (*C*), 1683 (1967).
298. H. Stamm and J. Hoenicke, *Ann. Chem.*, **748**, 143 (1971).
299. H. Stamm and J. Hoenicke, *Synthesis*, 145 (1971).
300. H. Stamm and H. Steudle, *Arch. Pharm.*, **309**, 935 (1976).
301. S. Zbaida and E. Breuer, *Tetrahedron*, **34**, 1241 (1978).
302. R. Huisgen and J. Wulff, *Chem. Ber.*, **102**, 746 (1969).
303. D. St. C. Black and V. C. Davis, *J. Chem. Soc., Chem. Commun.*, 416 (1975); *Australian J. Chem.*, **29**, 1735 (1976).
304. E. Breuer, S. Zbaida, J. Pesso and S. Levi, *Tetrahedron Letters*, 3103 (1975).
305. E. Breuer and S. Zbaida, *J. Org. Chem.*, **42**, 1904 (1977).
306. E. Breuer, S. Zbaida, J. Pesso and I. Ronen-Braunstein, *Tetrahedron*, **33**, 1145 (1977).
307. S. Zbaida and E. Breuer, *J. Chem. Soc., Chem. Commun.*, 6 (1978).
308. S. Zbaida, *Ph.D. Thesis*, The Hebrew University of Jerusalem, 1978.
309. S. Zbaida and E. Breuer, to be published.
310. S. Zbaida and E. Breuer, *Experientia*, **35**, 851 (1979).
311. E. G. Janzen and B. J. Blackburn, *J. Amer. Chem. Soc.*, **91**, 4481 (1969).
312. T. D. Lee and J. F. W. Keana, *J. Org. Chem.*, **43**, 4226 (1978).
313. T. D. Lee and J. F. W. Keana, *J. Org. Chem.*, **41**, 3237 (1976).
314. D. St. C. Black, V. M. Clark, R. S. Thakur and Lord Todd, *J. Chem. Soc., Perkin I*, 1951 (1976).
315. D. St. C. Black, N. A. Blackman and L. M. Johnstone, *Australian J. Chem.*, **32**, 2025 (1979).
316. C. Berti, M. Colonna, L. Greci and L. Marchetti, *Tetrahedron*, **31**, 1745 (1975).
317. C. Berti, M. Colonna, L. Greci and L. Marchetti, *Tetrahedron*, **32**, 2147 (1976).
318. C. Berti, M. Colonna, L. Greci and L. Marchetti, *J. Heterocycl. Chem.*, **16**, 17 (1979).
319. H. Stamm and J. Hoenicke, *Ann. Chem.*, **749**, 146 (1971).
320. H. Stamm and H. Steudle, *Tetrahedron*, **35**, 647 (1979).
321. M. Kinugasa and S. Hashimoto, *J. Chem. Soc., Chem. Commun.*, 466 (1972).
322. L. K. Ding and W. J. Irwin, *J. Chem. Soc., Perkin I*, 2382 (1976).
323. J. Satge, M. Lesbre, P. Riviere and S. Richelme, *J. Organometal. Chem.*, **34**, C18 (1972).
324. P. Riviere, S. Richelme, M. Riviere-Baudet, J. Satge, M. J. S. Gyanane and M. F. Lappert, *J. Chem. Res.* (*S*), 218 (1978).
325. P. Riviere, S. Richelme, M. Riviere-Baudet and J. Satge, *J. Chem. Res.* (*S*), 220 (1978).
326. H. Iwamura and N. Inamoto, *Bull. Chem. Soc. Japan*, **40**, 702 (1967); **43**, 856 (1970).
327. H. Iwamura and M. Iwamura, *Tetrahedron Letters*, 3723 (1970).
328. B. G. Ershov and I. E. Makarov, *Izv. Akad. Nauk SSSR, Ser. Khim.*, 2296 (1969); *Chem. Abstr.*, **72**, 37672h (1970).
329. M. Iwamura and N. Inamoto, *Bull. Chem. Soc. Japan*, **40**, 703 (1967); **43**, 860 (1970).
330. E. G. Janzen, *Accounts Chem. Res.*, **4**, 31 (1971).
331. A. L. Bluhm and J. Weinstein, *J. Org. Chem.*, **37**, 1748 (1972).
332. J. A. Howard and J. C. Tait, *Can. J. Chem.*, **56**, 176 (1978).
333. M. V. Merritt and R. A. Johnson, *J. Amer. Chem. Soc.*, **99**, 3713 (1977).

334. J. R. Harbour, V. Chow and J. R. Bolton, *Can. J. Chem.*, **52**, 3549 (1974).
335. E. G. Janzen, Y. Y. Wang and R. V. Shetty, *J. Amer. Chem. Soc.*, **100**, 2923 (1978).
336. G. R. Chalefont, M. J. Perkins and A. Horsfield, *J. Chem. Soc. (B)*, 401 (1970).
337. V. E. Zubarev, V. N. Belevski and L. T. Bugarenko, *Khim. Vys. Energ.*, **12**, 209 (1978); *Chem. Abstr.*, **89**, 138312x (1978).
338. J. G. Pacifici and H. L. Browning, Jr., *J. Amer. Chem. Soc.*, **92**, 5231 (1970).
339. E. G. Janzen, R. L. Dudley and R. V. Shetty, *J. Amer. Chem. Soc.*, **101**, 243 (1979).
340. A. T. Nielsen in *The Chemistry of the Nitro and Nitroso Groups*, Part I (Ed. H. Feuer), Wiley–Interscience, London, 1969, p. 349.
341. F. G. Bordwell, W. J. Boyle, Jr., J. A. Hautala and K. C. Yee, *J. Amer. Chem. Soc.*, **91**, 4002 (1969).
342. P. W. K. Flanagan, H. W. Amburn, H. W. Stone, J. G. Traynham and H. Shechter, *J. Amer. Chem. Soc.*, **91**, 2797 (1969).
343. M. Fukuyama, P. W. K. Flanagan, F. T. Williams, Jr., L. Frainier, S. A. Miller and H. Shechter, *J. Amer. Chem. Soc.*, **92**, 4689 (1970).
344. F. G. Bordwell, W. J. Boyle, Jr. and K. C. Yee, *J. Amer. Chem. Soc.*, **92**, 5926 (1970).
345. F. G. Bordwell and W. J. Boyle, Jr., *J. Amer. Chem. Soc.*, **93**, 511 (1971).
346. F. G. Bordwell and W. J. Boyle, Jr., *J. Amer. Chem. Soc.*, **94**, 3907 (1972).
347. F. G. Bordwell and W. J. Boyle, Jr., *J. Amer. Chem. Soc.*, **97**, 3446 (1975).
348. J. R. Keefe and N. H. Munderloh, *J. Chem. Soc., Chem. Commun.*, 17 (1974).
349. C. D. Slater and Y. W. D. Chan, *J. Org. Chem.*, **43**, 2423 (1978).
350. J. R. Keefe, J. Morey, C. A. Palmer and J. C. Lee, *Amer. Chem. Soc.*, **101**, 1295 (1979).
351. R. J. Sundberg and P. A. Bukowick, *J. Org. Chem.*, **33**, 4098 (1968).
352. F. G. Bordwell and K. C. Yee, *J. Amer. Chem. Soc.*, **92**, 5939 (1970).
353. S. J. Angyal and B. M. Luttrell, *Australian J. Chem.*, **23**, 1485 (1970).
354. M. J. Strauss and E. Weltin, *Tetrahedron Letters*, 692 (1971).
355. D. J. Sutor, F. J. Llewellyn and H. S. Maslen, *Acta Cryst.*, **7**, 145 (1954).
356. B. Klewe, *Acta Chem. Scand.*, **26**, 1049 (1972).
357. O. Simonsen, *Acta Cryst.*, **B29**, 2600 (1973).
358. O. Simonsen and J. P. Jacobsen, *Acta Cryst.*, **B33**, 3045 (1977).
359. R. Grée and R. Carrié, *Bull. Soc. Chim. Fr.*, 1314 (1975).
360. R. C. Kerber and A. Porter, *J. Amer. Chem. Soc.*, **91**, 366 (1969).
361. M. J. Brookes and N. Jonathan, *Spectrochim. Acta*, **25A**, 187 (1969).
362. M. J. Brookes and N. Jonathan, *J. Chem. Soc. (A)*, 1529 (1968).
363. A. C. Begbie, W. R. Bowman, B. T. Golding and W. P. Watson, *J. Chem. Soc., Chem. Commun.*, 645 (1974).
364. F. S. Alvarez and D. Wren, *Tetrahedron Letters*, 569 (1973).
365. F. Kienzle, G. W. Holland, J. L. Jernow, S. Kwoh and P. Rosen, *J. Org. Chem.*, **38**, 3440 (1973).
366. F. Freeman and A. Yeramyan, *Tetrahedron Letters*, 4783 (1978).
367. F. Freeman, A. Yeramyan and F. Young, *J. Org. Chem.*, **34**, 2438 (1969).
368. F. Freeman and A. Yeramyan, *J. Org. Chem.*, **35**, 2061 (1970).
369. F. Freeman and D. K. Lin, *J. Org. Chem.*, **36**, 1335 (1971).
370. J. E. McMurry, J. Melton and H. Padgett, *J. Org. Chem.*, **39**, 259 (1974).
371. P. Dubs and H. P. Schenk, *Helv. Chim. Acta*, **61**, 984 (1978).
372. J. R. Williams, L. R. Unger and R. H. Moore, *J. Org. Chem.*, **43**, 1271 (1978).
373. T. Severin and D. König, *Chem. Ber.*, **107**, 1499 (1974).
374. A. H. Pagano and H. Shechter, *J. Org. Chem.*, **35**, 295 (1970).
375. H. Shechter and R. B. Kaplan, *J. Amer. Chem. Soc.*, **75**, 3980 (1953).
376. K. Fukunaga, *Synthesis*, 55 (1978).
377. P. A. Bartlett, F. R. Green, III and T. R. Webb, *Tetrahedron Letters*, 331 (1977).
378. P. Noble, Jr., F. G. Borgardt and W. L. Reed, *Chem. Rev.*, **64**, 19 (1964).
379. V. Grakauskas and K. Baum, *J. Org. Chem.*, **33**, 3080 (1968).
380. M. J. Kamlet and H. G. Adolph, *J. Org. Chem.*, **33**, 3073 (1968).
381. L. T. Eremenko and F. Ya. Natsibullin, *Izv. Akad. Nauk SSSR*, 912 (1968).
382. K. Baum, *J. Org. Chem.*, **35**, 846 (1970).

383. J. E. McMurry, *Accounts Chem. Res.*, **7**, 281 (1974).
384. T. L. Ho and C. M. Wong, *Synthesis*, 196 (1974).
385. R. Kirchoff, *Tetrahedron Letters*, 2533 (1974).
386. T. Atika, M. Inba, H. Uchida and A. Ohta, *Synthesis*, 792 (1977).
387. T. Severin and D. König, *Chem. Ber.*, **107**, 1499 (1974).
388. D. Seebach, E. W. Colvin, F. Lehr and T. Weller, *Chimia*, **33**, 1 (1979).
389. J. Kimura, A. Kawashima, M. Sugizaki, N. Nemoto and O. Mitsunobu, *J. Chem. Soc., Chem. Commun.*, 303 (1979).
390. V. A. Tartakovskii, I. E. Chlenov, S. L. Ioffe, G. V. Lagodzinskaya and S. S. Novikov, *Zh. Org. Khim.*, **2**, 1593 (1966).
391. V. A. Tartakovskii, Z. Ya. Lapshina, I. A. Savost'yanova and S. S. Novikov, *Zh. Org. Khim.*, **4**, 236 (1968).
392. (a) H. Sato, T. Kusumi, K. Imaye and H. Kakisawa, *Bull. Chem. Soc. Japan*, **49**, 2815 (1976).
(b) R. Grée and R. Carrié, *J. Heterocycl. Chem.*, **14**, 965 (1977).
393. K. Torsell and O. Zeuthen, *Acta Chem. Scand.*, **B32**, 118 (1978).
394. S. L. Ioffe, M. V. Kashutina, V. M. Shitkin, A. Z. Yankelevich, A. A. Levin and V. A. Tartakovskii, *Izv. Akad. Nauk SSSR, Ser. Khim.*, 1341 (1972).
395. V. A. Tartakovskii, O. A. Duk'yanov, N. I. Shlykova and S. S. Novikov, *Zh. Org. Khim.*, **4**, 231 (1967).
396. R. Grée and R. Carrié, *Tetrahedron*, **32**, 683 (1976).
397. R. Grée, F. Tonnard and R. Carrié, *Tetrahedron*, **32**, 675 (1976).
398. R. Grée, F. Tonnard and R. Carrié, *Bull. Soc. Chim. Fr.*, 1325 (1975).
399. R. Grée and R. Carrié, *Bull. Soc. Chim. Fr.*, 1319 (1975).
400. I. S. Levina, E. I. Mortikova and A. V. Kamernitzky, *Synthesis*, 562 (1974).
401. J. E. Baldwin, R. G. Pudussery, A. K. Qureshi and B. Sklarz, *J. Amer. Chem. Soc.*, **90**, 5325 (1968).
402. R. Grée and R. Carrié, *J. Amer. Chem. Soc.*, **99**, 6667 (1977).
403. O. v. Schikh, G. Apel, H. G. Padeken, H. H. Schwarz and A. Segnitz in *Houben—Weyl: Methoden der Organischen Chemie*, Vol. X (Ed. E. Müller), Georg Thieme Verlag, Stuttgart 1971.
404. M. V. Kashutina, S. L. Ioffe and V. A. Tartakovskii, *Dok. Akad. Nauk SSSR*, **218**, 109 (1974).
405. E. W. Colvin and D. Seebach. *J. Chem. Soc., Chem. Commun.*, 689 (1978).
406. E. Kaji, A. Igarashi and S. Zen, *Bull. Chem. Soc. Japan*, **49**, 3181 (1976).
407. T. Mukaiyama and H. Nambu, *J. Org. Chem.*, **27**, 2201 (1962).
408. N. Kornblum and P. A. Wade, *J. Org. Chem.*, **38**, 1418 (1973).
409. D. Seebach, V. Ehrig, H. F. Leitz and R. Henning, *Chem. Ber.*, **108**, 1946 (1975).
410. A. McKillop and R. J. Kobylecki, *Tetrahedron*, **30**, 1365 (1974).
411. K. Kaji, K. Harada and S. Zen, *Chem. Pharm. Bull. Tokyo*, **26**, 3254 (1978).
412. D. Seebach and F. Lehr, *Angew. Chem.* (*Internat. Ed.*), **15**, 505 (1976).
413. N. Kornblum, *Angew. Chem.* (*Internat. Ed.*), **14**, 734 (1975).
414. N. Kornblum, 'Radical anion reactions of nitro compounds', this volume.
415. S. J. Etheredge, *Tetrahedron Letters*, 4527 (1965).
416. S. Gabriel, *Chem. Ber.*, **36**, 570 (1903).
417. S. Zen and E. Kaji, *Bull. Chem. Soc. Japan*, **43**, 2277 (1977).
418. E. Kaji and S. Zen. *Bull. Chem. Soc. Japan*, **46**, 337 (1973).
419. S. Zen and E. Kaji, *Chem. Pharm. Bull. Tokyo*, **22**, 477 (1974).
420. A. R. Katritzky, G. De Ville and R. C. Patel, *J. Chem. Soc., Chem. Commun.*, 602 (1979).
421. D. Seebach, R. Henning, F. Lehr and J. Gonnermann, *Tetrahedron Letters*, 1161 (1977).
422. H. L. Finkbeiner and G. W. Wagner, *J. Org. Chem.*, **28**, 215 (1963).
423. K. E. Koenig and W. P. Weber, *Tetrahedron Letters*, 2275 (1974).
424. M. E. Childs and W. P. Weber, *J. Org. Chem.*, **41**, 3486 (1976).
425. J. F. Normant and C. Piechucki, *Bull. Soc. Chim. Fr.*, 2402 (1972).
426. D. C. Baker and S. R. Putt, *Synthesis*, 478 (1978).
427. J. Gostelli, *Helv. Chim. Acta*, **60**, 1980 (1977).

428. J. Bagli and T. Bogri, *Tetrahedron Letters*, 3815 (1972).
429. P. Bakuzis, M. L. F. Bakuzis and T. F. Weingarten, *Tetrahedron Letters*, 2371 (1978).
430. E. Kaji, H. Kohno and S. Zen, *Bull. Chem. Soc. Japan*, **50**, 928 (1977).
431. W. E. Noland, *Chem. Rev.*, **55**, 137 (1955).
432. M. Langrenée, *Compt. Rend., Ser. C*, **284**, 153 (1977).
433. D. St. C. Black, *Tetrahedron Letters*, 1331 (1972).
434. R. M. Jacobsen, *Tetrahedron Letters*, 3215 (1974).
435. P. Dubs and R. Stüssi, *Helv. Chim. Acta*, **61**, 990 (1978).
436. E. Keinan and Y. Mazur, *J. Amer. Chem. Soc.*, **99**, 3861 (1977).
437. J. L. Hogg, T. E. Goodwin and D. W. Nave, *Org. Prep. Proced. Int.*, **10**, 9 (1978).
438. H. Lerche, D. König and T. Severin, *Chem. Ber.*, **107**, 1509 (1974).
439. M. T. Shipchandler, *Synthesis*, 666 (1979).

Nitrones, Nitronates and Nitroxides
Edited by S. Patai and Z. Rappoport

CHAPTER **3**

Nitrones and nitronic acid derivatives: an update†

ELI BREUER

Department of Pharmaceutical Chemistry, The School of Pharmacy, The Hebrew University of Jerusalem, Jerusalem, Israel

†The material in this appendix is divided in the same manner as in the body of the original Chapter 13 in Supplement F (see Chapter 2). The section numbers in the appendix are preceded by an asterisk; occasionally, some section numbers are omitted or added.
The numbers of structures, schemes and references run continuously in the original chapter and the appendix.

*II. NITRONES

The present update includes the additional sections II.C and II.D, which present some biological aspects and newer results regarding the synthesis of nitrones, respectively. These topics were not included in the original chapter.

*A. Structure of Nitrones

*1. Theoretical calculations

Ab initio molecular orbital calculations have been used to calculate the potential energy surface connecting nitrosomethane and its isomers, the *syn*- and *anti*-formaldoximes. Formaldonitrone ($CH_2{=}N(H)O^-$) is the favoured intermediate for the *syn–anti* isomerization, and from the results it was concluded that this nitrone might even be amenable to experimental observation[440].

CNDO/2 calculations were carried out on a series of (*E, E*) and (*E, Z*) chalcone nitrones (**323**)[441], assuming coplanarity of the aryl(a) ring with the two coplanar double bonds (i.e. (*s-E*) conformation). The results confirm that ring b prefers an out-of-plane conformation; π, σ and total electron densities were calculated for the atoms of the nitrone function and for atoms 1 and 1′ of the two aryl rings. One of the conclusions to be made from the calculations for *C-p*-anisyl nitrone is that the *p*-methoxy group polarizes the styryl system, but is not a sufficiently strong electron donor to affect polarization of the nitrone function. This correlates quite well with what is said in Section II.A.2.c concerning the weak electron-withdrawing properties of the nitrone function. On the other hand the *p*-nitro group, a powerful electron acceptor, is able to polarize the whole conjugated system, thus increasing, in resonance terms, the importance of the resonance contributor of type **5**, also in accordance with what is said later regarding electrophilic substitution of *C*-aryl groups (see Section II.A.2.c).

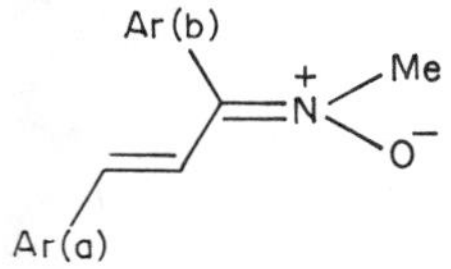

Ar(b), Ar(a), N+, Me, O−

(E,E-**323**)

Ar(b), Ar(a), N+, O−, Me

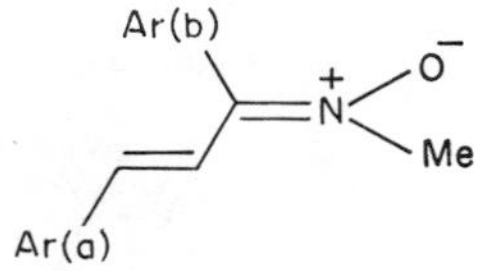

(E,Z-**323**)

A MINDO/3 study of C=N bond rotation was carried out on $CH_2{=}\overset{+}{N}(H)O^{-}$[442]. This study resulted in a rotational barrier of 40 kcal/mol, which is reasonably close to the experimental value. The results of this work also indicate that the **C**H=N—O carbon is negatively charged both in the ground state and in the 90° twisted state, thus lending theoretical support to resonance structure **5**. Additional calculations are reported in Section II.B.5.a.i, which deals with the mechanistic aspects of 1,3-dipolar cycloadditions.

*2. *Nonspectral physical methods*

a. X-ray studies.* The reaction products of *C,N*-diaryl nitrones with *N*-bromosuccinimide were found by X-ray crystallography to have the (*E*) structure **324[443]. The *C*-aryl group is coplanar with the plane of the C=N—O atoms, while the *N*-aryl and the *C*-succinimidyl rings are approximately orthogonal to that plane. The bond lengths and bond angles conform to those found in earlier works.

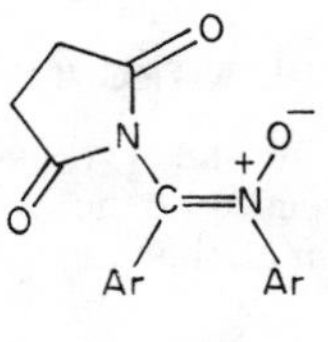

(**324**)

The structure of an *N*-glycosyl nitrone (derived from a protected mannofuranose), which served as a starting material for the asymmetric synthesis of α-aminophosphonic acid, was determined by X-ray crystallography[444] (see also Section II.B.8). The crystal structure of an unusual nitrone containing a cyclic phenylboronate moiety in the molecule (see **330**) was recently reported[450].

**c. Dipole moment and acid–base properties.* Molecular polarizability studies have been carried out on a few *C*-phenyl *N*-*p*-substituted phenyl nitrones. On the basis of model studies of (*Z*)-*C*-phenyl aldonitrones it was assumed that, because of steric interactions between the oxygen and the *ortho* hydrogens of the *C*-phenyl group, this ring is twisted out of the plane determined by the C=N—O atoms. This angle was assumed to be 30°. Following this, the angle between the plane of the *N*-ring and the nitrone function was determined by calculating the Kerr constants of the nitrones and comparing them with the experimental values. The angles thus obtained were around 25–30°, except for the *N*-*p*-dimethylaminophenyl *C*-phenyl nitrone, for which the angle was found to be 0°, presumably as a result of the powerful donor properties of the dimethylamino group[445].

Dipole moments were used, along with spectroscopic methods, to determine the structures of a series of mononitrones of cyclic α-diketones of type **325**. The (*Z*) isomer had a dipole moment of 5.3 D, somewhat higher than the value for simple nitrones. In contrast the (*E*) isomer showed the considerable lower value of ≈ 1.1 D[446].

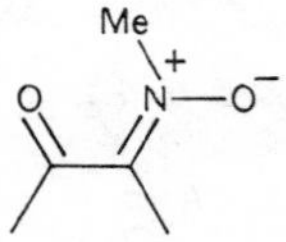

(*E*-**325**) 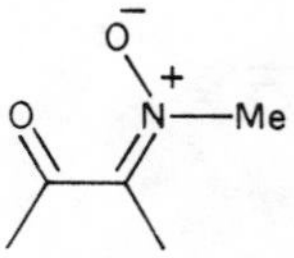 (*Z*-**325**)

The fact that the polarity and the polarizing ability of the nitrone function should not be overestimated, appears also from its *ortho-para* directing effect in the electrophilic substitution of an aromatic ring placed in the *C*-position[447]. Nitration of 4-phenyl-2,2,5,5-tetramethyl-3-imidazoline-3-oxide nitroxide yields the corresponding *p*-nitrophenyl derivative. This result can, presumably, be rationalized by invoking the contribution of a resonance structure of type **5**. In contrast to this, the nitrogen atom of the nitrone function has a positive charge that can help to stabilize an adjacent carbanion. Based on this, a new synthesis of *N*-hydroxy-α-amino acids was developed by alkylation of the carbanions (**326**) derived from the methyl esters of *N*-benzylidene-α-amino acid *N*-oxides[448].

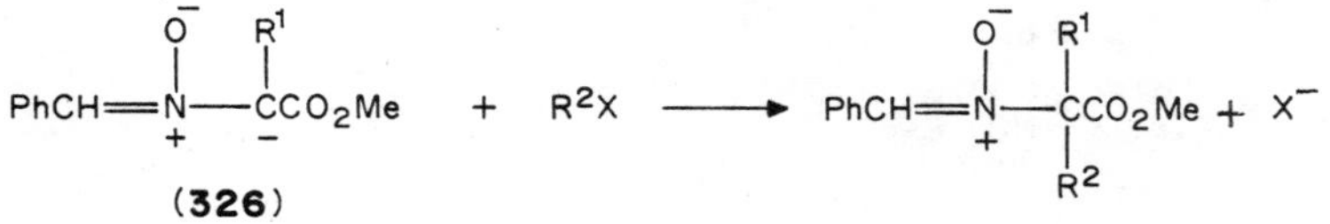

$$\mathrm{PhCH{=}\overset{+}{N}(O^-)\text{—}\overset{-}{C}(R^1)CO_2Me} \;+\; \mathrm{R^2X} \longrightarrow \mathrm{PhCH{=}\overset{+}{N}(O^-)\text{—}C(R^1)(R^2)CO_2Me} \;+\; \mathrm{X^-}$$

(**326**)

The donor properties of the nitrone function are usually assumed to be concentrated on the negatively charged oxygen. For this reason it was previously assumed that a metal would be bound by the two oxygens of the bidentate ligands, *C*-(2-hydroxy-1-naphthyl). *N*-methyl and *N*-*p*-chlorophenyl nitrones with the formation of seven-membered chelates. However ESR results of these complexes, determined more recently, are more consistent with structures in which the metal is bound by nitrogen, with the formation of six-membered complexes of type **327**[449]. In this respect, the nitrones investigated seem to resemble oximes, which are known to prefer to bond metals via nitrogen, rather than *N*-oxides that are known to bond via oxygen.

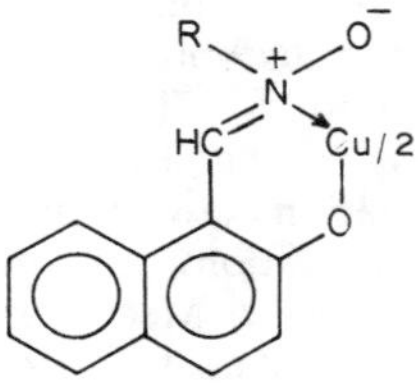

R, N⁺, O⁻, HC, Cu/2, O

(**327**)

The influence of the α-substituent upon the donor properties of the nitrone oxygen is shown in the difference in the behaviour of the *N*-bis(hydroxyalkyl) nitrones **328a** and **328b** towards phenylboronic acid[450]. While the reaction of the α-phenyl nitrone (**328a** yields the bicyclic structure **329**, through coordination of the nitrone oxygen with the boron atom, the weakening of the donor properties of the oxygen by the *p*-nitro substitution in the reaction product of **328b** stabilizes the nitrone phenylboronate **330**. Both structures have been confirmed by X-ray crystallography[450].

A series of papers report data such as stoichiometry, apparent formation constants and transition energies as well as solvent effects of charge-transfer complexes of nitrones with acceptors such as tetracyanoethylene, dichlorodicyanobenzoquinone and chloranil, based upon visible spectroscopic studies[451]. Such charge-transfer complexes have also been suggested to be involved as intermediates in the isomerization of 3*H*-indole 1-oxide to the corresponding lactam (see Sections II.B.1.c and *II.B.1.c)[452] and in 1,3-dipolar cycloaddition reactions (see Section *II.B.5.a.i)[490b].

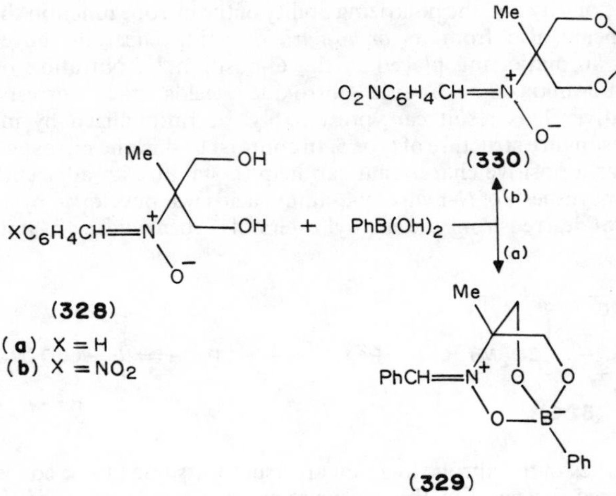

(328)
(a) X = H
(b) X = NO_2

(329)

(330)

3. Spectra of nitrones

**a. Photoelectron spectra.* A number of papers that deal primarily with mechanistic questions related to 1,3-dipolar cycloaddition of nitrones report photoelectron spectroscopic determination of nitrone ionization potentials (see Section *II.B.5.a).

b. Ultraviolet spectra.* The ultraviolet spectra of the series of chalcone nitrones mentioned above (323**) was measured[442]. From the results obtained it was concluded that only substituents on ring (a) shift the absorption maximum to longer wavelengths, while substituents on ring (b) have little or no effect. This is in contrast to the parent chalcones, in which substituents on both rings affect the absorption. From the results the conclusion was drawn that there is non conjugation between ring (b) and the rest of the molecule[442]. This confirms the results of the calculations presented in Section II.A.1, and is also in agreement with ^{13}C-NMR data (*vide infra*). Measurement of the ultraviolet spectra of the pairs of (*E*)- and (*Z*)-**325** nitrones revealed no differences in the spectra as a consequence of the geometric isomerism[446].

**c. Infrared spectra.* There is little to add in this area. The C═N vibration in α-ketonitrones appears at somewhat lower frequency than in simple nitrones, and differs for the two isomers, the (*E*) isomers absorbing at a higher frequency by 8–35 cm^{-1}. The values are (*E*): 1547–1586, (*Z*): 1524–1573. The vibration frequency of the N—O bond in these compounds is in the range of 1318–1360 cm^{-1}, higher than the usually observed value of 1200–1300 cm^{-1} for simple nitrones[446].

**d. Nuclear magnetic resonance spectra.* ^{1}H-NMR data are available now for some simple aliphatic nitrones[453], the structures of some of which were in question[452]. These include several *N*-alkyl *N*-alkylidene *N*-oxides that show the CH═N resonance at 6.57–6.86 ppm. In addition, ^{1}H-NMR data are presented for the *N*-methyl nitrones derived from acetone, cyclopentanone and cyclohexanone, which are isolable only as hydrochlorides[454].

Substituent effects on the ^{1}H and ^{13}C chemical shifts of a series of differently substituted (*Z*)-α, *N*-diaryl nitrones have been obtained. Correlations have been found between the chemical shifts and the Hammett σ parameters and the Swain and Lupton F and R parameters. These correlations confirm the dual behaviour of the nitrone group and the presence of 'through-resonance' in these nitrones[455].

^1H- and ^{13}C-NMR spectra of several (*E*) and (*Z*)-*N*-alkyl α-methoxycarbonyl nitrones (**331**) were determined[456]. It was found that the chemical shift of the CH=N proton is sensitive to the substituent, to the geometry around the C=N double bond and to the solvent. Such protons are deshielded in the (*E*) isomers, by the anisotropy of the *cis* oxygen, relative to the (*Z*) isomers. Large solvent shift differences between the two isomers ($\delta_{DMSO} - \delta_{benzene}$ values of about 0.9 for (*Z*) as compared to ≈ 0.4 for (*E*) were noted. The (*E*)/(*Z*) values were found to correlate linearly with solvent polarity and the slopes with the bulkiness of the N-substituent. It was also shown in this work that the ^{13}C resonances of the **C**—CH=N and CH=N carbons appear at lower field in the (*E*) isomer, in contrast to the N—**C** substituents which appear at higher field in these isomers.

$$\mathrm{MeO\overset{\overset{\displaystyle O}{\|}}{C}{-}CH{=}\overset{\overset{\displaystyle O^-}{|}}{\underset{+}{N}}R}$$

(331)

The ^{13}C chemical shifts of the α-carbon atom in the nitrone functions are around 125–150 ppm ($J_{CH} = 175$–180 Hz), with the large majority falling into the middle of this range[441,444,446,453–457,466,470].

The natural abundance ^{17}O-NMR spectroscopy of a series of (*Z*)-(substituted benzylidene) phenylamine *N*-oxides has recently been reported[458]. It was found that the chemical shift of the parent compound, α, *N*-diphenyl nitrone is down field from that of pyridine *N*-oxide, confirming that the N—O bond in this nitrone has more double-bond character than that of the aromatic *N*-oxide. The chemical shift range for these nitrones is 350–405 ppm in acetonitrile, relative to 2-butanone. A reasonable correlation was obtained when the ^{17}O chemical shifts were plotted against the corresponding Hammett σ values. ^{19}F-NMR data were reported for some polyfluorobenzophenone nitrone derivatives.

**e Mass spectra.* The mass spectra of some nitrones derived from oxomalonate and from camphorquinone were studied in comparison with the corresponding isomeric oxime ethers[460]. A characteristic feature in the mass spectra of nitrones is the absence of α-cleavage, which is proposed as a diagnostic tool to help distinguish between nitrones on the one hand and the isomeric oxime ethers or amides on the other. Another characteristic feature which is also absent in the isomeric compounds is the occurrence of the peaks $[M - O]^+$ and $[M - H]^+$, which are however found to be very weak (2–4% and 2–24%, respectively). A series of fluorenone nitrones was also examined by electron impact mass spectrometry[461]. The substituent on the nitrogen exerted a considerable influence on the fragmentation pattern, causing significant differences within this series.

**4. Geometrical isomerism of nitrones*

A set of substituent parameters was proposed which permits the prediction of (*E*/*Z*) equilibrium constants for nitrones along with other C=N and C=C double-bonded functional groups. The set includes values for the nonbonded electron pairs and the nitrone oxygen as well as for groups coplanar or orthogonal to the double bond[462].

It has been recognized that aldonitrones are usually more stable in the (*Z*) configuration. Recently it was found that when pure, crystalline (*Z*)-α-methoxycarbonyl *N*-alkyl nitrones are dissolved and their ^1H-NMR spectrum is examined, a time-dependent change in the spectrum is observed, which can be rationalized in terms of an (*E*–*Z*) equilibrium[463].

The final position of the equilibrium depends on the solvent polarity. Thus for α-methoxycarbonyl *N*-methyl nitrone (*E*/*Z*) values ranging from 6 in benzene to 0.67 in DMSO were observed. Reinvestigation of *C*-phenyl *N*-methyl nitrone by the same authors, however, did not show any evidence for the presence of the hitherto unknown (*E*) isomer. In this context it is worth mentioning a somewhat older report about the observation of two isomeric α-*p*-tolyl *N*-methyl nitrones in benzene and the assignment of their configuration by the use of lanthanide shift reagents[51].

*5. *Tautomerism of nitrones*

a. Nitrone–hydroxyenamine tautomerism.* Many acyclic, aliphatic and aromatic nitrones have been synthesized and studied[453]. It has been found that nitrones derived from aliphatic aldehydes and *N*-alkylhydroxylamines are converted to dimeric isoxazolidines as a result of 1,3-dipolar cycloaddition of the nitrone to the double bond of the tautomeric hydroxyenamine. In contrast, the nitrones of ketones or *N*-arylhydroxylamines are stable as such[453]. Additional examples of this type of tautomerism have been reported. One paper deals with a series of β-carbonyl nitrones in the imidazoline *N*-oxide series. The main conclusion of this work is that in all compounds the ketonitrone form (analogue of **40) is the disfavoured tautomer. In the β-formyl nitrone derivative the authors favour the hydroxyenaminoketone (of type **40b**) over the enol nitrone (**40a**) on the basis of the absence of coupling between the two hydrogens indicated: **=CH—O—H.** In the ketonitrones no such possibility exists, therefore the case between **40a** and **40b** is undecided[464].

Reactions of β-keto esters and β-diketones with N-substituted hydroxylamines, which constitute a new pyrrole synthesis, are reported to stop in certain cases at the nitrone stage. Thus the reaction of ethyl acetoacetate with *N*-cyclohexylhydroxylamine yields the nitrone as a mixture of (*E*) (82%) and (*Z*) (18%) isomers. In contrast, *N*-methylhydroxylamine reacts with benzoylacetone to give the hydroxyenamine ($\approx 80\%$) and the *C*-methyl *C*-phenacyl *N*-methyl nitrone as apparent from the ^{1}H-NMR spectrum of the product[465]. On the other hand, the majority of nitrones derived from α-formyl cyclic ketones exist in the enol nitrone tautomeric form[466].

**b. Behrend rearrangement.* The equilibrium in the triethylamine catalysed Behrend rearrangement of a series of *C*-aryl *N*-1-cyanoisobutyl nitrones was studied recently. It was found that in all cases the aldonitrone predominates in the equilibrium with constants ranging from 0.125 to 0.194[467]. The equilibrium constants were also temperature and solvent dependent. The final conclusion was that the most important influence on this equilibrium is solvation and the reason for the predominance of the aldonitrones is that these are more solvated that the corresponding ketonitrones[467].

$$\underset{\text{Aldonitrone}}{\mathrm{ArCH{=}\overset{\displaystyle O^-}{\overset{|}{N^+}}{-}\underset{\displaystyle CN}{\underset{|}{C}}HPr\text{-}\mathit{i}}} \rightleftarrows \underset{\text{Ketonitrone}}{\mathrm{ArCH_2\overset{\displaystyle O^-}{\overset{|}{N^+}}{=}\underset{\displaystyle CN}{\underset{|}{C}}Pr\text{-}\mathit{i}}}$$

c. Ring–chain tautomerism.* Additional papers have appeared on the subject of tautomerism in the series of *N*-(2-hydroxyalkyl) nitrones[468a]. The open-chain tautomers can be derivatized in the form of cyclic borinate derivatives of type **332, while treatment of the equilibrium mixture with acylating agents such as acyl halides or isocyanates affords the *N*-acyloxyoxazolidines (**333**) derived from the ring tautomers. In a number of cases the tautomeric equilibrium was detectable by ^{1}H-NMR spectroscopy and details of the spectra of the tautomers are reported[468a].

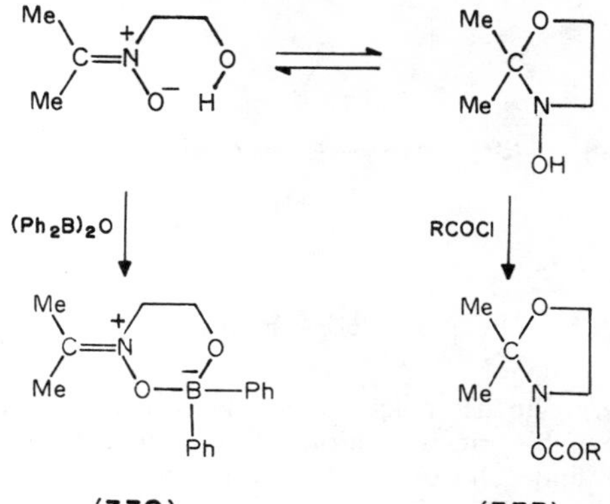

(332) (333)

Analogously, ring–chain tautomerism was observed in a series of *N*-(3-hydroxyalkyl) nitrones and the tautomers were characterized by the same methods[468b]. Another type of bifunctional nitrones which shows this kind of tautomerism is that of *C*-aryl *N*-carboxyalkyl nitrones. Although by NMR spectroscopy only the presence of the chain tautomer can be observed (which can be derivatized as a cyclic borinate derivative similar to **332**), acylation affords *N*-acyloxyoxazolidinones (**333**) derived from the ring tautomer[469]. *N*-(1-hydroxy-4-chloro-2-naphthylmethyl) nitrones derived from aliphatic aldehydes are in equilibrium with the corresponding naphtho-*N*-hydroxydihydrooxazines, analogous to **72**[470]. Both tautomers can be observed by ^{13}C-NMR spectroscopy. In the compounds derived from aromatic aldehydes, only the naphthol nitrones are observed.

*B. Reactions of Nitrones

*1. Rearrangements of nitrones

**b. Formation of oxaziridines.* Full details have been published of Reference 111[471]. Other cases of photoisomerization of a nitrone to an oxaziridine include the conversion of the dinitrone derived from 2,2,4,4-tetramethyl-1,3-cyclobutanedione to a bis-oxaziridine[472], and of some bicyclic 1-pyrroline 1-oxides[473a]. The oxaziridines described

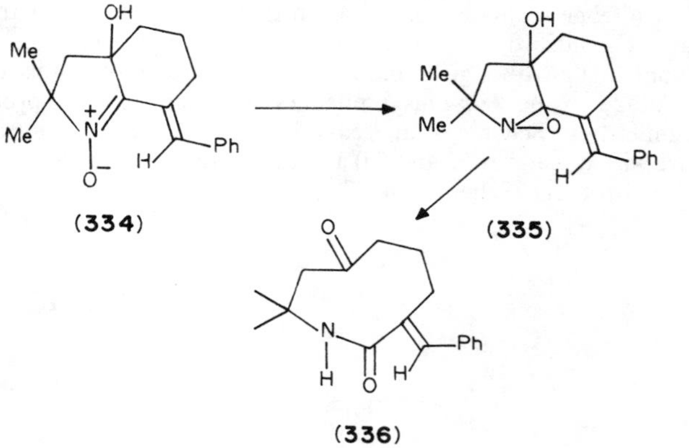

(334) (335)

(336)

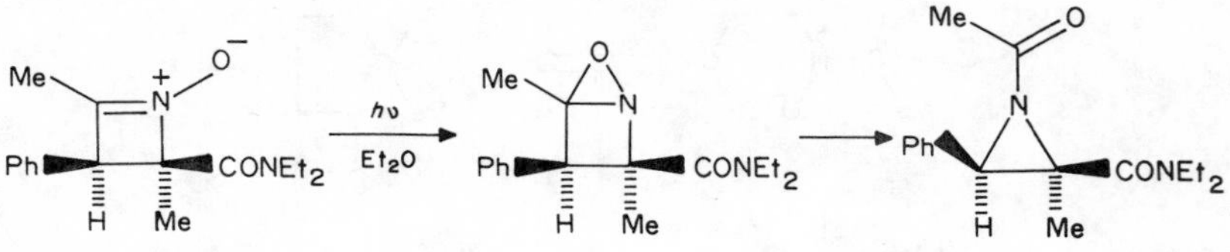

SCHEME 53

in these reports undergo further reactions ending with the formation of amides. An interesting application of this principle for the synthesis of medium-ring-sized oxolactams is described by Black and Johnstone[473b]. In this work, β-hydroxynitrones (**334**) are converted to oxaziridines (**335**). Treatment of these with iron(II) sulphate afforded ketolactams (**336**) in approximately 40% yield. This reaction constitutes a ring enlargement by three atoms.

A four-membered cyclic nitrone was converted to the bicyclic 5-oxa-1-azabicyclo[2.1.0]pentane derivative (**337**), which upon further irradiation rearranged to the aziridine (**338**) (Scheme 53)[474].

An attempt was made to achieve an asymmetric synthesis of oxaziridines by photocyclization of some *C*-aryl *N*-*t*-butyl nitrones in a chiral environment[475]. In the presence of cyclodextrin, optical yields of 0.06–0.41% were observed, which could be raised to about 1% upon the addition of an equimolar amount of optically active amino acid.

c.* *Formation of amides.* In an attempt to apply the ring-expansion reaction of the steroidal nitrones (*E*)- and (*Z*)-21**, which leads to seven-membered lactams (**111**)[50], to simple cyclic conjugated ketones, it was found that it is not applicable to these in any efficient manner[476]. The mechanism of the conversion of nitrones to amides in their reaction with aroyl chlorides was studied[477a]. The result of the reaction of a diaryl nitrone with O-labelled benzoyl chloride in the presence of a base in Scheme 54 is that the label is equally divided between the two oxygens of the diacylamine. In the absence of base the reaction brings about isomerization of the nitrone to amide. The mechanism proposed to account for these results (Scheme 55)[477a] is preferred over that in Scheme 7. In the absence of base, the nitrilium ion (**339**) is attacked by benzoic acid, followed by chloride ion, resulting in a rearranged amide containing 50% of the label, while the other 50% remains in the recovered benzoyl chloride[477a].

However, the nitrone → amide isomerization does not occur when the nitrone's *N*-phenyl ring is substituted with an electron-releasing group[477b]. *N*-*p*-Dimethylaminophenyl nitrones give imidoyl chlorides instead of amides upon treatment with an acyl halide, while *N*-*p*-anisyl nitrones are converted to products of *ortho* substitution (similarly to Scheme 34 and equation 51) that can be converted to benzoxazoles (compare with equations 53 and 54) and the corresponding aldehyde derived from the nitrone's C-substituent (Scheme 56)[477b].

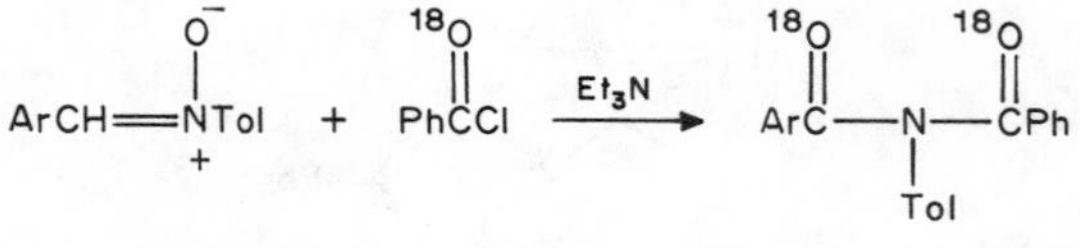

$$\mathrm{ArCH{=}\overset{+}{N}(O^-)Tol} + \mathrm{Ph\overset{^{18}O}{\overset{\|}{C}}Cl} \xrightarrow{Et_3N} \mathrm{Ar\overset{^{18}O}{\overset{\|}{C}}{-}N(Tol){-}\overset{^{18}O}{\overset{\|}{C}}Ph}$$

SCHEME 54

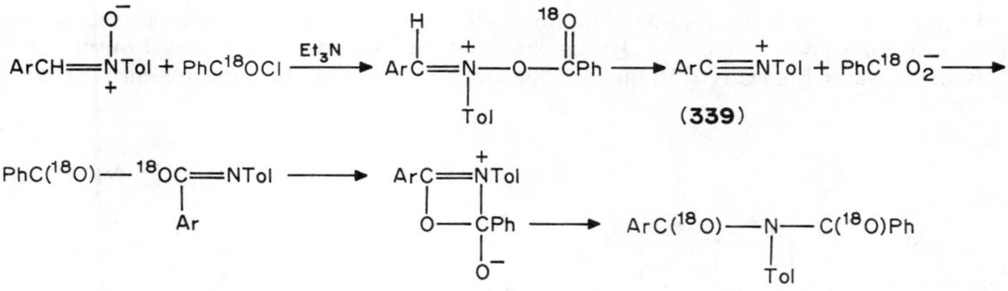

SCHEME 55

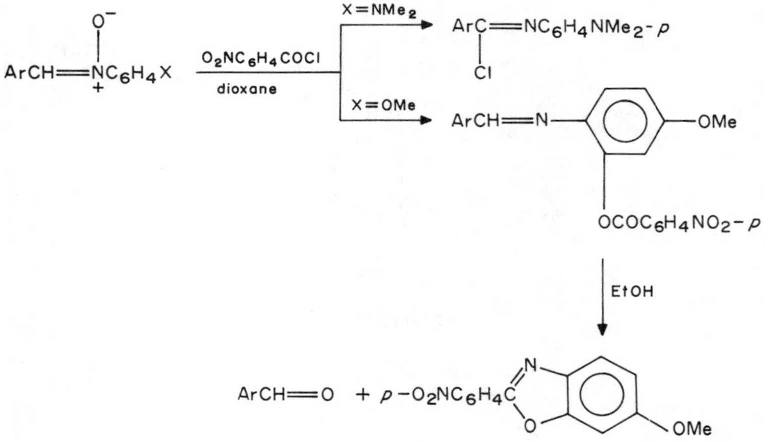

SCHEME 56

The *p*-toluenesulphonyl-chloride-induced isomerization of nitrones to amides shows different pH dependence in different systems. The treatment of 3,4-dihydroisoquinoline *N*-oxide with *p*-toluenesulphonyl chloride in the presence of sodium hydroxide gave 1-phenylisoquinoline, while in the presence of sulphuric acid a high yield of 2-phenylisocarbostyryl (**340**) was obtained[478].

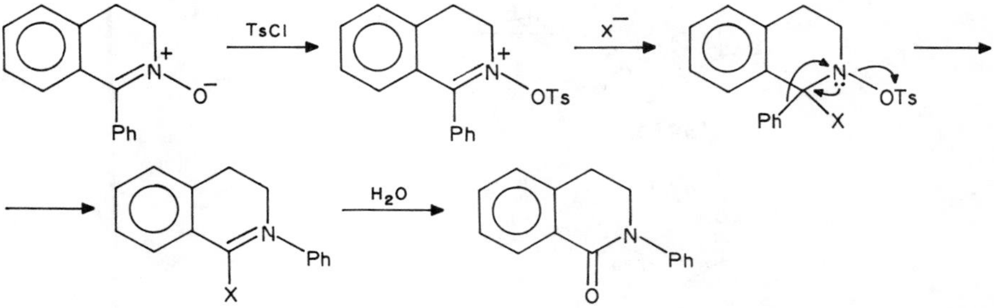

(**340**)

In contrast, the action of *p*-toluenesulphonyl chloride (TsCl) upon nitrones derived from 17-ketosteroids gives rearranged *D*-ring-expanded lactams (**341**) when carried out in the presence of water, but leads to 16-functionalized steroids of type **342** in the presence of base (Scheme 57)[479].

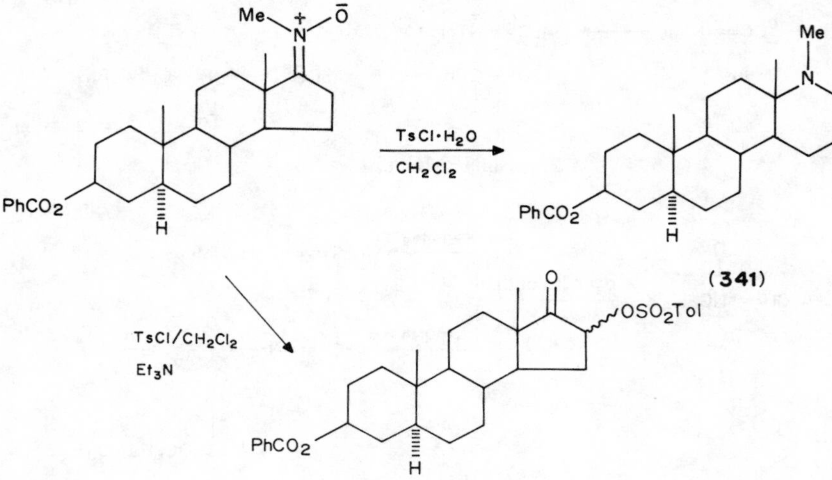

Me N O TsCl·H2O CH2Cl2 PhCO2 H Me N O PhCO2 H (341) TsCl/CH2Cl2 Et3N O OSO2Tol PhCO2 H (342)

SCHEME 57

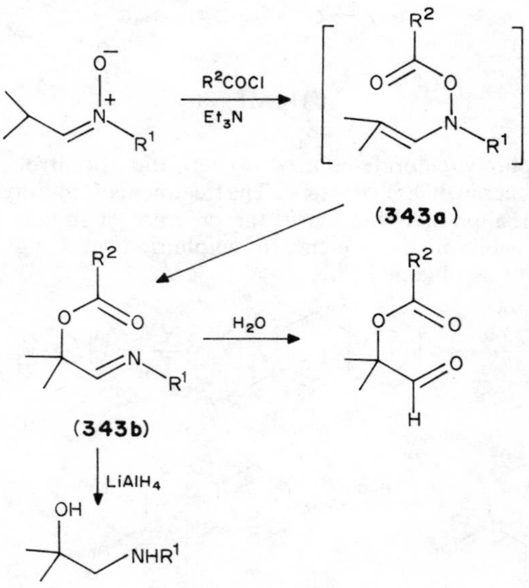

O N R1 R2COCl Et3N R2 O O N R1 (343a) R2 O O N R1 (343b) H2O R2 O O O H LiAlH4 OH NHR1

SCHEME 58

This reaction appears to have been discovered and applied independently in a series of keto-and aldo-nitrones and in proposed as a novel method for α-oxygenation of ketones and aldehydes[480]. The reaction is suggested to proceed by way of a spontaneous [3,3]sigmatropic rearrangement of the intermediate *N*-vinyl-*O*-acylhydroxylamine (**343a**) to the α-acyloxyimine (**343b**)[480]. This can either hydrolyse to the α-acyloxycarbonyl compound or may be reduced by hydrides to the corresponding amino alcohol or, in special cases, to the esters (Scheme 58).

A certain type of *C*-3-(1,2,4-oxadiazolyl) *N*-*p*-dimethylaminophenyl nitrone was found to rearrange to the corresponding amide, simply by heating in ethanol[481]. In this rearrangement a stable, dipolar oxadiazolium intermediate could be isolated. A charge-transfer complex has also been suggested to be involved as intermediate in the fluoranil-catalysed isomerization of 3*H*-indole 1-oxide to the corresponding lactam[452]. However, addition of fluoranil to open-chain nitrones did not give the corresponding amides[452].

e. Miscellaneous.* Four-membered cyclic nitrones of type **344 have been shown to undergo concerted conrotatory ring opening to yield the corresponding *N*-vinylnitrones (**345**) (Scheme 59)[482]. On the other hand, treatment of the same nitrones (**344**) with potassium *t*-butoxide results in a different kind of ring opening which leads to mixtures of stereoisomeric oximes (Scheme 60). In this reaction the first step is the abstraction of a proton by the base to yield the anion of *N*-hydroxy-1,2-dihydroazete, which then undergoes electrocyclic ring opening to the final products, (*E* + *Z*)-**346**[483].

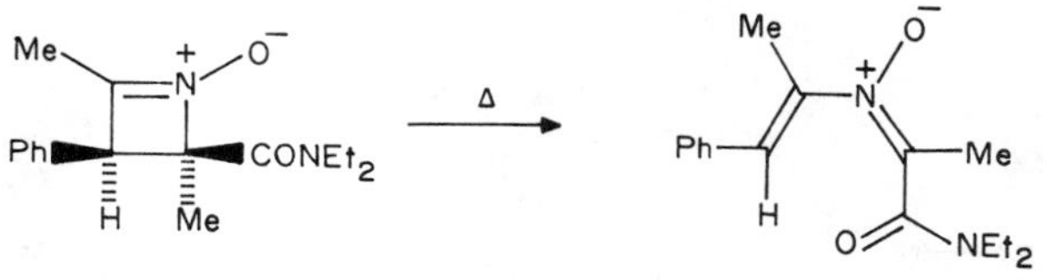

(**344**) (**345**)

SCHEME 59

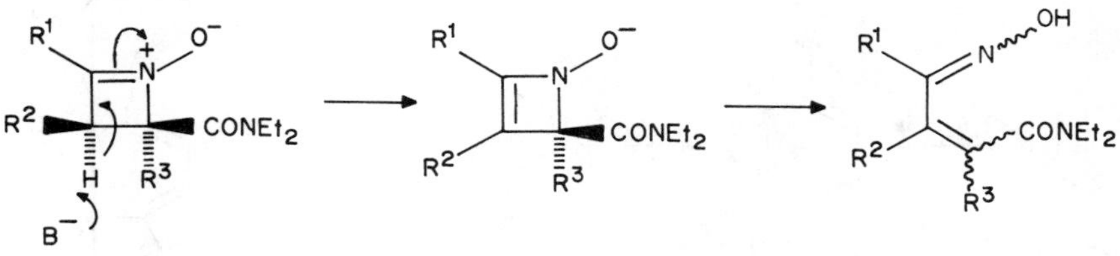

(**344**) (*E* + *Z*-**346**)

SCHEME 60

Another remarkable case of isomerization is shown by the highly reactive butenynyl nitrones of type **347** which give a number of rearrangement products in short duration thermolysis (350 °C, 10 sec) in a substituent-dependent manner (Scheme 61)[484].

These results are rationalized in terms of the formation of a ketocarbene intermediate (**348**), which can be formed either directly from the seven-membered cyclization product of the starting nitrone or from its secondary rearrangement products as shown in Scheme 62. This reaction may become applicable for the synthesis of special pyrrole or pyridine derivatives.

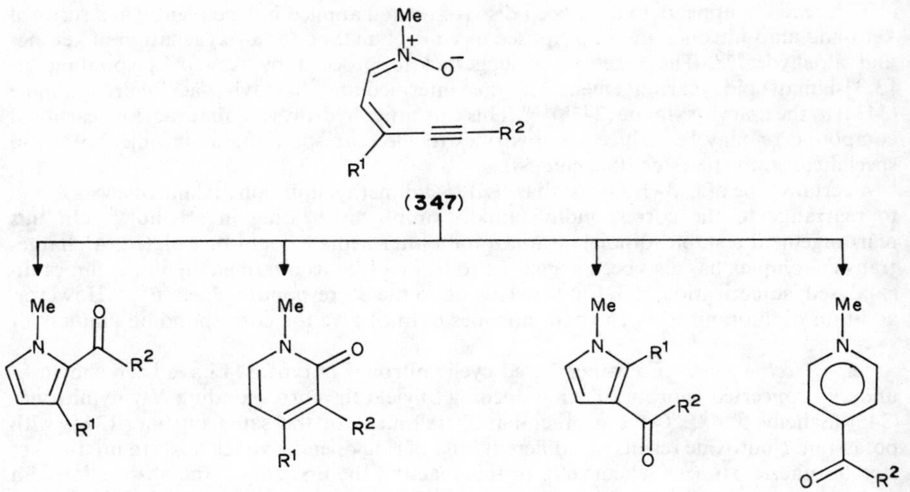

(347)

SCHEME 61

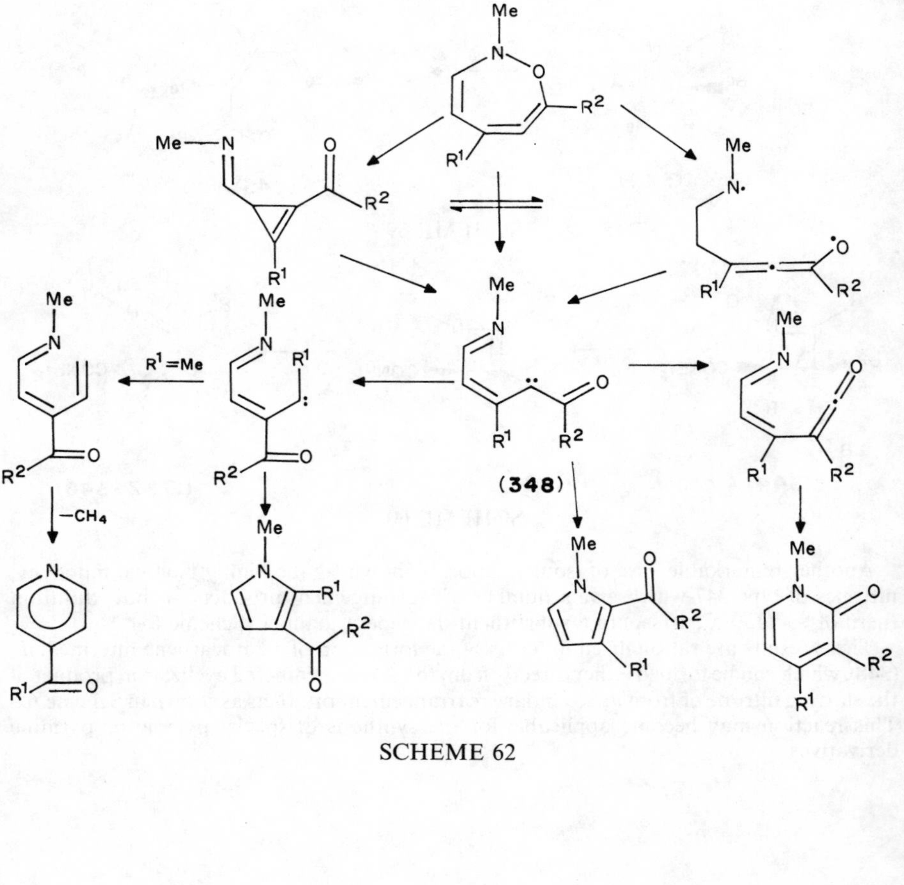

(348)

SCHEME 62

*3. Oxidation of nitrones

a. Lead tetraacetate (LTA).* Oxidation of the four-membered cyclic nitrone **349 gives *N*-acetoxyazetidinone (**350**), which can be hydrolysed to the corresponding *N*-hydroxy compound which in turn is reduced by $TiCl_3$ to the corresponding *β*-lactam (**351**)[485].

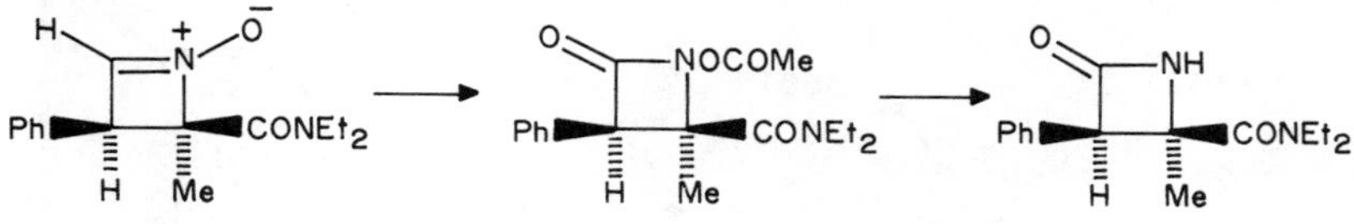

(349) (350) (351)

The oxidation of a series of enol nitrones derived from α-formyl cyclic ketones with lead dioxide should also be mentioned in this context: the initial products of these reactions were vinylaminyl oxides that could in some cases be detected by ESR spectroscopy, before they dimerized, either by *β*-C, *β*-C- or *β*-C,O bond formation[466].

d. Halogens.* Reaction of a series of α,*N*-diaryl nitrones with *N*-bromosuccinimide gave the α-aryl α-(*N*-succinimidyl) *N*-aryl nitrones **324 that were mentioned in Section *II.A.2.A[443].

**f. Ozone.* *N*-Glycosyl *C*-aryl nitrones can be prepared in high yields from sugar oximes that are in equilibrium with the ring tautomer, *N*-glycosylhydroxylamine, and an aromatic aldehyde. These can be oxidized by ozone to the corresponding 1-deoxy-1-nitro sugars[486]. An advantage of this method is that it is compatible with unprotected hydroxy groups that are present in the sugar molecule.

h. Miscellaneous.* Nitric acid oxidation of the tricyclic nitrone **352 was reported to afford the diketonitronitrone **353**[473a].

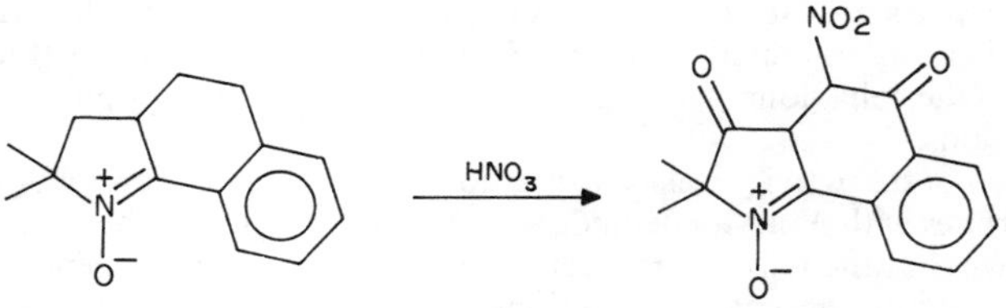

(352) (353)

C-Benzyl nitrones (**354**) show sensitivity to air and on storing most of them are observed to form pyrroles in low yields. The yields can be increased by carrying out the reactions with excess oxygen[487]. In the first step the reaction leads to the linear dimers **355** which undergo cyclization as shown in Scheme 63.

*4. Reduction of nitrones

**b. Deoxygenation.* Sodium hydrogen telluride is reported to reduce nitrones in a pH-dependent manner. At alkaline pH (10–11) deoxygenation is obtained, while at lower pH (6) the imine obtained in the first step is reduced to the secondary amine[488]. Three

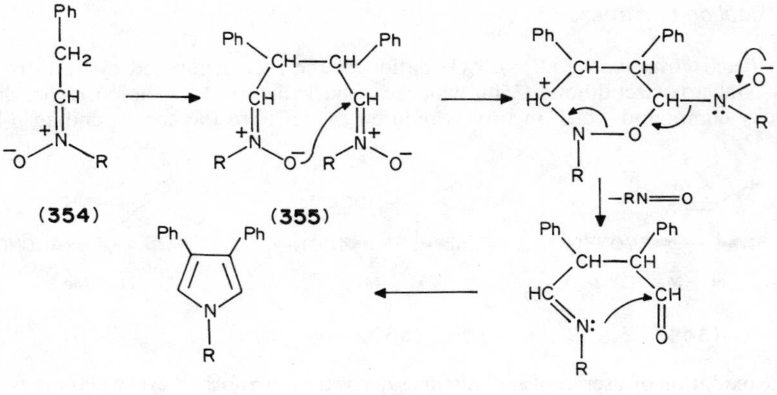

SCHEME 63

examples are given: 1-phenyl-3,4-dihydroisoquinoline 2-oxide and two steroidal nitrones (see also Section *II.B.4.d).

**c.* *Reduction to hydroxylamines and amines.* See preceding section.

**d.* *Nitrones as oxidizing agents.* Aliphatic nitrones act as oxygen donors and are converted to imines. Thus they react with benzylisothiocyanate to give the unstable intermediate benzyl isothiocyanate *S*-oxide which is trapped by a second molecule of isothiocyanate to form 2,4-dibenzyl-3-thiono-1,2,4-thiazolidin-5-one[489].

*5. Cycloaddition reactions of nitrones

**a.* *1,3-Dipolar cycloadditions.* This aspect is the most intensively studied in the chemistry of nitrones and was recently reviewed[490]. Because of the large volume of work done in recent years, this section will be subdivided, in contrast to the original chapter.

(i) Mechanistically oriented studies. The more recent work is focused on the examination of the behaviour of special types of nitrones or dipolarophiles, or both, under special conditions.

The question of the activity–selectivity ratio in the 1,3-dipolar cycloaddition of *C*-aryl *N*-phenyl nitrones with *N*-arylmaleimides and tetracyanoethylene has been studied[491]. It has been shown that the relation between activity and selectivity depends on the relative influence of the donor–acceptor interactions of the dipole with the dipolarophile on one hand or their localization energies on the other. The mode of addition of nitrones to the four-membered ring thiete 1,1-dioxide is markedly different from that to the acyclic vinyl sulphone[492]. While in the acyclic series a mixture of the two regiomers 4- or 5-sulphonylisoxazoline, is formed, in the cyclic compound only 4-sulphonylisoxazoline is obtained. It is concluded that frontier orbital interactions, calculated by CNDO/2 approximations, alone are not sufficient to account for the experimentally observed behaviour and that the participation of lower lying orbitals needs to be considered. Neither can calculations explain the stereoselectivity in the reactions of α, *N*-diphenyl nitrone with the homologous ring size benzo[*b*]thiophene *S*-oxide and *S*, *S*-dioxide series[493]. In contrast, in this case it was possible to discuss the regioselectivity in terms of frontier orbital interactions on the basis of CNDO/S calculations and photoelectron spectral ionization potentials in a satisfactory manner. The regiochemical behaviour of some simple cyclic nitrones such as pyrroline *N*-oxide and tetrahydropyridine *N*-oxide[494] and a β-carboline-derived nitrone[495] follow the predictions of frontier orbital theory; however, the explanation of reactivity differences between the five- and the six membered nitrones (found by kinetic studies) requires the consideration of strain-related influences in the transition state[494]. There are conflicting reports on the formation of charge-transfer complexes in the course of interactions of nitrones with electron-deficient dipolarophiles. While the interaction between *C,N*-diaryl nitrones with tetracyanoethylene is assumed to result in the formation of a charge-transfer complex on the basis of absorption bands that appear in the visible spectrum, and cannot be assigned either to reactants or products[491b], there was no indication of such complex formation in the reactions of nitrones with β-nitrostyrene or phenyl vinyl sulphone[452]. A thermochemical study was carried out on the reaction of α-benzoyl *N*-phenyl nitrone with a number of norbornene derivatives[496]. The reactions of this nitrone with olefins that are more electron rich than norbornene, such as styrenes and vinyl selenides, are known to take place by 1,3-dipole donor–dipolarophile acceptor interactions. Hence the fact that there was no effect of the substituents in the norbornene on the reaction was taken as an indication that the influence of donor–acceptor interactions is cancelled out by localization effects, resulting in the leveling of reactivities in the substituted norbornenes.

The cycloaddition of a series of *C,N*-diaryl nitrones to 'captodative' alkenes (e.g. α-*t*-butylthioacrylonitrile and related olefins containing both electron-withdrawing and -donating groups) was examined. Such systems are considered as good tools to differentiate between concerted and stepwise diradical mechanisms. The regio- and stereo-chemical results in this work were completely in accord with the accepted principles of frontier orbital control and concerted mechanism and could rule out any diradical intermediates[497]. In contrast, the possibility of a two-step ionic mechanism was raised for the reactions of some oxazoline *N*-oxides with acetylene dicarboxylate and methyl propiolate, because of the observation of isoxazoles in these reactions[498b]. The formation of these isoxazoles could be rationalized by zwitterionic intermediates.

It was found possible to change the known regioselectivity in nitrone–olefin cycloadditions to give 4-substituted isoxazolines by introducing electron-releasing substituents in the α-position. Thus *C*-cyclopropyl *N*-alkyl nitrones and *C,C*-dicyclopropyl *N*-methyl nitrone give significant amounts of 4-substituted products with electron-deficient dipolarophiles. This behaviour was expected from the observed low ionization potentials measured for these nitrones and the results of calculations that indicate that their HOMOs are heavily localized on the nitrone moiety[499]. Another way of achieving the same goal is by introducing the electron-donating alkoxy group into the nitrone α-position. This approach also yields predominantly 4-isoxazoline: however, because of their low reactivity[500] these nitrones are limited to the most active dipolarophiles such as acetylene carboxylates, acrylates and phenyl isocyanate[498]. This is rationalized by assuming that the α-oxygen substitution on the nitrone releases electrons and raises the HOMO of the nitrone, thereby narrowing the HOMO–LUMO gap between the dipole and dipolarophile. The structure of the dipolarophile can also influence the regioselectivity, as for example in the reactions of nitrones with allyl ethers and allyl silanes. While reactions of silanes give predominantly the usual 5-substituted isoxazolidines, those of the allyl ethers lead mainly to 4-(alkoxymethyl)isoxazolidines[501].

The reactivities of α-ketonitrones towards different types of dipolarophiles have been studied. Thus, the rate of reaction of *C*-benzoyl *N*-phenyl nitrone with styrenes[502] and aryl vinyl ethers[503] increases with the electron-attracting character of the substituent on the dipolarophile. On the other hand, with more nucleophilic dipolarophiles the rate increases for electron-donating substituents, α-piperidinostyrenes[503] and furans[504a,b] behaving analogously, and correlations can be found between the kinetic data and the ionization potentials of the dipolarophile. These results and those obtained for other heterocyclic α-ketonitrones[494,505] can be rationalized by considering that the carbonyl group lowers the LUMO of the nitrone and therefore the interaction between LUMO (nitrone) and HOMO (dipolarophile) is the one that controls the reaction.

In the reactions of *C*-benzoyl *N*-phenyl nitrone with heterocycles such as *N*-methylpyrrole, thiophene and their benzo derivatives, side-products derived from elimination, substitution and decomposition are observed together with the usual cycloaddition products[504c,d].

The reactions of nitrones with cumulenes are interesting. *C,N*-Diphenyl nitrone reacts with allene to yield three products (Scheme 64)[506]. The 4-methylene isoxazolidine derivative **356** (22%) is the minor product of the 1,3-dipolar cycloaddition, while the 3-pyrrolidinone **357** (23%) and the benzazepinone **358** (31%) come from rearrangement of the unstable major regiomer, 5-methylene isoxazolidine, via a diradical species.

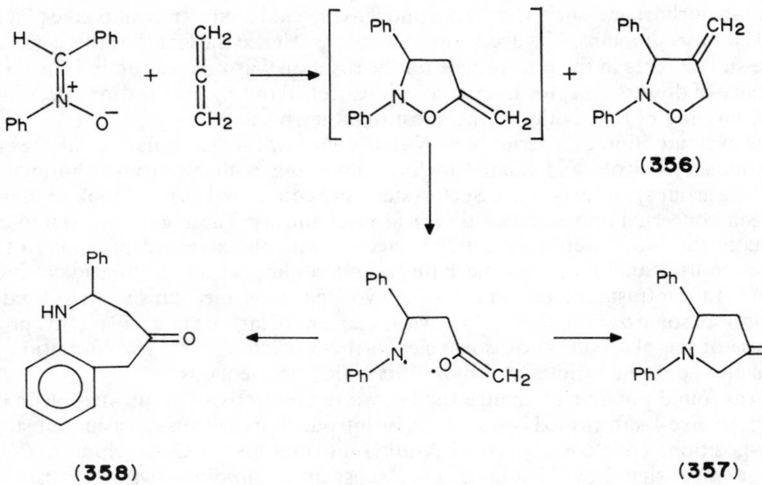

(**356**)

(**358**) (**357**)

SCHEME 64

In contrast, the reaction of fluoroallene with nitrones gives high yields of the corresponding 4-fluoromethyleneisoxazolidines (**359**). In this case, however, an additional effect came to light, namely the significant (> 80%) preference for the formation of the *syn* isomer (**359**) (Scheme 65)[507]. There is no satisfactory explanation for this phenomenon. The reaction of nitrones with substituted butatrienes also leads to 4-alleneisoxazolidines although in this case the low yields observed do not permit any conclusion to be made about the absence of the other regiomer[508].

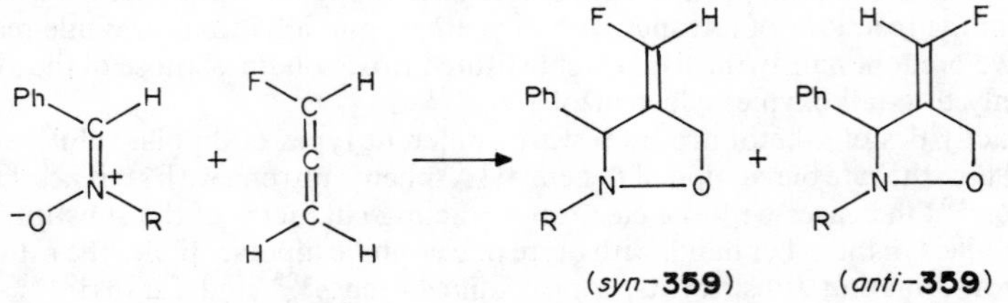

(*syn*-**359**) (*anti*-**359**)

SCHEME 65

(ii) The influence of external factors upon the reaction. Apart from the obvious possibility of heating the reaction mixture in order to increase its rate, other possibilities do exist:

(*a*) Some very interesting results have been reported on the influence of pressure on rate as well as on stereoselectivity. The application of pressures of 2–4 kbar make it possible to reach close to quantitative yields in reactions in which either the nitrones or the cycloadducts are heat sensitive and which therefore cannot be carried out at elevated temperatures[509]. The improvement in the stereoselectivity is exemplified by the reaction of *C*-phenyl *N*-methyl nitrone with vinylene carbonate. In the thermal reaction a 1:1 ratio of the *exo* and *endo* products is formed, while in the pressure-induced reaction the *endo* product predominates to the extent of 2:1[509]. A high-pressure kinetic study has been made of the 1,3-dipolar cycloaddition reaction of α-benzoyl *N*-phenyl nitrone with a number of dipolarophiles. Activation volumes have been determined, and the experimentally determined volume changes compared with the theoretically calculated changes in the van der Waals' volume for a prototype 1,3-dipolar cycloaddition between HC≡N—O and HC≡CH[510].

(*b*) Gallium chloride, a Lewis acid, has been found to have a spectacular effect in enhancing the rate of cycloaddition of *C*-phenyl *N*-methyl nitrone with maleic anhydride by a factor of $> 10^6$. The products are identical to those of the uncatalysed reaction[511]. In contrast, in a spectroscopic investigation of the effect of aluminium chloride on the 1,3-dipolar cycloaddition of a series of nitrones with *N*-phenylmaleimide, a decrease of reactivity and increase of stereoselectivity has been reported[512].

(*c*) In contrast to α-benzoyl *N*-phenyl nitrones, discussed previously, ketonitrones bearing electron-withdrawing substituents in the α-position, such as α-cyano α,*N*-diphenyl and α-methoxycarbonyl α,*N*-diphenyl nitrone, show lack of reactivity in the ground state[513]. Irradiation of these nitrones gives products which are quite different from the usual cycloadducts. The primary products of these reactions, which are not always stable enough to be isolated, are oxazolidines (**360**). Their formation is proposed to take place by way of the corresponding oxaziridines and their subsequent homolytic ring fission to diradicals that attack the olefin in a stepwise manner (Scheme 66)[513].

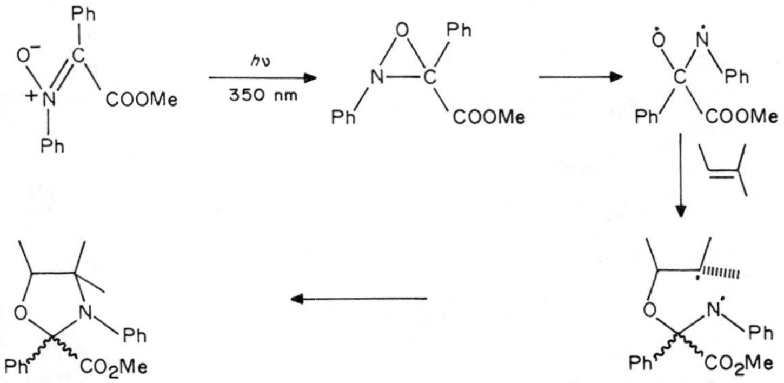

(**360**)

SCHEME 66

(iii) Synthetically oriented studies. It is hard to predict the steric outcome of intermolecular cycloadditions of acyclic nitrones to open-chain olefins. The addition of a series of *N*-benzyl *C*-alkyl and *C*-β-alkoxyalkyl nitrones to methyl crotonate gave predominantly the 3,5-*trans*-substituted isoxazolidines in 3:1 selectivity, while *N*-benzyl *C*-α-alkoxynitrones gave 4:1 excess of the *cis* product[514]. In contrast, the hindered *C*-tetramethyldioxolanyl nitrone **361** gave only the *trans* product (Scheme 67). A less hindered *C*-dioxolanyl nitrone showed little stereoselectivity[514,515].

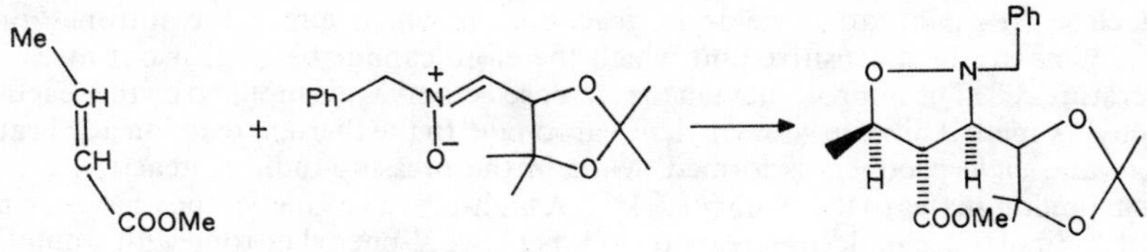

(**361**)

SCHEME 67

The behaviour of nitrones derived from benzaldehyde, bearing in the *ortho* position olefinic side-chains suitable for intramolecular cycloaddition (e.g. **362**), was examined[516]. It was found that such compounds react preferentially with acetylene dicarboxylate (intermolecularly!) to give isoxazolines that rearrange to azomethine ylids which are finally captured by the side-chain double bonds (Scheme 68).

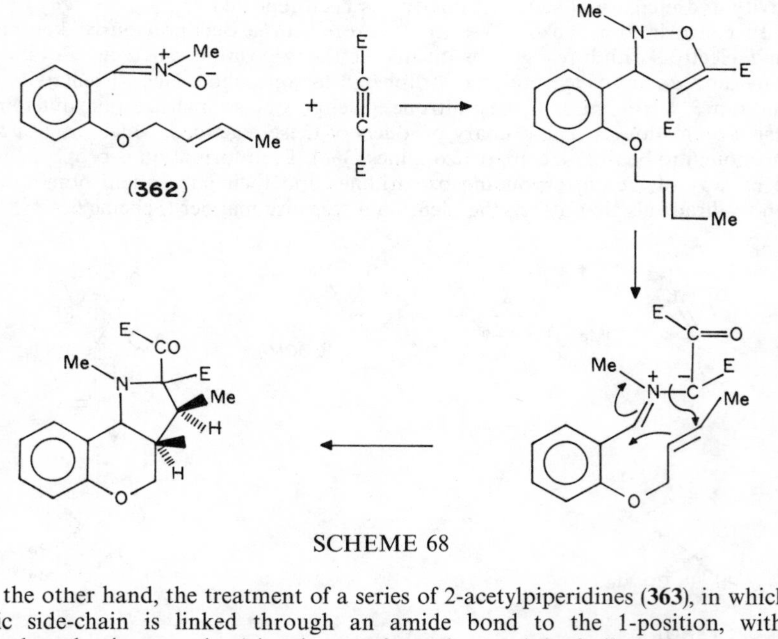

(**362**)

SCHEME 68

On the other hand, the treatment of a series of 2-acetylpiperidines (**363**), in which the olefinic side-chain is linked through an amide bond to the 1-position, with *N*-alkylhydroxylamines results (via nitrones formed *in situ*) in facile stereo- and regio-selective intramolecular cycloaddition, yielding δ-lactams (Scheme 69)[517].

An interesting utilization of the 1,3-dipolar cycloaddition is the preparation of amino and azabicyclic diols by reacting nitrones with cyclic dienes. The cycloaddition reaction is coupled with the *N*-oxidation of the isoxazolidine products (by *m*-chloroperbenzoic acid, MCPBA to afford new nitrones, which can further cycloadd intramolecularly to the remaining double bond. This is illustrated in Scheme 70 using *N*-methyl nitrone and 1,4-cyclohexadiene[518].

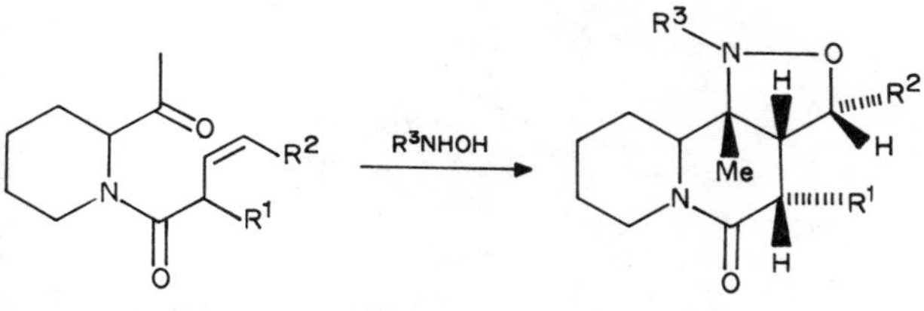

(363)

SCHEME 69

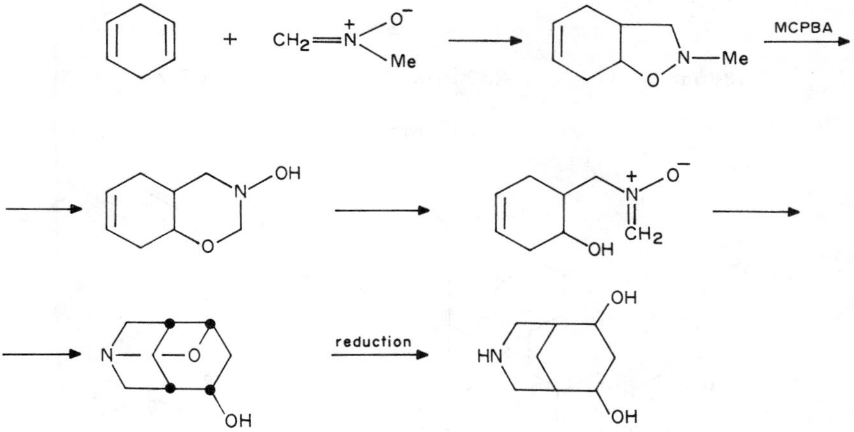

SCHEME 70

Nitrones **364**, derived from a series of 5-alkenyl and 6-alkenyl aldehydes, undergo intramolecular 1,3-dipolar cycloaddition. All 5-alkenyl nitrones and most 6-alkenyl nitrones afford the fused products **365**. Only the 6-alkenyl nitrones bearing an aryl group at C(6) give predominantly the bridged isoxazolidines of type **366** (Scheme 71)[519].

The relationship between reactivity in intramolecular dipolar cycloaddition and the distance between the nitrone and the double bond was also studied in a series of *N*-methylcyclohexylidenimine oxides having an unsaturated side-chain at position 2[520]. It was found that only compounds in which the double bond is in position 5 or 6 undergo intramolecular cycloaddition, the former being the more reactive. The 2-allylcyclohexylideneimine oxide, in which the double bond is in position 4 relative to the nitrone, underwent rearrangement, while there was no reaction at all in the compound in which the double bond was more remote, e.g. at the end of a decyl chain[520]. It was recently pointed out that, in addition to the factors mentioned, nonbonded interactions might also be decisive in determining the steric outcome of the cycloaddition[521]. Thus intramolecular cycloaddition of the *C*-cyclohexenyloxycarbonyl nitrone **367** affords only product **368**. The formation of product **369**, which would be the one predicted on stereoelectronic grounds, is presumably disfavoured due to nonbonded interactions of the carbonyl oxygen and the quasi-axial allylic hydrogen, as indicated in Scheme 72[521].

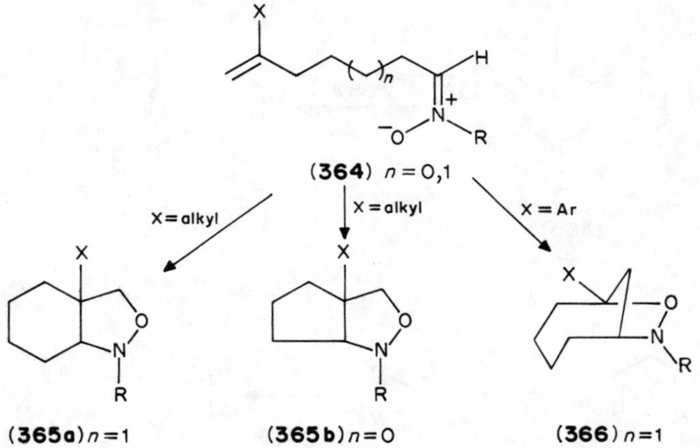

SCHEME 71

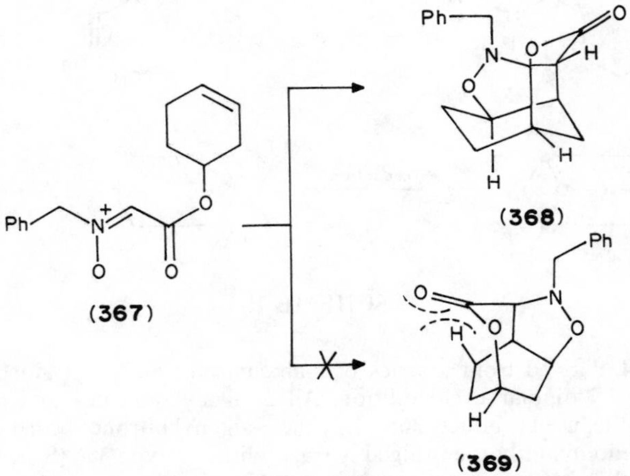

SCHEME 72

Intramolecular reactions of bicycloalkenyl nitrones provide a convenient entry into noradamantane, protoadamantane and adamantane systems[522]. Thus at room temperature the *C-endo*-bicyclo[3.2.1]oct-6-enyl nitrone **370** affords the epoxyiminonoradamantane derivative **371**.

An unusual kind of intramolecular cycloaddition was noted when tricyclic nitrones of type **372** were heated. The formation of the pentacyclic cage compounds **373**, which were obtained in high yields, required the hitherto unprecedented intramolecular addition of a cyclic 'ene' to a cyclic nitrone[522].

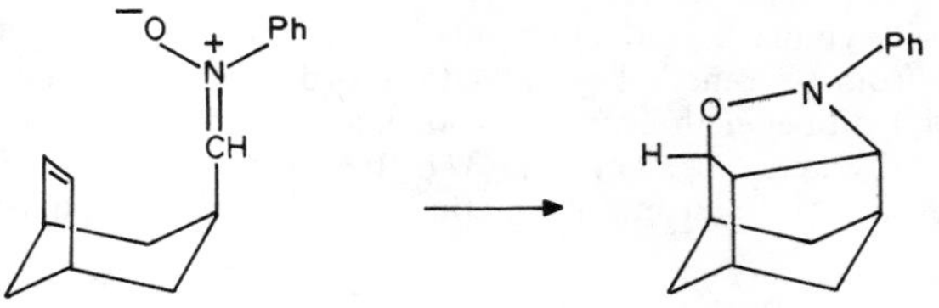

(370) (371)

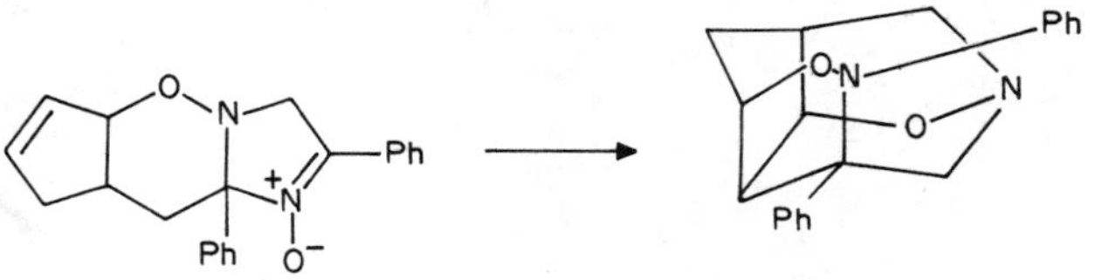

(372) (373)

In intramolecular additions distinction is made between two types of starting materials—those having the double bond in the C-substituent and others having it linked to the nitrogen. The examples discussed previously in the update belong to the first class, although Schemes 13 and 15 (Section II.B.5.a) illustrate synthetic applications of the second type. A systematic study was made recently of *N*-3-alkenyl and *N*-4-alkenyl nitrones, and of some cycloalkenylalkyl derivatives, in an attempt to understand and to be able to predict their regio- and stereo-chemistry[524]. In a series of straight-chain *N*-3-butenyl, *N*-4-pentenyl and *N*-5-hexenyl nitrones dependence of the product structure upon the chain length was observed (see Scheme 73)[524]. Similarly, nitrones derived from *N*-cyclohexylmethyl-, *N*-cyclohexylethyl-, *N*-cyclopentylmethyl- and *N*-cyclopentylethylhydroxylamines were examined and found to give products, often, but not in all cases, in high degree of regioselectivity, depending on the distance between the reacting groups[524,525].

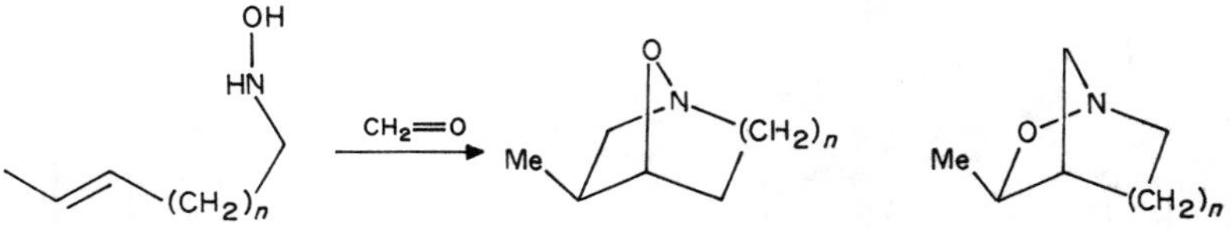

$n = 1$	100%	0%
$n = 2$	0%	95%
$n = 3$	25%	75%

SCHEME 73

N-Bicycloalkenyl nitrones derived from hydroxylamines **374**, **375** and **376** have been shown to yield azatricyclo ring systems in good yields and with regioselectivity[526].

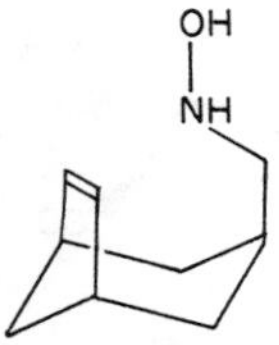

(374) 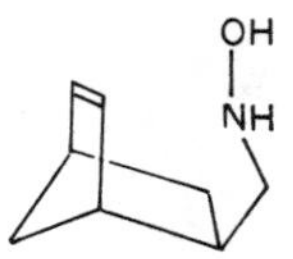 (375) 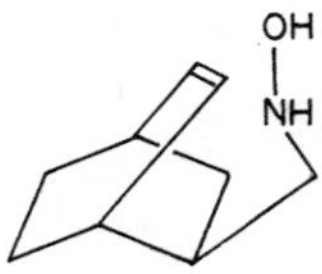 (376)

The question of asymmetric induction involving nitrones is being further pursued. Nitrones derived from (−)-methyl glyoxalate afford products of cycloaddition with styrene and with 1,1-diphenylethylene giving an enantiomeric excess of up to 20%[527]. In contrast, the optically active spironitrone **377** gives approximately 85% of the major product **378**, in addition to small amounts of the other three possible diastereoisomers[528].

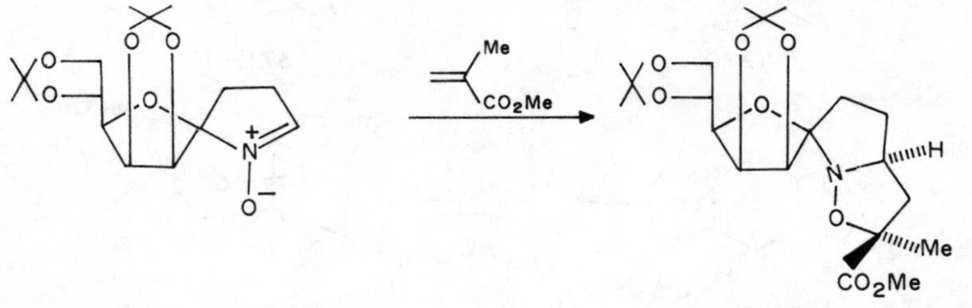

(377) (378)

On the other hand the addition of achiral nitrones (e.g. *C*-phenyl *N*-methyl nitrone and *C*,*N*-diphenyl nitrone) to the double bond in (*R*)-(+)-*p*-tolyl vinyl sulphoxide results in 4-(*p*-tolylsulphinyl)isoxazolidines as a mixture of diastereoisomers. After removal of the tosylsulphinyl function by desulphurization, and opening of the isoxazolidine ring, the final amino alcohol products are obtained in enantiomeric excess in the range of 80–90%[529].

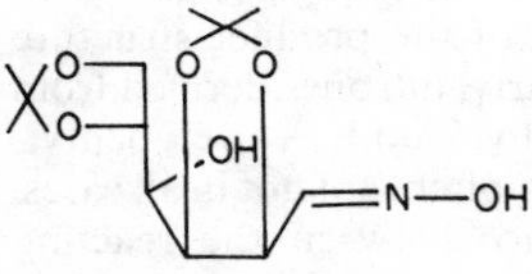

(379)

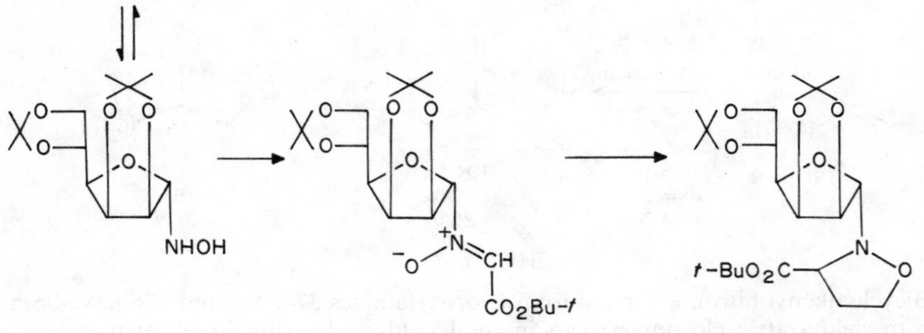

(380) (381)

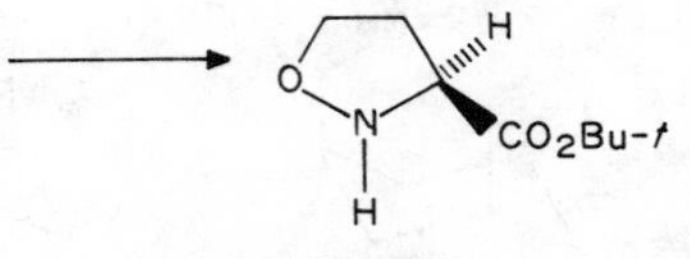

(382)

SCHEME 74

Asymmetric induction in cycloaddition has been exploited for the synthesis of some α-amino acid analogues and α-aminophosphonic acids. As an asymmetric handle the oxime of diisopropylidene-D-mannose (**379**) was used. Its reaction with *t*-butyl glyoxalate gave nitrone **380** which underwent cycloaddition with ethylene to isoxazolidine **381** in over 90% yield and 54% diastereomeric excess. After separation of the diastereoisomers the pure *t*-butyl ester of (L)-5-oxaproline (**382**) could be obtained (Scheme 74)[530].

In analogy to this, cycloaddition of the *N*-glycosyl *C*-phosphonyl nitrone **383** with ethylene gave the isoxazolidine **384** in a diastereoisomeric excess of 50%, of which the major isomer could be converted to (L)-5-oxa-1-phosphaproline (**385**) (Scheme 75).

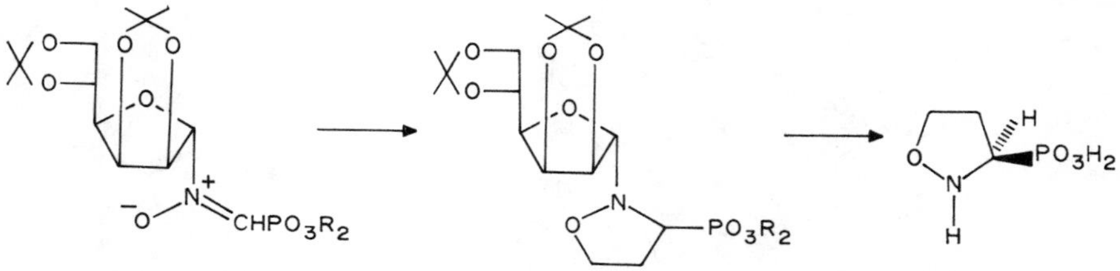

(**383**) (**384**) (**385**)

SCHEME 75

As is apparent from the previous examples, the 1,3-dipolar cycloaddition reaction is eminently suited for the synthesis of natural products, in particular where stereoselectivity and enantioselectivity are required. Although the first step of the cycloaddition reaction always leads to an isoxazolidine, this heterocyclic ring can be elaborated in a variety of ways to yield other functional groups or just a chiral centre, or none at all. The subject of natural product synthesis based on nitrones has been recently reviewed[490,532,533], and therefore only a few recent examples will be described to present the different uses of this methodology.

A short synthesis of 4-hydroxyproline was achieved when a series of nitrones derived from *N*-benzylhydroxylamines and methyl glyoxylate were condensed with acrolein. Hydrogenolysis of isoxazolidines **386** opened the ring by cleaving the N—O bond, and *in situ* recyclization of the amino aldehyde **387** and reduction of the imine, followed by hydrolysis of the ester gave the final products (Scheme 76)[534]. For other examples of amino acid synthesis see Schemes 74 and **75**.

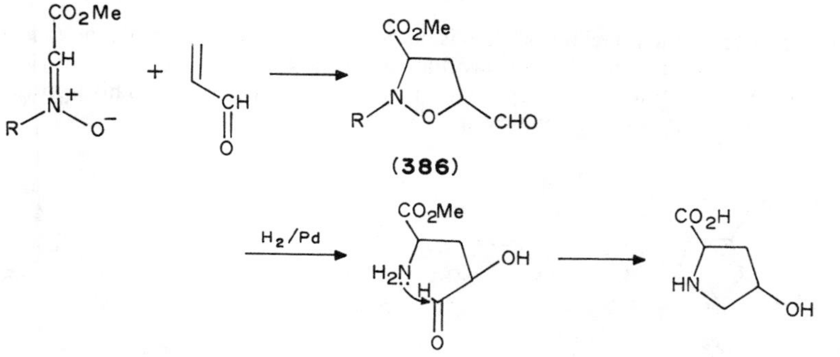

(**386**)

H_2/Pd

(**387**)

SCHEME 76

Negamycin (**390**), a peptide-like natural product, which has interesting antibacterial properties, has been synthesized by the nitrone cycloaddition methodology[535]. Nitrone **388**, derived from D-gulono-γ-lactone and methyl glyoxylate, was added to *N*-benzyloxycarbonylallylamine. Removal of the chiral auxiliary group by acid hydrolysis gave two diastereoisomeric isoxazolidines, one of which (**389**) was converted to the optically active natural product **390** (Scheme 77)[535].

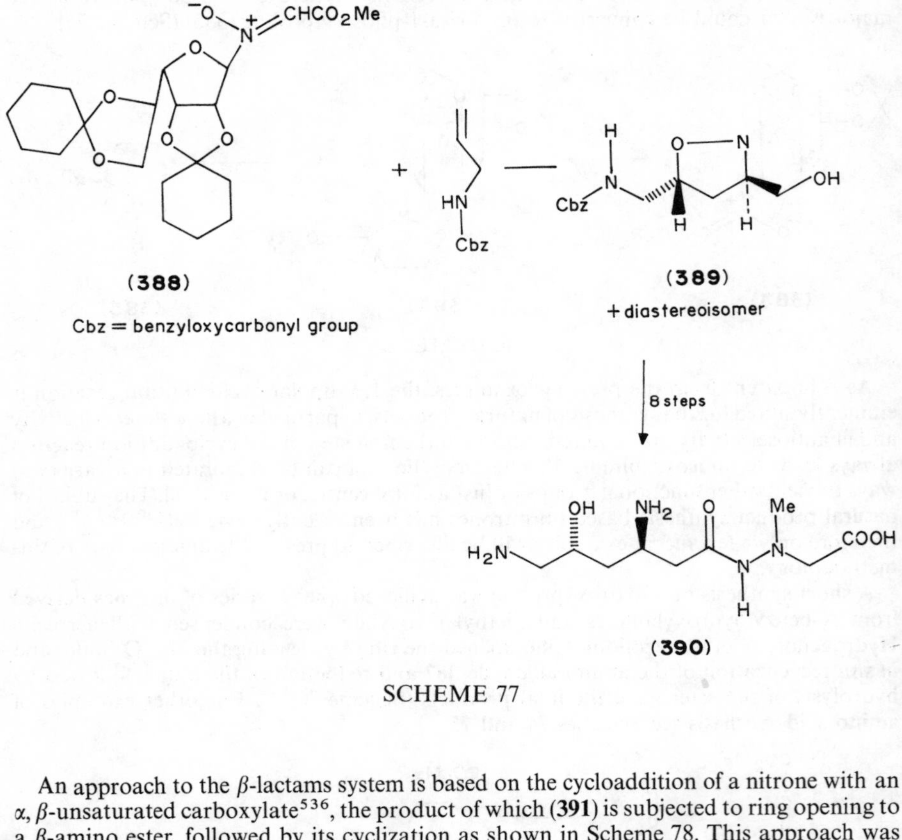

(388)

Cbz = benzyloxycarbonyl group

(389)

+ diastereoisomer

8 steps

(390)

SCHEME 77

An approach to the β-lactams system is based on the cycloaddition of a nitrone with an α,β-unsaturated carboxylate[536], the product of which (**391**) is subjected to ring opening to a β-amino ester, followed by its cyclization as shown in Scheme 78. This approach was applied to the synthesis of thienamycin[537].

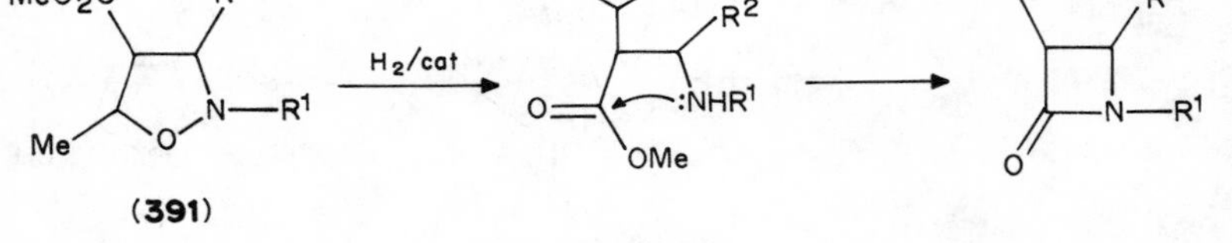

H_2/cat

(391)

SCHEME 78

Another way to obtain β-lactams from nitrones was discovered when it was observed that 4-nitro-5-cyanoisoxazolidines (**392**), obtained in dipolar cycloaddition, undergo thermal reorganization to β-lactams **394** (Scheme 79)[538]. The mechanism suggested for this reaction involves the opening of the isoxazolidine by homolytic fission of the N—O bond to an acylnitro intermediate **393** which cyclizes by the expulsion of nitrous acid.

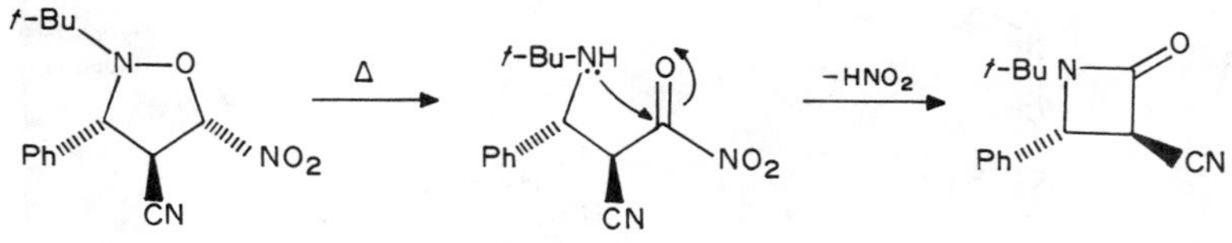

(**392**) (**393**) (**394**)

SCHEME 79

Full details have appeared of the biotin synthesis described in Scheme 14[539]. Another biotin synthesis, based on cysteine, involves intramolecular cycloaddition between the nitrone function constructed on the carboxyl carbon of cysteine and the double bond of thiovinyl ether attached at the other end of the chain[540]. It was found that the stereoselectivity in the cyclization of an acyclic compound (**395**) was disappointingly low. However, confinement of the thioenol ether side-chain in a ring, by linking it to the α-amino group, blocked the formation of the undesired stereoisomer and the cycloaddition resulted in a product that had the required two additional chiral centres, and could be converted to biotin (Scheme 80)[540].

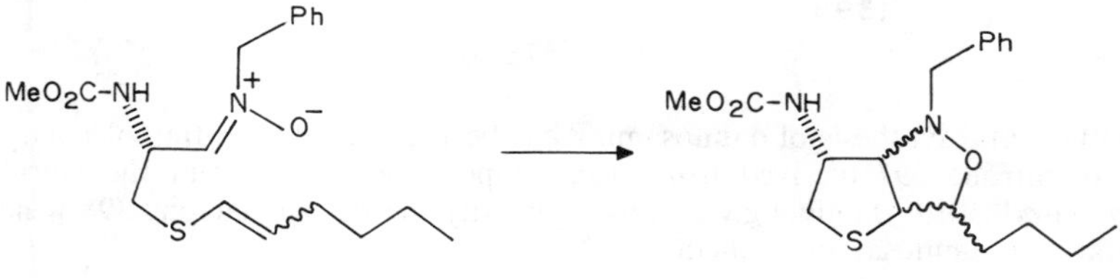

(**395**)

SCHEME 80

Cycloaddition is particularly suited for the synthesis of amino sugars containing several chiral centres, among them one nitrogen. The various groups can be inserted by judicious choice of starting materials or reactants.

The synthesis of nojirimycin, 5-amino-5-deoxyglucose (an antibiotic), was achieved by cycloadding an *N*-glycosyl nitrone to furan, followed by further elaboration as shown in Scheme 81[541].

The amino sugar daunosamine, a constituent of the important antitumour drug, adriamycin, was the target of two synthetic approaches. In one, the isoxazolidine **396**, derived from a nitrone based on tartaric acid and ethyl vinyl ether, was elaborated to daunosamine in high yield (Scheme 82)[542].

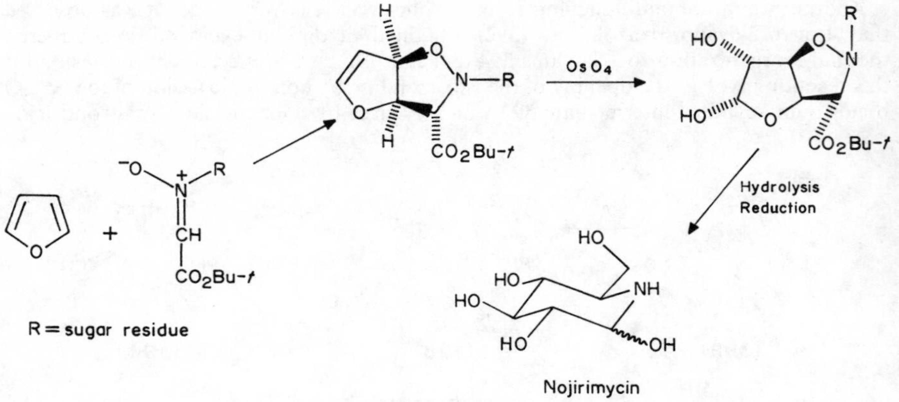

SCHEME 81

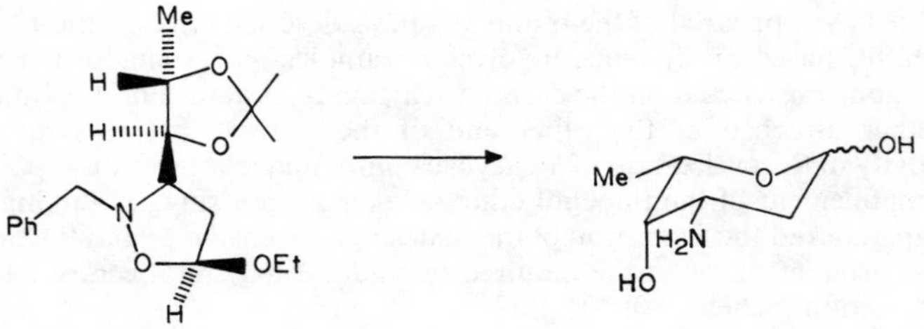

SCHEME 82

Another total synthesis of daunosamine has been achieved by intramolecular cyclization of nitrone **397** (derived from *trans*-propenyl glyoxylate and the chiral *N*-α-methylbenzylhydroxylamine) giving predominantly the bicyclic lactone **398** which was elaborated to daunosamine (Scheme 83)[543].

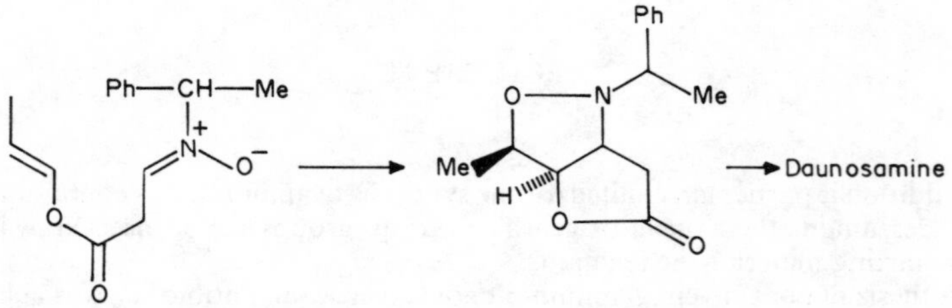

SCHEME 83

Further syntheses of pyrrolizidine alkaloids—*dl*-retronecine[544] and croalbinecine[545]—have been accomplished by adaptations of the methodology shown in

Scheme 11. Utilization of the dipolar cycloaddition to achieve (eventually) α,α'-dialkylation of 1-pyrroline *N*-oxide (**32**) for the synthesis of solenopsis ant venoms (e.g. **399**) is displayed in Scheme 84[546].

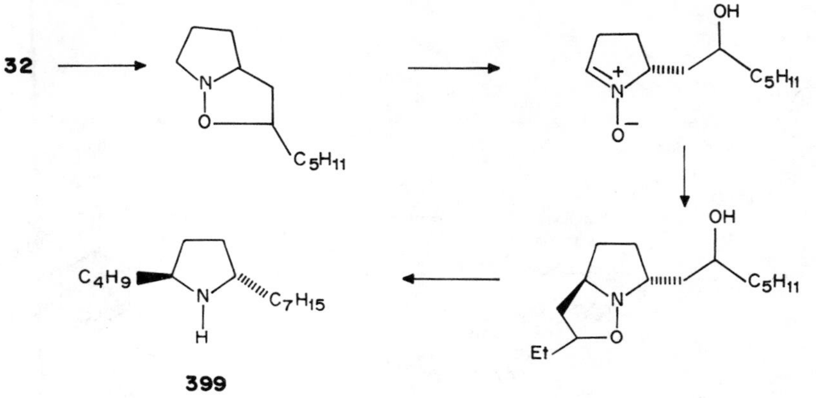

32

C_5H_{11} OH C_4H_9 C_7H_{15} Et

399

SCHEME 84

Nitrone **32** also served as a starting material in the synthesis of anatoxin-a, the 'very fast death factor'[547]. In this work **32** was reacted with 1,3-hexadien-5-ol to give **400**, which was oxidized to a new nitrone that underwent spontaneous intramolecular cycloaddition to **401**, which in turn could be converted to the target molecule (Scheme 85)[547].

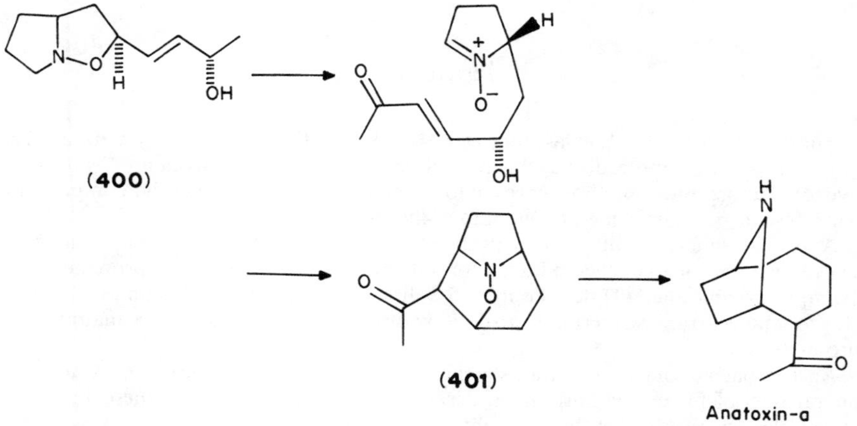

(400) **(401)** Anatoxin-a

SCHEME 85

A number of strategies for the synthesis of alkaloids having nitrogen in a bridgehead position have been described. Examples of these compounds are the indolizidine, quinolizidine, phenanthroindolizidine and phenanthroquinolizidine alkaloids. One approach is represented by the synthesis of a group of elaeocarpus alkaloids (e.g. elaeocarpine, **402**) by elaborating the β-aminoketone **403**, obtained from the cycloadduct of nitrone **32** with a styrene (Scheme 86)[548]. An analogous approach was employed for the synthesis of the pentacyclic alkaloids, tylophorine, cryptopleurine, septicine and julandine[549].

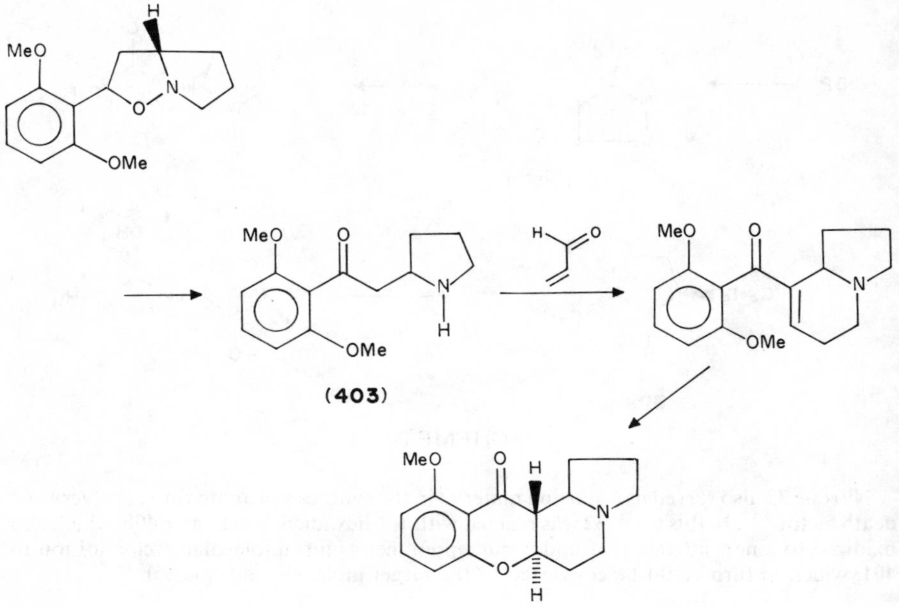

(403)

(402)

SCHEME 86

The quinolizidine alkaloid lasubine II (**404**) was recently synthesized by a quite different type of route. Intramolecular cyclization of nitrone **405**, derived from an *N*-homoallylic hydroxylamine and 4-methoxycarbonylbutyraldehyde gave the bicyclic compound **406** which could be transformed to the target **404** (Scheme 87)[550].

A totally different entry to bridgehead nitrogen compounds is provided by the rearrangement of the cycloadducts of nitrones to methylenecyclopropane[551]. For example, cycloadduct **407** derived from 5,5-dimethylpyrroline 1-oxide and methylenecyclopropane rearranges thermally to indolizidinone **408**; there are also enaminone by-products[551].

An approach to the total synthesis of the indole derivatives, clavine alkaloids, also based on nitrone–olefin cycloadditions, is described[552]. The key step in these involves an intramolecular reaction of the transient nitrone **409** as demonstrated by the synthesis of (±)-chanoclavin (Scheme 88)[552].

Representatives of another class of indole alkaloids, the Aristotelia alkaloids (−)-hobartine and (+)-aristoteline, have been synthesized from indole and (*S*)-1-*p*-menthen-8-ylamine via nitrone **410**, which underwent intramolecular cycloaddition. Further elaboration of the cycloadduct resulted in the two alkaloids indicated (Scheme 89)[553].

Another class of natural products that was the target of such a synthetic approach is that of the guanidine-containing natural products, displaying a broad spectrum of biological activity, represented by (−)-ptilocaulin[554]. One of the key steps in this synthesis is also an

SCHEME 87

SCHEME 88

intramolecular dipolar nitrone–olefin cycloaddition, followed by cleavage of the N—O bond of the isoxazolidine **411** and construction of the cyclic guanidine ring (Scheme 90)[554].

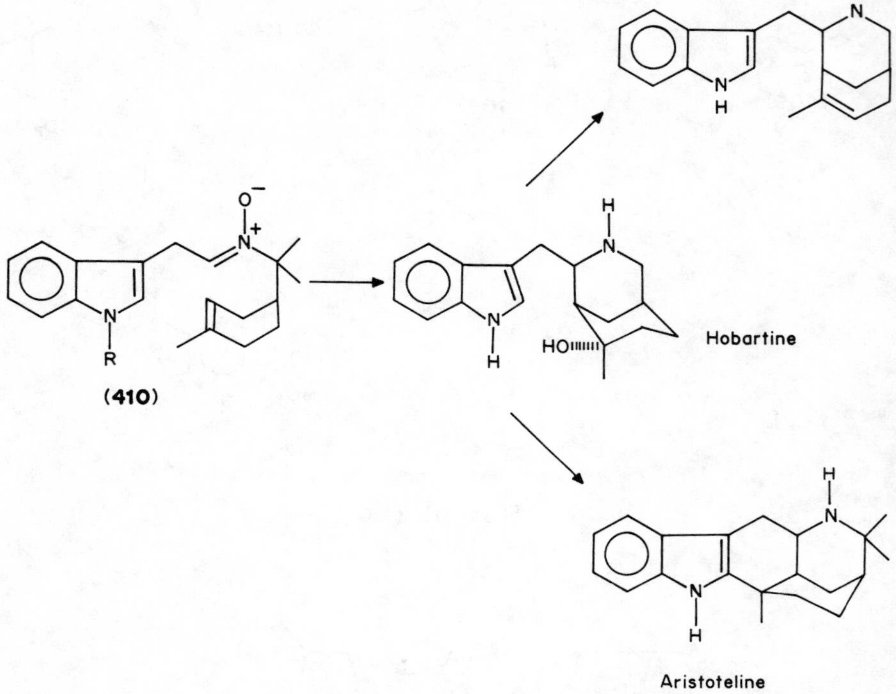

(410)

Hobartine

Aristoteline

SCHEME 89

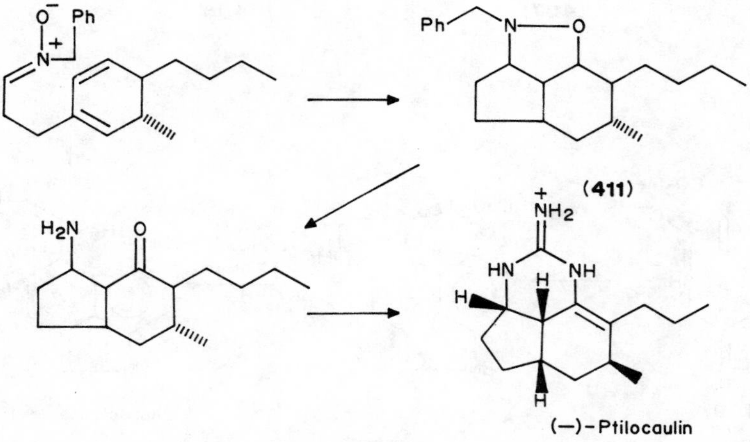

(411)

(−)-Ptilocaulin

SCHEME 90

An extensive study was carried out to examine the behaviour of nitrones derived from cyclic ketones having double-bond-containing side-chains[555]. 13 different systems derived from cyclohexanone and cyclopentanone with various substituents and side-chains were examined and found to yield bridged and fused polycarbocyclics. The exploitation of this methodology resulted in total syntheses of the sesquiterpenes (±)-secoishwaran-12-ol, (±)-hirsutene and (±)-coriolin[555].

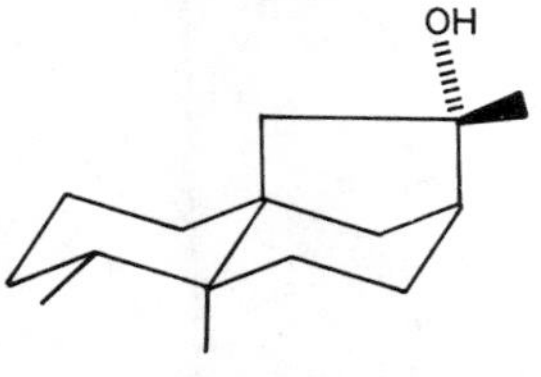

(±)-Secoishwaran-12-ol 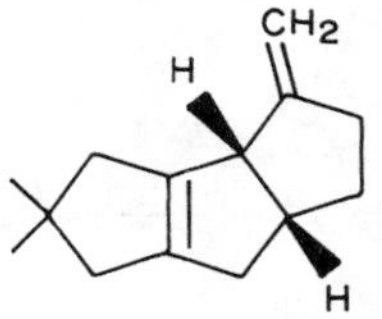 (±)-Hirsutene 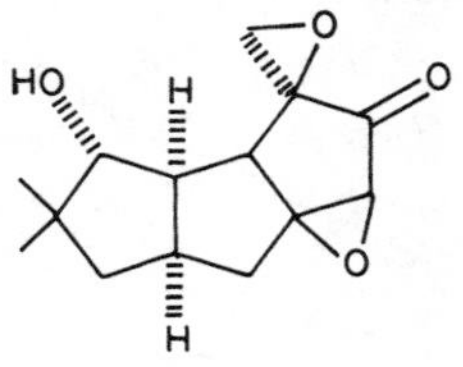 (±)-Coriolin

The nitrone cycloaddition was also used as a means of introducing the chiral centre to position 25 in the side-chain of trihydroxycholecalciferol. Reaction of nitrone **412** with methyl methacrylate gave isoxazolidine **413** as the major product, which was transformed via ketone **414** to trihydroxycholecalciferol (Scheme 91)[556].

A new approach to prostanoid synthesis was recently presented. It was found that nitrones derived from γ,δ-unsaturated aldehydes (of type **415**) undergo intramolecular cycloaddition to give the oxaza[2.2.1]bicycloheptane system **416** which is an analogue of prostaglandin H, and may serve as starting material for other prostaglandin systems[557].

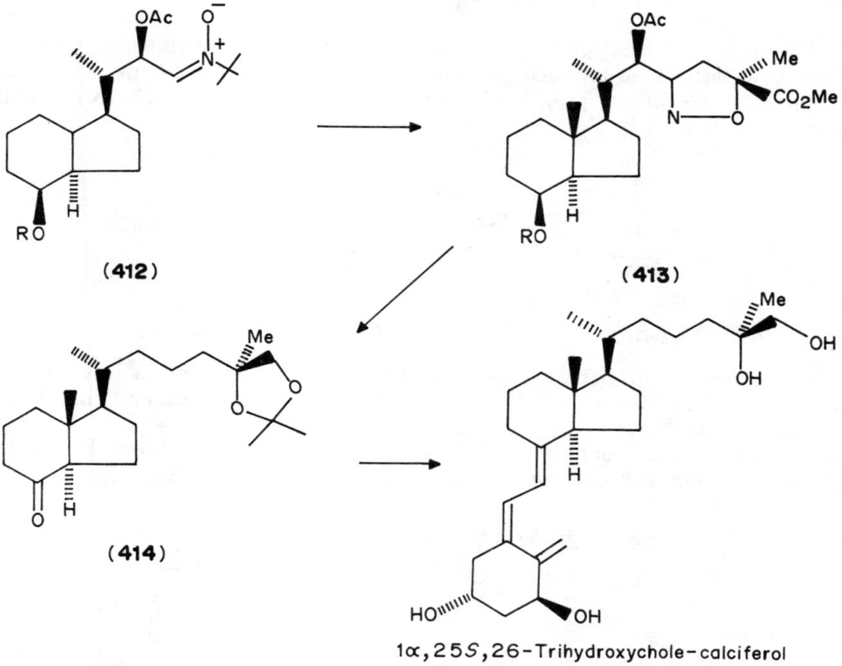

(**412**) (**413**)

(**414**)

1α,25*S*,26-Trihydroxychole-calciferol

SCHEME 91

Nitrones have been shown to react with vinylphosphonates to form 4-phosphonoisoxazolidines[558]. The cycloaddition of nitrones to vinylsilanes is suggested as a method for the homologation of aldehydes to α,β-unsaturated aldehydes, an alternative to the Wittig reaction. The 5-trimethylsilylisoxazolidines (**417**), which are the primary products in this cycloaddition, undergo ring opening followed by elimination, to give the final products in high yields, upon treatment by HF in acetonitrile at room temperature for one minute (Scheme 92)[559].

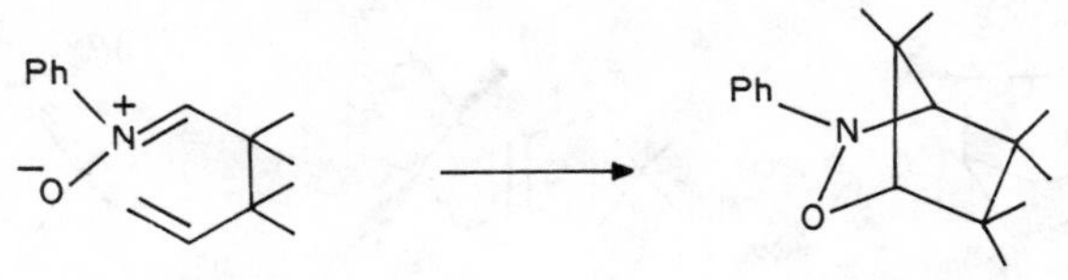

(**415**) (**416**)

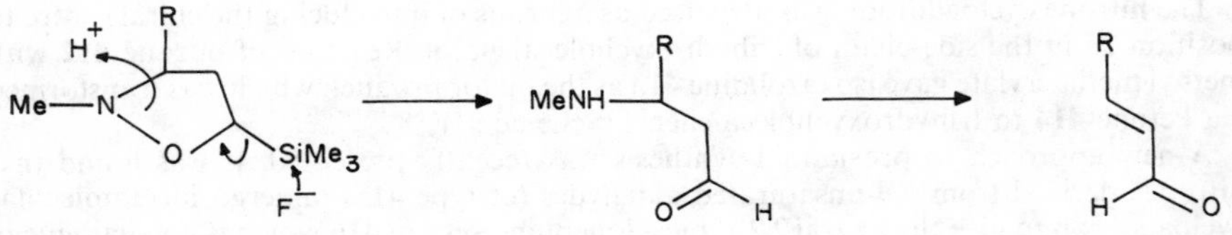

(**417**)

SCHEME 92

A different type of behaviour is seen in the reaction of nitrones with dibenzoylacetylene. In this case the acyclic products 3-anilino-1,4-diphenylbutane-1,2,4-trione (**419**) and aldehyde **420** were formed, presumably by fragmentation of the isoxazoline **418** initially obtained[560].

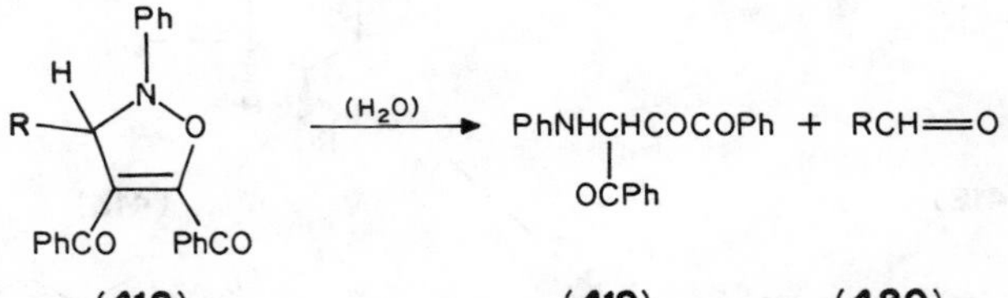

(**418**) $\xrightarrow{(H_2O)}$ PhNHCH(COPh)COCOPh (**419**) + RCH=O (**420**)

The reaction of nitrones with α,β-unsaturated azo compounds has been found to proceed selectively at the C=C (and not the N=N) double bond, to yield 5-azoisoxazolidines[561].

Nitrones have been found to react with tetracyanoethylene at the C≡N bond rather than the C=bond with the formation of Δ^4-1,2,4-oxadiazolines (**421**)[491b].

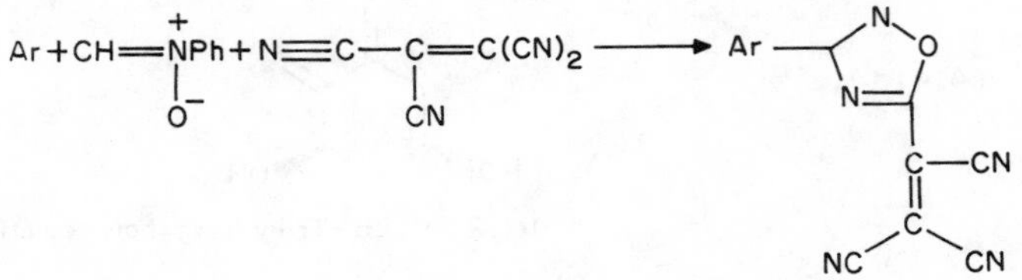

$$\text{Ar}\,\text{CH}{=}\overset{+}{\text{N}}(\text{O}^-)\text{Ph} + \text{N}{\equiv}\text{C}{-}\text{C}(\text{CN}){=}\text{C}(\text{CN})_2 \longrightarrow$$

(**421**)

The reactions of aldonitrones with isocyanides are rather complicated[562]. Under the influence of boron trifluoride etherate 4-imino-1,2-oxazetidines (**422**) and 4-imidazolidinones (**423**) predominantly are formed. Further work has revealed that the four-membered ring products **422** can undergo ring expansion to the five-membered **423**. In certain cases a rarer type of product a derivative of 1,2,4-oxadiazolidi-5-one (**424**) is formed[562].

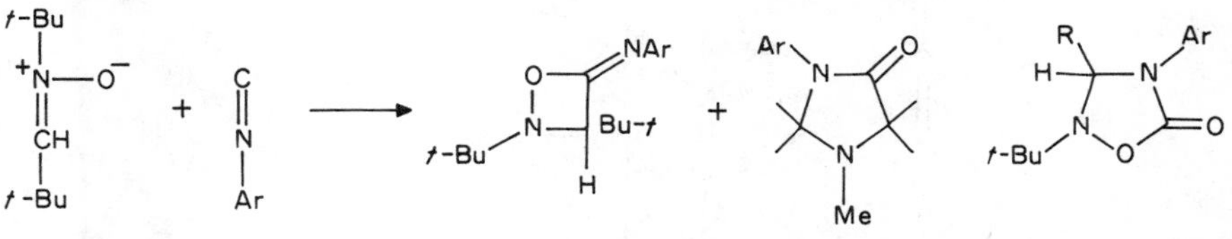

(**422**) (**423**) (**424**)

b. Cycloaddition to other 1,3-dipoles.* The reaction of 2,3-dihydro-2,2,4-trimethyl-1*H*-1,5-benzodiazepine (425**) with *C*-benzoyl *N*-phenyl nitrone (**426**) is rather unusual[563]. The formation of the final product, which was identified by X-ray crystallography as a pyrrolobenzodiazepinone derivative (**427**), was explained by a multistep route (Scheme 93). In the first step the nitrone acts as an oxygen donor (see Section II.B.4.d) and converts **425** to the corresponding *N*-oxide (partial structure **428**) which was assumed to cycloadd to another molecule of unreacted nitrone **426** to give **429**, which in turn was suggested to rearrange via **430** to the final product. Although the structure of the final product **427** seems to have been solidly established, the mechanism depicted in Scheme 93 is purely speculative[563].

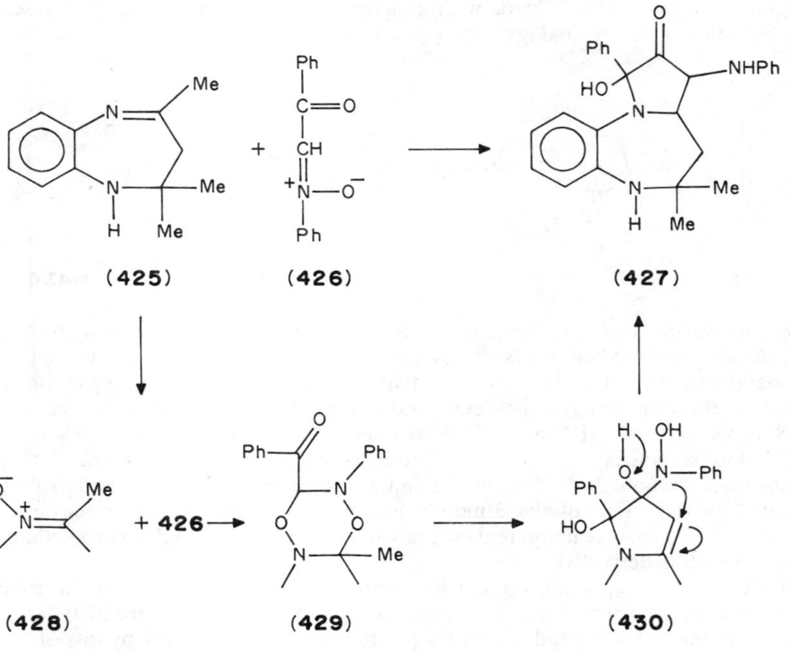

(**425**) (**426**) (**427**)

(**428**) (**429**) (**430**)

SCHEME 93

Another unusual observation is the formation of 1,2,4-Δ^2-triazolines in the reaction of aldonitrones with the 1,3-dipolar *C*-acetylnitrile imine **431**[564]. From the product **432** it is clear that in some stage during the reaction deoxygenation of the nitrone took place, followed by an apparent 1,3-dipolar cycladdition between **431** and the azomethine.

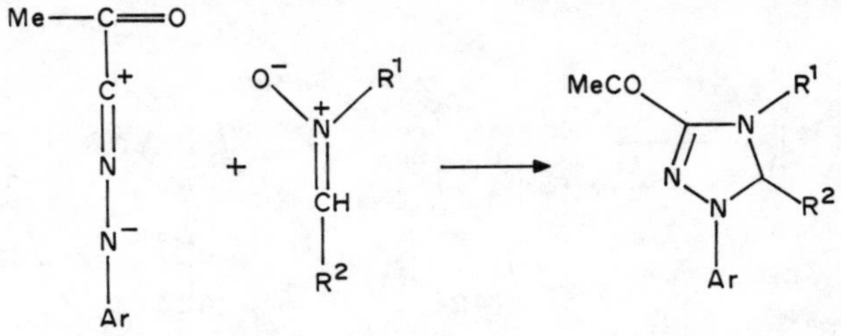

(431) (432)

*6. Reactions of nitrones with heterocumulenes

**a. Ketenes.* Further details have been published on the reactions of ketenes with nitrones derived from 9-fluorenone, reporting results similar to those in Scheme 26[565]. The subject has also been reviewed[566].

b. Ketene imines.* Results in a further report on the reaction of nitrones with ketene imines largely conform to those described earlier (see Scheme 27 and structures **221–224). One novel result is the formation of spiranic and fused products **433** and **434** in the reaction of 5,5-dimethylpyrroline *N*-oxide with dimethylketene *N*-phenylimine[567]. These results may be rationalized by analogy with known ketene chemistry[567].

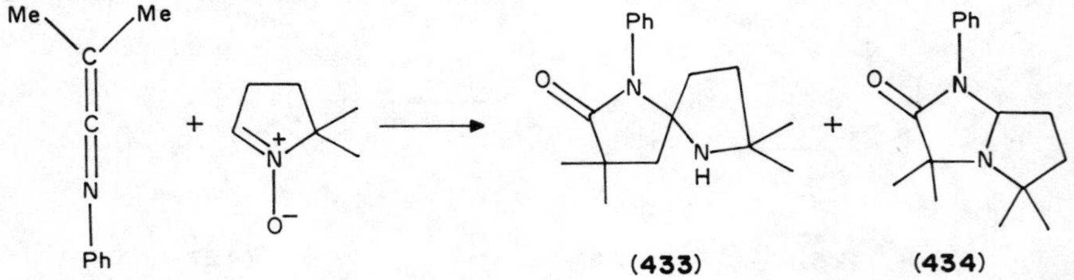

(433) (434)

**c. Isocyanates and isothiocyanates.* Reactions of dinitrones derived from glyoxal and *N*-alkylhydroxylamines with isocyanates have been shown to give either monoadducts, oxadiazolidinones see equation 46), or bis-adducts, depending on the nature of the nitrone's *N*-substituent and the isocyanate[568]. When heated with aryl isocyanates, α-(3-indolyl-*N*-aryl(alkyl) nitrones are converted into *N*-aryl-*N*-*aryl*(*alkyl*)-*N*-(3-indolyl)amidines[569]. The latter products are presumably formed through the extrusion of carbon dioxide from the initially obtained, expected oxadiazolidinones, followed by migration of the 3-indolyl group from carbon to the electron-deficient nitrogen[569]. This new reaction makes it possible to transform 3-indolecarboxaldehyde to a 3-aminoindole derivative.

Further papers reporting on the interaction of isothiocyanates with nitrones and dinitrones, derived from glyoxal, have appeared[570,571]. The products are 1,2,4-oxadiazolidine-5-thiones and secondary products formed from these by loss of COS and further reaction of the amidine with isothiocyanate. In view of this previously reported conversion of nitrones to thioamides by isothiocyanates (equation 47) it is relevant to mention that a similar conversion could be effected by the use of 1,1-thiocarbonyldiimidazole[572].

*7. *Reactions of nitrones with electrophiles*

**a. Nitrones as oxygen nucleophiles.* The reaction of a 1,3-oxazoline with *N*-phenylbenzimidoyl chloride has been shown to give substitution of the nitrone in the β-position (Scheme 94). This approach is suggested as a new general synthesis of substituted α-amino acids[573].

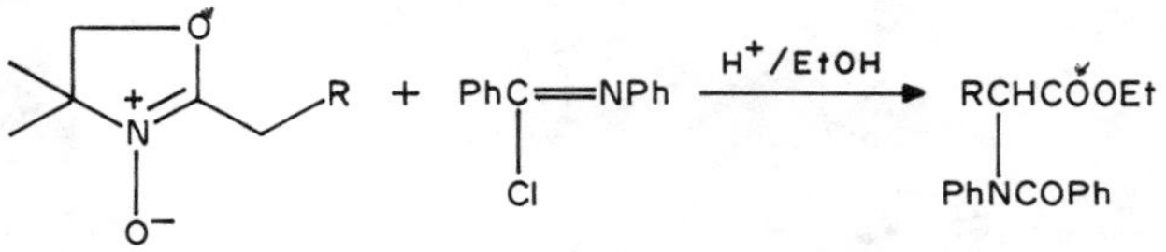

SCHEME 94

N-Carboxyalkyl nitrones are converted by the action of acyl halides to 2-oxazolin-5-ones (azlactones)[574]. This reaction may be rationalized either by assuming a rearrangement of the nitrone to an amide, followed by cyclization (route a) or by tautomerization of the nitrone to *N*-hydroxyoxazolidinone followed by elimination (route b, Scheme 95)[574]. This reaction is reminiscent of the acid-catalysed conversion of *N*-β-oxoalkyl nitrones to oxazoles[575].

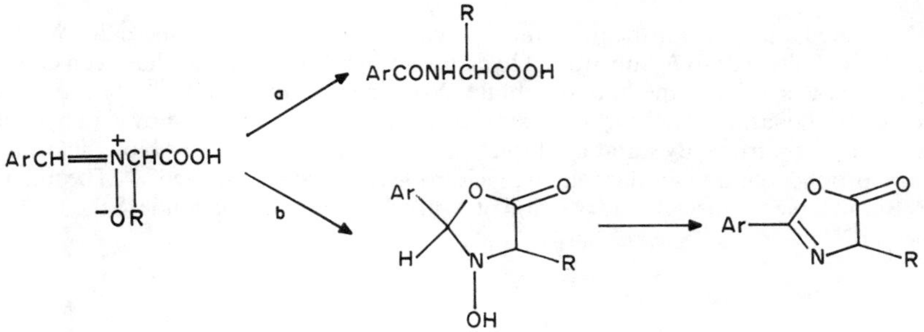

SCHEME 95

Full details have appeared on the conversion of *N*-aryl nitrones to 2-arylbenzothiazoles by phenylphosphonothioic dichloride (equation 53)[576]. Treatment of *N*-alkyl nitrones with the same reagent, or with its related analogues, diphenylphosphinothioic chloride or thiophosphoryl trichloride, gives amides and thioamides[576]. The full paper of the related conversion of nitrones to 2-arylbenzoxazoles (equation 54) by *O*-methyl diphenylphosphinothioate has also been published[577].

**b. Nitrones as carbon nucleophiles.* The carbanion derived from 1,2,2,4,5,5-tetramethyl-3-imidazolin-3-oxide by phenyllithium treatment has been added to various electrophiles (e.g. amyl nitrite, benzaldehyde, benzophenone and a number of carboxylates) to afford the expected products[464].

The unusual result that the reaction of *C*-aroyl nitrones with α-arylidene-β-dicarbonyls such as benzylidene Meldrum's acid or benzylidenebarbituric acid give 2-aroyl-α-indolenines (**435**) in unreported yields was reported recently (Scheme 96)[578]. The mechanism of this reaction is not known, but it is clear from the end-product that the elements of arylaldehyde (from the arylidene group and the nitrone oxygen) are lost in the reaction. The bond formed at the *ortho* position of the nitrone's *N*-aryl ring is not without precedent (e.g. Schemes 24, 25, 27, 31 and 35, and equations 52–54).

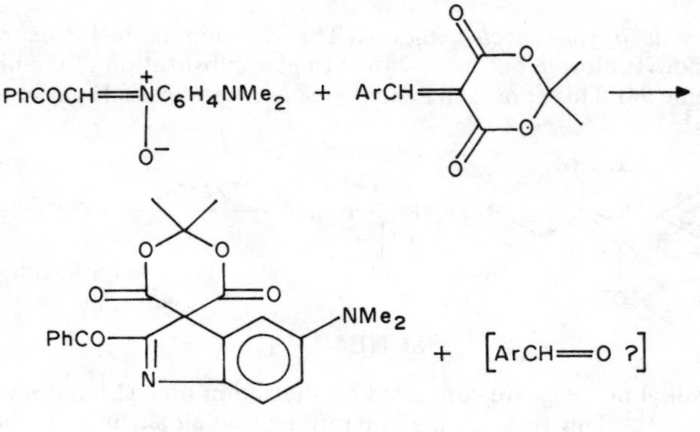

$$PhCOCH{=}\overset{+}{N}(O^-)C_6H_4NMe_2 + ArCH{=}\text{(Meldrum's acid)} \longrightarrow$$

(435) + [ArCH=O ?]

SCHEME 96

8. Reactions of nitrones with nucleophiles

As pointed out previously the simplest reaction belonging to this category is the hydrolysis of nitrones to *N*-substituted hydroxylamines. This principle has been exploited for a general synthetic method to obtain *N*-hydroxyamino acids[579]. The procedure involves *N*-alkylation of (*Z*)-furfural oxime by an α-halo acid, followed by acid hydrolysis of the resulting nitrone. By suitable adaptation of this methodology, the biologically active natural products emimicyn (**436**) and hadacidin (**437**) can be synthesized[579]. The full paper on the imidazole synthesis in the reaction of *N*-alkyl nitrones with cyanide (Scheme 37) has appeared[580].

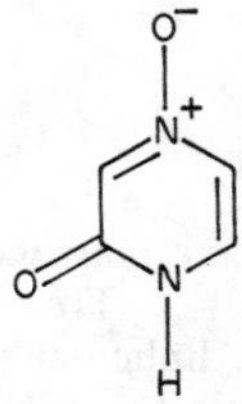

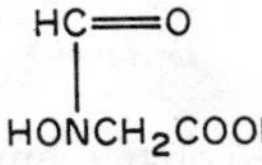

(436) **(437)** $HC{=}O$ / $HONCH_2COOH$

The reactions of trimethylsilyl cyanide with *C,N*-diaryl nitrones give the expected α-cyano-*O*-trimethylsilyl products (**438**)[581]. Thermal decomposition of these gives a variety of products—azoxybenzene, diarylsuccinonitrile, iminonitriles and benzanilides[581]. Compounds of type **438** can be utilized for the blocking of the nitrone function; their treatment with silver fluoride regenerates the original nitrone[582]. When the addition of trimethylsilyl cyanide to nitrone is carried out in the presence of a base (triethylamine) α-

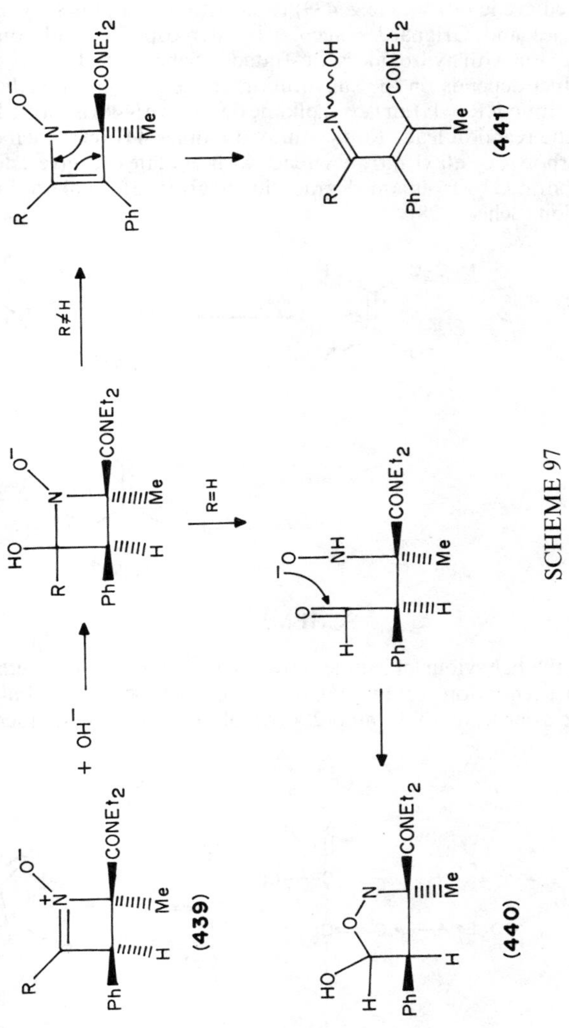

SCHEME 97

iminonitriles are formed in high yields through the elimination of Me_3SiOH (cf. equation 64)[583].

$$ArCH{=}\overset{+}{N}(O^-)Ph + Me_3SiCN \longrightarrow ArCH(CN)-N(OSiMe_3)Ph$$

Four-membered cyclic nitrones (e.g. **439**) react with nucleophiles (cyanide, hydroxide, alkoxides, hydrides and Grignard reagents) by stereospecific addition to the C=N bond[584]. The reaction with hydroxide is illustrated in Scheme 97. In this case the structure of the final product depends on the substitution at the α-position of the nitrone. If the nitrone is unsubstituted (R = H) an isoxazolidine derivative (**440**) is formed. In cases where R = Me or Ph, the reaction leads to unsaturated oximes **441** (cf. Scheme 60)[584].

C-Methoxycarbonyl *N*-alkyl nitrones react with enolates to give addition products which can be elaborated by Hofmann degradation to give the formal products of a directed aldol condensation (Scheme 98)[585].

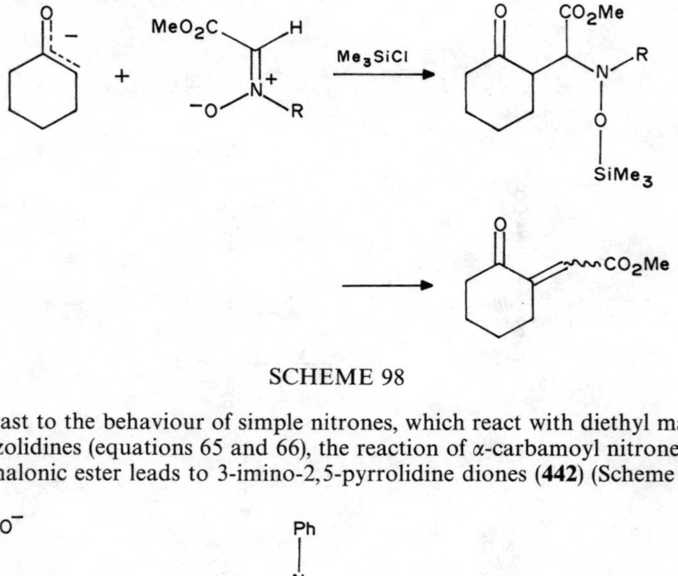

MeO2C, H, N+, −O, R, Me3SiCl, O, CO2Me, R, N, O, SiMe3, O, CO2Me

SCHEME 98

In contrast to the behaviour of simple nitrones, which react with diethyl malonate to give isoxazolidines (equations 65 and 66), the reaction of α-carbamoyl nitrones with the anion of malonic ester leads to 3-imino-2,5-pyrrolidine diones (**442**) (Scheme 99)[586].

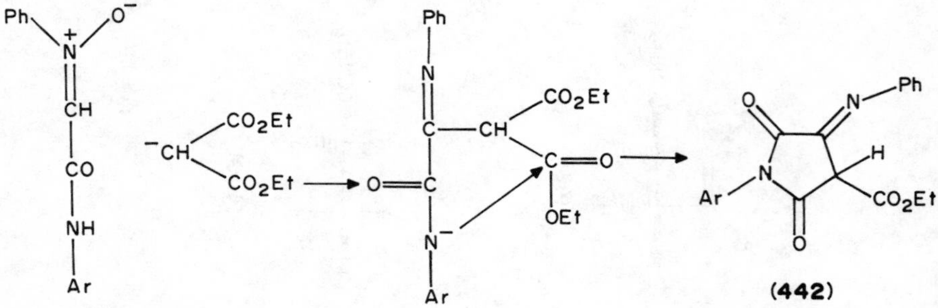

Ph, O−, N+, CH, CO, NH, Ar, −CH, CO2Et, CO2Et, Ph, N, C, CH, CO2Et, O=C, C=O, OEt, N−, Ar, O, N, Ph, H, Ar, N, CO2Et, O, **(442)**

SCHEME 99

The reaction of α-3-indolyl nitrones with the anion of malonic ester is unusual. The final product is a β-*N*-(3-indolyl)-*N*-aryl(alkyl)aminopropenoic acid (**443**). This indicates that in the course of the reaction C to N migration of the indole moiety has taken place. The mechanism suggested to account for this assumes the initial formation of the expected isoxazolidinone product (**444**) and its subsequent ring opening (Scheme 100)[587].

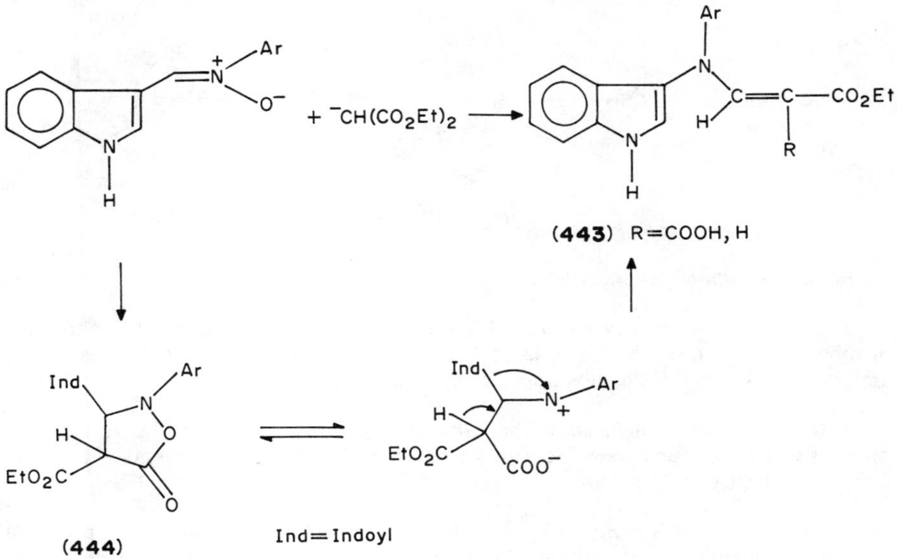

(**443**) R=COOH, H

(**444**)

Ind=Indoyl

SCHEME 100

As in the reactions of nitrones with α-phosphoryl carbanions (equation 68 and Scheme 39), carbanions derived from trimethylsilyl compounds give aziridines[588].

Nitrones have been found to be valuable starting materials for the synthesis of α-aminophosphonic acids, which are of considerable current interest. The addition of lithium diethylphosphite, and in cases where this did not react, of tris(trimethylsilyl) phosphite was shown to proceed with some diastereoselectivity and can be used as a synthetic route to these compounds[444,589].

N-2-(β-Indolyl)ethyl nitrones (e.g. **445**) have been shown to act as electrophiles in intramolecular Friedel–Crafts-like electrophilic substitution of the indole nucleus affording tetrahydrocarbolines[590].

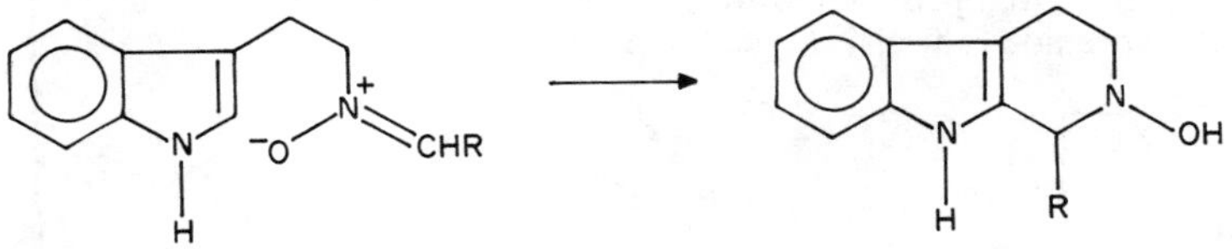

(**445**)

N-1-Cyanoalkyl *C*-phenyl nitrones react with thiocarboxylic acids to give thiazoles in 40–70% yields[591]. This reaction constitutes a relatively simple and convenient thiazole synthesis. Although the mechanism is not clear, the reaction probably involves a nucleophilic attack of the thiocarboxylate at the nitrone's α-position in the first step. In the ring-closure step the acyl group is transferred from the sulphur to the nitrogen by a mechanism which is not understood (Scheme 101)[591].

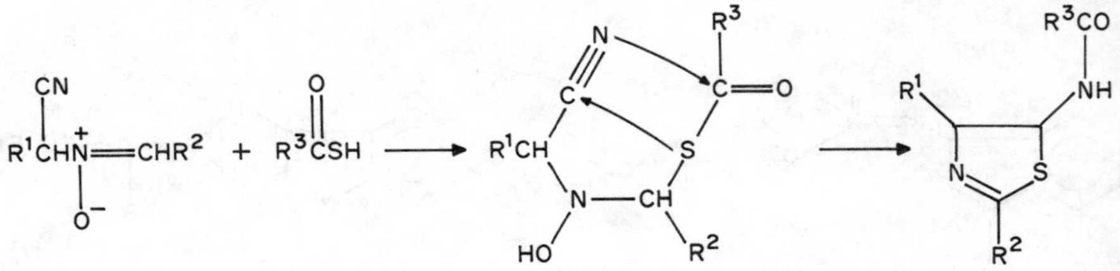

SCHEME 101

*9. *Reactions with organometallic compounds*

a. Organomagnesium compounds.* The Grignard reaction has been applied to four-membered cyclic nitrones (type **439)[584], to α,β-unsaturated nitrones[592a] and to α-aminonitrones[592b] to yield substituted hydroxylamines.

**b. Organocopper compounds.* The β-lactam synthesis based on the reaction of nitrones with copper acetylides (Scheme 42) has been extended to a variety of *C*-aryl and *C*-heteroaryl nitrones[593].

f. Boron derivatives of nitrones.* Boron derivatives of nitrones have been prepared from salicylaldehyde nitrones having seven-membered N—O—B dipolar heterocyclic structures with charged N^+ and B^- atoms[594]. On the other hand, six-membered N—O—B heterocyclic structures (also with dipolar structure) could be formed when nitrones were made from *o*-formylbenzeneboronic acid[595]. The crystal structures of some boron derivatives of nitrones (328–330**) have already been mentioned[450].

*10. *Reactions of nitrones with free radicals*

The spin trapping of *t*-butylperoxy radicals and tetralylperoxy radicals by *C*-phenyl *N*-*t*-butyl nitrone and *C*-methyl *N*-duryl nitrone was investigated. The peroxy radicals were generated by the alkoxy-radical-induced decomposition of hydroperoxides. The spin adducts of the alkoxy radicals could clearly be distinguished from the adducts of peroxy radicals[596]. A number of special nitrones of increased lipophilicity (5,5-dipropyl-1-pyrroline 1-oxide, 5-butyl-5-methyl-1-pyrroline 1-oxide and 2-aza-2-cyclopentenespirocyclopentane (**446**)) were designed for intracellular trapping of superoxide and hydroxide radicals[597].

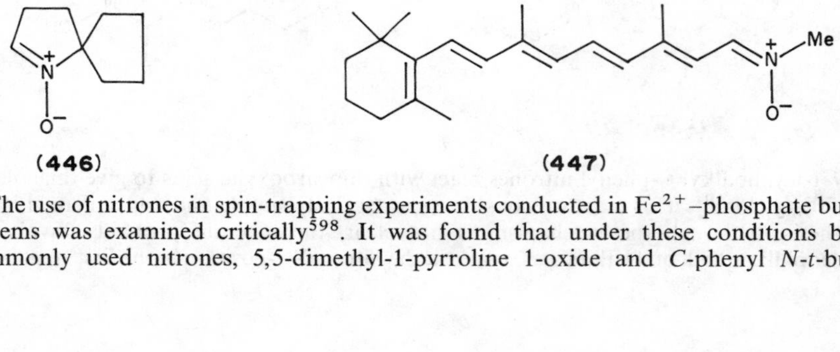

(446) **(447)**

The use of nitrones in spin-trapping experiments conducted in Fe^{2+}–phosphate buffer systems was examined critically[598]. It was found that under these conditions both commonly used nitrones, 5,5-dimethyl-1-pyrroline 1-oxide and *C*-phenyl *N*-*t*-butyl nitrone, generate spin adducts without added substrate. It was suggested that decomposition of the nitrones is the source for these radicals[598]. The spin-trapping technique was applied to the detection of free radicals in NADPH-dependent lipid peroxidation of rat ischaemic brain homogenate[599].

The question of the applicability of nitrones to serve as antioxidants in polymer stabilization was recently studied[600].

C. Biological Aspects

A series of retinoids containing the nitrone function (e.g. **447**) have been found to be effective in reversing the keratinization in hamster tracheal organ cultures[601]. Nitrone **447** also showed inhibitory activity against platelet aggregation induced by ADP or collagen and potentiates the submaximal platelet aggregation induced by a low concentration of arachidonic acid[601].

D. Synthesis of Nitrones

The following new section consists mainly of a survey of recent improvements in synthetic methods and some applications for the syntheses of special nitrones. The reader is directed to earlier sources[1–9] for a general review of the subject of nitrone synthesis.

1. Oxidative methods

a. Oxidation of secondary amines. A series of *C*-aryl *N*-*t*-butyl nitrones were synthesized by hydrogen peroxide oxidation of substituted benzyl-*t*-butylamines catalysed by sodium tungstate[602]. Subsequently this method was applied to purely aliphatic secondary amines, both acyclic (diisopropylamine) and cyclic (piperidine derivatives)[603]. The catalytic effect of other metals (V, Ti and Mo) has also been studied, but found to be inferior to tungsten[603]. An improved method for the preparation of *C, N*-diaryl nitrones is based on the oxidation of *N*-benzylanilines by *m*-chloroperbenzoic acid in acetone at low temperatures[604].

b. Oxidation of N,N-*dialkylhydroxylamines.* These are quite susceptible to oxidation and therefore even very mild reagents are capable of performing the task.

(i) Electrochemical oxidation. *N*-Hydroxy derivatives of cyclic amines such as piperidine, pyrrolidine etc. were oxidized to the corresponding nitrone electrochemically, using halide ions as mediators, in high yields[605].

(ii) Catalytic oxidation. The dehydrogenation of *N,N*-disubstituted hydroxylamines could be carried out by heating the starting material with palladium black. The yields were 60–90% and the reaction is applicable to dialkylhydroxylamines (cyclic and acyclic) as well as *N*-benzyl-*N*-alkylhydroxylamines[606].

(iii) Oxidation by peroxides. Seven-membered cyclic nitrones of the dibenzazepine system were prepared by hydrogen peroxide oxidation of the corresponding *N*-benzyloxyamines in acetic acid[607].

(iv) Oxidation by quinones. Dichlorodicyanobenzoquinone was used for the preparation of nitrone **448** from the corresponding *N*-hydroxy-β-carboline[495]. An *N*-glycosyl-*N*-hydroxy-α-aminophosphonate was oxidized by *p*-benzoquinone to the corresponding nitrone, which was trapped *in situ* by ethylene[531].

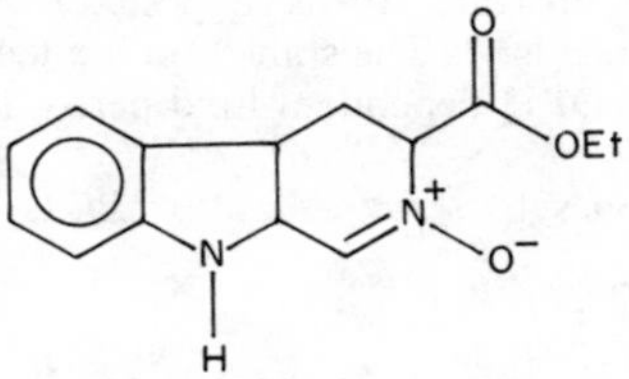

(448)

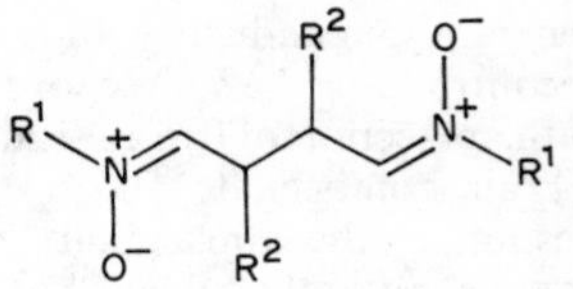

(449)

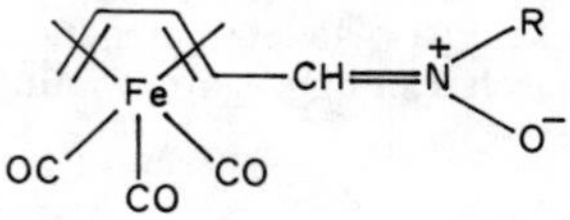

(450)

(v) Oxidation by metal oxides. *N,N*-Diethylhydroxylamine was oxidized to *C*-methyl *N*-ethyl nitrone, in quantitative yield, by silver oxide in ether[608]. 1-Hydroxyazetidines could be oxidized to four-membered cyclic nitrones by yellow mercuric oxide in dichloromethane[609]. Subsequently active lead(IV)oxide was found to be milder and better suited for the preparation of such sensitive and unstable nitrones[610,611]. This reagent was also used to oxidize acyclic dialkylhydroxylamines; however, in some cases the initially obtained nitrones underwent further oxidation to vinylaminyl oxide radicals, which in turn dimerized to 1,4-dinitrones of type **449**[612].

c. Oxidation of imines. The reaction of *N*-methylhydroxylamine-*O*-sulphonic acid has been found to be a mild method for the preparation of *N*-methyl nitrones from imines[613]. The original nitrogen of the imine along with its substituent is replaced by the nitrogen and the methyl group of the reagent. Thus the reaction of fluorenylideneanil gives fluorenone methyl nitrone. The yields are reportedly high[613]. The anil of perfluorobenzophenone could not be oxidized by peracetic acid; however treatment by trifluoroperacetic acid yielded a series of desired polyfluoroaryl nitrones[459]. The imine–peroxy acid reaction may lead either to oxaziridines or to nitrones. It was found that both steric and mesomeric effects have marked influence on the course of the reaction[614].

2. Condensation reactions of N-substituted hydroxylamines

A convenient 'one-pot' method was recently reported for the preparation of *N*-substituted hydroxylamines by the borane/tetrahydrofuran reduction of nitroalkenes[615]. Alternatively, pyridine–borane in ethanol was also shown to be applicable for the same kind of reduction[616].

a. Condensation with aldehydes and ketones. This is one of the most widely used synthetic entries to nitrones and has been used for the preparation of a variety of special nitrones, e.g. *N*-cyclohexyl chloral nitrone[616], the butadiene–irontricarbonyl-derived nitrone **450**[617], *C*-cyclopropyl *N*-methyl nitrone[499], epoxynitrone **203**[232] and *N*-benzyl α-ethoxycarbonyl nitrone derived from ethyl glyoxalate and *N*-benzylhydroxylamine[618].

The preparation of *C*-unsubstituted nitrones requires the condensation of hydroxylamines with formaldehyde. Such syntheses can take place either by heating paraformaldehyde with an N-alkylhydroxylamine[518,526,619] or by the introduction of gaseous formaldehyde, prepared by heating paraformaldehyde, into the solution of N-alkylhydroxylamine[524].

A modification of this reaction, using N-methyl-N,O-bis(trimethylsilyl)hydroxylamine is reported to be extremely efficient[620]. This reagent is prepared from N-methylhydroxylamine hydrochloride and chlorotrimethylsilane at room temperature in the presence of triethylamine in ether and reacts with carbonyl compounds under mild conditions giving nitrones in high yields, along with the volatile by-product hexamethyldisiloxane[620].

It was found possible to avoid the somewhat undesirable step of isolating N-substituted hydroxylamine, by carrying out the reduction of the nitro compound in the presence of an aldehyde (by zinc/ammonium chloride for aromatic nitro compounds[580] and by activated zinc dust/acetic acid in ethanol for 2-methyl-2-nitropropane[621]) and obtaining the nitrone in one step.

A nitrone could also be obtained by this approach from ethyl acetoacetate and N-cyclohexylhydroxylamine[465]. In contrast, the analogous nitrones derived from N-methyl- and N-benzyl-hydroxylamine proved to be too reactive to be isolated and reacted further to give pyrroles as secondary reaction products[465].

b. Condensation with acetylenes. A study of the reaction of hydroxylamines with a variety of acetylenes has been carried out[622]. The nitrones initially formed in this reaction are usually too reactive to be isolated and they react further with a second molecule of acetylene (Scheme 102). However, in some cases (e.g. the reaction of N-methylhydroxylamine and phenylacetylene) a mixture of nitrone and isoxazoline was isolated[622]. When acetylenes having an additional double bond were used, the nitrones formed were trapped intramolecularly[622].

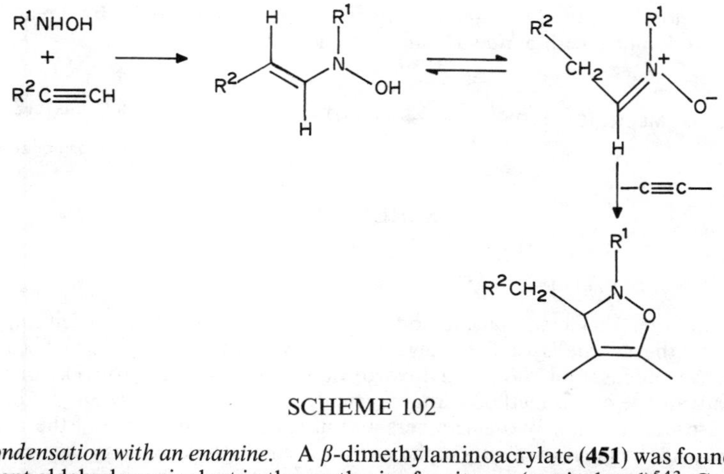

R^1NHOH + $R^2C{\equiv}CH$ → $R^2CH{=}CH{-}N(R^1)OH$ ⇌ $R^2CH_2{-}CH{=}N^+(R^1){-}O^-$ —($-C{\equiv}C-$)→ isoxazoline (R^1 on N, R^2CH_2 at C-3)

SCHEME 102

c. Condensation with an enamine. A β-dimethylaminoacrylate (**451**) was found to be a convenient aldehyde equivalent in the synthesis of a nitrone (not isolated)[543]. Compound **451** was prepared by heating propenyl acetate with bis-dimethylamino-t-butoxymethane. The nitrone underwent *in situ* intramolecular cycloaddition to the double bond in the molecule (see Scheme 83).

d. Intramolecular ketone alkylhydroxylamine condensations. The reduction of γ-nitroketones has been used recently for the preparation of some bicyclic[623] and bridgehead hydroxylated[624] pyrroline *N*-oxide derivatives. Improved conditions for such reductions have been developed, using activated zinc dust in ethanol and glacial acetic acid[625].

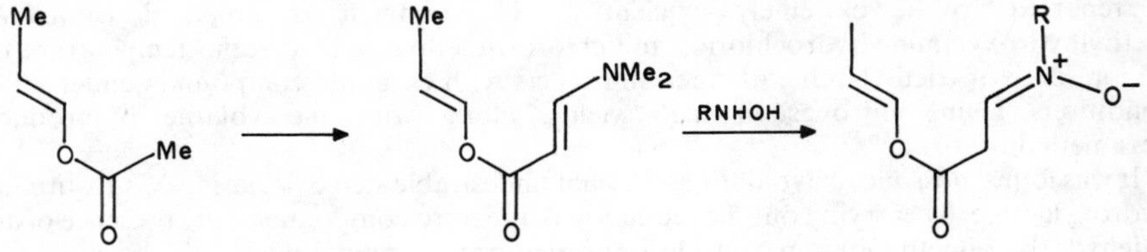

(451)

e. Condensation with an orthoester or amide acetal. Treatment of 2-hydroxylamino alcohols with orthoesters[498a] or with amide acetals[498b] gave cleanly and efficiently oxazoline *N*-oxides, which were previously accessible from the isomeric oxaziridines in low yields (e.g. Scheme 103).

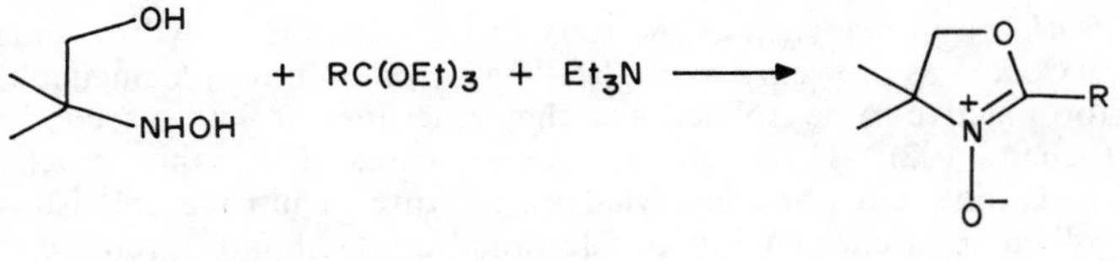

$$\text{HOCH}_2\text{C(Me)}_2\text{NHOH} + \text{RC(OEt)}_3 + \text{Et}_3\text{N} \longrightarrow \text{oxazoline } N\text{-oxide}$$

SCHEME 103

The analogous acyclic α-alkoxynitrones were prepared by reacting *N*-alkylhydroxylamines with amide acetals (Scheme 104)[498b].

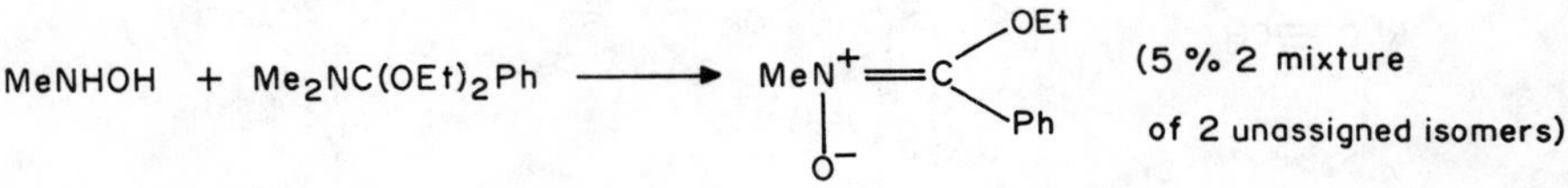

$$\text{MeNHOH} + \text{Me}_2\text{NC(OEt)}_2\text{Ph} \longrightarrow \text{MeN}^+(\text{O}^-){=}\text{C(OEt)Ph}$$

(5 % 2 mixture of 2 unassigned isomers)

SCHEME 104

3. N-*Alkylation of oximes*

This is another very common method for nitrone synthesis. It has been applied, for example, to the synthesis of *C*,*C*-dicyclopropyl *N*-methyl nitrone, which cannot be prepared by condensation of *N*-methylhydroxylamine with dicyclopropyl ketone[499]. The main disadvantage of this method is that it usually gives mixtures of *O*-alkylation and *N*-alkylation products, namely oxime ethers and nitrones. The outcome of the reaction is correlated with the stereochemistry of the starting oxime; (*Z*)-aldoximes give preferably *N*-alkylation, whereas (*E*)-oximes suffer *O*-alkylation[579]. There is some success in influencing the *N*-vs. *O*-alkylation ratio by reaction conditions. It has recently been found that when alkylation of chalcone oximes is carried out using methyl iodide in benzene aqueous sodium hydroxide using benzyltriphenylphosphonium bromide as phase-transfer catalyst, 60–80% *N*-methylation versus ca. 10% *O*-methylation is observed[441]. Another way to suppress *O*-alkylation has been devised by protecting the oxime as the *O*-trimethylsilyl derivative. The alkylation can be carried out using trimethyl- or triethyloxonium fluoroborate, or methyl or allyl trifluoromethanesulphonate (triflate)[626]. This method is equally applicable to ketoximes or to aldoximes. In the case of aldoximes, the usually pure (*E*)-*O*-trimethylsilyloximes are used and in rapid ice-cold work-up the presence of the less stable (*E*)-aldonitrones can be detected. However, purification always gives the more stable (*Z*)-aldonitrone[626].

$$\mathrm{PhCH{=}NOH + Me_3SiCH_2OSO_2CF_3} \longrightarrow \mathbf{(452)}$$

(452): H(Ph)C=N⁺(CH₂SiMe₃)OH

$$\mathbf{(452)} \xrightarrow{\mathrm{PhCH{=}O/CsF}} \mathrm{PhCH{=}\overset{+}{N}(O^-)CH_2CH(OH)Ph} \quad \mathbf{(453)}$$

$$\mathbf{(452)} \xrightarrow{\mathrm{CsF}} \mathrm{PhCH{=}\overset{+}{N}(OH)\overset{-}{C}H_2} \quad \mathbf{(454)}$$

The reaction of oximes with trimethylsilylmethyl triflate gives the iminium derivative **452**[627]. When this compound is treated with benzaldehyde, in the presence of caesium fluoride as the desilylating agent, nitrone **453** is formed. Treatment of **452** in the presence of a dipolarophile gives a cycloadduct indicating the intermediacy of an azomethine ylide **454**[627].

4. Nitrones from nitroso compounds

The best known reaction in this class is the Kröhnke reaction, which involves the conversion of benzyl halides into *N*-benzylpyridinium salts, and the reaction of the latter with *p*-nitroso-*N*,*N*-dimethylaniline to give *C*-aryl *N*-*p*-dimethylaminophenyl nitrones[278]. This reaction has now been extended to primary amines as possible starting materials[628]. Primary amines react with 4,6-diphenyl-2-(ethoxycarbonyl)pyrylium cation, to give the corresponding pyridinium cation, which can be reacted, after hydrolysis of the ester function, with *p*-dimethylaminonitrosobenzene to give the nitrone. Nitrosobenzenes may act as nucleophilic reagents at other electrophilic sites with the formation of nitrones. Thus reaction of nitrosoarenes with α-bromoheterocyclic carbonyl compounds (2-pyrazolin-5-one, isoxazolin-5-one or oxindole) lead to nitrones (e.g. **455**)[505b,629].

$$\text{3-bromooxindole} + \mathrm{ArN{=}O} \longrightarrow \mathbf{(455)}$$

(455): 3-(Ar–N⁺(O⁻)=)oxindole

The opposite kind of (namely electrophilic) behaviour is exhibited by nitrosoarenes in their reaction with silyl enol ethers, which leads to adducts **456**, which can be transformed to α-aroyl nitrones **457** by desilylation followed by silver oxide oxidation[630].

$$\underset{\displaystyle\text{OSiMe}_3}{\text{ArC}}{=}\text{CH}_2 + \text{PhN}{=}\text{O} \longrightarrow \underset{\displaystyle\text{O}}{\text{ArC}}\text{CH}_2\underset{\displaystyle\text{OSiMe}_3}{\text{NPh}} \longrightarrow \text{ArCOCH}{=}\underset{\displaystyle\text{O}^-}{\overset{+}{\text{N}}}\text{Ph}$$

(456) **(457)**

This synthesis provides an alternative to the Kröhnke reaction using haloketones. In the present method[630], electron-donating substituents on the aryl ring of the enolate promote the reaction, contrasting with the classical Kröhnke synthesis. Nitrogen analogues of α-aroyl nitrones, also called bifunctional 1,4-diazabutadiene nitrones (**458**), are available by the related, base-catalysed condensation of acetophenone anils (presumably via the enamines) with nitrosoarenes[631].

$$\text{Ar}^2\text{N}{=}\underset{\displaystyle\text{Ar}^1}{\text{C}}\text{Me} + \underset{\text{excess}}{\text{Ar}^3\text{N}{=}\text{O}} \longrightarrow \text{Ar}^2\text{N}{=}\underset{\displaystyle\text{Ar}^1}{\text{C}}\text{CH}{=}\underset{\displaystyle\text{O}^-}{\overset{+}{\text{N}}}\text{Ar}^3$$

(458)

The reaction of α-chloronitrosoalkanes with aliphatic and aromatic Grignard reagents has also been shown to lead to nitrones, usually in quite good yields (Scheme 105)[632]. By this method a series of nitrones derived from adamantanone as well as other aliphatic, cyclic and acyclic ketones could be prepared[632] (α-chloronitrosoalkanes are made by chlorination of ketoximes).

$$\text{R}^1_2\text{ClCN}{=}\text{O} \longrightarrow \text{R}^1_2\text{ClC}{-}\underset{\displaystyle\text{OMgCl}}{\text{N}}\text{R}^2 \longrightarrow \text{R}^1_2\text{C}{=}\underset{\displaystyle\text{O}^-}{\overset{+}{\text{N}}}\text{R}^2$$

SCHEME 105

5. *Miscellaneous*

Nitrocycloalkenes undergo cycloaddition to ynamines with the formation of four-membered cyclic nitrones (**459**) as the main products[482]. These nitrones undergo isomerization by ring opening to *C*-carbamoyl *N*-vinyl nitrones (**460**) upon standing in solution (Scheme 106).

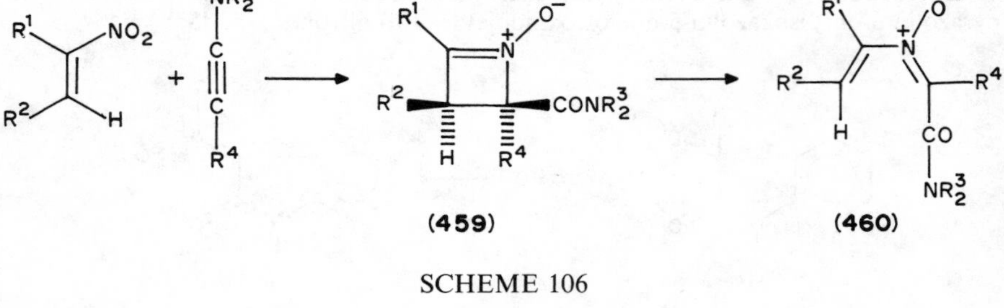

(459) **(460)**

SCHEME 106

*III. NITRONIC ACID DERIVATIVES

*A. Nitroalkane–Nitronic Acid Tautomerism

The subject of nitro-activated carbon acids, which are tautomers of nitronic acids, has been reviewed[633].

There are new opportunities to study the stereochemistry of the nitronate $\rightleftharpoons$ nitroalkane tautomerism in the transformation of the adducts of Grignard reagents of various aromatic nitro compounds. When the adduct **461**, derived from 4-methoxy-1-nitronaphthalene, is treated with acetate buffer the nitronic acid **462**, which is formed initially, tautomerizes predominantly to the thermodynamically less stable *cis* nitro compound **463**. The predominant formation of the less stable isomer is rationalized in terms of steric control of the approach of the proton in the nonplanar cyclohexadienic system as shown in structure **464**[634].

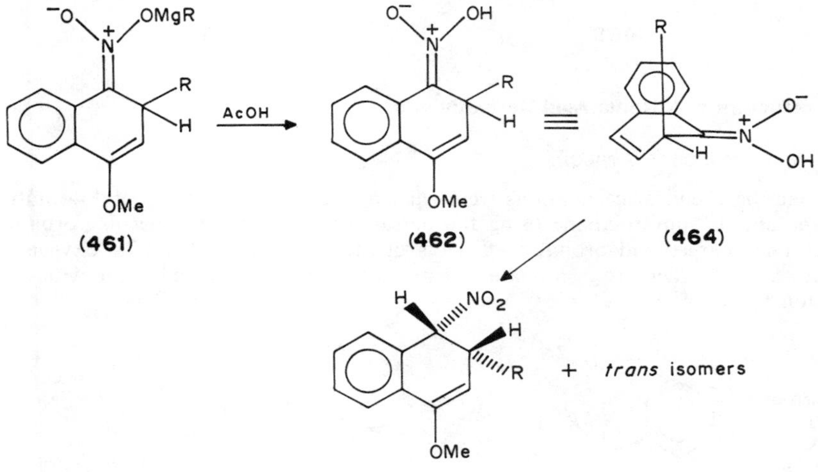

(461) (462) (463) (464)

AcOH

+ *trans* isomers

Similar effects are seen in 2-methoxy-1-nitronaphthalene[635a], in which the Grignard reagent enters position 4, and in 9-nitroanthracene[635b] in which the site of attack is position 10. However, contrary to the above, in both of these cases the *cis* isomers are the more thermodynamically stable. All dihydroaromatic compounds can be converted to the corresponding aromatics by oxidation with DDQ; consequently this reaction is an efficient procedure for nucleophilic alkylation of aromatic nitro compounds.

Although the nitronic acid is usually the less stable tautomer, and the majority of cases in which it has been isolated belong to the arylalkyl series, the purely aliphatic dinitronic acid **465**, derived from 1,4-dinitro-2-butene, was isolated as a crystalline compound and was characterized by infrared spectroscopy in the solid state. Other spectroscopic studies (ultraviolet and NMR) carried out in solution indicated rapid tautomerization to the more stable 1,4-dinitro-2-butene[636].

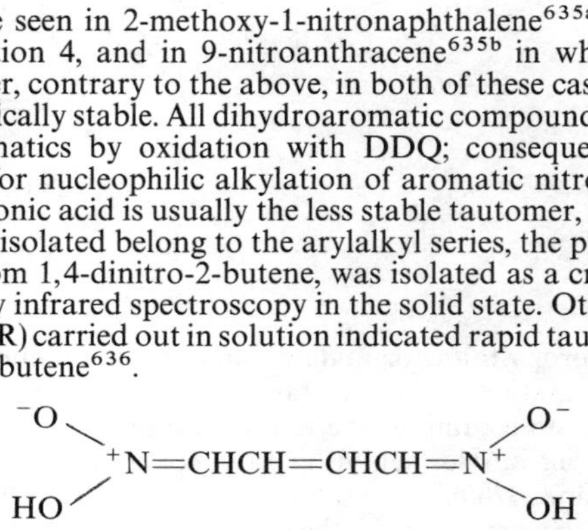

$$(^{-}O)(HO)\overset{+}{N}{=}CHCH{=}CHCH{=}\overset{+}{N}(O^{-})(OH)$$

(465)

A different type of ring–chain tautomerism, involving equilibrium between a nitro compound and a cyclic nitronate ester, has been observed in the cycloaddition product of 1-phenyl-2-nitropropene and 4-*t*-butyl-1-aminocyclohexene (**466**), which was found to exist in solution as an equilibrium mixture containing the nitroenamine **467**[637]. NMR and infrared spectral data are reported for both tautomers.

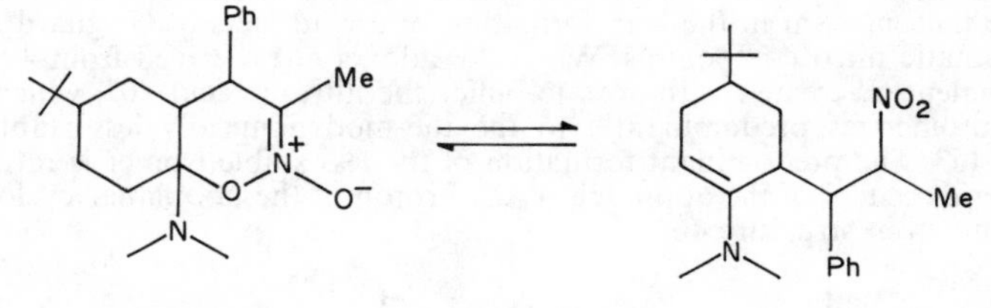

(**466**) (**467**)

*B. Structure of Nitronic Acid Derivatives

*1. Theoretical considerations

Quantum-chemical calculations were carried out for the dianions of 1,4-dinitro-2-butene and 1,2-dinitroethane using the perturbation theory of molecular orbitals to determine charges and orbital coefficients of the potential nucleophilic oxygens and carbons, in order to correlate them with their observed reactivities (see Section *III.C.4)[638].

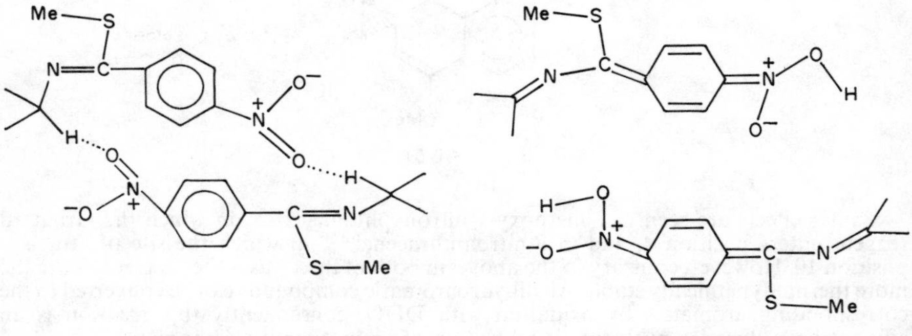

(**468**) (**469**)

*2. X-ray studies

Methyl 4-nitro-*N*-isopropylthiobenzimidate (**468**) exhibits photochromism in the crystalline state. Determination of its crystal structure confirmed the possibility of intermolecular prototropism leading to the tautomeric nitronic acid **469**[639].

The crystal structures of *t*-butyldimethylsilyl esters of *aci*-α-nitrotoluene (**470a**) and of *aci*-α-nitrodiphenylmethane (**470b**) have been determined at low (101 K) temperature[640]. The interesting feature in these structures is the relatively small distance between Si and O^2 (≈ 2.8 Å), which is considerably shorter than the sum of the van der Waals' radii of the two atoms (3.62 Å). This appears to be in accordance with the phenomenon of bond switching

(O^1—Si $O^2 \rightleftharpoons O^1$ Si—O^2) which was indicated by the magnetic equivalence of the carbon substituents in a series of silyl nitronates examined (see also Section *III.B.4)[640]. The *trans* structure of the zwitterionic nitronate **471** was in question because of the small olefinic H—H coupling constant (10.6 Hz) until it was proved by X-ray crystallography[641].

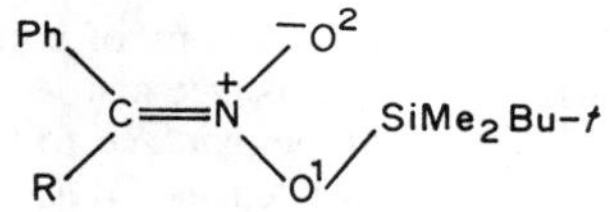

(470 a) R=H

(470 b) R=Ph

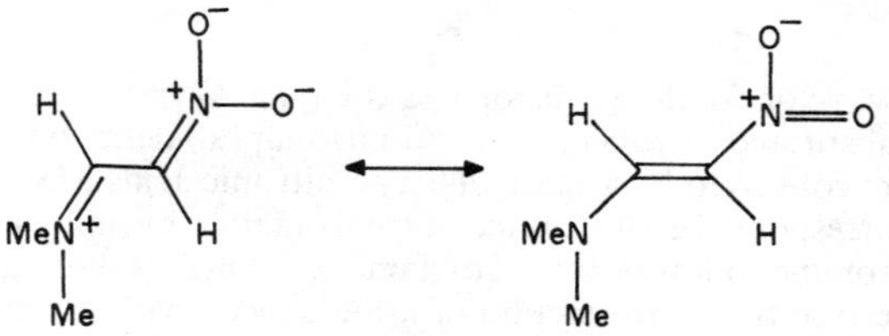

(471)

**4. Spectra*

The ^{13}C- and ^{1}H-NMR spectra of a series of aliphatic and alicyclic nitroalkanes and their nitronate anions have been determined in methanol and dimethyl sulphoxide[642]. The conclusions of this study agree with those summarized in Section III.B.4 regarding the influence of the nature of the solvent upon the structure of the nitronate and the distribution of charge in it, which were based on other spectral data and a preliminary account from the same laboratory[363].

A series of silyl nitronates (prepared by reacting primary or secondary nitroalkanes with lithium diisopropylamide, followed by quenching with trialkylsilyl chloride) was examined by NMR[640]. All of the silyl nitronates thus prepared appeared to be single compounds. Those derived from primary nitroalkanes show only one triplet for the olefinic proton in their ^{1}H-NMR spectra—around 6 ppm. Those derived from secondary nitroalkanes show no evidence of nonequivalence of the α-and α'-positions. For example the silyl nitronate derived from 2-nitropropane shows a singlet at 2.0 ppm in its ^{1}H-NMR spectrum and one at 17.6 ppm in its ^{13}C-NMR spectrum, corresponding to both methyl groups. Other examples are given in the paper[640]. These results are in marked contrast to those obtained from alkyl nitronates. For example methyl nitronate derived from 1-nitropropane shows the N=CH protons as two triplets at 5.75 and 6.06 ppm (see also the data in Chapter 1, Section IV.B). However, magnetic nonequivalence in these compounds could be demonstrated at low temperature, on the nitronate derived from nitrocyclopentane in Frigen 22 at $-75\,^{\circ}$C with a coalescence at $-65\,^{\circ}$C corresponding to an activation energy for a first-order process of 10 kcal/mol[640]. This phenomenon is rationalized in terms of a bond-switching process, moving the silicon from one oxygen to the other through a transition state in which both oxygens are bound to the silicon (involving the formation of a pentacoordinated silicon), thus proceeding with retention at silicon[640]. X-Ray crystallography lends support to this hypothesis (see previous section).

*C. Reactions of Nitronic Acid Derivatives

**1. Oxidation of nitronates*

2-Nitroalkylphosphonates (**472**) have been converted to β-aldehydophosphonates (**473**) by oxidation of the corresponding nitronate anions (in methanol in the presence of sodium methoxide with singlet oxygen[643].

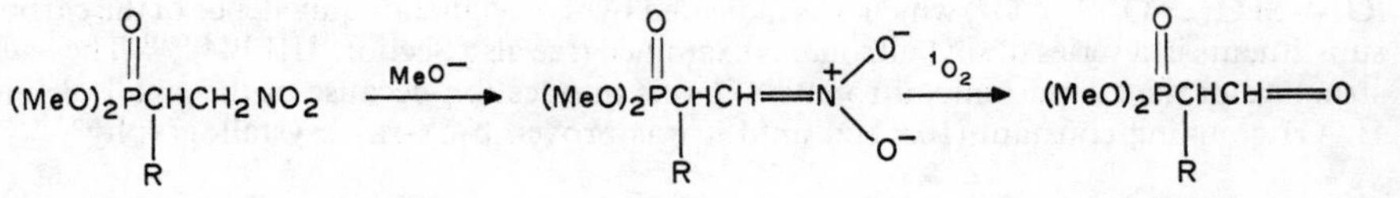

(472) (473)

Cyclohexadienyl nitronates (of type **474**), which are formed by treatment of some substituted nitrobenzenes and nitronaphthalenes with Grignard reagents (see also **461**), are converted by acetate buffer to nitronic acids **375**[644,645]. These can be oxidized to the corresponding substituted nitroaromatics by mild oxidizing agents such as DDQ or bromine followed by triethylamine. Similar results have been obtained with nitronates derived from the reaction of lithiumalkyls with nitroarenes[644]. Subsequently a modification has been developed that makes it possible to oxidize the initial adduct of type **474** directly, by adding a solution of potassium permanganate to the reaction mixture[646].

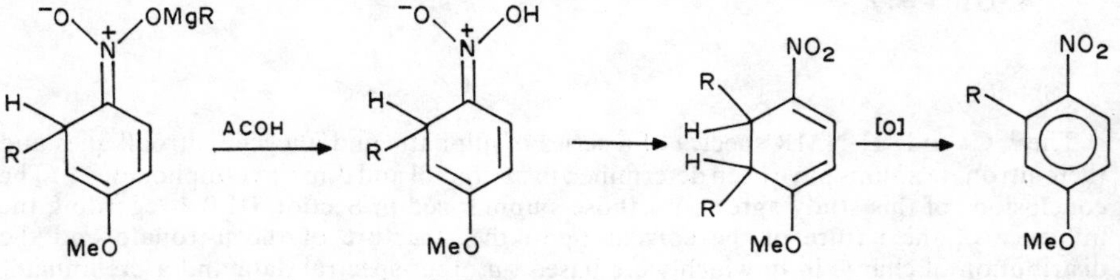

(474) (475)

2. Reduction of nitronates

The addition of Grignard reagents to aromatic nitro compounds, which has already been mentioned in previous sections, is a versatile reaction that may lead, *via nitronates*, to

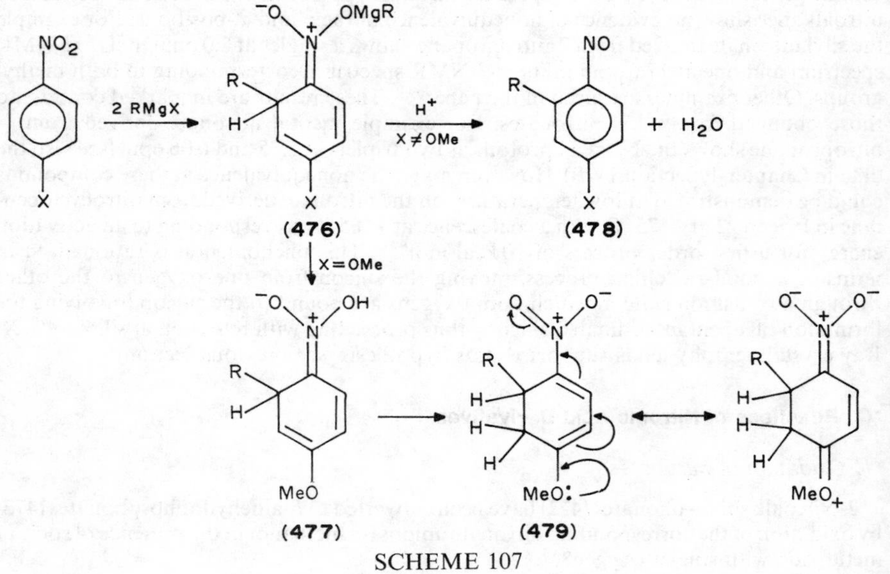

(476) (478) (477) (479)

SCHEME 107

aromatic nitrogen copounds in varying oxidation states. The basic reaction is illustrated in Scheme 107. In this scheme it can be seen that acidification of the primary product of addition (**476**) yields nitronic acid **477**, which in most instances will lose a molecule of water to give the corresponding nitrosoarene **478**[647]. However, in cases in which there is an *ortho* or *para* methoxy substituent on the aromatic ring, this will stabilize **477** and suppress the elimination leading to **478**. Nitronic acid **477** will tautomerize to the dihydronitroarene **479**[644]. It was found that by carrying out the reaction in the presence of cuprous iodide, the corresponding anilines were obtained in reasonable yields[648].

In analogy to equation (79), treatment of nitronates of type **461** with hexamethylphosphorus triamide causes their deoxygenation, and leads to the corresponding (*E*)-dihydronaphthalene-1-oximes (**480**) or, by acid treatment, to (*E*)-dihydronaphthoquinone monoximes (**481**)[649]. On the other hand, the action of phosphorous trichloride followed by acid on **461** gives the completely reduced, aromatic 2-alkyl-1-amino-4-methoxynaphthalene **482**[649].

461

1. $P(NMe_2)_3$ 2. NH_4Cl → (480)

1. $P(NMe_2)_3$ 2. HCl/MeOH → (481)

1. PCl_3 2. HCl → (482)

(480) **(481)** **(482)**

Nitronic acids (**483**), formed by the addition of β-dicarbonyl compounds to nitro olefins, were reduced by sodium sulphide to γ-keto oximes, which spontaneously underwent cyclization to dihydro-1,2-oxazines (**484a**) and to 1,2-oxazines (**484b**)[650a]. Other reducing agents (Sn^{2+}, Zn/H^+, Fe^{2+}) give pyrroles (**485**)[650b].

The fragmentation reaction of nitronate esters to carbonyl compounds and oximes (Chapter 1, equations 174–176 and Scheme 44) has been extended. Ethyl nitroacetate, in the presence of diethyl azodicarboxylate and triphenylphosphine oxidizes alcohols to aldehydes and ketones in good yields, under neutral conditions and in one step, without the need to isolate the intermediate nitronate ester **486** (Scheme 108)[651].

*3. *1,3-Dipolar cycloaddition reactions of nitronates*

The *N*-trimethylsilyloxyisoxazolidine products (**487**) of cycloaddition of silyl nitronates to olefins have proved to be versatile synthetic intermediates that can be transformed to isoxazolines, isoxazoles, pyridazines, pyridazones, hydroxyfurans, hydroxy-1,4-diketones and cyclopentenones[652,653]. It has been found that it is advantageous to carry out the sequence of silylation, cycloaddition and silanol elimination to isoxazoline **488** as a one-pot reaction[652]. Reduction of **488** with Ti^{2+} gave hydroxyketone **489**, or 2-hydroxy-1,4-diketone when R^1 = MeC=O. Treatment of hydroxyketones of type **489** (or even **488**) with hydrazine gave 3,6-disubstituted pyridazine **490**. In the case of R^1 = MeOC=O, **488** gave 6-substituted 3-pyridazones **491**[652].

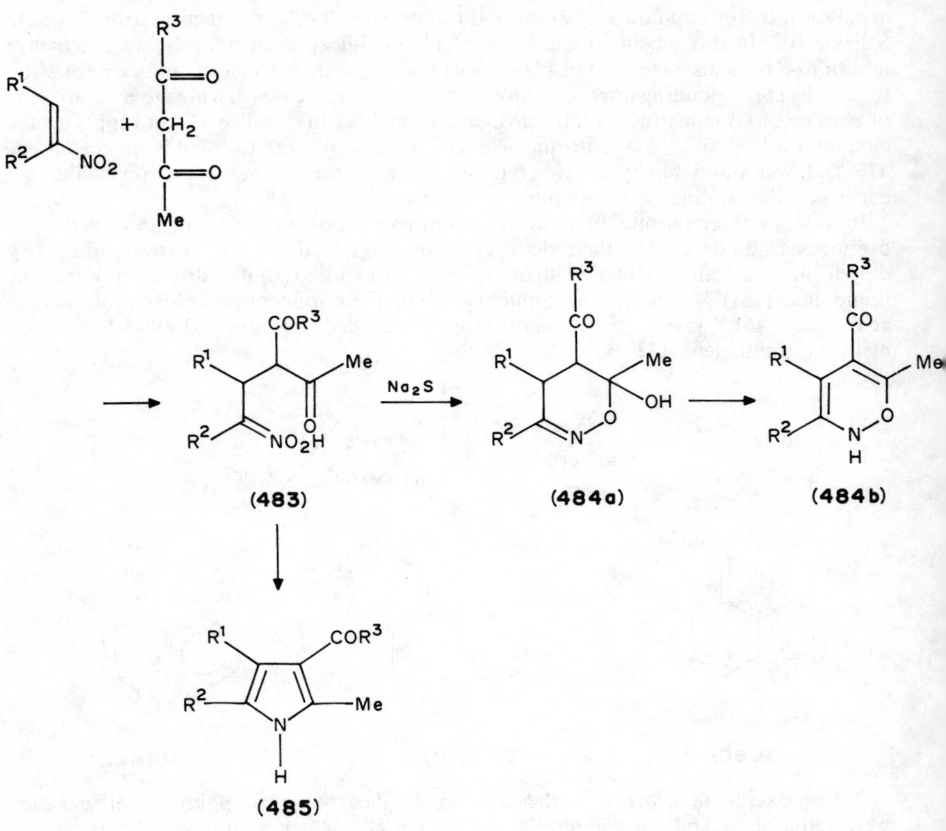

$$\mathrm{RCH_2OH + O_2NCH_2COOEt \xrightarrow[Ph_3P]{RO_2CN{=}NCO_2R} RCH_2ON^{+}(O^{-}){=}CHCOOEt \ (486) \rightarrow RCH{=}O}$$

SCHEME 108

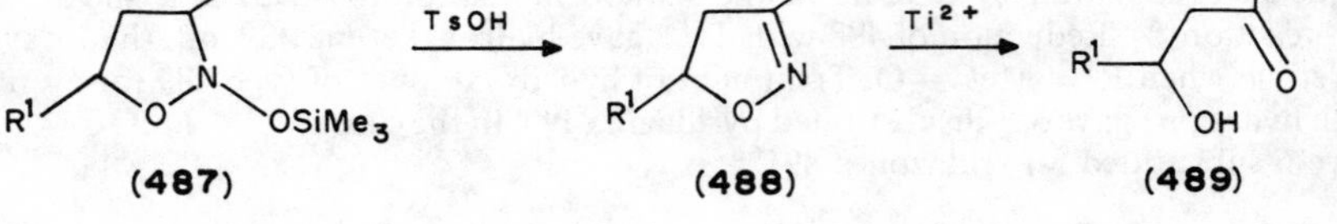

489 —(1. Ti^{2+}; 2. H_2NNH_2; R^1 = alkyl)→ **(490)**

→(H_2NNH_2)→ **(491)**

Further work on hydroxydiketones **489** (R^1 = MeC=O) revealed that these can be converted with acid or acetic anhydride to α,β-unsaturated ketones **492** or, where $R^1 = MeO_2C$—, to β-acylated acrylates. In one case, however, the furan derivative **493** was formed[652].

(493) ←(R^1=MeCO)— **(489)** —(R^1=alkyl)→ **(492)**

An additional synthetic option was discovered upon treatment of hydroxydiketones **489** with base. This reaction gave hydroxycyclopentenones **494**, which can serve as valuable synthetic intermediates to natural products in the terpene field and in addition provide a new approach to prostaglandins[653].

(489) —(OH^-)→ **(494)**

Another interesting transformation is shown by the adducts of acrylonitrile with silyl nitronates **495**. These compounds rearrange, when treated by fluoride, to 5-silyloxy derivatives **496**, presumably by the mechanism indicated in Scheme 109. Subsequent treatment of **496** by *p*-toluenesulphonic acid gave isoxazoles **497**[652].

The 1,3-dipolar cycloaddition product of the unstable cyclic nitronate ester **498** and an acetylene dipolarophile has the same type of structure (**499**) as the adducts obtained from four-membered cyclic nitrones (**459**)[482]. The formation of products **499** from nitronates **498** was rationalized by assuming the initial formation of adducts **500**, followed by their rearrangement as shown in Scheme 110[482].

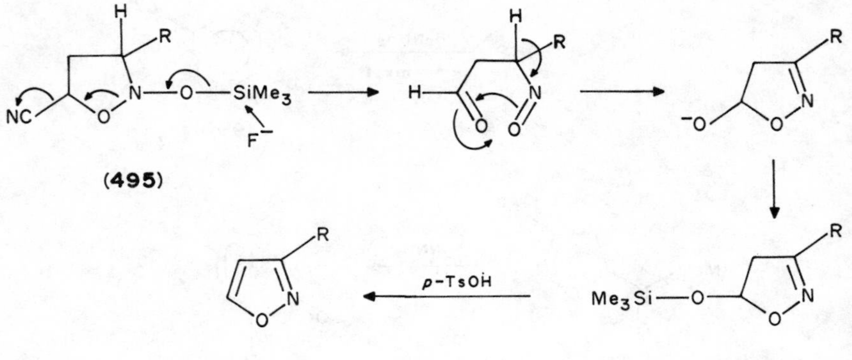

(495)

(497)

(496)

SCHEME 109

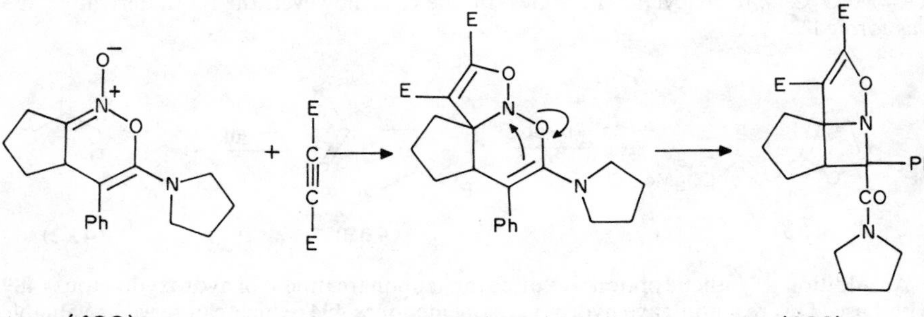

(498)

(500)

(499)

SCHEME 110

*4. *Reactions of nitronates with electrophiles*

In contrast to the nitronates derived from simple nitro compounds, which follow the rules of hard and soft acids and bases (HSAB) in their reactions with electrophiles (reacting with hard electrophiles such as protons, Meerwein's salts etc., on the oxygen and with soft electrophiles such as bromine or diazonium salts on the carbon), dinitro dianions derived from dinitroalkenes and dinitrodienes behave somewhat differently. The products from oxygen attack (by Meerwein's reagent) in this series are considerably more stable than those obtained from simple nitronates and the resulting aryl nitronates can be isolated[654]. It has also been found that in this series even the relatively soft arenediazonium cations react at the oxygen, with the formation of *O*-aryl nitronates (through the loss of nitrogen). Frontier orbital calculations have been carried out to correlate the reactivities of the two potential nucleophilic sites (C vs. O) of these bidentate nucleophiles[638].

a. Nitronates as oxygen nucleophiles.* Full experimental details have appeared of an improved procedure for the silylation of nitronate anions using trialkylsilyl halides (equation 86)[640,652]. However, when the silylation is carried out with an excess of trimethylsilyl trifluoromethanesulphonate (TMSTf) the initially formed trimethylsilyl nitronate is further silylated on the second oxygen to give *N,N*-bis(trimethylsilyloxy)aminoalkenes (501**)[655].

$$R^1CH_2CH(R^2)NO_2 \xrightarrow[Et_3N]{TMSTf} R^1CH_2C(R^2){=}N^+(O^-)OTMS \xrightarrow[Et_3N]{TMSTf} R^1CH{=}C(R^2)N(OTMS)_2$$

(**501**)

Enamines **501** undergo thermal or acid-catalysed rearrangement to 2-(trimethylsilyloxy)oxime *O*-trimethylsilyl ethers (**502**)[655].

$$R^1CH{=}C(R^2)N(OSiMe_3)_2 \xrightarrow{\text{heat or acid}} R^1CH(OSiMe_3)C(R^2){=}NOSiMe_3$$

(**501**) (**502**)

Enamines **501** can also add amines to their double bond to form, with the loss of one molecule of trimethylsilanol, α-aminooxime *O*-trimethylsilyl ethers (**503**) as products. Compounds **503** can be easily hydrolysed to α-aminooximes[655].

$$R^1_2\ddot{N}H + (R^2)(H)C{=}C(R^3){-}N(OSiMe_3)_2 \xrightarrow{-Me_3SiOH} R^1_2NCH(R^2)C(R^3){=}NOSiMe_3$$

(**501**) (**503**)

In analogy to nitrones, the influence of acetic anhydride and of acetyl chloride on lithium nitronate derived from 2-phenylnitroethane was also examined[656]. It appears that in the first step, the reaction leads to mixed nitronic–acetic anhydride **504**, which loses acetic acid to a nitrile oxide (**505**), which in turn can dimerize to a furoxan (**506**). Other products that have been observed in the reaction mixture are *O*-acetylbenzhydroxamic acid ($PhCH_2CONHOCOMe$) and the chlorooxime acetate $PhCH_2C(Cl){=}NOCOMe$[656].

b. Nitronates as carbon nucleophiles.* The full paper reporting the fluoride-catalysed addition of silyl nitronates to aldehydes (equation 96)[405] has appeared[657]. The resulting vicinal silyloxynitroalkanes can be smoothly reduced to amino alcohols, thus constituting an efficient synthetic route to a wide range of substituted amino alcohols[657]. Further work revealed that by carefully following a list of precautions during the experimental work, the 2-(*t*-butyldimethylsilyloxy)nitroalkanes may be obtained in high degree of diastereoselectivity (> 95% *erythro*)[658]. In addition to this, conditions were defined to direct the Henry reaction to proceed diastereoselectively to give either *threo*- or *erythro* enriched nitroaldols. It was found that the dilithio adducts **507 obtained from the addition of doubly deprotonated nitroalkanes undergo stereoselective protonation in the presence of HMPA or DMPU (1,3-dimethyl-2-oxohexahydropyrimidine) to give, predominantly, the *threo*- nitroaldol **508**[658].

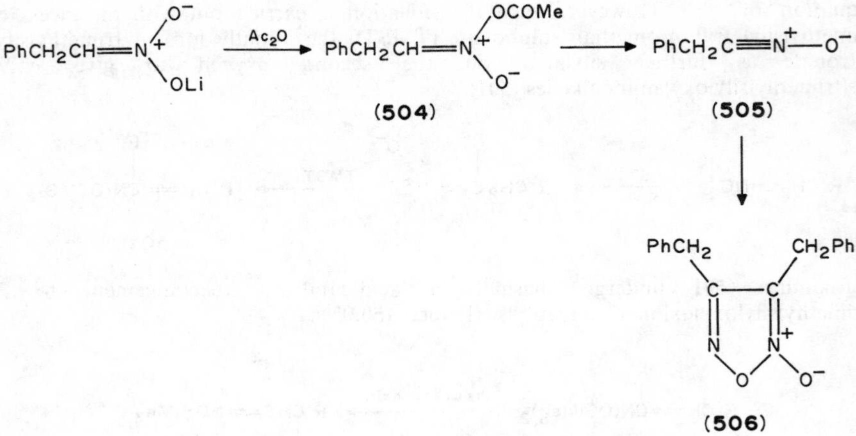

(504) (505)

(506)

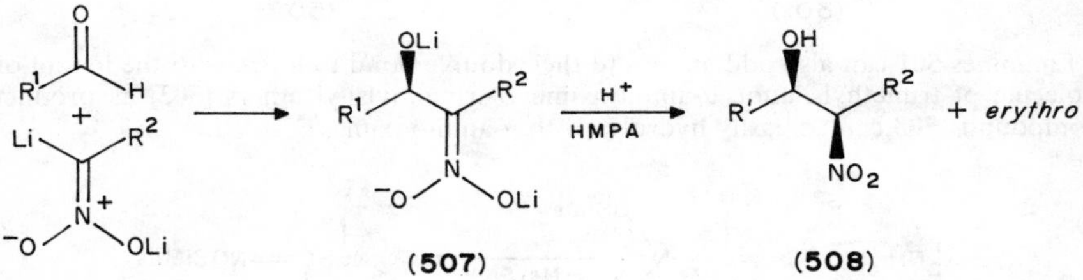

(507) (508)

On the other hand, when a diastereomeric mixture of *O*-*t*-butyldimethylsilyl-protected 2-nitro alcohols (**509**) is treated with lithium diisopropylamide at dry-ice temperature, a solution of nitronate **510** is formed. This nitronate can be protonated with high diastereoselectivity to give a considerably (> 90% d.e.) *erythro* enriched product (**511**)[658].

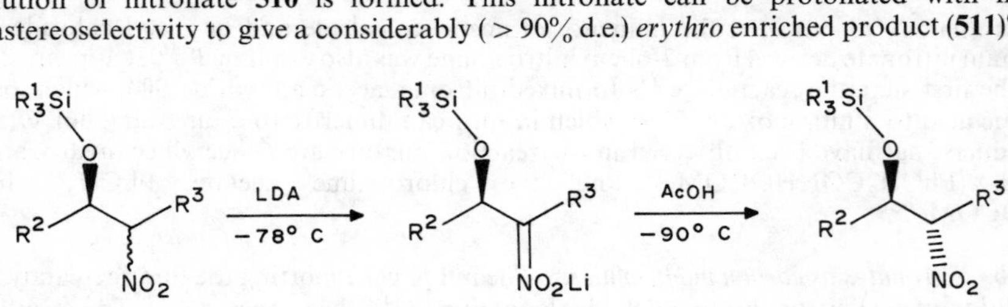

(509) (510) (511)

Some of the *C*-alkylation reactions of nitronates proceed by a single-electron-transfer ($S_{NR}1$) mechanism involving radical anions and free radicals[659]. The influence of branching at the proximity of the reacting sites on the rate and regiochemistry of $S_{NR}1$ reaction of *aci*-nitronates with a series of *p*-nitrobenzylic substrates has been studied, and a set of rules that makes it possible to predict the outcome of the reaction has been formulated[660]. In addition to the usual products of *C*-alkylation, other products such as *p*-nitrobenzyl alkyl ketones, oximes and *p*-nitrophenol may also be formed in these reactions[660]. The following two examples of synthetically useful reactions belong, mechanistically, to this class. Nitronates of type **513**, which are generated in the base-catalysed retro-Henry reaction of 5-hydroxymethyl-5-nitro-1,3-oxazines (**512**), are alkylated on the carbon to form 5-substituted 1,3-oxazines (**514**)[661].

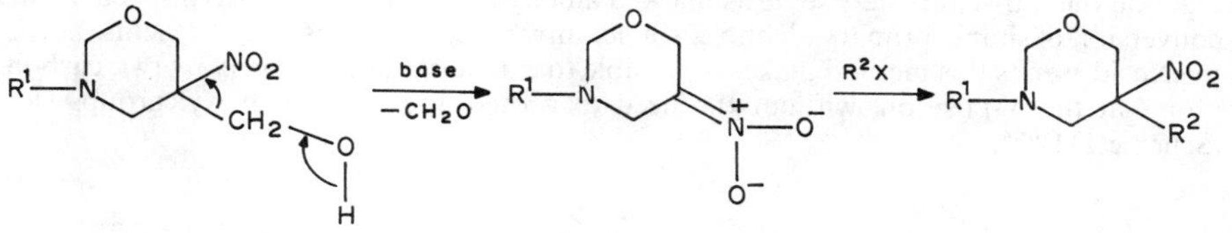

(512) (513) (514)

Reactions of this kind do not necessarily give nitro compounds as final products. Such a case is the synthesis of 2-vinylimidazole derivatives (**517**) in the reaction of 2-chloromethyl-5-nitroimidazoles (**515**) with nitronates derived from *s*-nitroalkanes (e.g. 2-nitropropane)[662]. The addition products (**516**) formed in the first step of this process suffer elimination of nitrous acid, to afford the final products (**517**) having unsaturated side-chains[662].

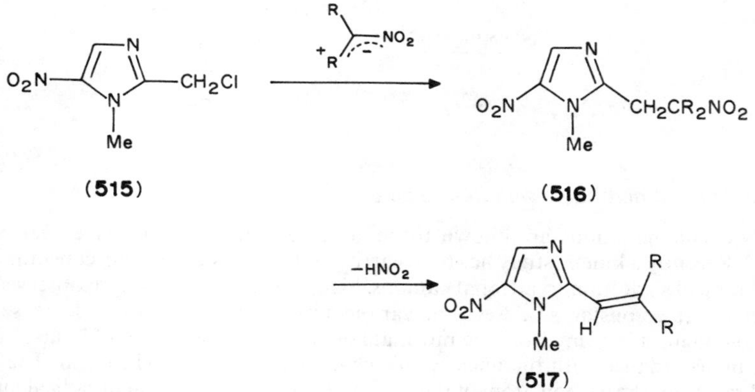

(515) (516)

(517)

**5. Reactions of nitronates with nucleophiles*

Nitronic acids **483**, obtained from the addition of active methylene compounds to nitro olefins (Section *III.C.2), react with a second molecule of active methylene compound to yield 4,5-dihydro-5-methyleneaminofuran-3-carboxylates (**519**) presumably via the dihydroxyamino intermediate **518**, which is formed by intramolecular nucleophilic cyclization of the 4-enolnitronic acid **483**[663].

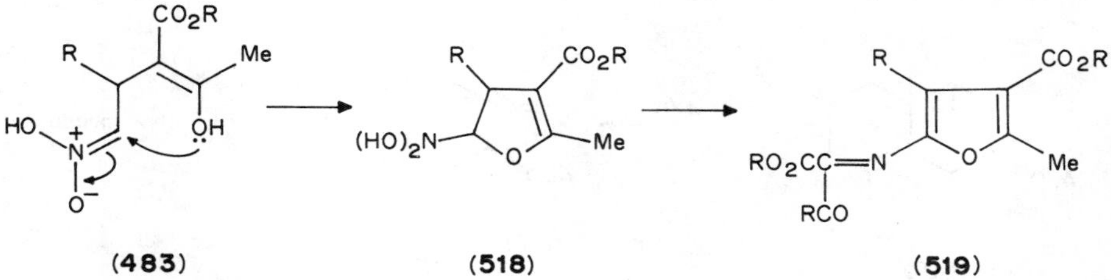

(483) (518) (519)

aci-Nitro esters obtained from the treatment of alcohols with 2,6-di-*t*-butyl-4-nitrophenol in the presence of diethyl azodicarboxylate and triphenylphosphine (Scheme 44) react with stabilized Wittig reagents to yield, directly, olefins[664]. In this reaction the nitronate esters serve as marked aldehydes. This reaction was applied to the conversion of simple primary alcohols, via *aci*-nitronates, to olefins in good yields. In the nucleoside series this method makes it possible to extend the sugar moiety by two carbon atoms at the 5′-position, without the need to protect the other hydroxyl groups (see (Scheme 111)[664].

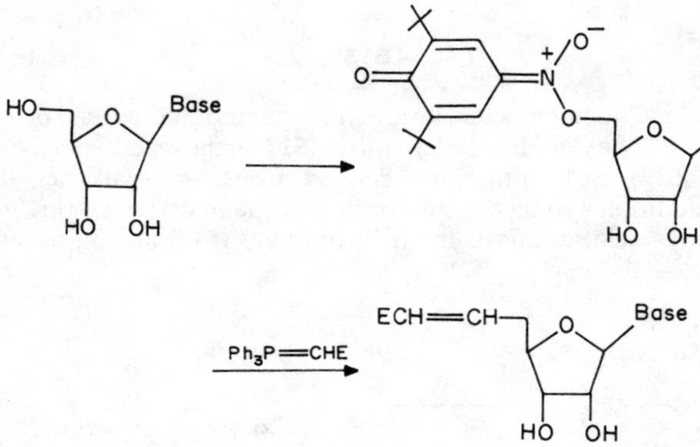

6. Reactions of nitronates with free radicals

aci-Nitroalkane anions are known to be good radical traps and have been used as such[665]. Recently a kinetic study has been carried out to measure the rate constants for the methyl radical's addition to nitronate anions. Significant differences were observed in the bimolecular rate constants between the various nitronates. There was a decrease in two orders of magnitude going from the nitronate of nitromethane to that of 2-nitropropane. This is in accordance with the nucleophilic character of the methyl radical. The methyl radical reacts on the carbon atom of the nitronate with the formation of radical anions of type $R^1R^2MeCNO_2^-$ these were characterized by absorption in the ultraviolet at 270 nm[666].

7. Photochemistry of nitronic acids

Irradiation of certain α,β-unsaturated nitro compounds, such as 6-nitrocholest-5-ene-3β-yl acetate (**520**) was observed to give, among others, the corresponding α,β-unsaturated ketones: 6-oxocholest-4-ene-3β-yl acetate (**521**)[667].

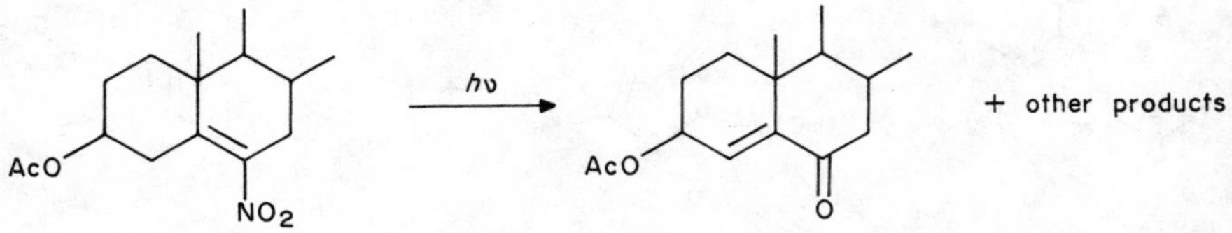

+ other products

(**520**) (**521**)

The formation of the ketone (of type **525**) was rationalized in terms of a γ-hydrogen abstraction leading, via a dipolar diradical intermediate (**522**), to an unsaturated nitronic acid (**523**). This undergoes a photochemical cyclization to hydroxyoxaziridine (**524**) (in analogy to the behaviour of nitrones), which in turn decomposes to the corresponding ketone. Support for this assumption can be derived from the results obtained by examination of the behaviour of the stable, saturated nitronic acids 4-*t*-butyl-*aci*-nitrocyclohexane and 9-*aci*-nitrofluorene[667,668].

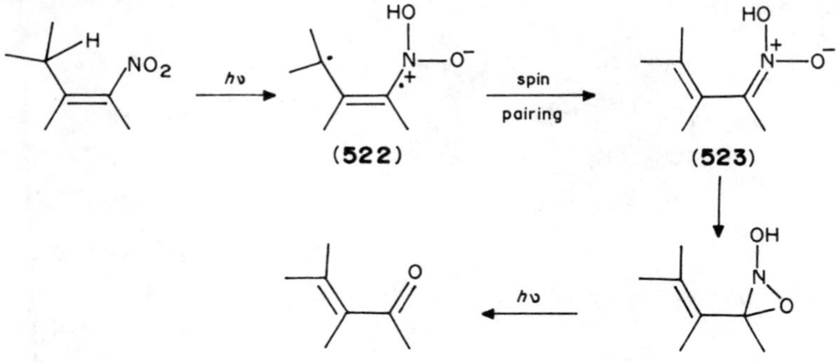

(**522**) (**523**) (**524**) (**525**)

8. Miscellaneous

a. Rearrangement of a nitronic ester. Cyclic nitronates of type **526** (derived from the cycloaddition of nitroalkenes to enaminones) undergo thermal rearrangement to hexahydropentalenone derivatives (**529**). The mechanism suggested for this transformation

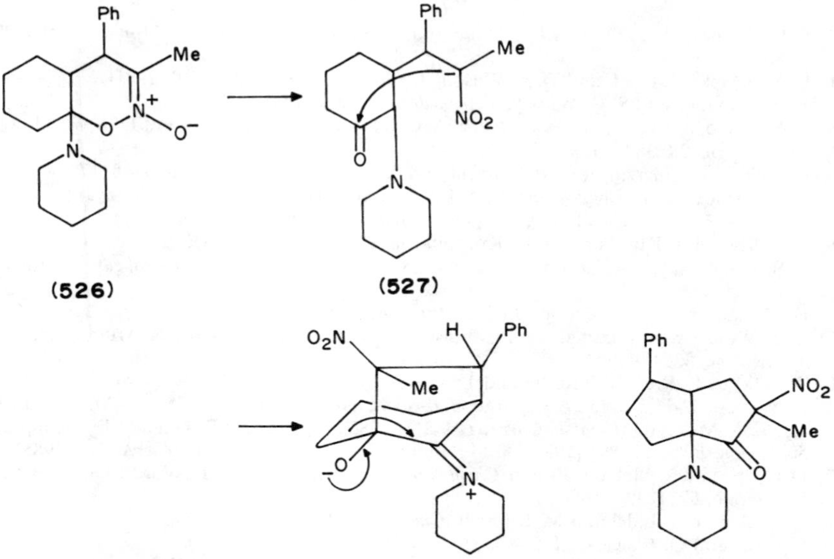

(**526**) (**527**) (**528**) (**529**)

involves opening of the nitronate ring, its reclosure by the nucleophilic attack of the α-carbon of the nitronate anion on the carbonyl in **527** and finally a ring contraction of the bridged intermediate (**528**) leading to the final product (**529**)[669].

a. Autocondensation of nitronic acids. When secondary polynitro compounds were treated with concentrated sulphuric or trifluoroacetic acid they were found to undergo dimerization: e.g. 2,2,6,6-tetranitroheptane-3-nitronic acid (**530**) was converted to *O*-α-nitroalkyloxime (**531**)[670].

$$2CH_3C(NO_2)_2CH_2C(=N^+(OH)O^-)CH_2C(NO_2)_2CH_3 \xrightarrow{H^+} CH_3C(NO_2)_2CH_2C(NO_2)(O\!-\!N\!=\!C[CH_2C(NO_2)_2CH_3]_2)CH_2C(NO_2)_2CH_3$$

(**530**) (**531**)

*IV. ACKNOWLEDGEMENTS

I wish to thank my wife Aviva D. Breuer and my sons Jonathan and Oded for their patience and help during the trying period when this chapter was written.

*V. REFERENCES

440. P. D. Adeney, W. J. Bouma, L. Radom and W. R. Rodweel, *J. Amer. Chem. Soc.*, **102**, 4069 (1980).
441. D. W. Clack, N. Khan and D. A. Wilson, *J. Chem. Soc., Perkin II*, 860 (1981).
442. W. B. Jennings and S. D. Worley, *Tetrahedron Letters.*, 1435 (1977).
443. A. M. Lobo, S. Prabhakar, H. S. Rzepa, A. C. Skapski, M. R. Tavares and D. A. Widdowson, *Tetrahedron*, **39**, 3833 (1983).
444. R. Huber, A. Knierzinger, J.-P. Obrecht and A. Vasella, *Helv. Chim. Acta*, **68**, 1730 (1985).
445. S. I. Adamova, S. B. Bulgarevich, L. A. Litvinenko, I. M. Andreeva, E. A. Medyantseva, V. A. Kogan, R. I. Nikitina and O. A. Osipov, *J. Gen. Chem.*, **49**, 745 (1979).
446. L. L. Rodina, I. Kuruts and I. K. Korobitsyna, *J. Org. Chem.* (*USSR*), **17**, 1711 (1981).
447. L. B. Volodarsky, I. A. Grigor'ev, L. N. Grigor'eva and I. A. Kirilyuk, *Tetrahedron Letters*, **25**, 5809 (1984).
448. H.-H. Lau and U. Schoellkopf, *Liebigs Ann. Chem.*, 1378 (1981).
449. D. X. West, S. Sivasubramanian, P. Manisankar, M. Palaniandavar and N. Arumugam, *Trans. Met. Chem.*, **8**, 317 (1983).
450. W. Kliegel, L. Preu, S. J. Rettig and J. Trotter, *Can. J. Chem.*, **63**, 509 (1985).
451. A. M. Nour El-Din and E. A. Mourad, *Bull. Soc. Chim. Belg.*, **91**, 539 (1982); A. M. Nour El-Din and E. A. Mourad, *Monatsh. Chem.*, **114**, 211 (1983); A. M. Nour El-Din and D. Doepp, *Bull. Soc. Chim. Belg.*, **93**, 891 (1984); A. M. Nour El-Din, *Spectrochim. Acta*, **41A**, 721 (1985).
452. D. Doepp and A. M. Nour El-Din, *Chem. Ber.*, **111**, 3952 (1978); A. M. Nour El-Din, *Bull. Chem. Soc.* Japan, **59**, 1239 (1986).
453. H. G. Aurich, J. Eidel and M. Schmidt, *Chem. Ber.*, **119**, 18 (1986).
454. B. Princ and O. Exner, *Coll. Czech. Chem. Commun.*, **44**, 2221 (1979).
455. (a) N. Arumugam, P. Manisankar, S. Sivasubramanian and D. A. Wilson, (a) *Org. Magn. Reason.*, **22**, 592 (1984).

(b) N. Arumugan, P. Hanisankar, S. Sivasubramanian and D. A. Wilson, *Reson. Chem.*, **23**, 246 (1985).
456. Y. Inouye, K. Takaya and H. Kakisawa, *Magn. Reson. Chem.*, **23**, 101 (1985).
457. J. Moskal and P. Milart, *Magn. Reson. Chem.*, **23**, 361 (1985).
458. M. Thenmozhi, S. Sivasubramanian, P. Balakrishnan and D. W. Boykin, *J. Chem. Res. (S)*, 340 (1986).
459. N. I. Petrenko, T. N. Gerasimova and E. P. Fokin, *Bull. Natl. Acad. USSR Ser Chem.*, 1268 (1984).
460. R. G. Kostyanovskii, A. P. Pleshkova, V. N. Voznesenskii, A. I. Mishchenko, A. V. Prosyanik and V. I. Markov, *Bull. Natl. Acad. USSR Ser. Chem.*, 240 (1980).
461. (a) M. A. Abou-Gharbia and M. M. Joullie, *Org. Mass. Spectrom.*, **15**, 489 (1980).
(b) M. A. Abou-Gharbia, T. Terwilliger, M. M. Joullie and R. M. Srivastava, *J. Chem. Eng. Data*, **26**, 216 (1981).
462. R. Knorr, *Chem. Ber.*, **113**, 2441 (1980).
463. (a) Y. Inouye, J. Hara and H. Kakisawa, *Chem. Letters*, 1407 (1980).
(b) Y. Inouye, K. Takaya and H. Kakisawa, *Bull. Chem. Soc. Japan*, **56**, 3541 (1983).
464. V. V. Martin and L. B. Volodarskii, *Bull. Natl. Acad. USSR Ser. Chem.*, 956 (1980).
465. S. Saeki, H. Honda and M. Hamana, *Chem. Pharm. Bull. Tokyo*, **31**, 1474 (1983).
466. H. G. Aurich, O. Bubenheim and M. Schmidt, *Chem. Ber.*, **119**, 2756 (1986).
467. K. Suda, E. Sekizuka, Y. Wakamatsu, F. Hino and C. Yijima, *Chem. Pharm. Bull. Tokyo*, **33**, 1297 (1985).
468. (a) W. Kliegel, B. Enders and H. Becker, *Chem. Ber.*, **116**, 27 (1983); *Liebigs Ann. Chem.*, 1712 (1982).
(b) W. Kliegel and L. Preu, *Liebigs Ann. Chem.*, 1937 (1983); *Chem. Ztg.*, **108**, 283 (1984).
469. W. Kliegel and J. Graumann, *Liebigs Ann. Chem.*, 1545 (1984).
470. H. Moehrle and K. Troester, *Arch. Pharm.*, **314**, 836 (1981).
471. D. R. Boyd, R. M. Campbell, P. B. Coulter, J. Grimshaw, D. C. Neill and W. B. Jennings, *J. Chem. Soc., Perkin I*, 849 (1985).
472. A. J. Boyd, D. R. Boyd, J. F. Malone, N. D. Sharma, K. Dev and W. B. Jennings, *Tetrahedron Letters*, **25**, 2497 (1984).
473. (a) D. St. C. Black and L. M. Johnstone, *Australian J. Chem.*, **37**, 577 (1984).
474. M. L. M. Pennings, G. Okay, D. N. Reinhoudt, S. Harkema and G. J. van Hummel, *J. Org. Chem.*, **47**, 4413 (1982).
475. I. Nakamura, T. Sugimoto, J. Oda and Y. Inouye, *Agric. Biol. Chem.*, **45**, 309 (1981).
476. R. H. Prager, K. D. Raner and A. D. Ward, *Australian J. Chem.*, **37**, 381 (1984).
477. (a) H. W. Heine, R. Zibuck and W. J. A. VandenHeuvel, *J. Amer. Chem. Soc.*, **104**, 3691 (1982).
(b) H. W. Heine, R. Zibuck and J. Helder, unpublished results.
478. M. Cherest and X. Lusinchi, *Tetrahedron*, **38**, 3471 (1982).
479. M. Cherest and X. Lusinchi, *Bull. Soc. Chim. Fr.*, II-227 (1984).
480. C. H. Cummins and R. M. Coates, *J. Org. Chem.*, **48**, 2070 (1983); R. M. Coates and H. Cummins, *J. Org. Chem.*, **51**, 1383 (1986).
481. L. Baiocchi, G. Picconi and G. Palazzo, *J. Heterocycl. Chem.*, **16**, 1479 (1979).
482. M. L. M. Pennings and D. N. Reinhoudt, *J. Org. Chem.*, **47**, 1816 (1982).
483. P. J. S. S. van Eijk, D. N. Reinhoudt, S. Harkema and R. Visser, *Rec. Trav. Chim.*, **105**, 103 (1986).
484. W. Eberbach and J. Roser, *Tetrahedron Letters*, **28**, 2689 (1987).
485. M. L. M. Pennings, D. N. Reinhoudt, S. Harkema and G. J. van Hummel, *J. Org. Chem.*, **48**, 486 (1983).
486. B. Aebischer and A. Vasella, *Helv. Chim. Acta*, **66**, 789 (1983); R. Julina and A. Vasella, *Helv. Chim. Acta*, **68**, 819 (1985).
487. F. DeSarlo, A. Brandi and A. Guarna, *J. Chem. Soc., Perkin I*, 1395 (1982).
488. D. H. R. Barton, A. Fekih and X. Lusinchi, *Tetrahedron Letters*, **26**, 4603 (1985).
489. E. Eghtessad and G. Zinner, *Arch. Pharm.*, **312**, 1027 (1979).
490. (a) J. J. Tufariello in 1,3-Dipolar Cycloaddition Chemistry (Ed. A. Padwa), Vol. 2, Wiley-Interscience, New York, 1984, p. 83.
(b) A. Padwa in 1,3-Dipolar Cycloaddition Chemistry (Ed. A. Padwa), Vol. 2, Wiley-Interscience, New York, 1984, p. 277.
491. (a) Ya. D. Samuilov, S. E. Solov'eva and A. I. Konovalov, *J. Gen. Chem. USSR*, **49**, 555 (1979).

(b) Ya. D. Samuilov, S. E. Solov'eva and A. I. Konovalov, *J. Gen. Chem. USSR*, **50**, 117 (1980).
492. P. G. DeBenedetti, S. Quartieri, A. Rastelli, M. De Amici, C. De Micheli, R. Gandolfi and P. Gariboldi, *J. Chem. Soc., Perkin II*, 96 (1982).
493. A. Bened, R. Durand, D. Pioch, P. Geneste, C. Guimon, G. Pfister Guillouzo, J.-P. Declercq, G. Germain, P. Briard, J. Rambaud and R. Roques, *J. Chem. Soc., Perkin II*, 1 (1984).
494. Sk. A. Ali and I. M. Wazeer, *J. Chem. Soc., Perkin II*, 1789 (1986).
495. R. Plate, P. H. H. Hermkens, J. M. M. Smits and H. C. J. Ottenheijm, *J. Org. Chem.*, **51**, 309 (1986).
496. Ya. D. Samuilov, A. I. Movchan, S. E. Solov'eva and A. I. Konovalov, *J. Org. Chem. USSR*, **21**, 1618 (1984).
497. D. Doepp and M. Henseleit, *Chem. Ber.*, **115**, 798 (1982); D. Doepp and J. Walter in *Substituent Effects in Radical Chemistry*, (Ed. H. G. Viehe), D. Reidel Publishing Co., Dordrecht, 1986, pp. 375, 379.
498. (a) S. P. Ashburn and R. M. Coates, *J. Org. Chem.*, **49**, 3127 (1984).
(b) S. P. Ashburn and R. M. Coates, *J. Org. Chem.*, **50**, 3076 (1985).
499. A. Z. Bimanand and K. N. Houk, *Tetrahedron Letters*, **24**, 435 (1983).
500. J. B. Hendrickson and D. A. Pearson, *Tetrahedron Letters*, **24**, 4657 (1983).
501. S. Niwayama, S. Dan, Y. Inouye and H. Kakisawa, *Chem. Letters*, 957 (1985).
502. Y. D. Samuilov, S. E. Slov'eva, A. I. Konovalov and T. C. Mannafov, *Zh. Org. Khim.*, **15**, 279 (1979).
503. Y. D. Samuilov, S. E. Slov'eva and A. I. Konovalov, *Zh. Org. Khim.*, **16**, 1228 (1980).
504. (a) L. Fisera, J. Kovac, J. Poliacikova and J. Lesko, *Monatsh. Chem.*, **111**, 909 (1980).
(b) L. Fisera, A. Gaplovsky, H.-J. Timpe and J. Kovac, *Coll. Czech. Chem. Commun.*, **46**, 1504 (1981).
(c) L. Fisera, J. Kovac and J. Patus, *Chem. Zvesti*, **37**, 819 (1983).
(d) L. Fisera, P. Mesko, J. Lesko, M. Dandarova, J. Kovac and I. Goljer, *Coll. Czech. Chem. Commun.*, **48**, 1854 (1983).
505. (a) J. Tacconi, P. P. Righetti and G. Desimoni, *J. Prakt. Chem.*, **322**, 679 (1980).
(b) R. Commisso, A. Corsico Coda, G. Desimoni, P. P. Righetti and G. Tacconi, *Gazz. Chim. Ital.*, **112**, 483 (1982).
506. J. J. Tufariello, Sk. A. Ali and H. O. Klingele, *J. Org. Chem.*, **44**, 4213 (1979).
507. W. R. Dolbier, C. R. Burkholder, G. E. Wicks, G. J. Palenik and M. Gawron, *J. Ammer. Chem. Soc.*, **107**, 7183 (1985).
508. H. Gotthardt and R. Jung, *Chem. Ber.*, **119**, 563 (1986).
509. C. M. Dicken and P. DeShong, *J. Org. Chem.*, **47**, 2047 (1982).
510. Y. Yoshimura, J. Osugi and M. Nakahara, *J. Amer. Chem. Soc.*, **105**, 5414 (1983).
511. V. D. Kiselev, D. G. Khuzyasheva and A. I. Konovalov, *J. Org. Chem. USSR*, **19**, 782 (1983).
512. A. M. Nour El-Din, A. F. El-Said Mourad and H. AbdelNabi, *J. Chem. Eng. Data*, **31**, 259 (1986).
513. G. W. Griffin, D. C. Lankin, N. S. Bhacca, H. Terasawa and R. K. Sehgal, *Tetrahedron Letters*, **23**, 2753 (1982).
514. M. J. Fray, R. H. Jones and E. J. Thomas, *J. Chem. Soc., Perkin I*, 2753 (1985).
515. P. DeShong, C. M. Dicken, J. M. Leginus and R. R. Whittle, *J. Amer. Chem. Soc.*, **106**, 5598 (1984).
516. O. Tsuge, K. Ueno and S. Kanemasa, *Chem. Letters*, 797 (1984).
517. R. Brambilla, R. Friary, A. Ganguly, M. S. Puar, B. R. Sunday, J. J. Wright, K. D. Onan and A. McPhail, *Tetrahedron*, **37**, 3615 (1981).
518. J. T. Bailey, I. Berger, R. Friary and M. S. Puar, *J. Org. Chem.*, **47**, 857 (1982).
519. S. W. Baldwin, J. D. Wilson and J. Aube, *J. Org. Chem.*, **50**, 4432 (1985).
520. S. Takahashi, T. Kusumi, Y. Sato, Y. Inouye and H. Kakisawa, *Bull. Chem. Soc. Japan*, **54**, 1777 (1981).
521. D. M. Tschaen, R. R. Whittle and S. M. Weinreb, *J. Org. Chem.*, **51**, 2604 (1986).
522. T. Sasaki, S. Eguchi and T. Suzuki, *J. Org. Chem.*, **47**, 5250 (1982).
523. D. Mackay and K. N. Watson, *J. Chem. Soc., Chem. Commun.*, 777 (1982).
524. W. Oppolzer, S. Siles, R. L. Snowden, B. H. Bakker and M. Petrzilka, *Tetrahedron*, **41**, 3497 (1985).
525. B. B. Snider and C. P. Cartaya-Marin, *J. Org. Chem.*, **49**, 1688 (1984).
526. S. Eguchi, Y. Furukawa, T. Suzuki, K. Kondo, T. Sasaki, M. Monda, C. Katayama and J. Tanaka, *J. Org. Chem.*, **50**, 1895 (1985).

527. I. Panfil and C. Belzecki, *Polish J. Chem.*, **55**, 977 (1981).
528. B. Bernet, E. Krawczyk and A. Vasella, *Helv. Chim. Acta*, **68**, 2299 (1985).
529. T. Koizumi, H. Hirai and E. Yoshii, *J. Org. Chem.*, **47**, 4004 (1982).
530. A. Vasella and R. Voeffray, *J. Chem. Soc., Chem. Commun.*, 97 (1981); A. Vasella, R. Voeffray, J. Pless and R. Huguenin, *Helv. Chim. Acta*, **66**, 1241 (1983).
531. A. Vasella and R. Voeffray, *Helv. Chim. Acta*, **65**, 1953 (1982).
532. N. Balasubramanian, *Org. Prep. Proc. Internat.*, **17**, 25 (1985).
533. W. Oppolzer (Ed.), 'Tetrahedron symposia in print', *Tetrahedron*, **41**, 3447 (1985).
534. J. Hara, Y. Inouye and H. Kakisawa, *Bull. Chem. Soc. Japan*, **54**, 3871 (1981).
535. H. Iida, K. Kasahara and C. Kibayashi, *J. Amer. Chem. Soc.*, **108**, 4647 (1986).
536. J. J. Tufariello, G. E. Lee, P. A. Senaratne and M. Al-Nuri, *Tetrahedron Letters*, 4359 (1979).
537. (a) R. V. Stevens and K. Albizati, *J. Chem. Soc., Chem. Commun.*, 104 (1982).
(b) T. Kametani, S. P. Huang, A. Nakayama and T. Honda, *J. Org. Chem.*, **47**, 2328 (1982).
538. A. Padwa, K. F. Koehler and A. Rodriguez, *J. Amer. Chem. Soc.*, **103**, 4974 (1981).
539. P. N. Confalone, G. Pizzolato, D. Lollar Confalone and M. R. Uskokovic, *J. Amer. Chem. Soc.*, **102**, 1954 (1980).
540. E. G. Baggiolini, H. L. Lee, G. Pizzolato and M. R. Uskokovic, *J. Amer. Chem. Soc.*, **104**, 6460 (1982).
541. A. Vasella and R. Voeffray, *Helv. Chim. Acta*, **65**, 1134 (1982).
542. P. DeShong and J. M. Loginus, *J. Amer. Chem. Soc.*, **105**, 1686 (1983).
543. P. M. Wovkulich and M. R. Uskokovic, *Tetrahedron*, **41**, 3455 (1985).
544. J. J. Tufariello and G. E. Lee, *J. Amer. Chem. Soc.*, **102**, 373 (1980).
545. J. J. Tufariello and K. Winzenberg, *Tetrahedron Letters*, **27**, 1645 (1986).
546. J. J. Tufariello and J. M. Puglis, *Tetrahedron Letters*, **27**, 1489 (1986).
547. J. J. Tufariello, H. Meckler and K. P. A. Senaratne, *Tetrahedron*, **41**, 3447 (1985).
548. J. J. Tufariello and Sk. A. Ali, *J. Amer. Chem. Soc.*, **101**, 7114 (1979).
549. H. Iida, Y. Watanabe, M. Tanaka and C. Kibayashi, *J. Org. Chem.*, **49**, 2412 (1984).
550. R. W. Hoffmann and A. Endesfelder, *Liebigs Ann. Chem.*, 1823 (1986).
551. A. Brandi, A. Guarna, A. Goti and F. DiSarlo, *Tetrahedron Letters*, **27**, 1727 (1986).
552. W. Oppolzer and J. I. Grayson, *Helv. Chim. Acta*, **63**, 1706 (1980); W. Oppolzer, J. I. Grayson, H. Wegman and M. Urrea, *Tetrahedron*, **39**, 3695 (1983).
553. G. W. Gribble and T. C. Barden, *J. Org. Chem.*, **50**, 5900 (1985).
554. A. E. Watts and W. R. Roush, *Tetrahedron*, **41**, 3463 (1985).
555. R. L. Funk, G. L. Bolton, J. U. Daggett, M. M. Hansen and L. H. M. Horcher, *Tetrahedron*, **41**, 3479 (1985).
556. P. M. Wovkulich, F. Barcelos, A. D. Batcho, J. F. Sereno, E. G. Baggiolini, B. M. Hennessy and M. R. Uskokovic, *Tetrahedron*, **40**, 2283 (1984).
557. J. R. Hwu and J. A. Robl, *J. Chem. Soc., Chem. Commun.*, 704 (1986).
558. B. A. Arbuzov, A. F. Lisin, E. N. Dianova and Yu. Yu. Samitov, *Izv. Akad. Nauk SSSR, Ser. Khim.*, 2588 (1977); Engl. transl., p. 2314.
559. P. DeShong and J. M. Leginus, *J. Org. Chem.*, **49**, 3421 (1984).
560. A. M. Nour El-Din, A.-F. El-Said Mourad and R. Mekamer, *Heterocycles*, **23**, 1155 (1985).
561. L. I. Vasil'eva, G. S. Akimova and V. N. Chistokletov, *Zh. Org. Khim.*, **20**, 148 (1984); Engl. transl., p. 133.
562. D. Moderhack, M. Lorke and D. Schomburg, *Liebigs Ann. Chem.*, 1685 (1984).
563. M. C. Aversa, P. Giannetto, A. Feriazzo and G. Bruno, *J. Chem. Soc., Perkin II*, 1533 (1986).
564. D. Prajapati and J. S. Sandhu, *Heterocycles*, **23**, 1143 (1985).
565. M. A. Abou-Gharbia and M. M. Joullie, *J. Org. Chem.*, **44**, 2961 (1979).
566. M. A. Abou-Gharbia and M. M. Joullie, *Heterocycles*, **12**, 819 (1979).
567. O. Tsuge, H. Watanabe, K. Masuda and M. M. Yousif, *J. Org. Chem.*, **44**, 4543 (1979).
568. G. Zinner, H. Blass, E. Eghtessad and J. Schmidt, *Chem.-Ztg.*, **104**, 145 (1980).
569. V. S. Velezheva, I. S. Yaroslavskii and N. N. Suvorov, *Zh. Org. Khim.*, **21**, 601 (1985); Engl. transl., p. 544.
570. G. Zinner and E. Eghtessad, *Arch. Pharm.*, **312**, 907 (1979).
571. G. Zinner and E. Eghtessad, *Arch. Pharm.*, **313**, 357 (1980).
572. D. N. Harpp and J. G. MacDonald, *Tetrahedron Letters*, **24**, 4927 (1983).
573. D. A. Abramovitch, R. A. Abramovitch and H. Benecke, *Heterocycles*, **23**, 25 (1985).
574. J. Graumann and W. Kliegel, *Chem.-Ztg.*, **107**, 136 (1983).

575. L. B. Volodarskii and T. K. Sevast'yanova, *Zh. Org. Khim.*, **7**, 1687, 1752 (1971); T. K. Sevast'yanova and L. B. Volodarskii, *Zh. Org. Khim.*, **7**, 1974, 2046 (1971).
576. M. Yoshifuji, R. Nagase, T. Kawasima and N. Inamoto, *Bull. Chem. Soc. Japan*, **55**, 870 (1982).
577. M. Yushifuji, R. Nagase and N. Inamoto, *Bull. Chem. Soc. Japan*, **55**, 873 (1982).
578. E. Liedl and P. Wolschann, *Monatsh. Chem.*, **113**, 1067 (1982).
579. G. Goto, K. Kawakita, T. Okutani and T. Miki, *Chem. Pharm. Bull.*, **34**, 3202 (1986).
580. E. Cawkill and N. G. Clark, *J. Chem. Soc., Perkin I*, 244 (1980).
581. O. Tsuge, S. Urano and T. Iwasaki, *Bull. Chem. Soc. Japan*, **53**, 485 (1980).
582. A. Padwa and K. F. Koehler, *J. Chem. Soc., Chem. Commun.*, 789 (1986).
583. D. K. Dutta, D. Prajapati, J. S. Sandhu and J. N. Baruah, *Synth. Commun.*, **15**, 335 (1985).
584. M. L. M. Pennings, D. N. Reinhoudt, S. Harkema and G. J. van Hummel, *J. Org. Chem.*, **47**, 4419 (1982).
585. M. F. Schlecht, *J. Chem. Soc., Chem. Commun.*, 1239 (1985).
586. N. A. Akmanova, D. Ya. Mukhametova, Kh. F. Sagitdinova and F. A. Akbutina, *Zh. Org. Khim.*, **15**, 2061 (1979); Engl. transl., p. 1863.
587. V. S. Velezheva, I. S. Yaroslavskii and N. N. Suvorov, *Zh. Org. Khim.*, **18**, 2001 (1982); Engl. transl., p. 1785; V. S. Velezheva, I. S. Yaroslavskii, L. N. Kurkovskaya and N. N. Suvorov, *Zh. Org. Khim.*, **19**, 1518 (1983); Engl. transl., p. 1367.
588. O. Tsuge, K. Sone, S. Urano and K. Matsuda, *J. Org. Chem.*, **47**, 5171 (1982).
589. B. Bernet, E. Krawczyk and A. Vasella, *Helv. Chim. Acta*, **68**, 2299 (1985); A. Vasella in *Organic Synthesis: an Interdisciplinary Challenge*, (Ed. J. Streith, H. Prinzbach and G. Schill), Blackwell Scientific Publications, Oxford, 1985, p. 255.
590. S.-Y. Han, M. V. Lakshmikantham and M. P. Cava, *Heterocycles*, **23**, 1671 (1985).
591. K. Suda, H. Fuke, F. Hino and C. Yijima, *Chem. Letters*, 1115 (1985).
592. (a) N. Khan and D. A. Wilson, *J. Chem. Res.*, (S), 150, (M) 1531 (1984).
(b) H. Moehrle and B. Schmidt, *Arch. Pharm.*, **316**, 47 (1983).
593. D. K. Dutta, R. C. Boruah and S. Sandhu, *Indian J. Chem.*, **25B**, 350 (1986).
594. W. Kliegel and D. Nanninga, *J. Organometal. Chem.*, **243**, 373 (1983).
595. W. Kliegel and D. Nanninga, *J. Organometal. Chem.*, **247**, 247 (1983).
596. E. Niki, S. Yokoi, J. Tsuchiya and Y. Kamiya, *J. Amer. Chem. Soc.*, **105**, 1498 (1983).
597. M. J. Turner, III and G. M. Rosen, *J. Med. Chem.*, **29**, 2439 (1986).
598. S. Tero-Kubota, Y. Ikegami, T. Kurokawa, R. Sasaki, K. Sugioka, and M. Nakano, *Biochem. Biophys. Res. Commun.*, **108**, 1025 (1982).
599. T. Tominaga, S. Imaizumi, T. Yoshimoto, J. Suzuki and Y. Fujita, *Brain Research*, **402**, 370 (1987).
600. M. Schulz, B. Bach, M. Reinhardt, D. G. Pobedimskij and T. Kondrateva, *Plaste und Kautschuk*, **33**, 209 (1986).
601. V. Balogh-Nair and K. Nakanishi, *Pharm. Res.*, 93 (1984).
602. T. Markowicz, J. Skolimowski and R. Skowronski, *Polish J. Chem.*, **55**, 2505 (1981).
603. H. Mitsui, S. Zenki, T. Shiota and S.-I. Murahashi, *J. Chem. Soc., Chem. Commun.*, 874 (1984).
604. J. W. Gorrod and N. J. Gooderham, *Arch. Pharm.*, **319**, 261 (1986).
605. T. Shono, Y. Matsumura and K. Inoue, *J. Org. Chem.*, **51**, 549 (1986).
606. S.-I. Murahashi, H. Mitsui, T. Watanabe and S.-I. Zenki, *Tetrahedron Letters*, **24**, 1049 (1983).
607. R. Kreher and H. Morgenstern, *Chem. Ber.*, **115**, 2679 (1982).
608. H. F. Schmitthenner, K. S. Bhatki, R. A. Olofson and J. Heicklen, *Org. Prep. Proced. Int.*, **11**, 249 (1979).
609. M. L. M. Pennings and D. N. Reinhoudt, *Tetrahedron Letters*, **23**, 1003 (1982).
610. M. L. M. Pennings, D. Kuiper and D. N. Reinhoudt, *Tetrahedron Letters*, **24**, 825 (1983).
611. M. L. M. Pennings and D. N. Reinhoudt, *J. Org. Chem.*, **48**, 4043 (1983).
612. H. G. Aurich, M. Schmidt and T. Schwerzel, *Chem. Ber.*, **118**, 1105 (1985).
613. M. Abou-Gharbia and M. M. Joullie, *Synthesis*, 318 (1977); M. Abou-Gharbia and M. M. Joullie, *Org. Prep. Proced. Int.*, **11**, 95 (1979).
614. D. R. Boyd, P. B. Coulter and N. D. Sharma, *Tetrahedron Letters*, **26**, 1673 (1985).
615. R. S. Varma and G. W. Kabalka, *Org. Prep. Proced. Int.*, **17**, 254 (1985); M. Mourad, R. S. Varma and G. W. Kabalka, *J. Org. Chem.*, **50**, 133 (1985).
616. A. R. Gosh, *Indian J. Chem.*, **23B**, 449 (1984).
617. A. Monpert, J. Martelli and R. Gree, *J. Organometal. Chem.*, **210**, C45 (1981).

618. Y. Inouye, Y. Watanabe, S. Takahashi and H. Kakisawa, *Bull. Chem. Soc. Japan*, **52**, 3763 (1979).
619. S. Mzengeza and R. A. Whitney, *J. Chem. Soc., Chem. Commun.*, 606 (1984).
620. J. A. Robl and J. R. Hwu, *J. Org. Chem.*, **50**, 5913 (1985).
621. R. Huie and W. R. Cherry, *J. Org. Chem.*, **50**, 1531 (1985).
622. A. Padwa and G. S. K. Wong, *J. Org. Chem.*, **51**, 3125 (1986).
623. S. St. C. Black and L. M. Johnstone, *Australian J. Chem.*, **37**, 117 (1984).
624. D. St. C. Black and L. M. Johnstone, *Australian J. Chem.*, **37**, 587 (1984).
625. D. L. Haire, J. W. Hilborn and E. G. Janzen, *J. Org. Chem.*, **51**, 4298 (1986).
626. N. A. LeBel and N. Balasubramanian, *Tetrahedron Letters*, **26**, 4331 (1985).
627. A. Padwa, W. Dent and P. E. Yeske, *J. Org. Chem.*, **52**, 3944 (1987).
628. A. R. Katritzky, N. Dabbas, R. C. Patel and A. J. Cozens, *Rec. Trav. Chim.*, **102**, 51 (1983).
629. G. Tacconi, P. P. Righetti and G. Desimoni, *J. Prakt. Chem.*, **322**, 679 (1980).
630. T. Sasaki, K. Mori and M. Ohno, *Synthesis*, 279 (1985).
631. J. Moskal and P. Milart, *Synthesis*, 128 (1984); *Monatsh. Chem.*, **116**, 537 (1985); *Chem. Ber.*, **118**, 4014 (1985).
632. C. Schenk, L. Beekes and Th. J. deBoer, *Rec. Trav. Chim.*, **99**, 246 (1980); C. Schenk, L. Beekes, J. A. M. van der Drift and Th. J. deBoer, *Rec. Trav. Chim.*, **99**, 278 (1980).
633. E. S. Lewis in *Supplement F: The Chemistry of Amino, Nitroso and Nitro Compounds and their Derivatives* (Ed. S. Patai), Wiley–Interscience, Chichester, 1982, p. 715.
634. G. Bartoli, M. Bosco and G. Baccolini, *J. Org. Chem.*, **45**, 2649 (1980).
635. (a) G. Baccolini, G. Bartoli, M. Bosco and R. Dalpozzo, *J. Chem. Soc., Perkin II*, 363 (1984).
(b) G. Bartoli, M. Bosco, R. Dalpozzo and P. Sgarabotto, *J. Chem. Soc., Perkin II*, 929 (1982).
636. R. I. Bodina, E. S. Lipina and V. V. Perekalin, *Zh. Org. Khim.*, **15**, 875 (1979); Engl. transl., p. 784.
637. G. Pitacco and E. Valentin, *Tetrahedron Letters*, **26**, 2339 (1978).
638. E. S. Lipina, N. D. Stepanov, I. L. Bagal, R. I. Bodina and V. V. Perekalin, *Zh. Org. Khim.*, **16**, 2404 (1980); Engl. transl., p. 2059.
639. C. O. Meese and H. Guesten, *Z. Naturforsch.*, **41b**, 265 (1986).
640. E. W. Colvin, A. K. Beck, B. Bastani, D. Seebach, Y. Kai and J. D. Dunitz, *Helv. Chem. Acta*, **63**, 697 (1980).
641. A. Hazell and A. Mukhopadhyay, *Acta Cryst.*, **B36**, 747 (1980).
642. W. R. Bowman, B. T. Golding and W. P. Watson, *J. Chem. Soc., Perkin II*, 731 (1980).
643. M. Yamashita, H. Nomoto and H. Imoto, *Synthesis*, 716 (1987).
644. G. Bartoli, M. Bosco, A. Melandri and A. C. Boicelli, *J. Org. Chem.*, **44**, 2087 (1979).
645. F. Kienzle, *Helv. Chim. Acta*, **61**, 3436 (1978).
646. G. Bartoli, M. Bosco and G. Baccolini, *J. Org. Chem.*, **45**, 522 (1980).
647. G. Bartoli, R. Leardini, A. Medici and G. Rosini, *J. Chem. Soc., Perkin I*, 692 (1978).
648. G. Bartoli, A. Medici, G. Rosini and D. Tavernari, *Synthesis*, 436 (1978).
649. G. Bartoli, M. Bosco, G. Cantagalli and R. Dalpozzo, *Tetrahedron*, **40**, 3437 (1984).
650. (a) F. Boberg, M. Ruhr, K.-H. Garburg and A. Garming, *J. Heterocycl. Chem.*, **23**, 759 (1986).
(b) F. Boberg, K.-H. Garburg, K.-J. Goerlich, E. Pereit and M. Ruhr, *Liebigs Ann. Chem.*, **239** (1985).
651. O. Mitsunobu and N. Yoshida, *Tetrahedron Letters*, **22**, 2295 (1981).
652. S. H. Andersen, N. B. Das, R. D. Jorgensen, G. Kjeldsen, J. S. Knudsen, S. C. Sharma and K. B. G. Torssell, *Acta Chem. Scand.*, **B36**, 1 (1982).
653. N. B. Das and K. G. B. Torssell, *Tetrahedron*, **39**, 2227 (1983).
654. R. I. Bodina, E. S. Lipina and V. V. Perekalin, *Zh. Org. Khim.*, 2095 (1976).
655. (a) H. Feger and G. Simchen, *Liebigs Ann. Chem.*, 428 (1986).
(b) H. Feger and G. Simchen, *Liebigs Ann. Chem.*, 1456 (1986).
656. M. Cherest and X. Lusinchi, *Tetrahedron*, **42**, 3825 (1986).
657. E. W. Colwin, A. K. Beck and D. Seebach, *Helv. Chim. Acta*, **64**, 2264 (1981).
658. D. Seebach, A. K. Beck, T. Mukhopadhyay and E. Thomas, *Helv. Chim. Acta*, **65**, 1101 (1982).
659. N. Kornblum in *Supplement F: The Chemistry of Amino, Nitroso and Nitro Compounds and their Derivatives* (Ed. S. Patai), Wiley–Interscience, Chichester, 1982, p. 361.
660. R. K. Norris and D. Randles, *Australian J. Chem.*, **35**, 1621 (1982).
661. M. P. Crozet and P. Vanelle, *Tetrahedron Letters*, **26**, 323 (1985).
662. M. P. Crozet, J.-M. Surzur, P. Vanelle, C. Ghiglione and J. Maldonado, *Tetrahedron Letters*, **26**, 1023 (1985).

663. (a) F. Boberg, M. Ruhr and A. Garming, *Liebigs Ann. Chem.*, 223 (1984).
(b) F. Boberg, K.-H. Garburg, K.-J. Goerlich, E. Pereit and M. Ruhr, *Liebigs Ann. Chem.*, 911 (1984).
664. (a) O. Mitsunobu, J. Kimura, T. Shimizu and A. Kawashima, *Chem. Letters*, 927 (1980).
(b) J. Kimura, A. Kawashima and O. Mitsunobu, *Chem. Letters*, 1793 (1981).
(c) J. Kimura, H. Kobayashi, O. Miyahara and O. Mitsunobu, *Bull. Chem. Soc. Japan*, **59**, 869 (1986).
665. (a) B. C. Gilbert, R. O. C. Norman, G. Placucci and R. C. Sealy, *J. Chem. Soc., Perkin I*, 885 (1975).
(b) D. Behar and R. W. Fessenden, *J. Phys. Chem.*, **76**, 1706 1710 (1972).
666. D. Veltwisch and K.-D. Asmus, *J. Chem. Soc., Perkin II*, 1143 (1982).
667. J. T. Pinhey, E. Rizzardo and G. C. Smith, *Australian J. Chem.*, **31**, 97 (1978).
668. R. D. Grant, J. T. Pinhey, E. Rizzardo and G. C. Smith, *Australian J. Chem.*, **38**, 1505 (1985).
669. G. Barbarella, G. Pitacco, C. Russo and E. Valentin, *Tetrahedron Letters*, **24**, 1621 (1983).
670. G. V. Oreshko, M. A. Fadeev, G. V. Lagodzinskaya, I. Yu. Kozyreva and L. T. Eremenko, *Izv. Akad. Nauk SSSR, Ser. Khim.*, 2737 (1985); Engl. transl., p. 2533.

Nitrones, Nitronates and Nitroxides
Edited by S. Patai and Z. Rappoport

CHAPTER **4**

Nitroxides*

HANS GÜNTER AURICH
Fachbereich Chemie, Philipps-Universität Marburg, Marburg, Federal Republic of Germany

*Dedicated to Prof. Dr. K. Dimroth on the occasion of his 70th birthday.

I. GENERAL AND THEORETICAL ASPECTS

A. Stabilization of the Nitroxide Group

Nitroxides[1–4] are *N*,*N*-disubstituted NO radicals, the unpaired electron being delocalized between the nitrogen and oxygen atoms, as is indicated by the mesomeric formulae A and B. Following IUPAC rules these radicals should be named aminyl oxides, but aminoxyl would be more appropriate. In the Soviet literature the name nitroxyl or even iminoxyl is frequently used. Nevertheless, we prefer the term nitroxide in view of the vast amount of literature using that name.

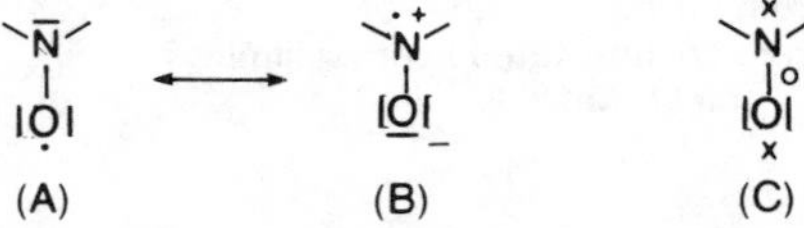

The resonance hybrid formulation A ↔ B and the Linnett formulation[5] C indicate the unique stabilization of these radicals in which three π-electrons are distributed over two atomic centres. A simple MO picture shows the remarkable gain in energy resulting from delocalization of the three π-electrons within the two molecular π-orbitals, obtained by a linear combination of the two atomic p_z-orbitals (Figure 1). Since the nitrogen atom of the skeleton has formally two positive charges, one has to assume that its atomic p_z-orbital lies below the atomic p_z-orbital of the singly charged oxygen atom. Two of the π-electrons occupy the low-lying bonding π-orbital, the third one is in the antibonding π*-orbital, thus yielding a net π-bonding of one electron. That the NO bond of the nitroxide group is in fact a one-and-a-half bond (one σ-bond and a half π-bond) is indicated by the bond energy of approximately 100 kcal/mol[6]. This is compared to 53 kcal/mol for the N—O single bond and 145 kcal/mol for the N=O double bond[6]. Further evidence of the bond order is found in the position of the IR frequencies for the NO valence vibration. For 2,2,6,6-tetramethylpiperidine-*N*-oxyl and 4-oxo-2,2,6,6-tetramethylpiperidine-*N*-oxyl, the delocalization energy is of the order of 30 kcal/mol[6], as shown by thermochemical studies. Thus, dimerization through formation of an O—O bond cannot occur (equation 1), because the gain of about 35 kcal/mol for

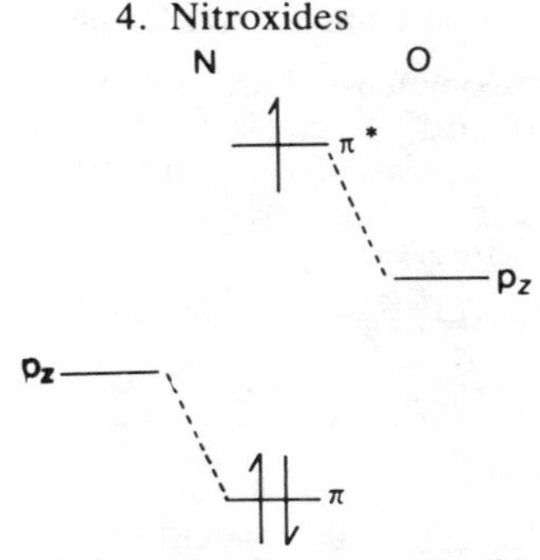

FIGURE 1. Molecular orbitals of nitroxides using linear combination of atomic orbitals.

the new bond cannot compensate for the loss of delocalization energy for the two nitroxide molecules. The gain in energy from delocalization compared to a hypothetical radical without delocalization, such as represented by formula A, is defined

$$2\ \rangle\overline{\underline{N}}-\overline{\underline{O}}\cdot \rightleftharpoons \rangle\overline{\underline{N}}-\overline{\underline{O}}-\overline{\underline{O}}-\overline{\underline{N}}\langle \tag{1}$$

according to Ingold's proposal[7] as stabilization of the nitroxide group. This means that every nitroxide radical is considerably stabilized. On the other hand, the persistence of nitroxides may be very different. There are, for example, nitroxides which have life-times of several years in the solid state. On the other hand, in solution some nitroxides exist for only a split second and therefore can be detected only under special conditions.

B. Types of Nitroxide Radicals

The variety of substituent groups which can be used in nitroxides is great[8]. Besides the parent compound, $R^1 = R^2 = H$, nitroxides with primary, secondary and tertiary alkyl groups, aryl, alkoxy and amino groups, and with substituents derived from other elements of the higher periods are known. Recently, nitroxides substituted by transition metal complexes have been described

$$R^1{-}\underset{\displaystyle \overset{|\cdot}{O}}{N}{-}R^2$$

R^1 and/or R^2 = H, Me, *n*-, *s*- and *t*-alkvl, aryl, OR, SR, NR_2, PR_2, SiR_3, ML_n
(M = transition metal, L = ligand)

Whereas the delocalization of the unpaired electron in nitroxides with alkyl groups is restricted to the nitroxide group, in aryl nitroxides and nitroxides with conjugated double bonds or substituents with a free electron pair further delocalization can occur. These nitroxides can be characterized as nitroxides with extended delocalization. Nitroxides with acyl, imino, nitrone and vinyl groups, belong to this class of radicals:

$$R^1{-}\underset{\displaystyle \overset{|\cdot}{O}}{N}{-}\overset{\displaystyle R^2}{\overset{|}{C}}{=}X \qquad X = {=}O,\ {=}NR,\ {=}\underset{\displaystyle O^-}{\overset{+}{\underset{|}{N}}}{-}R,\ {=}N{-}OH,\ {=}CR_2$$

One such nitroxide with extended delocalization, the porphyrexide **11**, is the first known isolable organic radical[9]. Radicals **1–12** are some representative nitroxides. Finally, many diradicals and polyradicals with nitroxide groups are known. In these radicals two or more nitroxide moieties are connected by various bridges[10], as shown for instance in structure **13**.

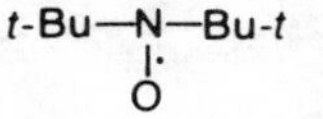

(**1**) t-Bu—N(O•)—Bu-t (**2**) CF_3—N(O•)—CF_3 (**3**) t-Bu—N(O•)—Ph (**4**) Ph—N(O•)—Ph (**5**) Ph—N(O•)—H

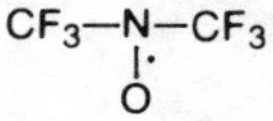

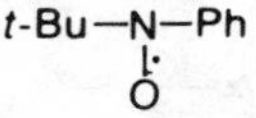

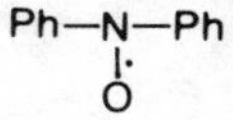

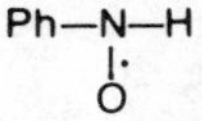

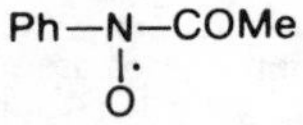

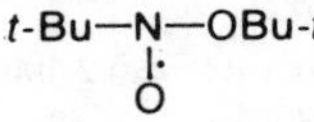

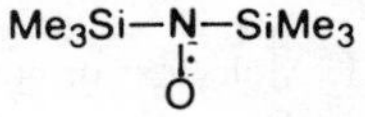

(**6**) Ph—N(O•)—COMe (**7**) t-Bu—N(O•)—OBu-t (**8**) Me_3Si—N(O•)—$SiMe_3$

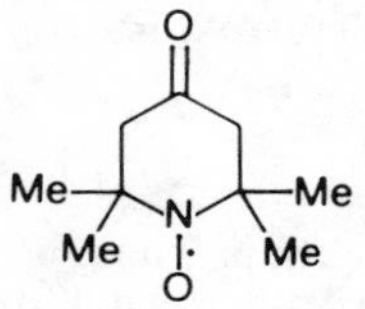

(**9**) 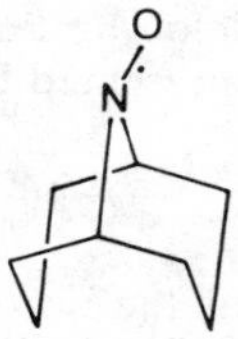 (**10**) 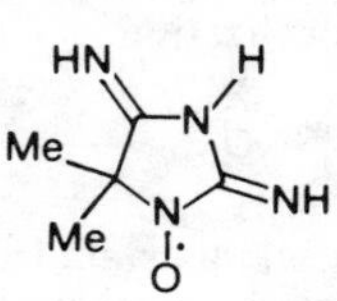 (**11**) 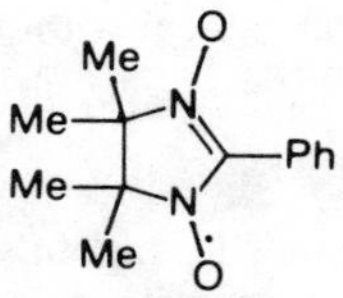 (**12**)

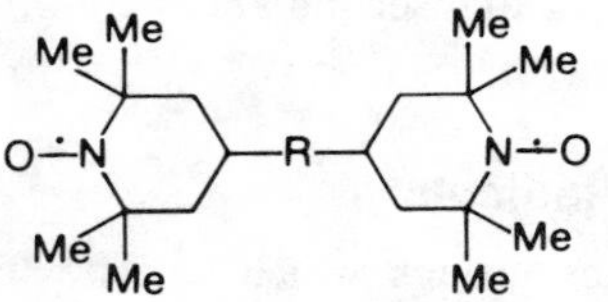

(**13**)

C. Related Radicals

1. Thionitroxides

Substitution of the oxygen in nitroxides by sulphur leads to thionitroxides. Since the electronegativity of sulphur is less than that of oxygen and the bond distance between nitrogen and sulphur is larger, the thionitroxides are less stabilized than nitroxides. Furthermore, dimerization is favoured in the case of thionitroxides due to the higher bond energy for formation of a S—S bond (equation 2). Therefore thionitroxides can be detected in only a few cases under favourable circumstances[11–13].

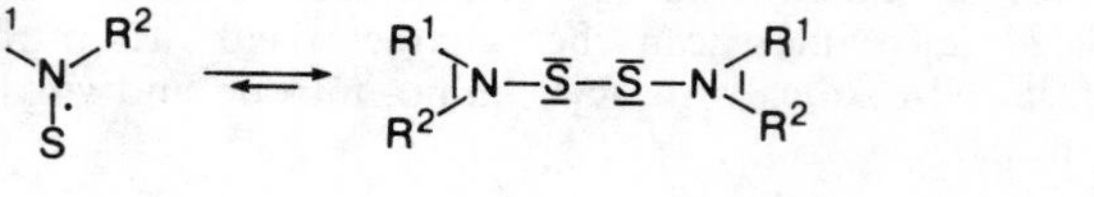

$$R^1R^2N\!-\!\dot{S} \rightleftharpoons R^1R^2N\!-\!S\!-\!S\!-\!NR^1R^2 \qquad (2)$$

2. Iminoxyls

The oxidation products of oximes, the iminoxyls, are fundamentally different from nitroxides. Whereas nitroxides are π-radicals, in which the unpaired electron

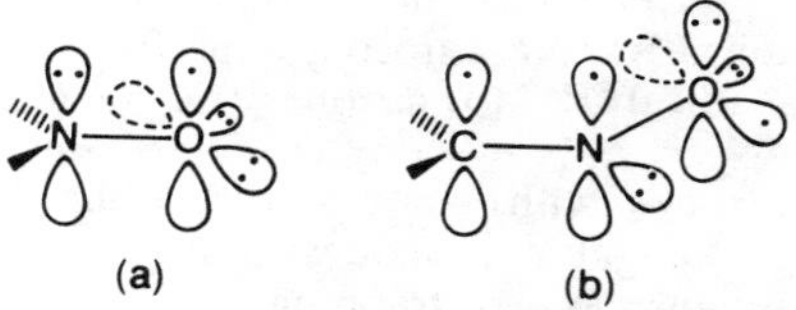

FIGURE 2. Electronic configuration of (a) nitroxides and (b) iminoxyls.

occupies the π^*-orbital in the z-plane, in iminoxyls the unpaired electron occupies a molecular orbital which results from a linear combination of the nitrogen sp^2-orbital and the oxygen p_y-orbital and is therefore orthogonal to the z-plane. Since this orbital has some s-character, iminoxyls are called σ-radicals. Obviously, an electron configuration joining four electrons in the three-centre molecular orbital system of the z-plane and three electrons in the two-centre molecular orbital system of the y-plane is more favourable than the alternative possibility with only three electrons in the three-centre molecular orbital system of the z-plane but four electrons in the two-centre molecular orbital system of the y-plane (see Figure 2).

II. FORMATION OF NITROXIDES

Because of their high stabilization nitroxides are easily formed. This is true for nitroxides which are persistent enough to be isolated as well as for those which are only intermediates in a reaction sequence. The most important methods of nitroxide formation are shown in Scheme 1:

(1) The oxidation of hydroxylamines or the corresponding anions, including cases where the hydroxylamino group is only an intermediate.

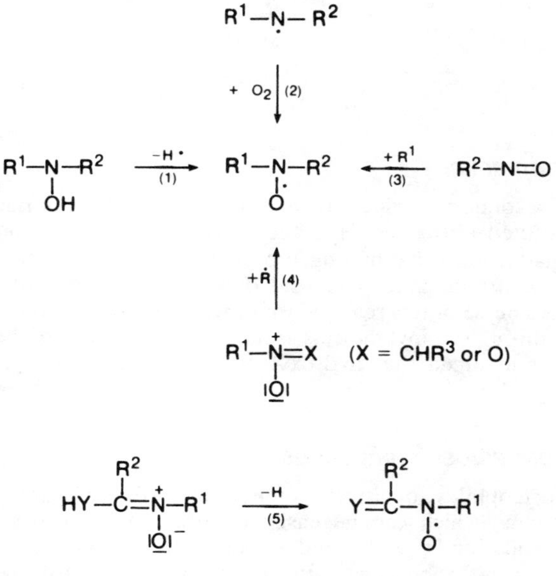

$$R^1{-}\dot{N}{-}R^2 \xrightarrow{+\,O_2\ (2)} R^1{-}N(O^\bullet){-}R^2$$

$$R^1{-}N(OH){-}R^2 \xrightarrow[(1)]{-H^\bullet} R^1{-}N(O^\bullet){-}R^2 \xleftarrow[(3)]{+R^1} R^2{-}N{=}O$$

$$R^1{-}\overset{+}{N}(|\underline{\overline{O}}|^-){=}X \xrightarrow{+\dot{R}\ (4)} R^1{-}N(O^\bullet){-}R^2 \qquad (X = CHR^3 \text{ or } O)$$

$$HY{-}C(R^2){=}\overset{+}{N}(|\underline{\overline{O}}|^-){-}R^1 \xrightarrow[(5)]{-H} Y{=}C(R^2){-}N(O^\bullet){-}R^1$$

SCHEME 1

(2) Addition of an oxygen atom to an aminyl radical.
(3) Addition of a radical R$^\cdot$ to a nitroso group.
(4) Addition of a radical R$^\cdot$ to nitrones (X = CHR^3) or nitro compounds (X = O).
(5) Oxidation of nitrones with heteroatoms Y in the β-position (especially Y = N), yielding conjugated nitroxides.

Routes (3) and (4) are known as spin trapping.

A. Oxidation of the Hydroxylamino Group

In hydroxylamines such as *N*-hydroxy-2,2,6,6-tetramethylpiperidine and *N*-hydroxy-4-oxo-2,2,6,6-tetramethylpiperidine, the bond energies for the OH bond are relatively low, 69.6 and 71.8 kcal/mol, respectively, compared to 80–85 kcal/mol for the corresponding oximes[14]. Thus, oxidation of hydroxylamines occurs readily. This method of preparation is very important because hydroxylamines synthesized from amines or nitroso compounds, respectively, can be oxidized directly without isolation of the intermediate.

1. Formation from hydroxylamines

Oxidation of hydroxylamines occurs with a variety of reagents[15,16]. Potassium hexacyanoferrate (III), silver oxide, lead dioxide, lead tetraacetate, nickel peroxide and mercury oxide are frequently used. In some cases oxidation can occur using oxygen itself either with or without a catalyst. Organic compounds such as nitroso compounds can also oxidize hydroxylamines[17]. In this connection, the formation of monoaryl[18]- or monoalkyl nitroxides[19] from hydroxylamines and nitroso compounds should also be mentioned (equation 3).

$$R^1{-}N(OH){-}H + R^2{-}N{=}O \rightleftharpoons R^1{-}N(\dot{O}){-}H + R^2{-}N(\dot{O}){-}H \rightleftharpoons R^1{-}N{=}O + R^2{-}N(OH){-}H \qquad (3)$$

2. Formation from amines

Oxidation of secondary amines is one of the most important methods of preparing disubstituted nitroxides. Hydrogen peroxide in the presence of tungstate, molybdate or vanadate, alkaline hydrogen peroxide, perbenzoic acid and substituted perbenzoic acids, hydroperoxides and lead dioxide are among the many oxidizing agents used[20]. The course of this reaction is not well known. In principle, oxidation of amines can occur through aminyl radicals as the direct precursors of the nitroxides, but in most cases it is assumed that hydroxylamines are the direct precursors of the nitroxides[21].

3. Formation from nitroso compounds

Addition of nucleophiles to nitroso compounds leads to hydroxylamines or the corresponding anions, which can be easily oxidized to nitroxides (equation 4)[22]. Frequently, the oxidation is performed by the nitroso compound itself which is partly reduced to azoxybenzene. In particular, Grignard compounds and organolithium compounds react according to equation (4)[23]. Sulphinic acids[24] and

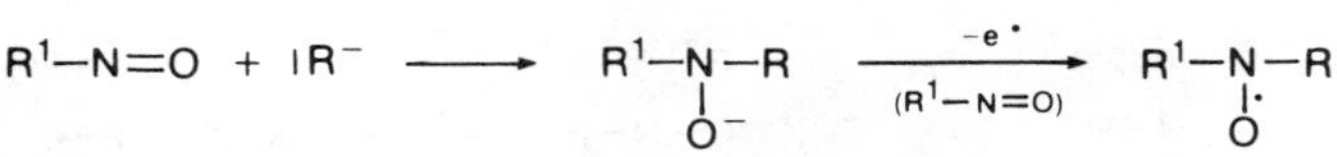

$$R^1{-}N{=}O + |R^- \longrightarrow R^1{-}N(O^-){-}R \xrightarrow[(R^1-N=O)]{-e^\cdot} R^1{-}N(\dot{O}){-}R \quad (4)$$

the hydrogen sulphite ion[25,26] also yield adducts with nitroso compounds, which are oxidized directly to the nitroxides **14** and **15**, respectively. Alkenes bearing at least one hydrogen atom at the allylic position may easily add to aromatic nitroso compounds forming nitroxides by the subsequent oxidation reaction (5)[27]. For instance, 2,3-dimethyl-2-butene yields **16** in this manner[28]. Radicals **17** and **18** are formed from imines[29] and ketones[30], respectively, which react in their tautomeric enamine and enolic forms in an analogous manner. Finally, 4-morpholino-1-cyclohexene and aromatic nitroso compounds yield nitroxides **19**[31].

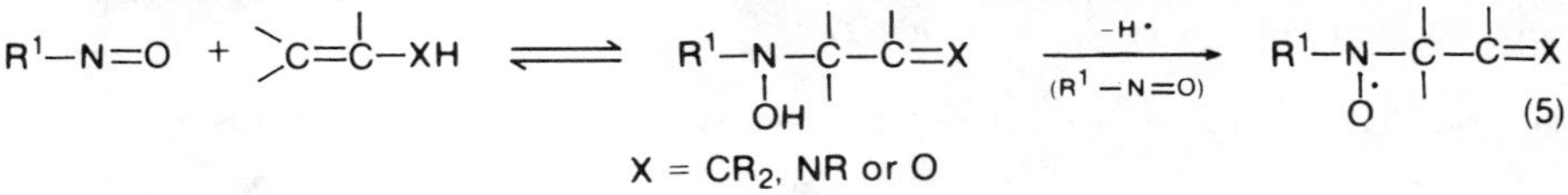

$$R^1{-}N{=}O + {>}C{=}C{-}XH \rightleftharpoons R^1{-}N(OH){-}C{-}C{=}X \xrightarrow[(R^1-N=O)]{-H^\cdot} R^1{-}N(\dot{O}){-}C{-}C{=}X \quad (5)$$

X = CR_2, NR or O

$R^1{-}N(\dot{O}){-}SO_2Ar$ (**14**)

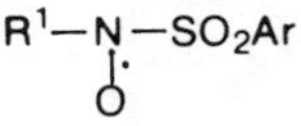

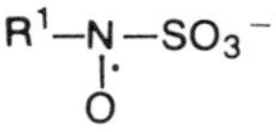

$R^1{-}N(\dot{O}){-}SO_3^-$ (**15**)

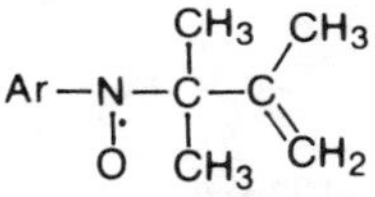

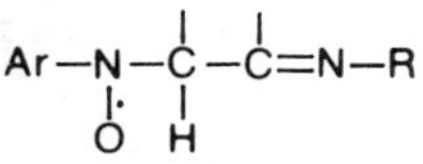

$Ar{-}N(\dot{O}){-}C(CH_3)_2{-}C(CH_3){=}CH_2$ (**16**)

$Ar{-}N(\dot{O}){-}C{-}C(H){=}N{-}R$ (**17**)

$Ar{-}N(\dot{O}){-}C{-}C{=}O$ (**18**)

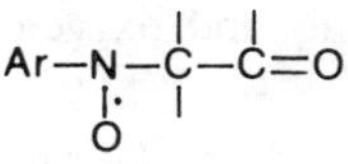

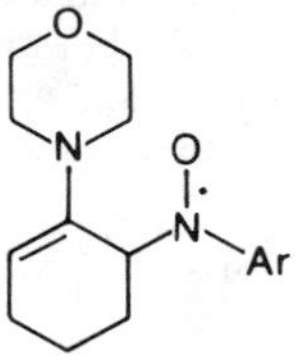

(**19**)

4. Formation from nitrones

In the same way, adducts from nucleophiles and nitrones may be converted to nitroxides (equation 6). Thus, Grignard compounds or organilithium compounds attack nitrones, the adducts **20** being subsequently oxidized to nitroxides **21**[32] (equation 7). The addition of amines to nitrones usually occurs to only a small extent giving an equilibrium mixture. Nevertheless, the adduct can be oxidized to nitroxides (equation 6, X = R_2N)[33,34]. The oxidation of nitrones **22**, yielding nitroxides **23** with extended delocalization[35,36] (see route 5 in Scheme 1), can formally be considered as a reaction of the tautomeric hydroxylamine (equation 8).

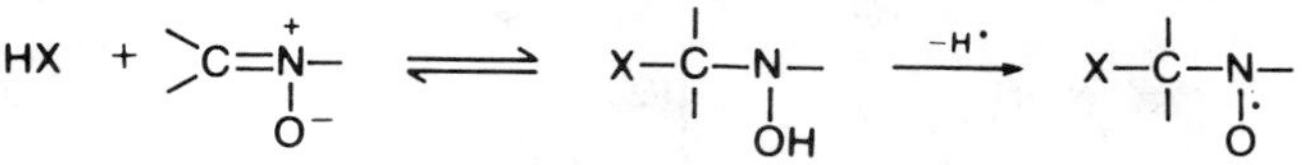

$$HX + {>}C{=}\overset{+}{N}(O^-){-} \rightleftharpoons X{-}C{-}N(OH){-} \xrightarrow{-H^\cdot} X{-}C{-}N(\dot{O}){-} \quad (6)$$

$$\xrightarrow[2.\ H_2O]{1.\ R^2MgX} \text{(20)} \xrightarrow{-H^\cdot} \text{(21)} \quad (7)$$

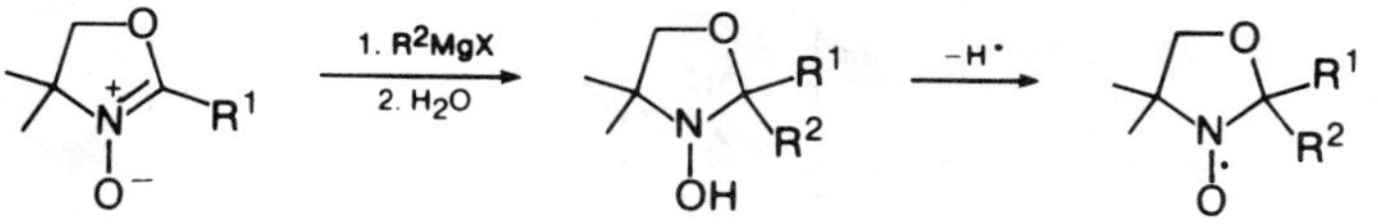

(**20**) (**21**)

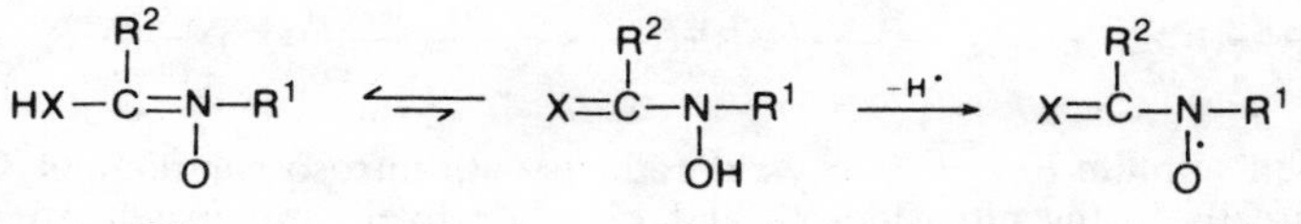

$$\underset{(22)}{HX{-}\overset{R^2}{\overset{|}{C}}{=}\underset{\underset{O}{|}}{N}{-}R^1} \rightleftharpoons X{=}\overset{R^2}{\overset{|}{C}}{-}\underset{\underset{OH}{|}}{N}{-}R^1 \xrightarrow{-H^\bullet} \underset{(23)}{X{=}\overset{R^2}{\overset{|}{C}}{-}\underset{\underset{O}{|\bullet}}{N}{-}R^1} \tag{8}$$

$$X = RN{=} \text{ or } RCH{=}$$

5. Formation from O-substituted hydroxylamines

O,N-Disubstituted hydroxylamines can, in special cases, be converted to nitroxides. Thus, deprotonation of **24** gives the anion **25** which exists in equilibrium with the rearranged anion **26** (equation 9). Oxidation of the latter, either by oxygen or electrochemically, gives the nitroxide **27**[37].

$$\underset{(24)}{RMe_2SiNH{-}OSiMe_2R} \xrightarrow{-H^+} \underset{(25)}{RMe_2Si\bar{\underline{N}}{-}O{-}SiMe_2R} \rightleftharpoons \underset{(26)}{(RMe_2Si)_2N{-}\bar{\underline{O}}|^-} \xrightarrow{-e} \underset{(27)}{(RMe_2Si)_2N\dot{-}O} \tag{9}$$

B. Nitroxides from Aminyl Radicals

Contrary to nitroxides, aminyl radicals which are substituted by two alkyl groups are not stabilized, and even when substituted by two aryl groups only to a small degree. Consequently, aminyl radicals are very reactive, forming in most cases nitroxides on reacting with oxygen (equation 10). Unambiguous evidence for the direct reaction of aminyl radicals and oxygen was obtained from the reaction of

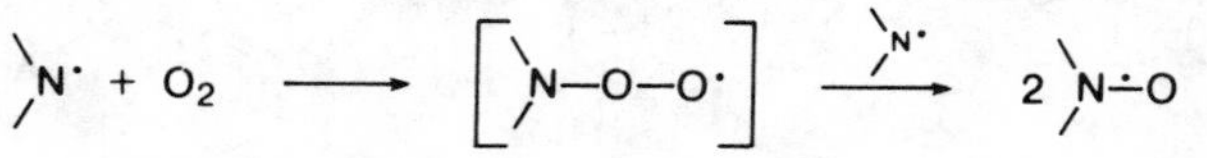

$$>N^\bullet + O_2 \longrightarrow \left[>N{-}O{-}O^\bullet\right] \xrightarrow{>N^\bullet} 2\ >N\dot{-}O \tag{10}$$

$>N{-}Cl \xrightarrow[h\nu]{-Cl^\bullet} >N^\bullet$

$>N{-}N{=}N{-}N< \xrightarrow[h\nu \text{ or } \Delta]{-N_2} >N^\bullet$

$>N{-}NO \xrightarrow[h\nu]{-NO^\bullet} >N^\bullet$

$>N{-}N< \xrightarrow[h\nu \text{ or } \Delta]{} >N^\bullet$

$>N{-}H \xrightarrow[(RO{-}OR)\ h\nu]{-H^\bullet} >N^\bullet$

$>N^\bullet \xrightarrow{O_2} >N\dot{-}O$

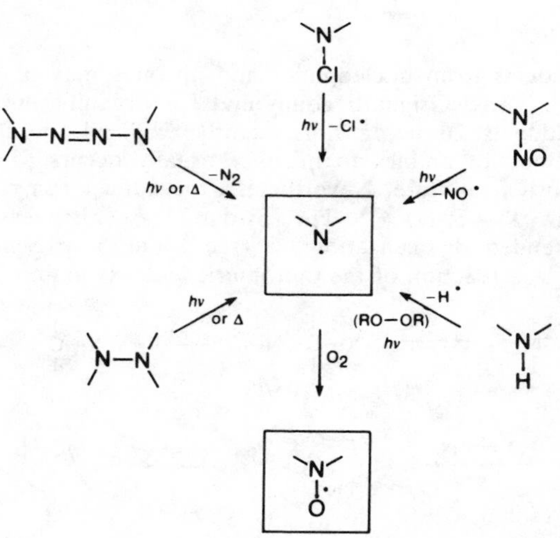

SCHEME 2

2,2,6,6-tetramethylpiperidyl radical and oxygen-17 at low temperatures[38]. The reaction rate is so fast that it cannot be measured. An adduct of the aminyl radical and an oxygen molecule is assumed to be an intermediate in the formation of the nitroxide. This method is extremely important in ESR spectroscopic investigations. It is especially suitable for the ^{17}O-labelling of nitroxides. The precursors of the nitroxides, the aminyl radicals, can be easily generated in the cavity of an ESR spectrometer by either photolysis or thermolysis of tetrazenes or hydrazines, by photolysis of *N*-chloroamines or *N*-nitrosoamines and by hydrogen abstraction from secondary amines[39] (see Scheme 2).

C. Spin Trapping

1. General remarks

The reaction of a short-lived radical with a nitroso compound, a nitrone or any other suitable compound, affording a new, persistent radical is called spin trapping because, in using this process the radical character, the spin, is retained[40,41]. This method has wide applications for two reasons. Firstly, it can be used to obtain nitroxides which are difficult to generate using other methods. Secondly, it can be used to study indirectly the mechanism of certain reactions, since it allows the detection of transient radicals. In the latter case, one must consider the possibility that the formation of a 'spin adduct' can also arise by the addition of a nonradical nucleophile to the spin trap, followed by oxidation of the adduct[42]. Furthermore, the high sensibility of the ESR method can easily lead to misinterpretations insofar as radicals arising from side-reactions are also trapped.

The exceptionally high reactivity of nitroso compounds in radical addition reactions was discovered many years ago. Szwarc found that nitrosobenzene reacts with methyl radicals about 10^5 times faster than benzene does[43]. This high reactivity can be understood if one assumes that the transition state itself is already stabilized by partial delocalization owing to the developing nitroxide moiety. Theoretical considerations suggest that the reaction begins with the transfer of electron density from the radical to the unoccupied antibonding π^*-orbital of the nitroso group. Consequently, the NO bond is stretched, reducing the bond energy and facilitating twisting of the NO group. During the progress of the reaction, one electron of the original NO double bond is used in the formation of the new bond with the attacking radical, the other one being localized at the oxygen. This decoupling of the electron pair is facilitated by the increasing twist of the stretched NO bond leading to an increase in overlap between the p_z-orbital of the oxygen with its single electron, and the original sp^2-orbital of the nitrogen with its electron pair. The latter gets more and more p-character which results in the final formation of the two molecular π-orbitals of the nitroxide group by linear combination with the single occupied orbital of oxygen:

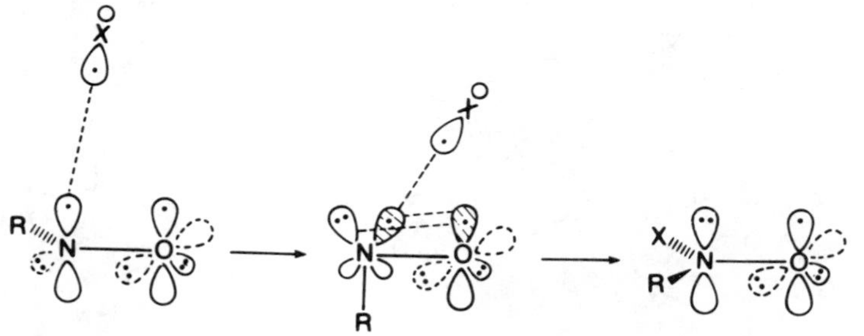

Similar arguments can be used to explain the high reactivity of nitrones in radical trapping reactions. In this case, the decoupling of the C=N double bond with the beginning of C—R bond formation is facilitated by the increasing gain in delocalization energy connected with the formation of the nitroxide moiety. Furthermore, it seems reasonable that the addition of radicals is rendered more difficult if the nitrone is stabilized by conjugation with other groups, such as the phenyl group.

On the other hand, decoupling of an electron pair of the considerably stabilized nitro group is much more difficult. Thus, nitro compounds do not trap alkyl radicals, except when the nucleophilic character of the radical is enhanced by electron-donating groups[44].

2. *The properties of spin traps*

Formulae **28–43** show a selection of spin traps. The use of nitroso compounds has the advantage over nitrones that the radical being trapped is directly attached

t-Bu—NO (**28**) [D_9]-*t*-Bu—NO (**29**) Me—C(=O)—C(Me)$_2$—NO (**30**) $(CF_3)_3C$—NO (**31**) PhNO (**32**)

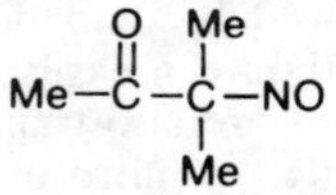

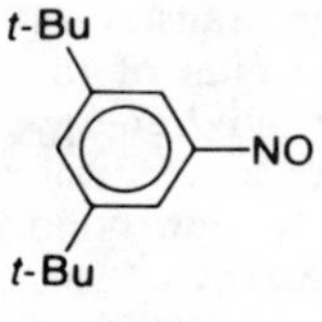

(**33**) 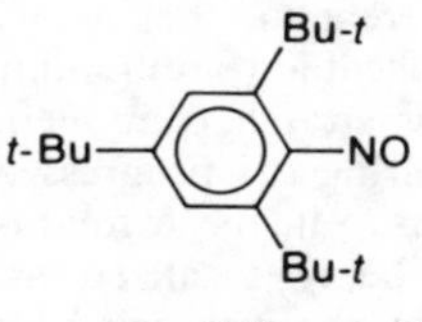 (**34**) 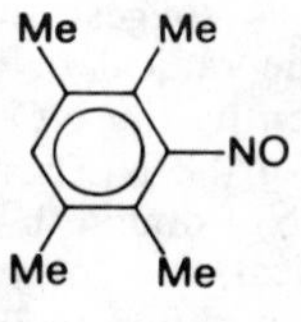 (**35**) 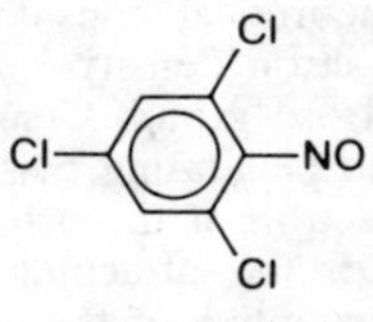 (**36**) 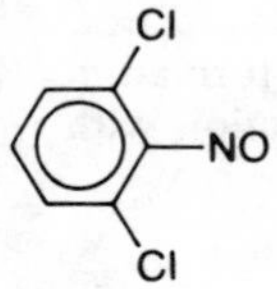 (**37**)

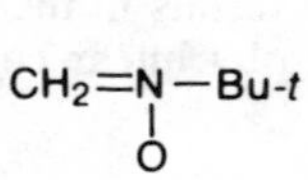

CH_2=N(O)—Bu-*t* (**38**) PhCH=N(O)—Bu-*t* (**39**) 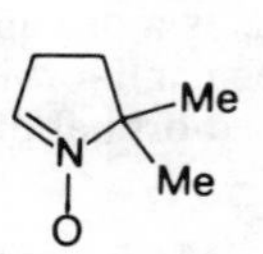 (**40**) 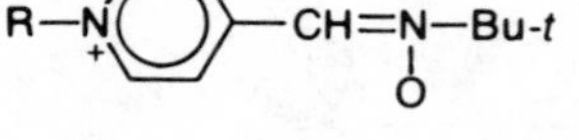 (**41**) 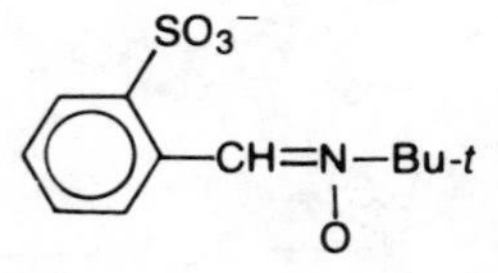 (**42**) 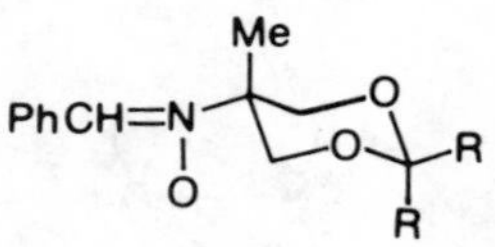 (**43**)

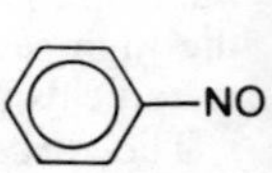

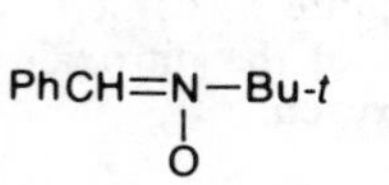

to the nitroxide group and thus makes a contribution to the splitting of the ESR spectrum, enabling, in most cases, unambiguous identification of the original radical. On the contrary, the ESR spectra of spin adducts with nitrones give less information. However, the ESR spectra arising from partial decomposition of the nitroso compounds are often superimposed on the spectra of spin adducts derived from nitroso compounds. Aliphatic (e.g. 2-methyl-2-nitrosopropane) as well as aromatic (e.g. nitrosobenzene) nitroso compounds may be cleaved to some extent thermally, even at room temperature, or photochemically, to nitrogen oxide and the corresponding alkyl or aryl radical which is trapped by intact nitroso compound, giving dialkyl or diaryl nitroxides, respectively[45]. This is not true for spin traps **34–37** which are finding increasing use. Nitrosodurene (**35**) easily gives spin adducts[46], though even in solution an equilibrium between dimeric and monomeric forms exists.[47,48]

As was shown recently, 2,6-dichloronitrosobenzene (**37**) and similar compounds (2,6-dibromonitrosobenzene, 2,4,6-tribromonitrosobenzene) react with compounds like *o*-xylene at slightly elevated temperature by abstraction of hydrogen and trapping of the benzylic radical, thus acting as a spin trap as well as a radical-generating reagent[49].

For nitrone spin traps the possibility for variations at the side-groups is greater. The compounds **41–43**, which have been synthesized only recently[50,51], are of special interest. The water-soluble compound **42**[50] is the first anionic spin trap. In nitrones **41** and **43** long-chain alkyl groups may also be introduced[51]. These spin traps are of particular interest for application in complex biphasic systems; as such they may be of value for the study of reactions in biological systems. The lifetime of the cationic spin adducts derived from **41** is longer than those of corresponding neutral spin adducts, since disproportionation to the corresponding nitrone and hydroxylamine is retarded by the positive charge[51].

3. *Examples of spin trap reactions*

In addition to the many different types of carbon-, nitrogen, oxygen-, sulphur- and phosphorus-centred radicals[52], an increasing number of transition metal complexes has recently been trapped by nitroso compounds[53]. A great number of radicals has also been trapped by nitrones.

Free-radical formation by the photolysis of aliphatic nitroso compounds in solution has been unambiguously confirmed by de Boer[54]. From compound **44** he has obtained nitroxide **47** which can arise only via the intermediate free radical **45** and its rearrangement product **46**. An example of intramolecular spin trapping is

$$\underset{\textbf{(44)}}{\text{CH}_3\overset{\text{c-C}_3\text{H}_5}{\text{CH}}\text{—NO}} \xrightarrow[-\text{NO}]{h\nu} \left[\underset{\textbf{(45)}}{\text{CH}_3\overset{\text{c-C}_3\text{H}_5}{\text{CH}}{}^{\bullet}} \longrightarrow \underset{\textbf{(46)}}{\text{CH}_3\text{CH}{=}\text{CHCH}_2\text{CH}_2{}^{\bullet}} \right] \xrightarrow{+\,\textbf{44}}$$

$$\underset{\textbf{(47)}}{\text{CH}_3\text{CH}{=}\text{CHCH}_2\text{CH}_2\text{—}\underset{\text{O}^{\bullet}}{\text{N}}\text{—}\overset{\text{c-C}_3\text{H}_5}{\text{C}}\text{HCH}_3} \qquad (11)$$

the photochemical conversion of **48** to the nitroxide **49**[55]. On the other hand, the acylnitroxides **52** may be obtained by a twofold spin trapping process between the intermediate diradicals **51**, generated by photolysis of cyclic ketones **50**, and nitrogen oxide[56].

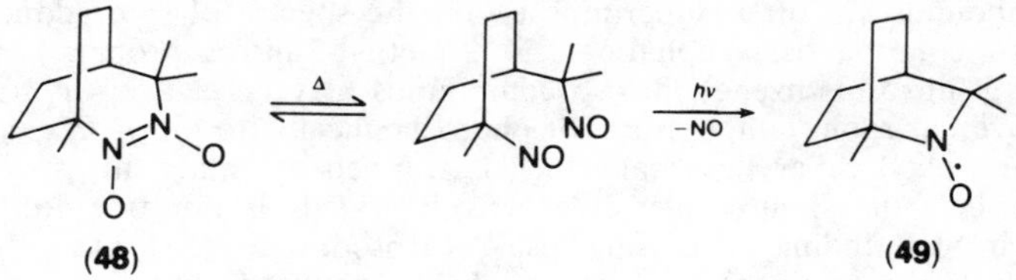

$$\textbf{(48)} \underset{}{\overset{\Delta}{\rightleftharpoons}} \text{(dinitroso)} \xrightarrow[-NO]{h\nu} \textbf{(49)} \tag{12}$$

$$\underset{\textbf{(50)}}{H_2C\overset{O}{\overset{\|}{C}}CHMe\;/\;H_2C{-}(CH_2)_n} \xrightarrow{h\nu} \left[\underset{\textbf{(51)}}{H_2C\overset{O}{\overset{\|}{C}}\dot{C}HMe\;/\;H_2\dot{C}{-}(CH_2)_n}\right] \xrightarrow{+\,NO^{\bullet}} \textbf{(52)} \tag{13}$$

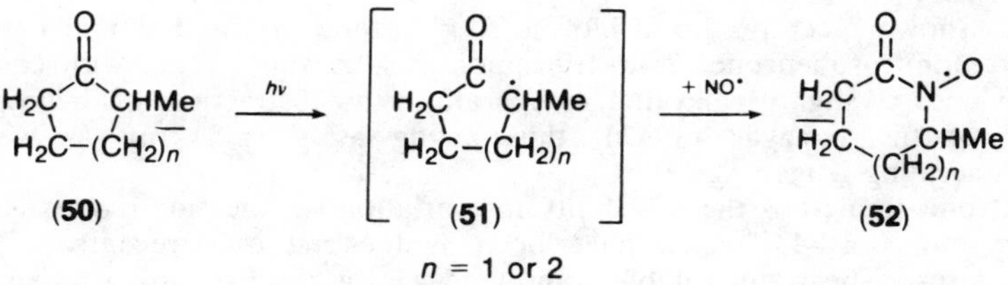

$n = 1$ or 2

After many unsuccessful attempts, peroxy radicals have been recently trapped by *t*-butyl phenyl nitrone **(39)** at low temperature, the adduct being converted to the corresponding alkoxy adduct at 0 °C[57]. Using 2-methyl-2-nitrosopropane, adducts of peroxy radicals have also been detected[57].

The following example shows that spin trapping can also be used in the investigation of biological processes. Radicals formed by γ-irradition of aqueous solutions of dipeptides or of bases such as uracil or thymine, may be trapped using 2-methyl-2-nitrosopropane (equation 14)[58].

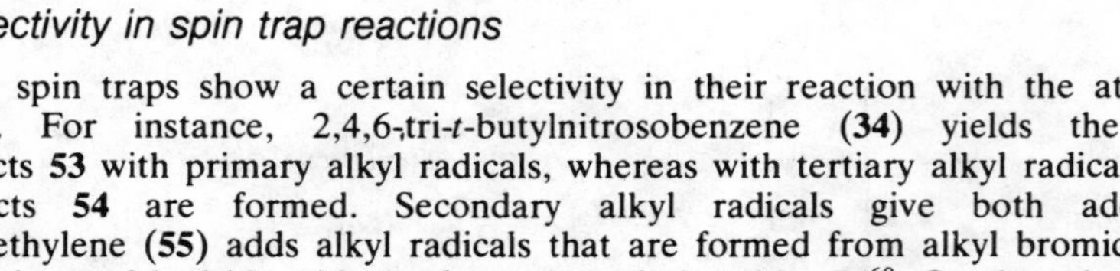

$$H_2O \longrightarrow HO^{\bullet} + \text{uracil} \xrightarrow{t\text{-BuNO}} \text{5-hydroxy-6-}(N\text{-}t\text{-butylnitroxyl})\text{-5,6-dihydrouracil} \tag{14}$$

4. Selectivity in spin trap reactions

Some spin traps show a certain selectivity in their reaction with the attacking radicals. For instance, 2,4,6-tri-*t*-butylnitrosobenzene **(34)** yields the usual *N*-adducts **53** with primary alkyl radicals, whereas with tertiary alkyl radicals, only *O*-adducts **54** are formed. Secondary alkyl radicals give both adducts[59]. Nitrosoethylene **(55)** adds alkyl radicals that are formed from alkyl bromides and tri-*t*-butylstannyl hydride with the formation of nitroxides **56**[60]. On the other hand, *t*-butyl radicals generated by the irradiation of 2,2-azoisobutane surprisingly yield spin adduct **57**[61].

Nitrone **58** allows for the differentiation between carbon- and oxygen-centred radicals[62]. Alkyl and aryl radicals add to the nitrone group, whereas oxygen-centred radicals abstract hydrogen from the hydroxy group thus affording a radical **59** that may be considered as both phenoxyl and nitroxide.

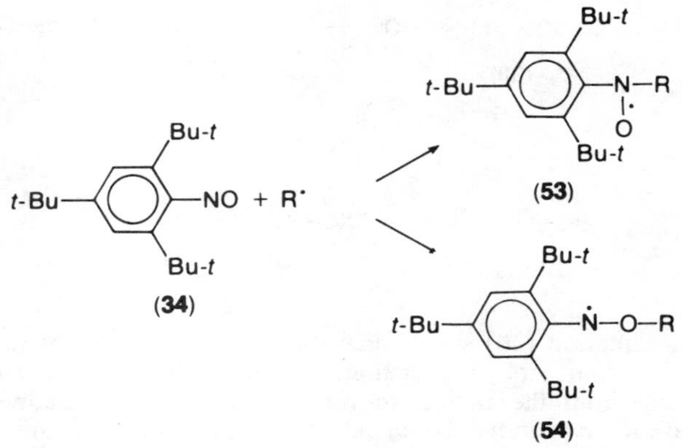

(15)

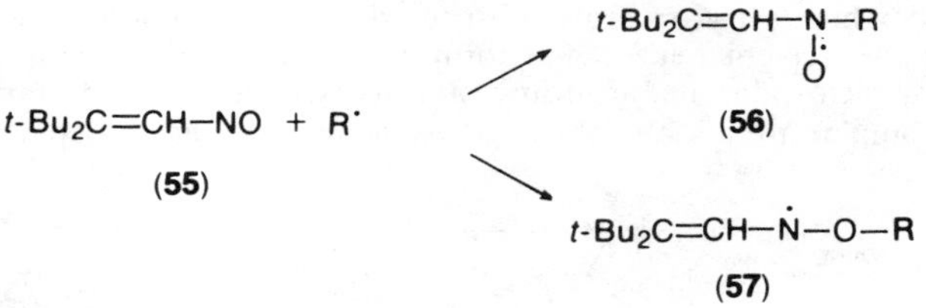

(16)

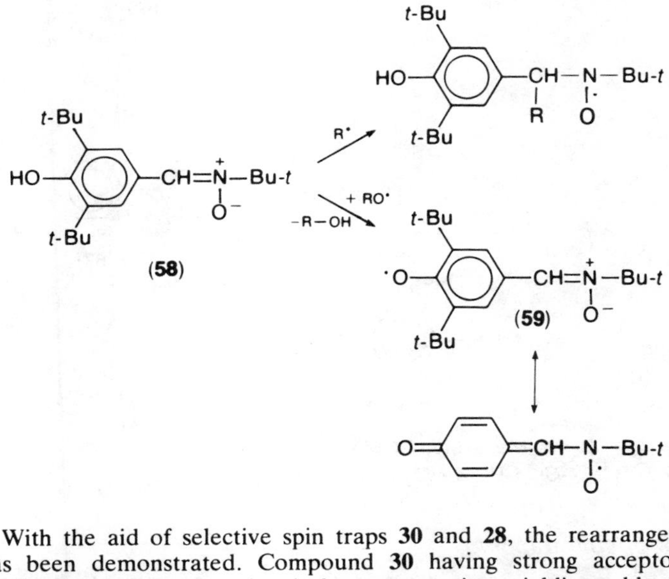

(17)

With the aid of selective spin traps **30** and **28**, the rearrangement of radical **60** has been demonstrated. Compound **30** having strong acceptor properties traps radical **60** in a fast reaction before rearranging, yielding adduct **62**. On the other hand, with compound **28** having weaker acceptor properties, rearrangement of the radical is faster and only the spin adduct **63** of the rearranged radical **61** is detected[63].

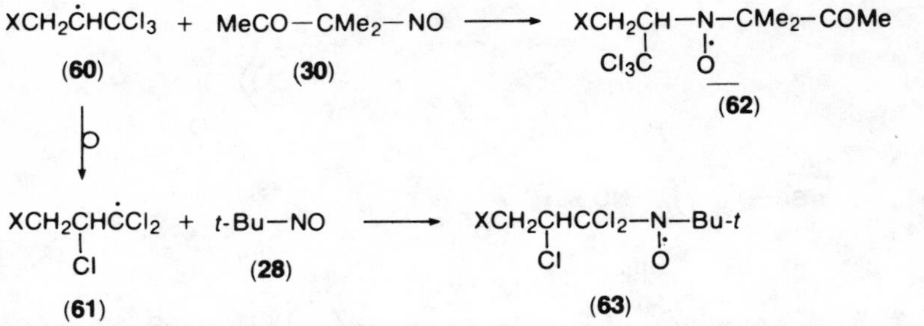

$$XCH_2\dot{C}HCCl_3 + MeCO{-}CMe_2{-}NO \longrightarrow XCH_2CH(CCl_3){-}N(\dot{O}){-}CMe_2{-}COMe \quad (18)$$

(60) (30) (62)

$$XCH_2CHCl\dot{C}Cl_2 + t\text{-}Bu{-}NO \longrightarrow XCH_2CHClCCl_2{-}N(\dot{O}){-}Bu\text{-}t \quad (19)$$

(61) (28) (63)

The example in equation (20) shows how the donor properties of the attacking radical influence the course of the reaction, indicating that a transfer of electron density is occurring from the radical to the spin trap before the new bond is formed. Aryl radicals substituted by an electron-donating group are sufficiently nucleophilic to attack nitrosodurene directly in benzene solution yielding radicals **64**. On the other hand, aryl radicals substituted by electron-accepting groups react primarily with the solvent benzene to form cyclohexadienyl radicals **65** which are then trapped by nitrosodurene, affording radicals **66**. These may be further oxidized to yield **67**[64]. Similar trends are observed using nitrones as the spin trap, but the effects are smaller[65].

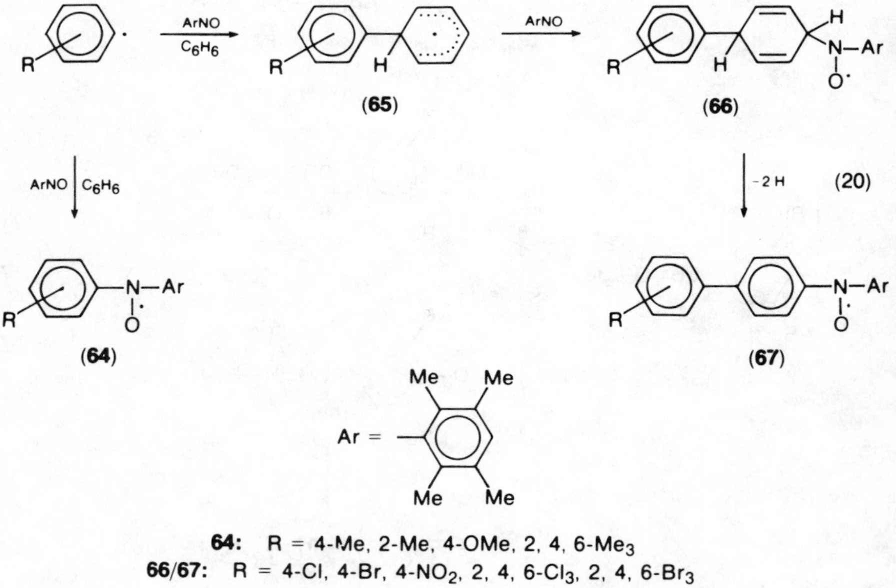

ArNO, C_6H_6 ArNO (65) (66) ArNO C_6H_6 −2 H (20) (64) (67)

Ar = 2,3,5,6-tetramethylphenyl (Me, Me, Me, Me)

64: R = 4-Me, 2-Me, 4-OMe, 2, 4, 6-Me_3
66/67: R = 4-Cl, 4-Br, 4-NO_2, 2, 4, 6-Cl_3, 2, 4, 6-Br_3

5. *Kinetic studies of spin trapping*

Ingold has determined the reaction rates for spin trapping for several spin traps. For this purpose, the competition between ring-closure of the ^{13}C-labelled 5-hexenyl radical $\mathscr{H}^{\cdot}$[66] or the 6-hepten-2-yl radical $\mathscr{S}^{\cdot}$[48] and the trap reaction was utilized.

$\mathscr{H}^{\bullet} \xrightarrow{\cdot\text{O}} \mathscr{C}^{\bullet}$ $\mathscr{S}^{\bullet} \xrightarrow{\text{O}} \mathscr{P}^{\bullet}$

$\mathscr{H}^{\bullet} \xrightarrow{+T} \mathscr{H}-T$, $\mathscr{C}^{\bullet} \xrightarrow{+T} \mathscr{C}-T$, $\mathscr{S}^{\bullet} \xrightarrow{+T} \mathscr{S}-T$, $\mathscr{P}^{\bullet} \xrightarrow{+T} \mathscr{P}-T$

From the rate of cyclization $\mathscr{H}^{\bullet} \rightarrow \mathscr{C}^{\bullet}$ or $\mathscr{S}^{\bullet} \rightarrow \mathscr{P}^{\bullet}$ and the proportion of spin adducts, $\mathscr{H}$–T/$\mathscr{C}$–T or $\mathscr{S}$–T/$\mathscr{P}$–T, absolute rate constants were calculated. The relative rate constants[48] are given in Table 1.

TABLE 1. Relative rate constants for the spin trapping of alkyl radicals in benzene at 40°C[48]

Spin trap	Primary ($\mathscr{H}^{\bullet}$)	Secondary ($\mathscr{S}^{\bullet}$)
2-Methyl-2-nitrosopropane (**28**)	100[a]	68
Nitrosodurene (**35**)	450	450
Tri-*t*-butylnitrosobenzene (**34**)	5.2	0.2
5,5-Dimethylpyrroline-1-oxide (**40**)	29	4.7
N-Methylene-*t*-butylamine-*N*-oxide (**38**)	34	14.5
N-Benzylidene-*t*-butylamine-*N*-oxide (**39**)	1.4	0.75

[a]Assumed for standard.

The relative values show that nitroso compounds, with the exception of the sterically hindered **34**, are superior to nitrones in reactions with alkyl radicals. Similar results have been obtained by Yoshida[47]. He has estimated the relative rates of spin trapping from competition experiments with 2-methyl-2-nitrosopropane and a second spin trap using *t*-butyl radicals.

The activation energy for the addition of primary alkyl radicals to nitroso compounds is in the range of 1–2 kcal/mol; for addition to nitrones it is somewhat higher[66]. With the exception of nitrosodurene, the rate of the trap reaction is in the order primary > secondary > tertiary alkyl radicals[48]. Aliphatic nitrones (e.g. **38** and **40**) react faster than aromatic nitrones (e.g. **39**). The reason is that addition to aromatic nitrones like **39** is not only sterically more hindered, but also disfavoured because conjugation between the aromatic ring and the nitrone group must be disrupted. Finally, it should be mentioned that in reactions with the electrophilic *t*-butoxy radical, nitrones are the better spin traps compared to nitroso compounds[67].

6. Spin trapping with nitro compounds

The low reactivity of nitro compounds in spin trap reactions requires a high degree of electron transfer in the transition state. Thus, only carbon-centred radicals substituted by electron-donating groups are sufficiently nucleophilic to be trapped by nitro compounds[68]. Frequently the reactions are performed photochemically. The primarily formed spin adduct **68** can be easily split to afford a nitroso compound and an alkoxy radical. The nitroso compound can then add to the original carbon-centred radical (equation 21a), or to the alkyl radicals being formed by cleavage of the alkoxy radical (equation 21b).

$$R{-}NO_2 + \cdot C(R^1)(R^2){-}OR^3 \longrightarrow \underset{(\mathbf{68})}{R{-}\dot{N}(O){-}O{-}C(R^1)(R^2){-}OR^3} \longrightarrow R{-}N{=}O + \cdot O{-}C(R^1)(R^2){-}OR^3 \quad (21)$$

$$\cdot\overline{O}{-}C(R^1)(R^2){-}OR^3 \longrightarrow R^2{-}CO_2R^3 + \dot{R}^1 \;\text{ or }\; R^1{-}CO_2R^3 + \dot{R}^2$$

$$R{-}N{=}O \xrightarrow{R^1R^2\dot{C}OR^3} R{-}\dot{N}(O){-}C(R^1)(R^2){-}OR^3 \quad (21a)$$

$$R{-}N{=}O \xrightarrow{\dot{R}^1 \text{ or } \dot{R}^2} R{-}\dot{N}(O){-}R^1 \text{ or } R{-}\dot{N}(O){-}R^2 \quad (21b)$$

Contrary to carbon-centred radicals, the highly nucleophilic silyl-, germanyl- and stannyl-centred radicals are easily trapped by nitro compounds[69,70] (equation 22), the reaction being further favoured by the gain of bond energy in the formation of the metal–oxygen bond.

$$R{-}NO_2 + \cdot MR^1_3 \longrightarrow R{-}\dot{N}(O){-}OMR^1_3 \quad (22)$$

$$M = Si, Ge, Sn$$

D. Further Modes of Formation of Nitroxides

1. From nitro compounds

The reduction of 2-methyl-2-nitropropane either with sodium or electrochemically, yields as the final product di-*t*-butyl nitroxide[71,72]. At first, the anion radical of 2-methyl-2-nitropropane (**69**) is formed, this readily decomposes to nitrite ion and *t*-butyl radical. The latter reacts further with additional anion radicals as indicated (equations 23a–c). Di-*t*-butyl nitroxide can then arise directly

$$t\text{-Bu}{-}NO_2 + e \longrightarrow \underset{(\mathbf{69})}{t\text{-Bu}{-}NO_2^{\cdot -}} \longrightarrow t\text{-Bu}^\cdot + NO_2^- \quad (23)$$

$$\mathbf{69} + t\text{-Bu}^\cdot \longrightarrow \underset{(\mathbf{70})}{(t\text{-Bu})_2\overset{+}{N}(O^-)_2} \quad (23a)$$

$$\mathbf{69} + t\text{-Bu}^\cdot \longrightarrow t\text{-Bu}{-}N(\underline{|O|}^-){-}O{-}\text{Bu-}t \longrightarrow t\text{-Bu}{-}N{=}O + {}^-O{-}\text{Bu-}t \quad (23b)$$

$$\mathbf{69} + t\text{-Bu}^\cdot \longrightarrow CH_2{=}C(CH_3)_2 + t\text{-Bu}{-}N(\underline{|O|}^-){-}OH \longrightarrow t\text{-Bu}{-}N{=}O + OH^- \quad (23c)$$

by the addition of *t*-butyl radicals to 2-methyl-2-nitrosopropane as well as by the hydrolysis of anion **70** during work-up. With the exclusion of water, the anion **70** can be isolated.

The reaction of phenylsodium with 2-methyl-2-nitropropane yields an analogous anion radical **71** (R^1 = Ph, R^2 = *t*-Bu), the hydrolysis of which gives *t*-butyl phenyl nitroxide[73]. A very important method for the preparation of nitroxides is the reaction of nitro-*t*-alkanes with aryl or *t*-alkyl Grignard compounds or of aromatic nitro compounds with *t*-alkyl Grignard compounds (equation 24). It is suggested that in these reactions also, the anion **71** is initially formed[74].

$$R^1R^2\overset{+}{N}(O^-)(O^-)$$

(**71**)

$$R^1NO_2 + R^2MgX \longrightarrow R^1-\underset{\underset{O}{|\cdot}}{N}-R^2 \qquad (24)$$

2. *From nitroso compounds*

The high tendency for the formation of nitroxides due to their high stabilization implies that nitroso compounds may not only trap free radicals but are also capable of oxidizing other compounds. Thus, nitroso compounds are converted into monoaryl or monoalkyl nitroxides, which can dimerize to give azoxy compounds or may subsequently be reduced further to hydroxylamines (equation 25). In this way not only hydroxylamines (see Sections II.A.1) but also amines can be oxidized by nitroxo compounds[75].

$$R-N{=}O \xrightarrow[(HX)]{+H} R-\underset{\underset{O}{|\cdot}}{N}-H \longrightarrow R-N{=}\underset{\underset{O}{|}}{N}-R \qquad (25)$$

$$R-\underset{\underset{O}{|\cdot}}{N}-H \xrightarrow{+H^\cdot} R-NHOH$$

$$HX = RNHOH,\ R_2NH,\ RNH_2$$

The formation of nitroxides in reactions of aromatic nitroso compounds and alkenes having no hydrogen atom in the allylic position has also been observed. It is assumed that diradicals are formed first and that these are converted to nitroxides of unknown structures in the course of the reaction (equation 26)[76].

$$Ar-N{=}O + \rangle C{=}C\langle \longrightarrow [Ar-\underset{\underset{O}{|\cdot}}{N}-\overset{|}{\underset{|}{C}}-\overset{|}{\underset{|}{C}}\cdot] \longrightarrow Ar-\underset{\underset{O}{|\cdot}}{N}-\overset{|}{\underset{|}{C}}-\overset{|}{\underset{|}{C}}-X \qquad (26)$$

Alkynes such as dimethylacetylenedicarboxylate react with nitrosobenzene in a similar manner affording nitroxides, for which a probable structure is **72** (equation 27)[77]. Reaction of aromatic nitroso compounds and nitrile oxides yielding *N*-hydroxybenzimidazole-*N*-oxides through the intermediates **73**[78] is accompanied by redox processes with the formation of nitroxide **74** or benzimidazole-3-oxyl-1-oxide **75** depending on the reaction conditions (equation 28)[79].

$$\mathrm{Ph{-}N{=}O + E{-}C{\equiv}C{-}E \longrightarrow [Ph{-}N(O^{\bullet}){-}C(E){=}\dot{C}(E)] \xrightarrow{+H} Ph{-}N(O^{\bullet}){-}C(E){=}C(E)H \xrightarrow{Ph-N=O} Ph{-}N(O){=}C(E){-}C(H)(E){-}N(O^{\bullet}){-}Ph} \quad (27)$$

E = CO_2Me **(72)**

$$\mathrm{PhN{=}O + ArC{\equiv}N{-}O \longrightarrow Ph{-}N(O){=}C(N{=}O)Ar \;(\mathbf{73})} \quad (28)$$

(73) $\xrightarrow{+H^{\bullet}}$ Ph—N(O•)—C(=N—OH)Ar **(74)**; (73) ⟶ 2-Ar-1-hydroxybenzimidazole 3-oxide $\xrightarrow{-H^{\bullet}}$ **(75)**

III. INVESTIGATION OF NITROXIDES BY PHYSICAL METHODS

A. X-ray Analysis and Electron Diffraction

Most physical methods of structure elucidation can be applied to nitroxides which are isolable in crystalline form. Obviously one of the most informative methods is X-ray analysis. X-ray analysis and electron diffraction data for some nitroxides are collected in Table 2. The electron diffraction results are congruent with the X-ray studies. For all nitroxides studied the NO bond lengths are in the range 1.23–1.29 Å, intermediate between the NO single bond (1.44 Å) and the NO double bond (1.20 Å). This confirms the existence of a two-centre-three-electron bond in nitroxides.

The other outstanding problem in discussing the structure of nitroxides is whether or not the nitroxide group is planar or pyramidal. As shown in Table 2, the out-of-plane angle α (dihedral angle between the NO bond and the C—N—C plane) is not uniform. The different values, such as for the similar piperidine-*N*-oxyls **9**, **79** and **80**, indicate that even small differences can cause a

(4-MeO-C_6H_4)$_2$N—O• **(76)** $(C_6Cl_5)_2$N—O• **(77)** t-Bu—N(O•)—Bu-t **(1)**

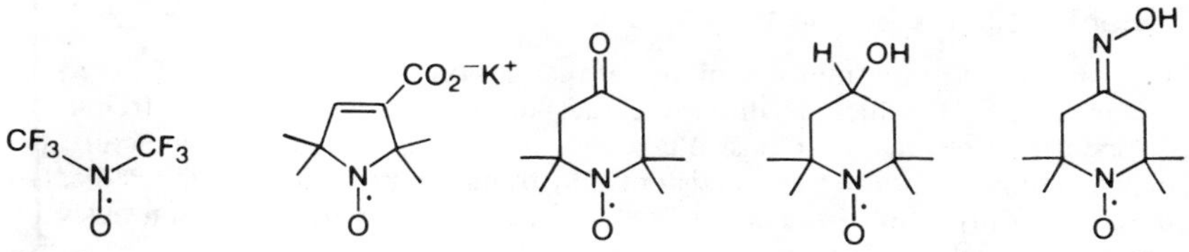

(2) (78) (9) (79) (80)

TABLE 2. Bond lengths (Å) and bond angles (degrees) for some nitroxides

	r_{NO}	r_{NC}	C—N̂—C	O—N̂—C	α^a	Reference
76	1.23	1.44	124[b]	—	0	80
77	1.272	1.435	119.9[c]	120.0	0	81
1[d]	1.28	1.51	136	—	0	80
2[d]	1.26	1.441	120.9	117.2	21.9	80
78	1.27	1.50	114.5	122.5	0	82
9	1.276	1.483	123.5	118.3	0	80
79	1.291	1.499	125.4	116.2	15.8	80
80	1.27	1.49	125	115	25.8	83

[a]See text.
[b]The phenyl rings are twisted about 33°.
[c]The phenyl rings are twisted about 57°.
[d]Electron diffraction.

considerable change in the geometry of the nitroxide group. In fact, several calculations show very little difference in the energy of an optimized pyramidal configuration and a planar configuration[84–87]. Thus it seems reasonable that results from studies in the crystalline or gaseous state cannot be easily applied to the structure of nitroxides in solution. There does seem, however, to be general agreement that the nitroxide group of diaryl nitroxides is planar in solution also[85].

B. Dipole Moment Measurements

The delocalization of the unpaired electron as shown by the two mesomeric formulae, a neutral one and a charged dipolar one, shows the probable existence of a relatively high dipole moment. This is in fact true, as a comparison of the dipole moments of nitroxides and the corresponding hydroxylamines shows[88]. The absolute dipole moment of 4-oxo-2,2,6,6-tetramethylpiperidine-*N*-oxyl is reduced, since the dipole moment of the carbonyl group is opposed to that of the nitroxide group (Table 3).

TABLE 3. Dipole moments of nitroxides and hydroxylamines[88]

Compound	μ (D)
2,2,6,6-Tetramethylpiperidine-*N*-oxyl	3.14
Diphenyl nitroxide	3.00
4-Oxo-2,2,6,6-tetramethylpiperidine-*N*-oxyl	1.36
N-Hydroxy-2,2,6,6-tetramethylpiperidine	1.76
N,*N*-Diphenylhydroxylamine	1.30

C. Electron Spin Resonance

The most important method in the investigation of nitroxides is Electron Spin Resonance (ESR), which is the preferred method for detection of free radicals. ESR spectroscopy makes it possible to get a more or less detailed picture of the structure of a vast number of persistent and transient nitroxides. Even radicals with a lifetime of only a split second can be studied when generated continuously in the cavity of the ESR spectrometer.

1. Anisotropic ESR spectra

Besides the ESR study of nitroxides in nonviscous solvents, nitroxides can be investigated under various special conditions. ESR spectra can be recorded from single crystals and from nitroxides included in host crystals, as well as in liquid crystals, from matrix-isolated radicals, and from nitroxides in the polycrystalline or powder state, in rigid glasses (frozen solutions) and in viscous solutions. The ESR spectra are characterized by three constants: the g-factor, the coupling constants of the nuclei with a nonzero nuclear spin, especially the nitrogen nucleus, and the line-width.

If a single crystal is orientated in the ESR spectrometer aligning the magnetic field with the principal axes of the nitroxide, three different spectra are obtained for the three principal axes (x-axis: parallel to NO bond, z-axis: parallel to the nitrogen p_z-orbital, y-axis: perpendicular to both the x- and z-axes).

For example the following are the values found for di-t-butyl nitroxide[90]:

$$g_{xx} = 2.00881 \quad g_{yy} = 2.00625 \quad g_z = 2.00271 \quad g_{iso} = 2.00592$$

$$a_{xx} = 7.59\ \text{G} \quad a_{yy} = 5.95\ \text{G} \quad a_{zz} = 31.78\ \text{G} \quad a_{iso} = 15.1\ \text{G}$$

In nonviscous solvents these anisotropic effects are nearly averaged to zero by the Brownian molecular motion, affording an isotropic spectrum with average values of g and a. However, small residual effects remain as time-dependent perturbations. Essential for the disappearance of anisotropy is a fast rotational motion. This rotational motion is dependent on several factors such as molecular size, temperature, viscosity of solvent and possible association with other molecules.

If the rotational motion of the nitroxides decreases, the spectra become increasingly asymmetric and change their line-shape[89]. Computer simulation of the line-form of the ESR spectra with the aid of known values of a_{xx}, a_{yy}, a_{zz} and the anisotropic g-factors, gives the rotational correlation time τ which is inversely related to the tumbling rate of the molecule. Thus, important information concerning the molecular environment of the molecules is obtained. This effect can therefore be used in the study of biological systems (see Section VI).

Anisotropic effects are also observed in ESR studies of nitroxide radicals and biradicals in nematic solvents[91]. In this case the solvent molecules first align themselves in the direction of the external magnetic field, then the radical molecules dissolved in the mesophase are forced to partially align with the field due to the solute–solvent interactions. This alignment causes anisotropy in the ESR spectrum. The observed anisotropic hyperfine splitting constants can be used to determine the orientation of the radicals in the liquid crystal. It has been found that the molecular shape of a molecule makes a major contribution to its ordering properties. Thus the degree of orientation increases with increasing molecular length and is more pronounced the narrower the molecular shape.

TABLE 4. Coupling constants a^{N} and $a^{^{17}O}$ for some nitroxides (in Gauss)

Compounds	a^{N}	$a^{^{17}O}$
Dialkyl nitroxides	14 –16.5	18.5–19.5
Alkyl aryl nitroxides	10.5–13.5	≈18.3
Diaryl nitroxides	9 –11	≈17.3
Monoalkyl nitroxides	12.0–13.5	—
Monoaryl nitroxides	8.5– 9.5	—
Acyl nitroxides	6.5– 8.5	≈20.4
Imino nitroxides	7.5–10.0	18.0–19.5
Nitronyl nitroxides	6.5– 7.5	≈12.2 (Ref. 95)
Alkyl alkoxy nitroxides	24 –29	≈18.5
Alkoxy aryl nitroxides	13 –15.5	—

2. *Isotropic ESR spectra*

The *g*-values in the isotropic spectra are usually of the order 2.0055 to 2.0065[92]. Heteroatoms in the vicinity of the nitroxide group shift the *g*-factor (for instance bistrialkylsilyl nitroxide: $g \approx 2.0093$). Since the *g*-factor of aminyl radicals lies generally between 2.0030–2.0045[93], an unambiguous differentiation between nitroxides and aminyls is possible on examination of the *g*-factor.

The isotropic coupling constants of the different nuclei of a radical yield even more information. The spectra of nitroxide radicals are best characterized by the nitrogen coupling constant a^{N}, but the proton coupling constants of neighbouring alkyl groups or conjugated aryl groups, as well as the coupling constants of the nuclei of conjugated groups, can also give interesting information. Only recently, some ^{17}O-labelled nitroxides were prepared to determine the oxygen coupling in the nitroxide group[39,94] (see Table 4).

D. NMR Spectroscopy and ENDOR Spectroscopy

In general, radicals can be studied by NMR spectroscopy[96]; however, a disadvantage of this method is that concentrated solutions of radicals (>0.1 M) must be used. On the other hand, resolution is better by one to two orders of magnitude compared to ESR. Thus NMR is particularly used in the determination of small coupling constants. In this way, coupling through several bonds in mono- and bi-cyclic nitroxides can be detected[97]. Not only the magnitude of hyperfine coupling but also its sign can be determined by NMR. Contrary to ESR spectra, NMR spectra are relatively simple since for every group of equivalent nuclei only a single line is obtained. The coupling constant for these nuclei can be determined from the shift of the line relative to its position in the corresponding diamagnetic molecule according to equation (29):

$$\Delta H = -a(\gamma_e/\gamma_n)(g\beta H/4\,kT) \tag{29}$$

A shift to a lower field corresponds to a positive coupling and vice versa. In dilute solutions the lifetime of the electron-spin state is too large, hence NMR spectra of radicals are not observed. However, the lifetime can be reduced through intramolecular spin exchange thus giving sharp lines. This can be attained by very high concentrations of the radical (>0.5M) or by using a paramagnetic solvent such as di-*t*-butyl nitroxide.

Concerning NMR spectroscopy, a very high concentration of radical is necessary for the application of the ENDOR method (Electron Nuclear Double Resonance). Even very small coupling constants can be detected by this method. For instance, in nitroxides **81** the coupling constants of R have been detected[98]. These nitroxides are formally spin adducts of radicals R$^{\bullet}$ and *N*-benzylidene-*t*-butylamine-*N*-oxide; they can, however, be prepared in sufficiently high concentration only by the addition of organometallic compounds to the nitrone and subsequent oxidation of the adduct.

Ph—CH—N—Bu-*t* (R on CH; O$^{\bullet}$ on N)

(**81**)

E. Other Spectroscopic Methods

In accordance with the proposed one-and-a-half bond of nitroxides, the characteristic IR absorptions for the NO bond lie in the range from 1340–1380 cm^{-1} (diphenyl nitroxide 1342[99], di-*t*-butyl nitroxide 1345[99], 4-oxo-2,2,6,6-tetramethylpiperidine-*N*-oxyl 1380[100] and 4-hydroxy-2,2,6,6-tetramethylpiperidine-*N*-oxyl 1371 cm^{-1})[100].

In the UV and visible spectra of dialkyl nitroxides, absorption maxima at 240 nm ($\varepsilon = 3000$) and 410–460 ($\varepsilon = 5$) were found[101]. Whereas the position of the first maximum is independent of the solvent, the second maximum is shifted to lower wavelengths with increasing polarity of solvent, according to a n $\rightarrow \pi^*$ absorption. The absorption is shifted to higher wavelengths by conjugation.

The absolute configuration of the chiral radical **82** has been determined using circular dichroism. The circular dichroism agrees with the octant rule if the molecule in solution adopts a half-chair conformation as it does in the crystalline state[105].

OH, Me, Me, Me, Me, N, O$^{\bullet}$ — *R* (+) form

(**82**)

The photoelectron spectra of di-*t*-butyl nitroxide and 2,2,6,6-tetramethylpiperidine-*N*-oxyl have been studied[106]. The vertical first ionization potentials were found to be 7.20 and 7.31 eV, arising from the unpaired electron in the antibonding π^*-orbital.

The mass spectra of several persistent nitroxides have been recorded[107]. Ionization by electron bombardment gives rise to an even-electron molecular ion

TABLE 5. Absorption maxima of conjugated nitroxides

Nitroxide	Absorption maxima (nm)	Reference
Bis(4-methoxyphenyl) nitroxide	493, 407, 333, 323, 262	102
2-Phenyl-4,4,5,5-tetramethylimidazoline-1-oxyl-3-oxide (**90**)	588, 360, 263, 238	103
N^2-Aryl-N^1-*t*-butylformamidine-N^1-oxyl	597, 542	104
N^1,N^2-Diarylformamidine-N^1-oxyl	607, 550	104

species. The fragmentation mode of the ions arising by this ionization is dependent on the structure of the different nitroxides in a specific manner.

IV. SPECIFIC PROPERTIES OF NITROXIDES AS STUDIED BY ESR

A. Spin Density Distribution

According to the polar character of the mesomeric formula B, the spin density distribution in nitroxides should be solvent-sensitive. In fact, an increasing spin density at nitrogen ρ^N and a decreasing spin density at oxygen ρ^O is observed with

(A) ⟷ (B)

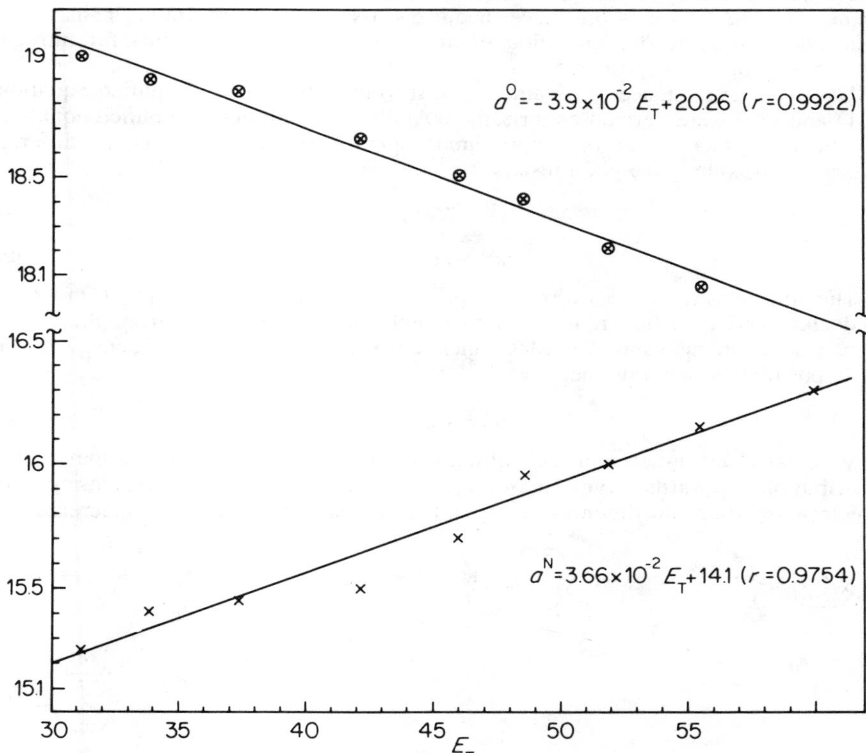

FIGURE 3. Values a^N and a^{17O} of tetramethylpiperidine-*N*-oxyl versus the solvent parameter E_T. Solvent (E_T): cyclohexane (31.2), toluene (33.9), tetrahydrofuran (37.4), acetone (42.2), acetonitrile (46.0), isopropyl alcohol (48.6), ethanol (51.9) and methanol (55.5). E_T is a solvent polarity paramter based on the transition energy (kcal/mol) for the largest wavelength solvatochromic absorption band of a pyridinum-*N*-phenol betaine[229]. Reproduced from H. G. Aurich, K. Hahn, K. Stork and W. Weiss, *Tetrahedron*, **33**, 971 (1977) by permission of Pergamon Press.

increasing polarity of the solvent as indicated by the increasing values of a^N and the decreasing values of a^O [94]. A linear correlation between a^N and the solvent polarity has been found by many workers[108]. The correlation between the E_T values of several solvents and the a^N and a^O values for tetramethylpiperidine-*N*-oxyl is shown in Figure 3.

Conversion of coupling constants to spin densities is performed with the aid of equations (30a) and (30b):

$$a^N = Q_N^N \cdot \rho^N + Q_{ON}^N \cdot \rho^O \quad (30a)$$

$$a^O = Q_O^O \cdot \rho^O + Q_{NO}^O \cdot \rho^N \quad (30b)$$

where Q_X^X or Q_{YX}^X is the contribution to the hyperfine splitting constant a^X expected for unit spin population at the atom X or Y, respectively. Since different Q values have been used by several authors, large discrepancies occur in the prediction of the spin densities. Thus for simple dialkyl nitroxides, spin densities from $\rho^N = 0.2$ to $\rho^N = 0.9$ have been discussed. More recently, it has been concluded from the determination of anisotropic coupling constants for nitrogen and oxygen that $\rho^N \approx \rho^O \approx 0.5$.

Using the dependence of a^N and a^O on solvent polarity, the simplified equations (31) and (32) were derived empirically[94]. Applications of these simplified equations to the determination of the approximate spin density distributions for different types of nitroxides gave good results.

$$a^N = 33.1\, \rho^N\, G \quad (31)$$

$$a^O = 35.3\, \rho^O\, G \quad (32)$$

The spin density at the various positions of the phenyl rings in diaryl, monoaryl and alkyl aryl nitroxide radicals, for example, may be detected from the proton hyperfine coupling using the McConnell equation (33). The spin density at the *meta*-positions is generally negative.

$$a^H = -27\, \rho^C\, G \quad (33)$$

Whereas electron-accepting substituents in the phenyl group shift the spin density distribution towards the mesomeric formula A, thus decreasing a^N, electron-donating substituents operate in the opposite direction, i.e. a^N increases.

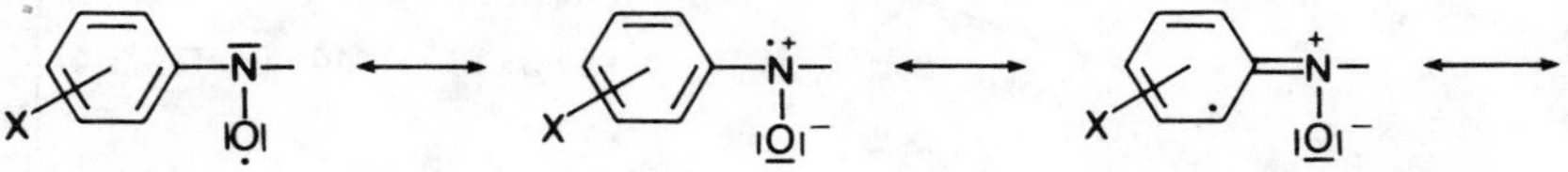

(A) (B) (C) 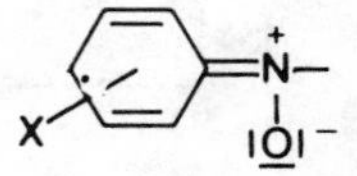 (D)

Hammett plots for a^N have been discussed together with other ESR-derived Hammett plots. In all cases the best agreement was obtained using σ^- values for *para* substituents[109,110]. The values in Table 6 indicate the increasing delocalization of the unpaired electron in the order **1** < **83** < **84** < **85**. The sum of the spin densities in the phenyl ring of **83** is 0.1 and in either of the phenyl groups of **84** is

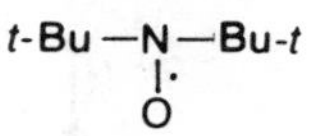

(1) 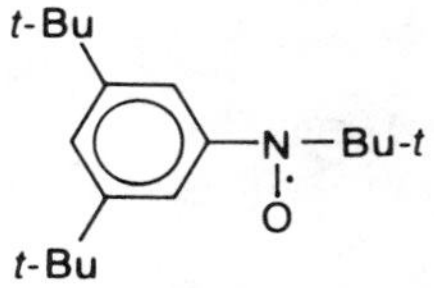 (83) 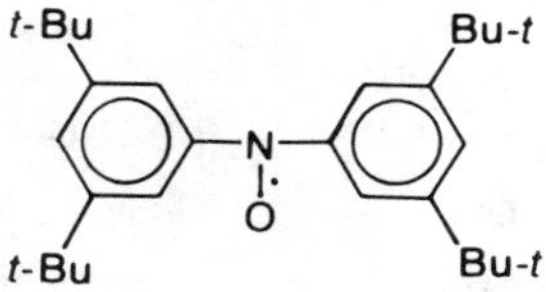 (84)

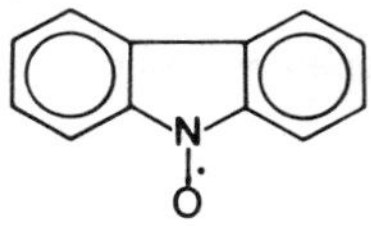

(85) 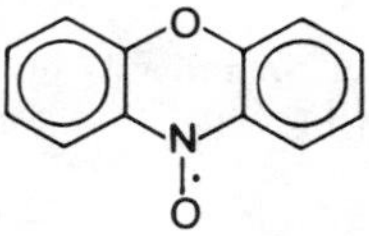 (86)

TABLE 6. Coupling constants a^{N} and a^{17O} (in Gauss) and spin densities ρ^{N} and ρ^{O} for some nitroxides[94]

	a^{N}	a^{17O}	ρ^{N}	ρ^{O}
1	15.0	19.1	0.455	0.54
83	12.6	18.3	0.38	0.52
84	9.9	17.3	0.30	0.49
85	6.65	16.5	0.20	0.47
86	9.50	13.1	0.29	0.37

0.12. The delocalization of the spin density into the phenyl rings occurs mainly at the expense of ρ^{N}. Another situation arises with radical **86**. In this case the additional π-electron pair of the ring-oxygen causes the spin density within the nitroxide group to be shifted more towards the oxygen atom.

In nitroxides with conjugated groups C=X, increasing electronegativity of X should increase ρ^{O} and ρ^{C} and decrease ρ^{N} and ρ^{X}. Inspection of Table 7 shows that this is true[39]. As expected, the spin density distribution is strongly modified if the π-system is extended by an additional oxygen atom as in **90**[95].

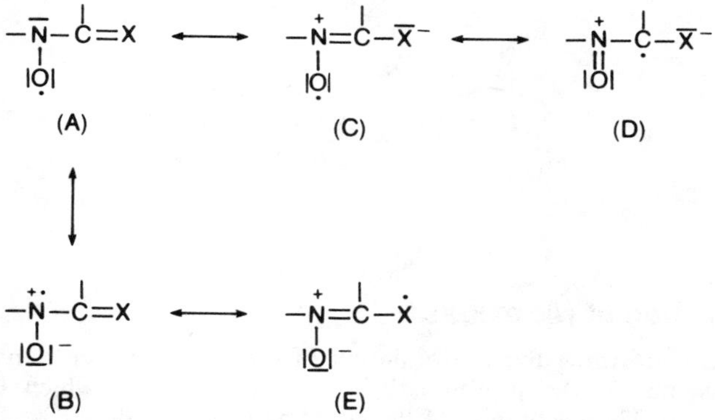

(A) (C) (D)

(B) (E)

The fact that the simplified equations (31) and (32) with unchanged values for Q may be applied to these different types of nitroxides is considered as evidence that

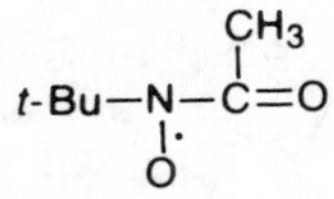

(87) 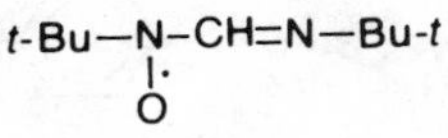 (88) 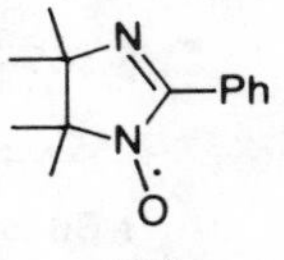 (89) 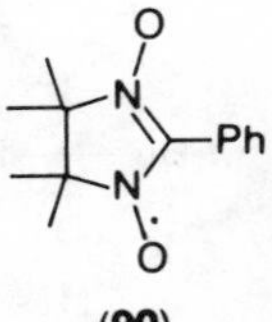 (90)

TABLE 7. Spin density distribution in nitroxides with conjugated groups[39,95]:

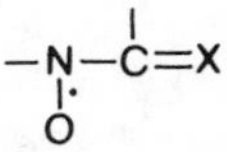

	ρ^{O}_{NO}	ρ^{N}_{NO}	ρ^{C}	ρ^{X}
87	0.58	0.23	0.07	0.12
88	0.55	0.28	−0.05	0.21
89	0.51	0.28	−0.06	0.28
90	0.34	0.22	−0.12	*a*

[a]See ρ^{O}_{NO} and ρ^{N}_{NO}.

all these radicals, including the dialkyl nitroxides, have virtually the same geometry of the nitroxide group in solution[111] and therefore should be approximately planar[39]. Since any deviation from planarity should introduce some s-character in the singly occupied orbital, an increased value of a^{N} compared to the planar structure should result. Thus, a total spin density considerably greater than one should be obtained on the application of equations (31) and (32). However, this has only been found to be true for alkoxy alkyl nitroxides[39] and for the bicyclic nitroxides **91–93**[112] for which a planar nitroxide group seems unreasonable due to considerable ring-strain. Nevertheless, discussions on the structure of the nitroxide groups, especially in dialkyl nitroxides, is continuing[113–115].

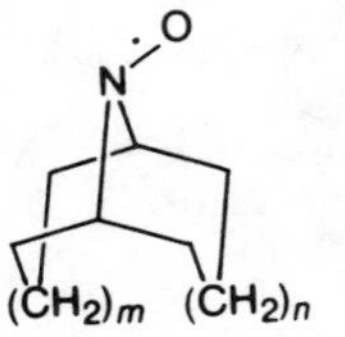

(**91**) $m = n = 1$
(**92**) $m = 1, n = 0$
(**93**) $m = n = 0$

B. Conformation of Nitroxides

Conclusions concerning the conformation of nitroxides can be derived from the coupling constants of the β-protons[116,117], the coupling of which is caused by hyperconjugation. The magnitude of the coupling constants depends on the angle φ between the single occupied orbital and the CH σ-orbital according to equation (34)[116].

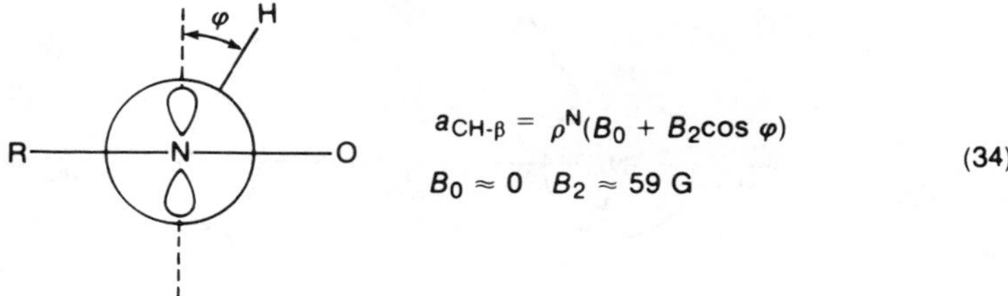

φ
H
R
N
O
$a_{CH\text{-}\beta} = \rho^N(B_0 + B_2\cos\varphi)$
$B_0 \approx 0 \quad B_2 \approx 59$ G
(34)

Assuming that ρ^N is approximately 0.45, the following values of a^H for some values of φ are given:

φ(degrees)	0	30	45	60	90
a^H (Gauss)	26	19.5	13	6.5	0

TABLE 8. Coupling constants a_β^H and a^N for some nitroxides[116]:

R—N(—O•)—CHR^{1}R^2

(94)–(99)

	R	R^1	R^2	a^N	a_β^H
94	*t*-Bu	H	H	16.4	12.7 (3H)
95	*t*-Bu	H	Me	15.7	10.7 (2H)
96	*t*-Bu	H	Ph	14.6	7.7 (2H)
97	*t*-Bu	Me	Me	16.8	1.8 (1H)
98	Et	Me	Me	16.0	4.8 (1H)
99	H	Me	Me	12.7	12.1 (1H)

In Table 8 some examples of β-proton couplings are given along with a^N values, which may be interpreted as follows. The methyl group in **94** rotates freely, thus for each individual proton a time-averaged angle of 45° may be assumed, according to a coupling constant of 13 G. In the case of methylene and methyne groups, rotation is restricted with the exception of **99**. Thus, molecules adopt preferred conformations, the energy of which is determined by the interactions of the various substituents. Steric effects have the largest influence, but hyperconjugation and the formation of hydrogen bridges can also contribute to these interactions. Therefore, quantitative conclusions are difficult and the discussion rests on a more qualitative basis.

For methyne group conformations A and B have the lowest energy, A being somewhat more favourable in the case of R = *t*-Bu. For both conformations, it would be expected that $a_\beta^H \approx 0$. The larger the substituent R, the more frequently the molecule should adopt conformation A or B, averaged over the time and the less the value of a_β^H. This effect explains nicely the trend in a_β^H values for nitroxides **97–99**. For **99**, free rotation of the isopropyl group is assumed.

The situation for the methylene group is even more complicated. Besides conformation C, conformations D–G must also be considered. In conformation C, a_β^H is expected to be about 6.5 G for both hydrogens. Thus, this conformation seems to be important, as for example in **96**, for the benzyl group only. With other groups such as ethyl, *n*-propyl and *n*-butyl, the a_β^H values are larger for two equivalent protons. In conformations D–G, a_β^H values ≈ 19.5 G for one proton and $a_\beta^H \approx 0$ G for the other proton are expected. However, a fast exchange between the enantiomeric conformations D and E, or F and G, would make both hydrogens equivalent so that a value of about 9.75 G for both protons would be expected. This is in agreement with the observed value of 10.7 G. Thus it may be concluded that radical **95** as well as other similar radicals exist preferentially in rapidly interchanging conformations similar to D–G.

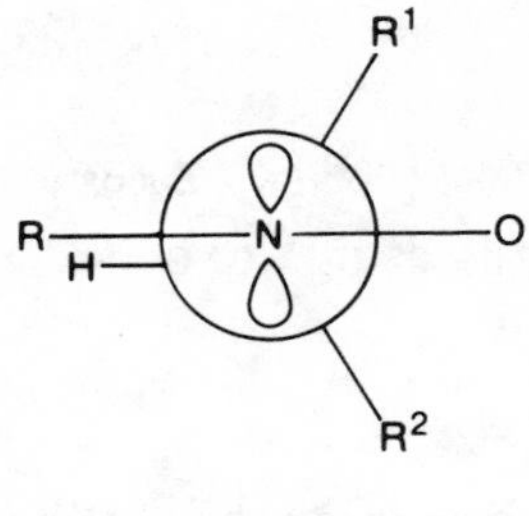

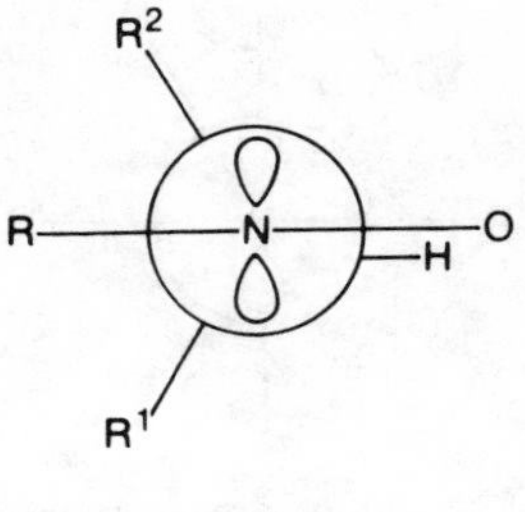

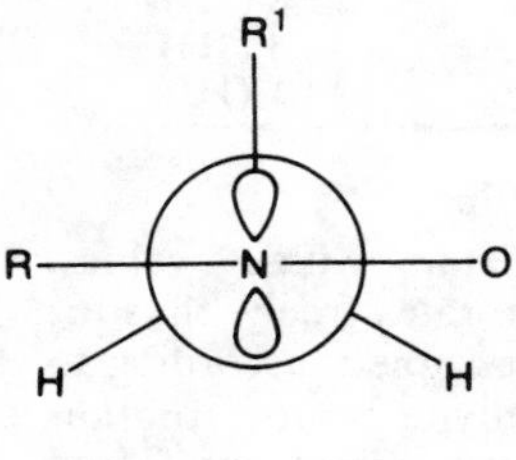

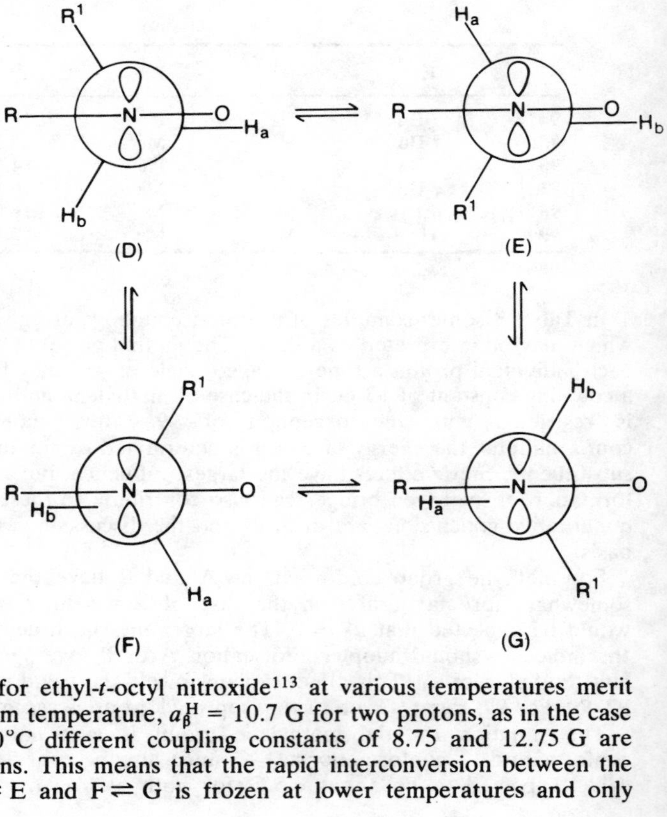

The results obtained for ethyl-*t*-octyl nitroxide[113] at various temperatures merit special comment. At room temperature, $a_\beta^H = 10.7$ G for two protons, as in the case of **95**, whereas at -100°C different coupling constants of 8.75 and 12.75 G are found for the two protons. This means that the rapid interconversion between the enantiomeric pairs D $\rightleftharpoons$ E and F $\rightleftharpoons$ G is frozen at lower temperatures and only interconversion of the diastereomeric pairs D $\rightleftharpoons$ F and E $\rightleftharpoons$ G can occur. In this case, the two protons are no longer equivalent and therefore exhibit different coupling constants.

The equivalence of both protons is also lost by the introduction of a chiral group R^1, when D and E, or F and G, are no longer enantiomers but diastereomers instead. For nitroxides of this type, even at room temperature, two different values for a_β^H have been found[119].

The fast ring-inversion in piperidine-*N*-oxyl at +110°C causes the equivalence of the four β-protons. At −100°C, however, the rate of the ring-inversion is sufficiently slow on the ESR time-scale that the molecule appears frozen in one conformation. Thus two different coupling constants are detected (a^H_{axial} = 26.3 G, $a^H_{equatorial}$ = 3.78 G)[120].

Coupling constants of protons distant from the nitroxide group (γ- or δ-protons), are usually very much smaller; γ- or δ-coupling constants can also be used to study conformational problems especially in cyclic and bicyclic nitroxides[97]. Frequently very small coupling constants can be detected by NMR spectroscopy only.

C. Dynamic Processes in Mono- and Poly-radicals

Dynamic processes in radicals can also be studied by ESR spectroscopy[121]. As already mentioned for nitroxides with a methylene group, the two β-protons are equivalent because of rapid interchange of the two enantiomeric conformations at room temperature, whereas at low temperature the conformational change is slow and two individual proton couplings are observed. In the intermediate temperature range, the change in position of the two protons causes broadening of the ESR lines. From the temperature dependence of the line-broadening, the activation energy for the conformational interchange can be determined. In this way, the activation energy for the rotation of the ethyl group in ethyl *t*-octyl nitroxide was estimated to be ≈7.5 kcal/mol[118]. Similar conformational changes in nitroxides have been studied by Janzen[122]. The study of ring inversion in piperidine-*N*-oxyl afforded an activation energy of 5–6 kcal/mol[123]. In radical **100**, a fast exchange of the hydrogen atom, as indicated, makes the two nitrogen atoms equivalent. For this process an activation energy of 4.5 kcal/mol was estimated from the temperature dependence of the line-widths[124].

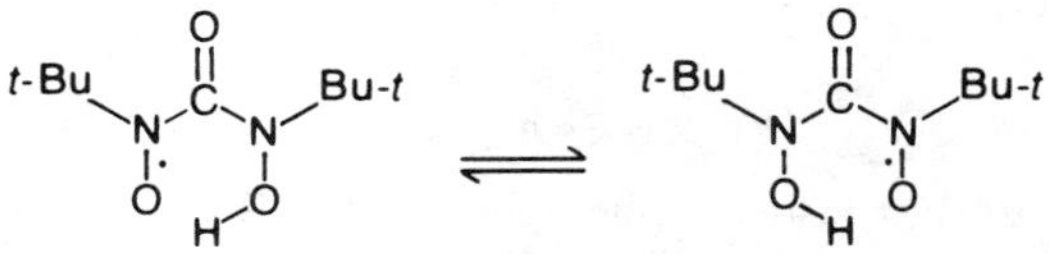

t-Bu–N(O·)–C(=O)–N(OH)–Bu-t $\rightleftharpoons$ t-Bu–N(OH)–C(=O)–N(O·)–Bu-t (35)

(**100**)

Similar dynamic processes are observed in di- and poly-radicals. As shown for diradicals of the type **101**, spin exchange between the two radical centres can occur[125]. In cases where spin exchange J is slow ($a \gg J$), the two halves of the

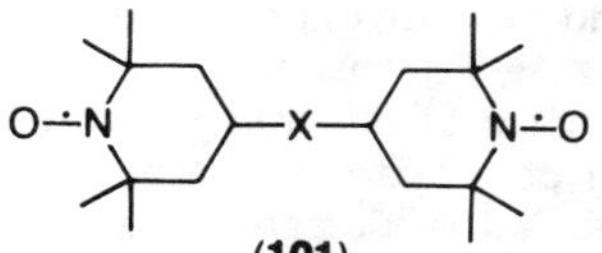

(**101**)

e.g. X = —O—CO—$(CH_2)_n$—CO—O—

diradical act as independent monoradicals and the hyperfine splitting is a triplet splitting, as has also been found for the analogous monoradical. When spin exchange is fast ($J \gg a$), each nitrogen nucleus interacts with both electrons and the hyperfine splitting drops to one half of the value of the coupling in the analogous monoradical, yielding a five-line spectrum with the intensity ratio 1:2:3:2:1. In the intermediate range, complicated spectra can result[126]. The magnitude of spin exchange J seems to be a complex function of the biradical structure being influenced by temperature and solvent as well as by the chemical nature of the connecting bridges[127]. Particularly rigid diradicals such as **102–105** have been proposed as structural probes in biological systems[128,129]. For polyradicals **106**, the

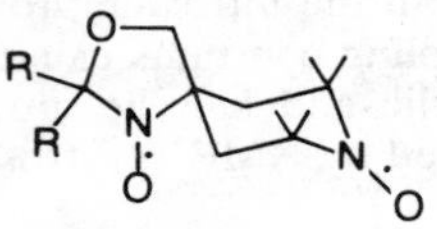

(102) 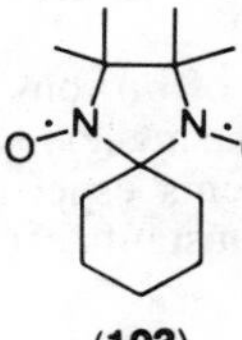 **(103)**

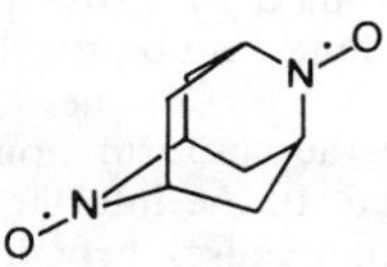

(104) 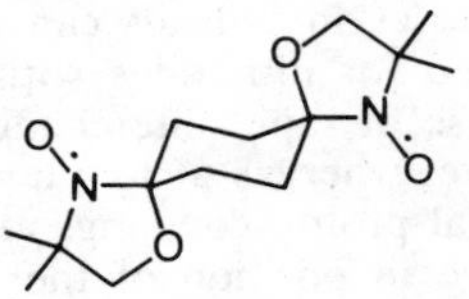 **(105)**

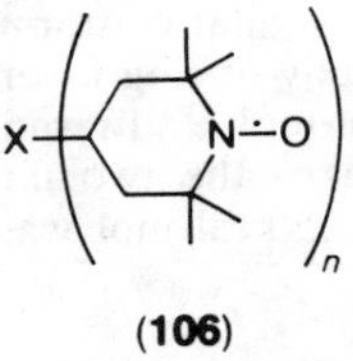

(106)

X = P, n = 3
X = Si, n = 4

situation is analogous. For instance, the ESR spectra of triradicals, for which $J \gg a$ is valid, show hyperfine splitting into seven lines, a^N being one third of the usual value[130].

V. REACTIONS INVOLVING THE NITROXIDE GROUP

In general most nitroxide group reactions can be divided into reductive processes yielding hydroxylamines and oxidative processes yielding nitroso compounds or nitrones (see Schemes 3 and 4). For instance, radicals can add to the oxygen atom of the nitroxide group affording O-substituted hydroxylamines. A special case of this reaction is the addition of a hydrogen atom. On the other hand, an electron can be transferred to the nitroxide group giving at first an anion, which subsequently can accept a cationic species, especially a proton, to yield a hydroxylamine. Protonation or complexation of the nitroxide group does not

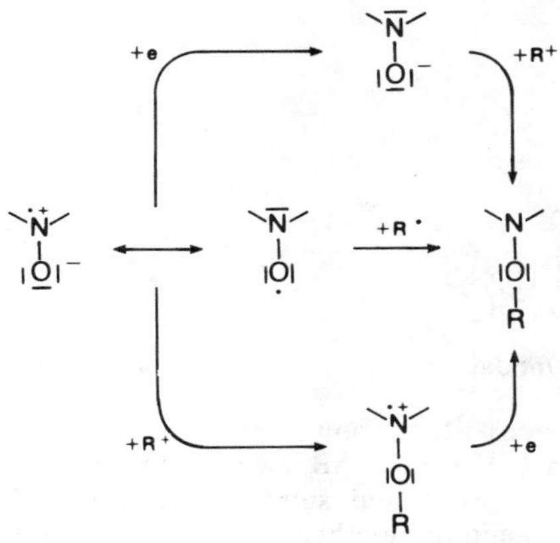

SCHEME 3

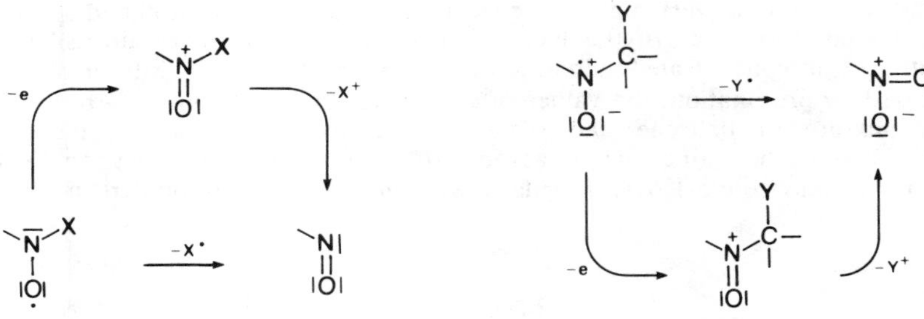

SCHEME 4

change the oxidation level, but stabilization due to delocalization of the unpaired electron is for the most part lost. Thus cationic species easily accept an electron, being reduced to hydroxylamines.

Removal of an electron leads to nitrosonium cations which are for the most part unstable. The nitrosonium cations can afford nitroso compounds or nitrones by splitting off the groups X^+ or Y^+, respectively. On the other hand, homolytic cleavage of the N—X bond (α-cleavage) or of the C—Y bond (β-cleavage) can give the nitroso or nitrone group directly.

A. Protonation and Complex Formation of Nitroxides

Nitroxides can interact with electrophiles via their free electron pair. The formation of hydrogen bridges in protic solvents such as alcohols, slightly increasing the spin density at nitrogen, is a weak interaction of this kind[131]. For hydrogen bridging, a σ-model and a π-model have been discussed, the former being favoured[132].

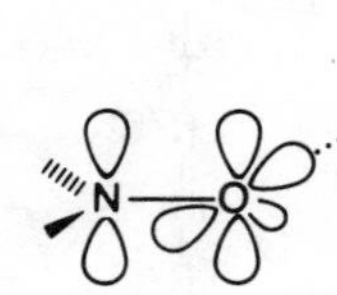

δ-model 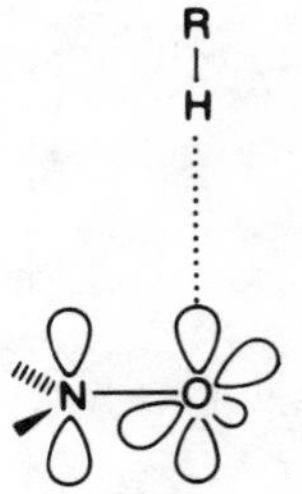 π-model

Protonation of nitroxides occurs only with strong acids. pK_A values of dialkyl nitroxides are in the range of −5.5. ESR spectra of protonated nitroxides, **107–109**, have been detected in concentrated sulphuric acid, in trifluoroacetic acid with added sulphuric acid[133], and in methylene chloride with added wet aluminium trichloride[134]. However, in dilute sulphuric acid (10%), the spectra of the unprotonated nitroxides are obtained. Increasing the acid concentration to 40% causes the lines to broaden, and at 55%, they disappear entirely, due to fast exchange of the protons between the nitroxide molecules. The fast exchange process stops only at very high acid concentration when the protonated species can be observed. Nevertheless, the yield of protonated nitroxide is surprisingly only about 0.3% in concentrated sulphuric acid[133]. Since the spin density at nitrogen is increased by protonation, the values of a^N increase. Protonation of carbazol-*N*-oxyl in a mixture of benzene and trifluoroacetic acid causes a^N to increase by approximately the same factor as for **107–109** to 8.8 G[135]. 2-Phenyl-4,4,5,5-tetramethylimidazoline-1-oxyl-3-oxide **(12)** can also be protonated in benzene

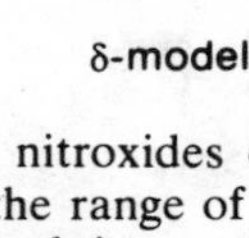

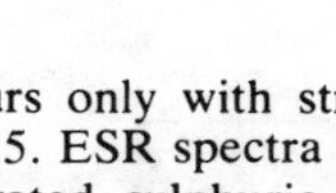

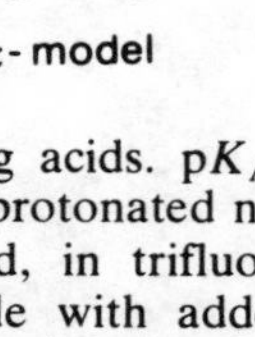

	(107)	**(108)**	**(109)**
a^N(G)	21.6	20.7	24.5

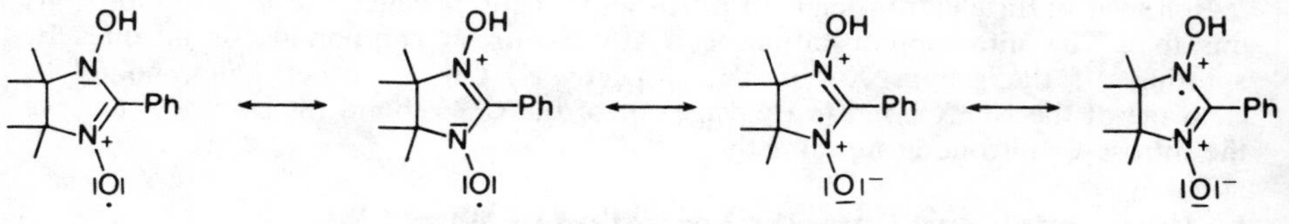

(110)

and trifluoroacetic acid[103] to give 110. However, in this case it is obvious that protonation leads to an increasing spin density at the oxygen atom of the unprotonated site of the molecule. Thus, the decrease in spin density at both the nitrogen atoms causes a^N to drop to 5.7 and 4.5 G, compared to a value of 7.5 G for both nitrogens in the unprotonated radical. In aqueous hydrogen chloride (1N) only the spectrum of the unprotonated form is obtained.

No individual spectra corresponding to the unprotonated and protonated species are obtained in the pH range from 7 to −1 for 2-phenyl-4,4,5,5-tetramethyl imidazoline-1-oxyl[136,137]. With increasing acidity, a continuous decrease in $a^{N}_{nitroxide}$ and increase in a^{N}_{imino} is observed. In this case, a fast proton transfer causes an averaging of the spectra of the two species. The pK_A for this radical is estimated to be 1.9[136] and 1.6[137].

With Lewis acids such as BX_3[138], AlX_3[139], SiX_4[140], GeX_4[140] and SnX_4[140] and others[141] formation of complexes (**111**) is observed, resulting in a more or less strong increase in a^N.

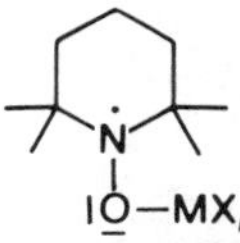

$\overline{O}$—MX_n^-

(**111**)

MX_n	BCl_3	$AlCl_3$	$SiCl_4$	$GeCl_4$	$SnCl_4$
a^N(G)	21.18	19.86	16.5	16.6	18.53

Nitroxides can act as electron donors and thus form electron donor–acceptor complexes (EDA) with various π-electron acceptors. For di-*t*-butyl nitroxide, complex formation with tetracyanoethylene, 2,3-dichloro-5,6-dicyano-*p*-benzoquinone and pyromellitic dianhydride has been confirmed[142]. The resulting complexes show charge-transfer absorption bands. From the effect of complex formation on the hyperfine splitting constant a^N, a contribution of electrostatic forces to complex formation has been demonstrated. However, in the case of complexes with strong acceptors such as tetracyanoethylene, the values of the hyperfine coupling constant have indicated a significant contribution of charge-transfer interaction to complex formation. Similarly, the formation of π-complexes from *t*-butyl mesityl nitroxide with various aromatic compounds and alkenes, substituted by electron-accepting groups, has been established[143]. The bond energy for the formation of these complexes is estimated to lie in the range 3–8 kcal/mol.

B. Addition of Radicals to the Nitroxide Group

Addition of a radical R$^\bullet$ to the nitroxide group (equation 36) occurs if the gain in energy for R—O bond formation overcomes the loss of delocalization energy which is of the order of 30 kcal/mol for the nitroxide group. This is true for most alkyl radicals. The kinetic studies of spin trapping have shown that the decay

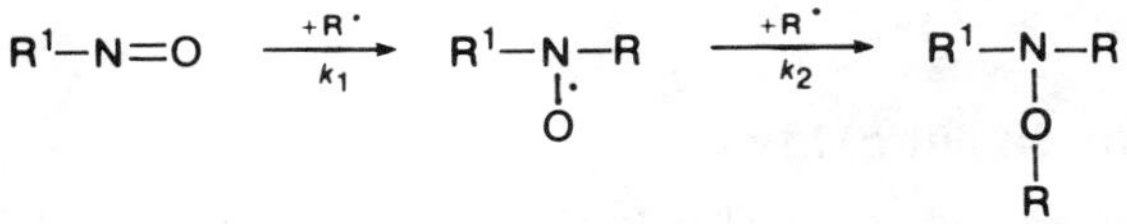

$$R^1\text{—N=O} \xrightarrow[k_1]{+R^\bullet} R^1\text{—N(—O}^\bullet\text{)—R} \xrightarrow[k_2]{+R^\bullet} R^1\text{—N(—O—R)—R} \qquad (36)$$

reaction of the nitroxides, caused by the addition of alkyl radicals, is even faster than the formation of nitroxides ($k_1 < k_2$)[47,66].

Furthermore, 4-oxo-2,2,6,6-tetramethylpiperidine-*N*-oxyl (**9**) readily adds 2-cyano-2-propyl radicals generated from the azoisobutyronitrile by thermal decomposition with formation of a stable adduct[144]. Even the stabilized 1,1-diphenylethyl radical is added to the nitroxide 9, but in solution this adduct is unstable. There exists an equilibrium with the precursor radicals (equation 37)

$$(\mathbf{9}) + \cdot C(R^1)(R^2)(R^3) \rightleftharpoons \text{piperidine-N-O-C}(R^1)(R^2)(R^3) \tag{37}$$

(**a**) $R^1 = R^2 = Me$, $R^3 = CN$
(**b**) $R^1 = R^2 = Me$, $R^3 = Ph$
(**c**) $R^1 = Me$, $R^2 = R^3 = Ph$

($\Delta H = -21.4$ kcal mol^{-1}, $\Delta S = -36$ cal deg^{-1} mol^{-1})[145] which obviously induces the decomposition to give 1-hydroxy-2,2,6,6-tetramethylpiperidine and 1,1-diphenylethylene. As expected, dissociation is far less in the case of the adduct of the cumyl radical, the adduct being much more stable. The high tendency for formation of such adducts is pertinent to the use of nitroxides in the inhibition of the autoxidation of hydrocarbons[146]. For the same reason, nitroxides can interrupt chain-reactions in polymerization and are thus frequently used as inhibitors[147].

Addition reactions of diaryl nitroxides have been known for a long time. Diphenyl nitroxide reacts with triphenylmethyl to give a 1:2 adduct, whereas from bis(4-nitrophenyl) nitroxide a 1:1 adduct has been obtained[148]. Furthermore, diphenyl nitroxide adds to the *para* position of tri-*t*-butylphenoxyl. Aminyl radicals can also add to the oxygen atom of nitroxides. For instance, diphenyl nitroxide and dianisylaminyl radicals react through an intermediate adduct with exchange of the oxygen atom, as indicated in equation (38)[148].

$$Ar_2N\cdot + O\dot{-}NPh_2 \longrightarrow Ar_2N{-}O{-}NPh_2 \longrightarrow Ar_2N\dot{-}O + Ph_2N\cdot \tag{38}$$

$$Ar = p\text{-}MeOC_6H_4$$

Bis(trifluoromethyl) nitroxide, the stabilization of which is reduced due to the electron-accepting properties of the trifluoromethyl groups, can even form stable adducts with nitrogen oxide or nitrogen dioxide[149]:

$$(F_3C)_2N{-}O{-}N{=}O \qquad (F_3C)_2N{-}O{-}NO_2$$

C. Dimerization of Nitroxides

Contrary to the addition of alkyl radicals to the nitroxide group, the dimerization of nitroxides to afford dimers **112** or **113** is improbable due to energy considerations (see Section I.A).

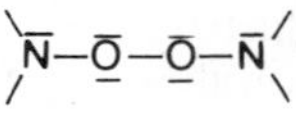

(112) 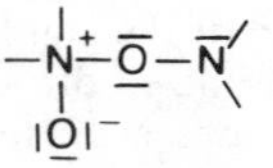 (113)

Nevertheless, some sterically unhindered nitroxides such as the Fremy salt[150], bis(trifluoromethyl) nitroxide[150], tetraphenylpyrrol-*N*-oxyl[150], 9-azabicyclo[3.3.1]-nonane-3-on-9-oxyl[150] and 8-azabicyclo[3.2.1]octane-8-oxyl[151], respectively, exist as dimers in the crystalline state. However, in the dimer of 9-azabicyclo[3.3.1]-nonane-3-on-9-oxyl, the distance between the nitroxide groups is estimated to be 2.278 Å by X-ray analysis, indicating that there is no genuine covalent bond between the groups[152]:

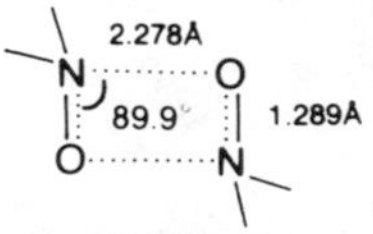

2.278Å 89.9° 1.289Å

In the case of sterically hindered nitroxides formation of diamagnetic dimers could no longer be observed. Instead, formation of radical pairs of tetramethyl piperidine-*N*-oxyl in the plastic phase of carbon tetrachloride[153] and of di-*t*-butyl nitroxide in the plastic phase of camphene[154] have been recently detected. In these radical pairs, the two nitroxide groups are at a distance of about 7.0 Å, but are in a relatively rigid orientation[153].

In solution, reversible dimerization is detected for methyl (2,4,6-tri-*t*-butyl) phenyl nitroxide[155], 8-azabicyclo[3.2.1]octane-8-oxyl[151] and 9-azabicyclo[3.3.1]-nonane-9-oxyl[151] at low temperatures, affording diamagnetic species. By comparison with the situation in the solid state it seems probable that these dimers are not covalently bonded, but rather are molecular complexes for which the geometry of the nitroxide group is only changed slightly. The bonding situation can arise by combination of the two singly occupied molecular orbitals to afford a bonding and an antibonding molecular-complex orbital, the former being capable of taking up the two originally odd electrons with gain of energy, provided that their spin is opposite. Similar dimers of nitrogen oxide have been detected in the gas phase, for which the van der Waals' bonding forces have been found to be only about 1% of the energy of the usual covalent bonds[156]. For the nitroxide dimers, however, calculations show that dipolar attraction forces of about 5 kcal/mol are operative between the two nitroxide groups. In this work has Ingold stated that the dipolar attraction can provide a very significant fraction of the enthalpy for dimerization of unhindered nitroxides in solution[151].

For nitroxides with extended delocalization there is an alternative way of dimerization. Phenyl-substituted nitroxides can form dimers of the type **114** which are obviously energetically less favourable than their monomers but can decompose

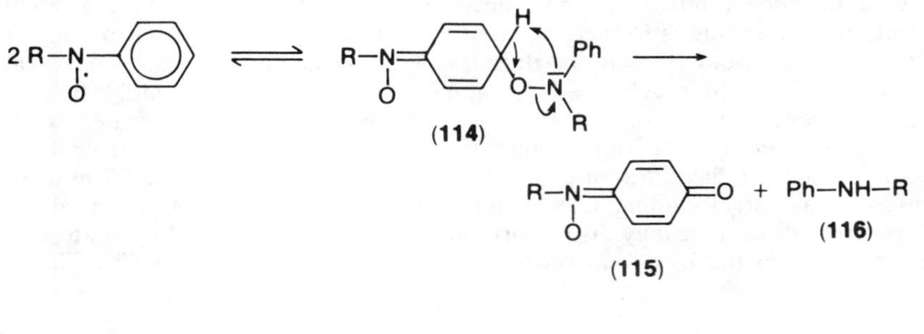

(114)

(115) + Ph—NH—R (116) (39)

irreversibly with formation of the stable products **115** and **116**[157]. Curiously, since the nitroxide is additionally stabilized through delocalization of the unpaired electron into the phenyl group, it becomes even less persistent than di-*t*-alkyl nitroxides, Bulky substituents at the *para*, or even at the *meta*, positions prevent the radical from dimerization and consequently from decomposition.

Radical **117** can be detected at only slightly elevated temperatures by ESR spectroscopy. The gain of about 80 kcal/mol for formation of the carbon–carbon bond in the dimerization to yield **118** causes the equilibrium to lie far to the right at room temperature[158,159].

$$2\ \mathrm{Ph{-}CH{=}N{-}\overset{\displaystyle O^{\bullet}}{N}{-}Ph} \rightleftharpoons \mathrm{Ph{-}C(H){-}N{=}N(O){-}Ph} \atop \mathrm{Ph{-}C(H){-}N{=}N(O){-}Ph} \tag{40}$$

(**117**) (**118**)

On the other hand, analogous dimerization of nitroxide **119** through formation of a nitrogen–nitrogen bond affords only about 40 kcal/mol and is therefore not sufficient to compensate the loss of delocalization energy for the two radicals. Thus, nitroxides like **119** are persistent for several hours or even days in solution,

$$\mathrm{Ph{-}N{=}CH{-}N(O^{\bullet}){-}Ph}$$

(**119**)

although they rapidly decompose when the solvent is removed[104]. In this way the persistence of acyl nitroxides can also be understood. One of the few known vinyl nitroxides dimerizes through formation of a C—C bond as does **117**, but in this case the dimer cannot dissociate reversibly to the monomeric radical but rather decomposes to afford other products[36].

D. Disproportionation of Nitroxides

Disproportionation is an alternative pathway for radical decay. Aryl nitroxides as well as alkyl nitroxides can easily disproportionate to yield hydroxylamines and nitroso compounds, or the corresponding oximes. Formally the oxidation reaction is an

$$2\ \mathrm{R{-}N(O^{\bullet}){-}H} \rightleftharpoons \mathrm{RNHOH} + \mathrm{RN{=}O} \tag{41}$$

α-scission. As kinetic studies have shown, this is a fast bimolecular reaction whose rate is barely influenced by the nature of the group R[160]. The reaction is reversible in the case of phenyl nitroxide and *t*-butyl nitroxide. This means that a small concentration of nitroxide is formed by mixing of the corresponding hydroxylamine and nitroso compounds (R = *t*-Bu: $\Delta H = -10.4$ kcal mol^{-1}; $\Delta S = +8.2$ cal mol^{-1} deg^{-1}; R = Ph: $\Delta H = -7.2$ kcal mol^{-1}; $\Delta S = -2.9$ cal mol^{-1} deg^{-1}). Whereas, however, the concentration of phenyl nitroxide decreases continuously according to the formation of azoxybenzene, *t*-butyl nitroxide is persistent for several days, because *t*-butylhydroxylamine and 2-methyl-2-nitrosopropane form the corresponding azoxy compound only extremely slowly, if at all.

Dialkyl nitroxides can usually disproportionate only if there is a hydrogen atom in the β-position, when the oxidation process affords a nitrone (equation 42). This is

$$2\ R{-}\underset{\underset{\displaystyle O}{|^{\bullet}}}{N}{-}\overset{\overset{\displaystyle H}{|}}{\underset{|}{C}}{-} \longrightarrow R{-}\underset{\underset{\displaystyle OH}{|}}{N}{-}\overset{\overset{\displaystyle H}{|}}{\underset{|}{C}}{-} + R{-}\underset{\underset{\displaystyle O}{|}}{N}{=}C\Big\langle \qquad (42)$$

also a bimolecular reaction, the carbon–hydrogen bond being broken in the rate-determining step. Isotope effects k_H/k_D between 6 and 14 have been found[161]. The disproportionation of diethyl nitroxide is calculated to be exothermic by 28 kcal/mol. The rate of disproportionation is very different for different nitroxides as can be seen from the relative values in Table 9[161].

It has been shown that the formation of a dimer precedes the disproportionation in the last three cases of Table 9, but the nature of this dimer is not known. Although Ingold has formulated a structure with a covalent oxygen–oxygen bond for the dimer, he himself has expressed doubts for thermodynamic reasons about this structure[162]. On the other hand, no dimer formation is observed in the disproportionation of diisopropyl nitroxide[161] or *t*-butyl isopropyl nitroxide[163], the reaction of which is also very slow.

The different reaction rates can be understood if one realizes that hydrogen abstraction from a nitroxide molecule affording a nitrone can proceed with a relatively small activation energy only if the molecule adopts a conformation **120** in which the hydrogen atom to be abstracted and the single occupied orbital are in the same plane. Diisopropyl nitroxide, *t*-butyl isopropyl nitroxide and other

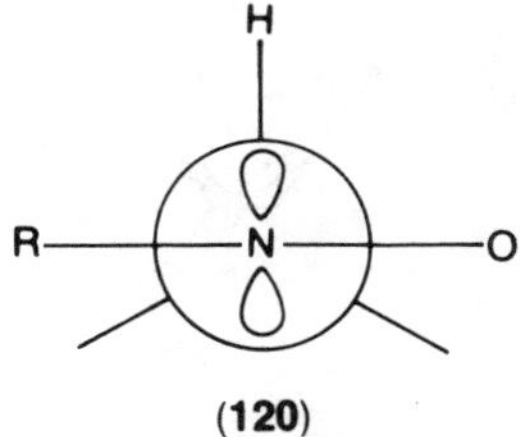

(**120**)

similar nitroxides, can adopt such a conformation only if supplied with an additional amount of energy, thus raising the activation energy for the reaction. The bicyclic nitroxides **10**, **121** and **92** can never adopt such a conformation. Thus **10** and **121** are very persistent radicals[151], whereas **92** surprisingly yields a dimer, **122**, which must have been formed from the corresponding nitrone[164]. Of course the formation of this dimer is very slow and the decay of **92** is about 10^9 times slower than that of diethyl nitroxide.

Isolation of the primary disproportionation products is often difficult, since secondary reactions can easily occur[165]. For example, the disproportionation of *t*-butyl isopropyl nitroxide affords products, the formation of which can be

TABLE 9. Relative rate of disproportionation of nitroxides[161]

Nitroxide	k_{rel}
Diisopropyl nitroxide	1
Diethyl nitroxide	1.3×10^3
Dimethyl nitroxide	3.9×10^3
Piperidine-*N*-oxyl	31×10^3

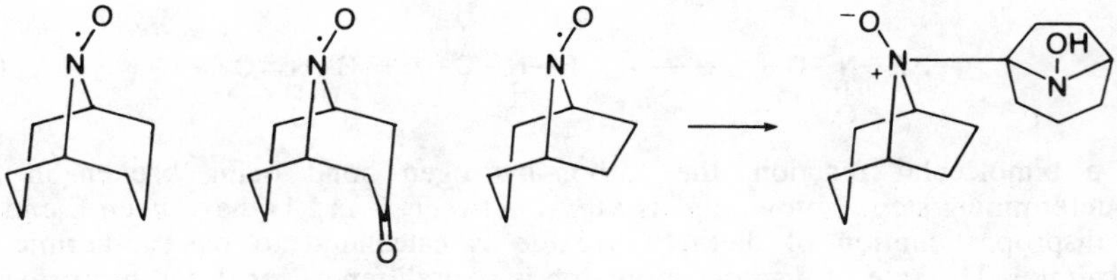

(10) (121) (92) (122) (43)

explained by the addition of unchanged nitroxide to the nitrone **123** formed in this reaction, to give adduct **124** (equation 44), followed by decomposition of this adduct to several fragments[163].

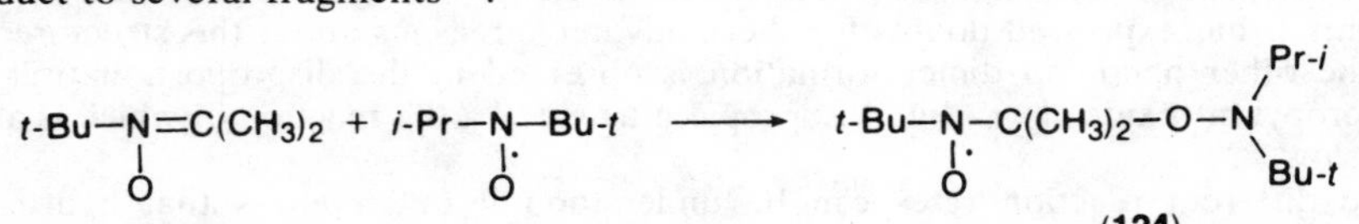

$$t\text{-Bu}-\underset{\mathrm{O}}{\mathrm{N}}=\mathrm{C(CH_3)_2} + i\text{-Pr}-\underset{\mathrm{O}}{\dot{\mathrm{N}}}-\mathrm{Bu}\text{-}t \longrightarrow t\text{-Bu}-\underset{\mathrm{O}}{\dot{\mathrm{N}}}-\mathrm{C(CH_3)_2}-\mathrm{O}-\mathrm{N}(\mathrm{Pr}\text{-}i)(\mathrm{Bu}\text{-}t) \quad (44)$$

(123) (124)

Only the nitrone **127** formed by the disproportionation of *N*-methyl (2,4,6-tri-*t*-butyl)phenyl nitroxide (**125**) was isolated directly, but the simultaneously formed hydroxylamine **126** was also detected (equation 45)[155]. As

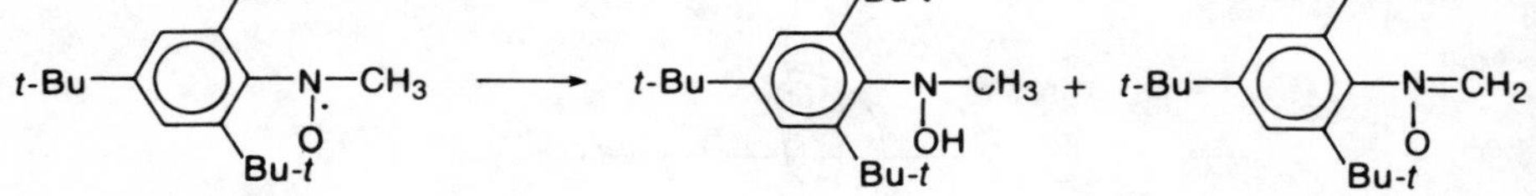

(125) (126) (127) (45)

shown by the isolation of the products **128** and **129**, *ortho*-substituted aryl *t*-butyl nitroxides can disproportionate in an analogous manner (equation 46)[166].

Contrary to nitroxides substituted by primary or secondary alkyl groups, di-*t*-alkyl nitroxides are usually very persistent. For instance, di-*t*-butyl nitroxide and 2,2,6,6-tetramethylpiperidine-*N*-oxyl remain unchanged over periods of several

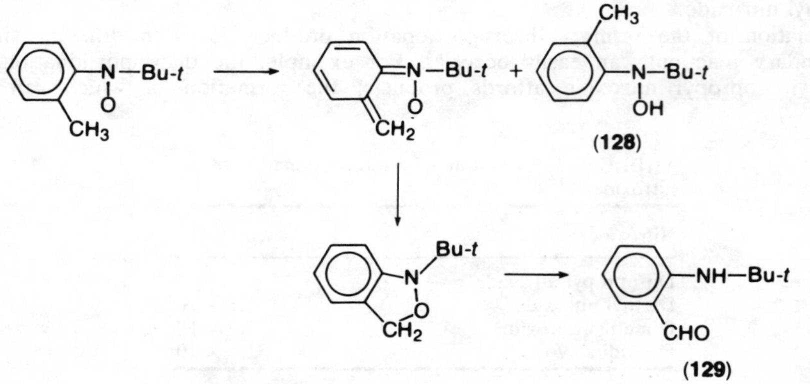

(128) (46) (129)

months or even years. On the other hand, 4-oxo-2,2,6,6-tetramethylpiperidine-*N*-oxyl (**9**) decomposes after six months, a dimer of structure **132** being formed. The diradical **131** is formulated as an intermediate in this decomposition reaction. Nitroxide **9** also decomposes yielding hydroxylamine **130** (60%), a polymer of unknown structure and phoron **134** (trace) by heating to 100 °C or by refluxing in benzene for several hours. It is suggested that the diradical **131** is also an intermediate in this reaction[167].

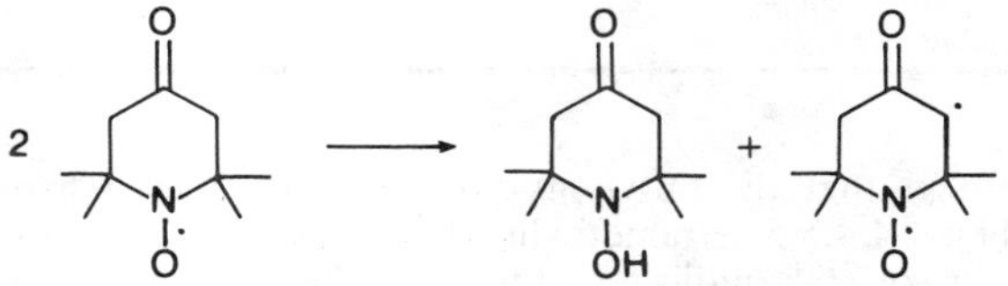

(47)

(9) **(130)** **(131)**

130 + **131** →

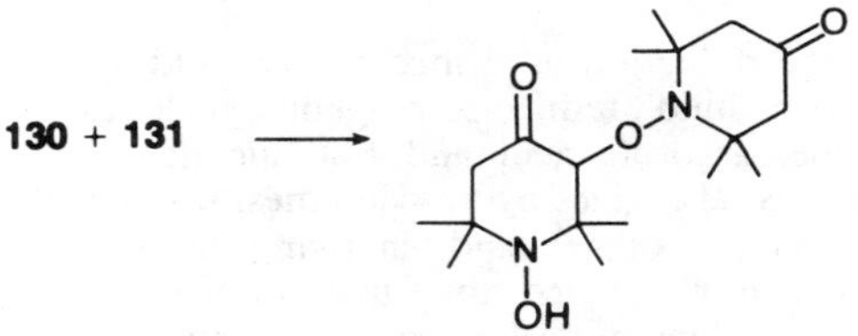

(132)

131 → ... $\xrightarrow[-H^{\bullet}]{-NO^{\bullet}}$

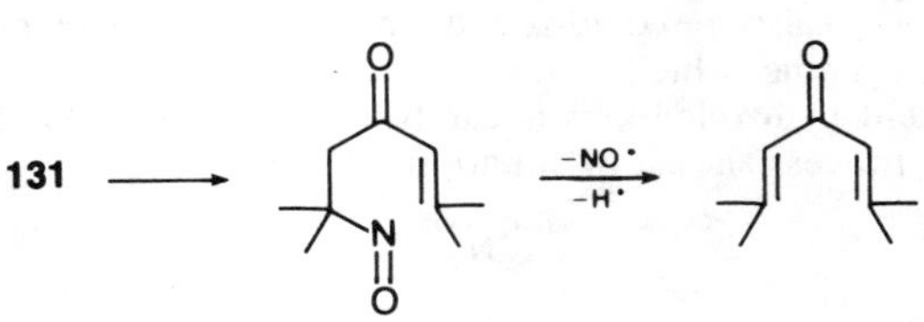

(133) **(134)**

E. Reduction of the Nitroxide Group

An increasing spin density at oxygen (increasing weight of contribution structure A) should enhance the electrophilic character of the nitroxide, enhancing the tendency for hydrogen abstraction (equation 48). This should be reflected by the O—H bond dissociation energy of the corresponding hydroxylamines. A high bond dissociation energy is indicative of a strong tendency for hydrogen abstraction in the nitroxide.

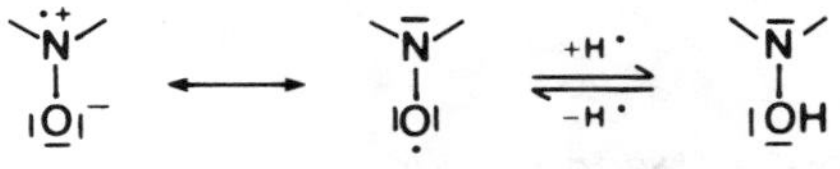

(48)

(B) **(A)**

Inspection of Table 10 shows that bicyclic nitroxides and acyl nitroxides should be reduced more easily than normal nitroxides. This is in fact found to be true. But the bistrifluoromethyl nitroxide shows an even higher tendency to be reduced. On

TABLE 10. O—H bond dissociation energies (D_{OH}) of hydroxylamines

Hydroxylamine	D_{OH} (kcal/mol)	Reference
N-Hydroxy-2,2,6,6-tetramethypiperidine	69.6	14
N-Hydroxy-4-oxo-2,2,6,6-tetramethylpiperidine	71.8	14
N-Hydroxy-8-azabicyclo[3.2.1]octane	77.0	14
N-Hydroxy-9-azabicyclo[3.3.1]nonane	76.2	14
N-*t*-Butylarylhydroxamic acids	76–80	168

the other hand, the fact that di-*t*-butyl nitroxide can abstract a hydrogen atom from 1-hydroxy-2-phenyl-4,4,5,5-tetramethylimidazoline-3-oxide giving the corresponding nitronyl nitroxide[103] indicates that in the nitronyl nitroxides the tendency to be reduced is very small. This order of reactivity for the reduction of nitroxides agrees with the spin density at oxygen as determined by the ^{17}O-coupling constants*.

In general, the reduction of a nitroxide to yield the corresponding hydroxylamine is possible with mild reducing reagents such as hydrazine, phenylhydrazine, hydrazobenzene, ascorbic acid and thio alcohols[169,170]. Catalytic reduction using a platinum catalyst also gives hydroxylamines, whereas the use of Raney nickel yields amines[169,170]. On the other hand, in many cases nitroxides can also be used as oxidizing agents. For instance, they may oxidize substituted phenylenediamines to give the corresponding radical cations, and hydroxybenzoquinones or phenols, to give quinones[169]. Heterocyclic compounds such as 1,5-dihydroflavines and 1,5-dihydrolumiflavines can also be oxidized by nitroxides[171]. The reaction of a nitroxide and a hydroxylamine which leads to an equilibrium between the two nitroxides and their parent hydoxylamines is catalysed by cupric ions. The catalysis can be explained by the process shown in equation (49)[172].

$$\text{>N–OH} + Cu^{2+} \rightleftharpoons \text{>N–O}\cdot + Cu^{+} + H^{+} \quad (49)$$

The strong electrophilic character of bis(trifluoromethyl) nitroxide is indicated by the fact that this nitroxide can abstract a hydrogen atom from hydrocarbons with the exception of methane. Reaction with ethane proceeds slowly, and the ethyl radical formed reacts with a second nitroxide molecule to yield **135**. Further reaction affords **136** and **137** (equation 50)[173] On the other hand, reaction with

$$\underset{\mathbf{(2)}}{(CF_3)_2NO\cdot} + CH_3CH_3 \longrightarrow (CF_3)_2N\text{—}OH + CH_3CH_2\cdot \xrightarrow{+2} \underset{\mathbf{(135)}}{CH_3CH_2\text{—}O\text{—}N(CF_3)_2}$$

$$\longrightarrow \longrightarrow \underset{\mathbf{(136)}}{CH_3CO\text{—}O\text{—}N(CF_3)_2} + \underset{\mathbf{(137)}}{(CF_3)_2N\text{—}O\text{—}CH_2CH_2\text{—}O\text{—}N(CF_3)_2} \quad (50)$$

*^{17}O-labelling of bistrifluoromethyl nitroxide has not yet been performed but in *t*-butyl trifluoromethyl nitroxide a^{17}O is larger than in the other nitroxides[39].

$$(CH_3)_3C{-}O{-}N(CF_3)_2 \quad \mathbf{(138)} \qquad (CH_3)_2C(CH_2{-}O{-}N(CF_3)_2){-}O{-}N(CF_3)_2 \quad \mathbf{(139)}$$

methylpropane occurs easily, the products **138** and **139** being formed. Formation of **139** occurs via the intermediate methylpropene[173]. Reaction of bis(trifluoromethyl) nitroxide (**2**) and toluene in the ratio 2:1 affords mainly **140**, whereas a large excess of nitroxide yields products **141**, **142** and **143**. These products are formed by the reaction sequences shown in equation (51)[174]. It seems remarkable that even the aldehydic hydrogen atom can be abstracted by the radical.

$$PhCH_3 + 2\,(CF_3)_2NO^{\cdot}\ \mathbf{(2)} \longrightarrow PhCH_2{-}O{-}N(CF_3)_2\ \mathbf{(140)} + (CF_3)_2NOH\ \mathbf{(141)}$$

$$\downarrow{+2}$$

$$Ph\dot{C}H{-}O{-}N(CF_3)_2 + \mathbf{141} \tag{51}$$

$$\downarrow$$

$$PhCOO{-}N(CF_3)_2\ \mathbf{(142)} + \mathbf{141} \xleftarrow{+2\times 2} PhCH{=}O + (CF_3)_2N^{\cdot} \xrightarrow{+2} (CF_3)_2N{-}O{-}N(CF_3)_2\ \mathbf{(143)}$$

Acyl nitroxides are also good hydrogen acceptors. Although benzoyl *t*-butyl nitroxide **144a** reacts with ethylbenzene only at elevated temperatures to yield **145** (equation 52), this reaction is faster by a factor of 10^3 than with dialkyl nitroxides,

$$ArCO{-}N(O^{\cdot}){-}Bu\text{-}t\ \mathbf{(144)} + H{-}R \longrightarrow ArCO{-}N(OR){-}Bu\text{-}t\ \mathbf{(145)} \tag{52}$$

(**a**) Ar = C_6H_5,
(**b**) Ar = 3, 5-$(NO_2)_2C_6H_3$, R = $C_6H_5CHCH_3$ or *c*-C_6H_{11}

The more reactive 3,5-dinitrobenzoyl *t*-butyl nitroxide **144b** can even attack cyclohexane[175]. **144b** can also oxidize alcohols, for instance, cyclohexanone is formed from cyclohexanol[175]. An aspect of special interest is the application of the chiral acyl nitroxide **146**, which can, for instance, oxidize benzoin in an enantioselective manner[176].

CH3 O ‖ C—N—Bu-*t*, N—O·

(**146**)

Whereas dialkyl nitroxides in the ground state do not usually react with hydrocarbons or the reaction is extremely slow, photochemical reaction with hydrocarbons occurs very readily. Thus, 4-hydroxy-2,2,6,6-tetramethyl piperidine-*N*-oxyl (**79**) is converted to hydroxylamine **147** and *O*-benzylhydroxylamine **148** in a 1:1 ratio when irradiated in toluene (equation 53)[177]. **148** is the addition product of nitroxide **79** and the benzyl radical. Hydrogen abstraction from alkanes

(**79**) + $PhCH_3$ $\xrightarrow[3500\ \text{Å}]{h\nu}$ (**147**) + (**148**) (53)

by photochemically excited dialkyl nitroxides occurs from the $n \rightarrow \pi^*$ excited state of the radical.

The activation energy for reactions of photochemically excited 2,2,6,6-tetramethyl piperidine-*N*-oxyl is estimated to be in the range 2–3 kcal/mol, indicating that the radical is very reactive in its excited state. The excitation causes the dipole moment to fall from 3.14 D for the ground state to 1.0 D in the excited state[178]. This reflects a shift in electron density from oxygen to nitrogen making the oxygen atom much more electrophilic.

Demethylation at a specific position has been achieved in a steroid molecule labelled with a nitroxide group by a photochemically induced intramolecular oxidation[179]. Another intramolecular hydrogen abstraction by a photochemically excited nitroxide group involves the photochemical conversion of nitronyl nitroxide **149** in aqueous solution, **152a** and **152b** being isolated. It has been suggested that this reaction occurs via the intermediates **150** and **151**[180].

(**149**) $\xrightarrow{h\nu}$ (**150**) $\longrightarrow$ (**151**) $\longrightarrow$ $\longrightarrow$ (**152**) (54)

(**a**) X = OH
(**b**) X = OOH

F. Oxidation of the Nitroxide Group

Oxidation of monoaryl and monoalkyl nitroxides affording nitroso compounds has already been discussed in Section V.D. This reaction is part of the oxidative conversion of hydroxylamines to nitroso compounds, in which the nitroxides are intermediates. An analogous situation arises for the oxidation of disubstituted nitroxides to give nitrones[181]. For instance, *N*-benzylidene benzylamine-*N*-oxide is formed by the oxidation of dibenzyl hydroxylamine via dibenzyl nitroxide (equation 55)[182]. Frequently, however, the resulting nitrones undergo secondary reactions and cannot be isolated[163,165]. Oxidation of nitroxides beyond the nitrone level yielding αβ-unsaturated nitroxides has been only observed for aminoalkyl nitroxides (equation 56)[183]. Oxidation of di-*t*-alkyl nitroxides gives the oxoammonium salts **153** (equation 57).

$$PhCH_2-N(OH)-CH_2Ph \xrightarrow{-H^\bullet} PhCH_2-N(O^\bullet)-CH_2Ph \xrightarrow{-H^\bullet} PhCH{=}N(O)-CH_2Ph \quad (55)$$

$$R^1NH_2 + R^2CH{=}N(O)-R^3 \rightleftharpoons R^1NH-CH(R^2)-N(OH)-R^3 \xrightarrow{-H^\bullet}$$

$$R^1NH-CH(R^2)-N(O^\bullet)-R^3 \xrightarrow{-H^\bullet} R^1NH-C(R^2){=}N(O)-R^3 \xrightarrow{-H^\bullet} R^1N{=}C(R^2)-N(O^\bullet)-R^3 \quad (56)$$

$$R_3C-N(O^\bullet)-CR_3 \xrightarrow{-e} R_3C-\overset{+}{N}({=}O)-CR_3 \quad (57)$$

(153)

In Table 11 the half-wave potentials determined by electrochemical oxidation of some nitroxides are listed[184] reflecting the trend for the formation of oxoammonium cations.

Oxidation occurs with a variety of oxidizing agents[185]. For instance, 2,2,6,6-tetramethylpiperidine-*N*-oxyl is oxidized by xenon difluoride to afford the corresponding oxoammonium fluoride[186]. Using bromine or chlorine in inert solvents, oxoammonium salts can be isolated in most cases[185]. Oxoammonium salts can also be obtained by the disproportionation of di-*t*-alkyl nitroxides after partial protonation (equation 58)[185].

Oxoammonium cations are electron-deficient compounds compared to nitroxides, thus easily undergoing secondary reactions. For instance, reaction with the solvent can occur. The hydroxylammonium cation **155** is formed from **154** in acetone[187].

TABLE 11. Half-wave potentials ($E_{1/2}$) of nitroxides[184]

Nitroxide	$E_{1/2}$(V)
Di-*t*-butyl nitroxide	0.207
Tetramethylpiperidine-*N*-oxyl	0.271
4-Hydroxytetramethylpiperidine-*N*-oxyl	0.331
4-Oxotetramethylpiperidine-*N*-oxyl	0.444
3-Oxo-9-azabicyclo[3.3.1]nonane-*N*-oxyl	0.473
1,5-Dimethyl-3-oxo-8-azabicyclo[3.2.1]octane-*N*-oxyl	0.448
3-Oxo-8-azabicyclo[3.2.1]octane-*N*-oxyl	0.509

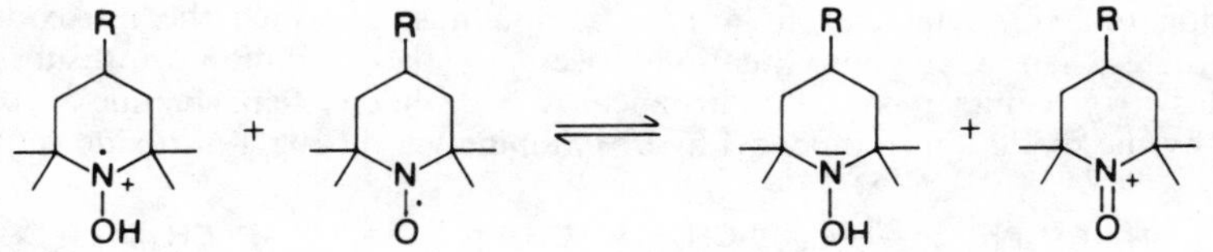

(58)

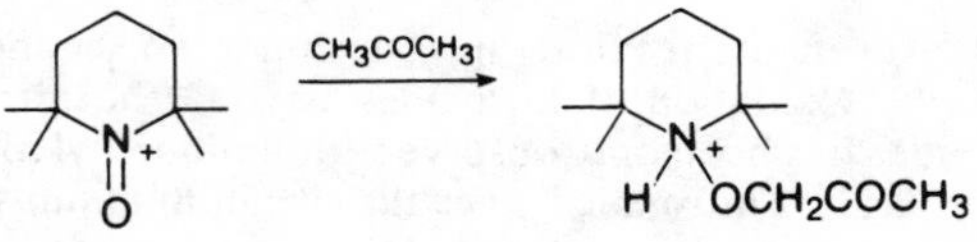

(59)

(154) (155)

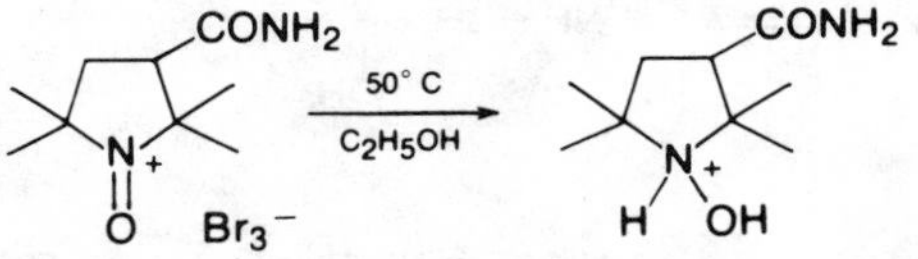

(60)

(156) (157)

The oxoammonium salt **156** is reduced by warming it in alcohol, to yield **157**, from which the corresponding free hydroxylamine can be obtained by treatment with potassium carbonate[187]. This reaction is used to oxidize secondary alcohols using a mixture of *m*-chloroperbenzoic acid and 2,2,6,6-tetramethylpiperidine-*N*-oxyl to yield ketones[188]. The oxidation does not occur in the absence of the nitroxide but proceeds with only a catalytic amount[189]. Therefore the reaction sequence shown in equation (61) is suggested[188]. The oxidation of methanol in the presence of di-*t*-butyl nitroxide and cupric phenanthroline complexes to yield formaldehyde appears to be more complicated[190].

In some oxoammonium salts, bond cleavage occurs, being favoured by the presence of suitable functional groups within the molecule. Thus the oxoammonium

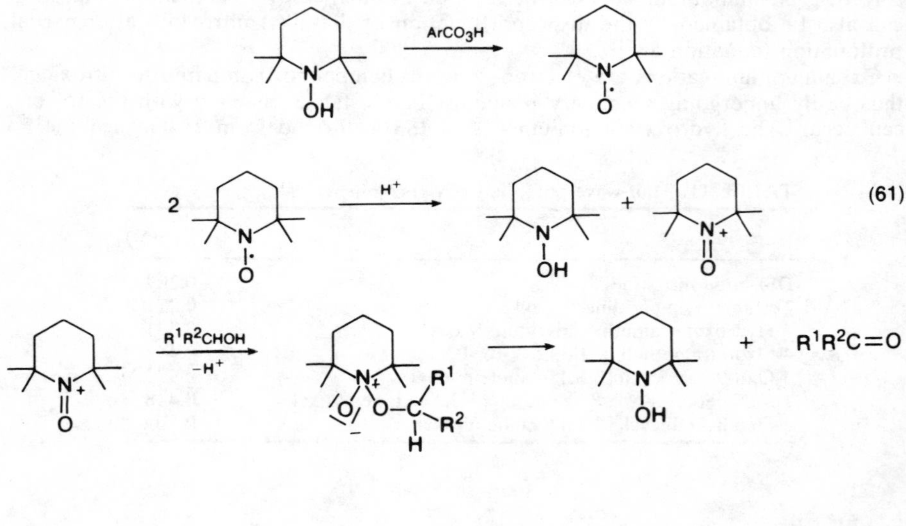

(61)

salt **158**, formed by oxidation of 4-oxo-2,2,6,6-tetramethylpiperidine-*N*-oxyl, is cleaved to afford the nitro compound **160** via the nitroso compound **159** (equation 62) whereas the cation obtained from 2,2,6,6-tetramethylpiperidine-*N*-oxyl is stable under these conditions[191].

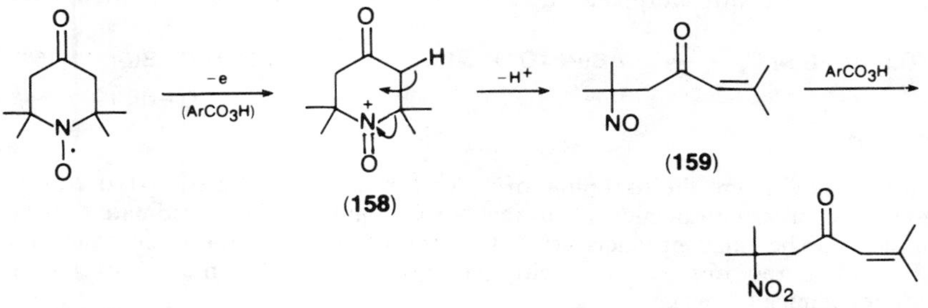

-e (ArCO$_3$H) → (**158**) → $-H^+$ → (**159**) NO → $ArCO_3H$ → NO$_2$ (**160**) (62)

Nitroxide **161** is oxidized by *m*-chloroperbenzoic acid or nitrogen dioxide, and decomposes via its oxoammonium cation to yield cyclohexanone, isobutylene and NO^+ (equation 63)[191]. Molecular rearrangements, accompanying the oxidation of bicyclic nitroxides by silver oxide, are assumed to be induced by cleavage of the primarily formed oxoammonium cations[192,193].

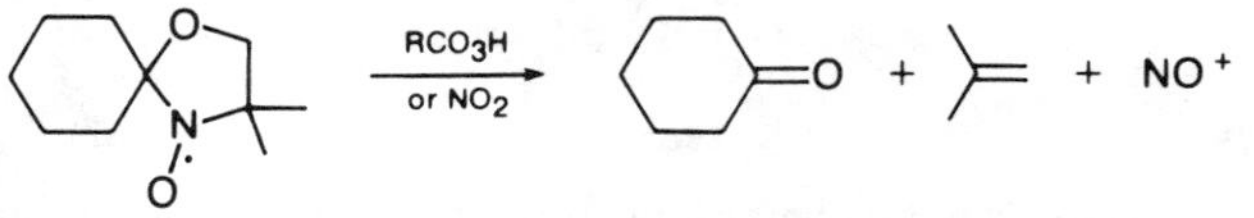

(**161**) $\xrightarrow[\text{or } NO_2]{RCO_3H}$ cyclohexanone + isobutylene + NO^+ (63)

G. α-Scission of Nitroxides

The reverse of spin trapping by nitroso compounds is the α-scission of a disubstituted nitroxide. The α-scission of the N—C bond occurs readily if the resulting radical $\dot{R}^2$ is stabilized (equation 64).

$$R^1{-}N(\dot{O}){-}R^2 \longrightarrow R^1N{=}O + \dot{R}^2 \qquad (64)$$

$$R^2 = Ph_3C^{\cdot},\ H_2C{=}CH{-}\dot{C}(CH_3)_2,\ HC{\equiv}C{-}\dot{C}(CH_3)_2$$

Thus, trityl nitroxides decompose yielding the triphenyl methyl radical and the nitroso compound[194]. This reaction occurs easily if the R^1 group is bulky, as for instance the *t*-butyl group. In solution, radical **162**, decomposes even at 25°C affording the allyl radical **163**, which further attacks the nitroxide **162** to yield the product **164** (equation 65)[195]. On the other hand, the corresponding

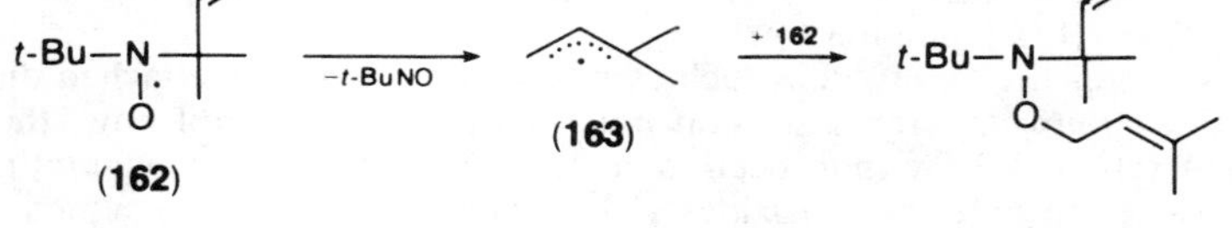

t-Bu—N(O·)— (**162**) → $-t$-BuNO → (**163**) → + **162** → *t*-Bu—N(O—)— (**164**) (65) *t*-butyl-(1,1-dimethyl) propyne-2-yl nitroxide is only cleaved in refluxing benzene solution, when subsequent reactions yield a bicyclic compound with a complicated structure[196].

Di-*t*-butyl nitroxide (**1**) is only decomposed at temperatures above 125°C, yielding 2-methyl-2-nitrosopropane and tri-*t*-butylhydroxylamine (equation 66),

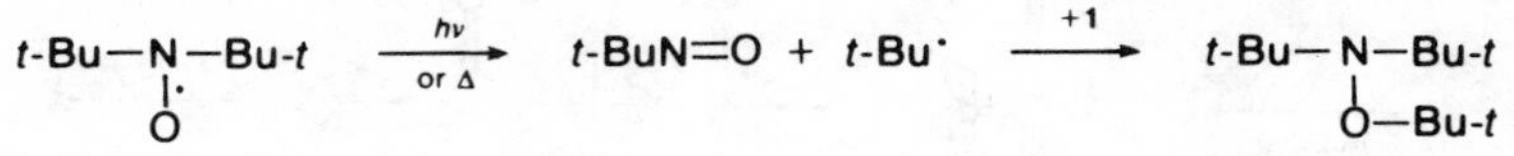

$$t\text{-Bu}-\underset{\underset{\text{O}}{|\cdot}}{\text{N}}-\text{Bu-}t \xrightarrow[\text{or }\Delta]{h\nu} t\text{-BuN=O} + t\text{-Bu}^{\cdot} \xrightarrow{+1} t\text{-Bu}-\underset{\underset{\text{O}-\text{Bu-}t}{|}}{\text{N}}-\text{Bu-}t \qquad (66)$$

(**1**)

the latter arising from the trapping of *t*-butyl radicals by unreacted di-*t*-butyl nitroxide. The dissociation energy for the N—C bond has been estimated to be 29 kcal/mol[197]. The same products are formed by the photochemical cleavage of **1** but the reaction requires $\pi \rightarrow \pi^*$ excitation ($\lambda < 320$ nm), the $n \rightarrow \pi^*$ excitation being photochemically inert[198].

The photochemical decomposition of nitroxide **165** is thought to proceed via the intermediate **166**, which is stabilized by its allylic group. Elimination of nitrogen oxide then affords product **167** (equation 67).

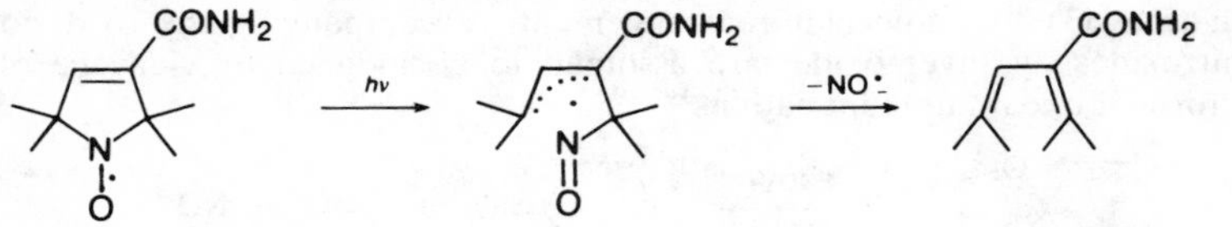

(67)

(**165**) $\xrightarrow{h\nu}$ (**166**) $\xrightarrow{-NO^{\cdot}}$ (**167**)

On the contrary, photochemically excited 4-hydroxy-2,2,6,6-tetramethylpiperidine-*N*-oxyl (**79**) does not decompose but abstracts hydrogen in the presence of toluene (equation 68) (see Section V.E). It is suggested that α-scission also occurs in this case, but that the radical **168** not being stabilized is converted reversibly to **79** by intramolecular spin trapping[198].

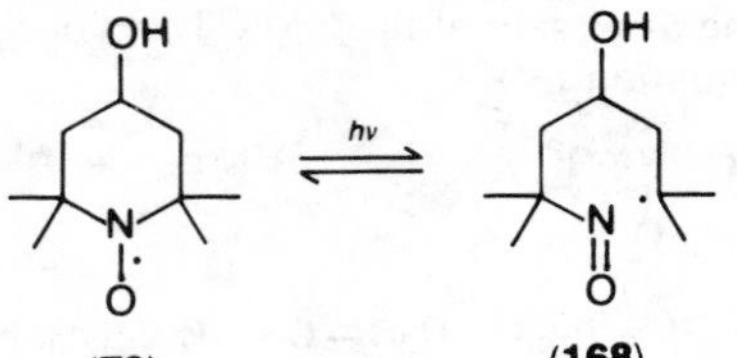

(68)

(**79**) $\underset{}{\overset{h\nu}{\rightleftharpoons}}$ (**168**)

The photochemically induced reaction of 2-*t*-butyl-4,4,5,5-tetramethylimidazoline-3-oxide-1-oxyl in aprotic solvents follows a rather complicated course[199]. Irradiation at low temperature causes the usual conversion of the nitrone group to an oxaziridine ring. At higher temperatures the NO bond of the oxaziridine ring is broken, inducing α-scission at the nitroxide moiety to afford the intermediate **169**. Finally this undergoes intramolecular spin trapping to yield the nitroxides **170** or **171** (equation 69).

Radicals in which a heteroatom such as nitrogen or oxygen is attached directly to the nitroxide group undergo α-scission much more readily. Until now, the acyclic triazene-*N*-oxyls **172** have not been detected. All attempts at generating these radicals have given only the nitroxides **173**[200]. Their formation is assumed to occur by α-scission of the transient triazeneoxyl giving the nitroso compound and a

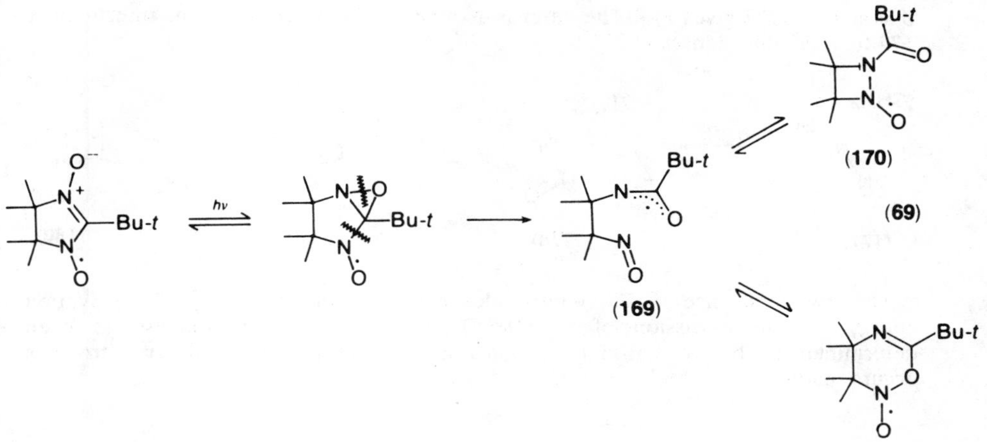

(170)

(69)

(169)

(171)

diazonium radical which further decomposes by elimination of nitrogen. The radical $\dot{R}^1$ is then trapped by the nitroso compound yielding the nitroxide **173** (equation 70).

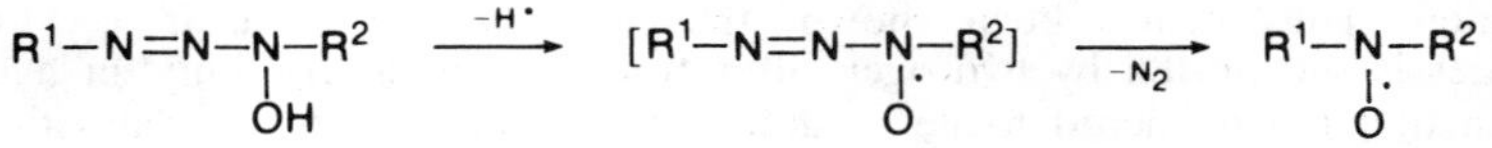

$$R^1{-}N{=}N{-}N(OH){-}R^2 \xrightarrow{-H^{\cdot}} [R^1{-}N{=}N{-}N(\dot{O}){-}R^2] \xrightarrow{-N_2} R^1{-}N(\dot{O}){-}R^2 \quad (70)$$

(172) (173)

On the other hand, benzotriazole-1-oxyl (**174**), is sufficiently persistent to be detected by ESR in several solvents[201]. In aromatic solvents, however, or even in the presence of aromatic compounds, **174** decomposes with the formation of the *o*-nitrosophenyl radical. In, for example, benzene, the reaction of the *o*-nitrosophenyl radical with the solvent affords the carbazole-*N*-oxyl **176** (equation 71). Substituent effects on the persistence of radicals of the type **174** makes it

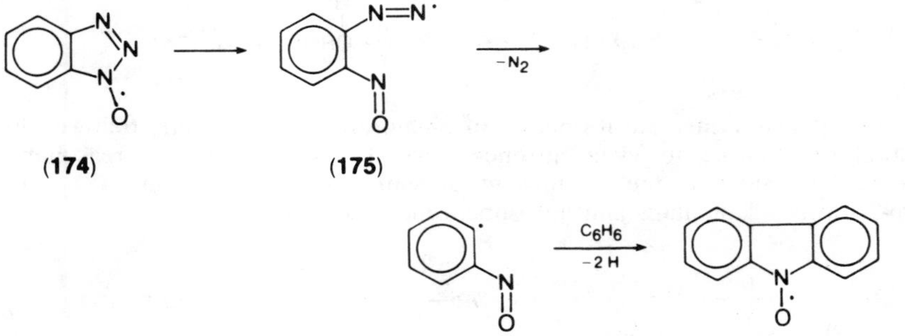

(174) (175)

C_6H_6, $-2\,H$

(176) (71)

probable that in the first step of the reaction α-scission occurs to yield the α-nitrosodiazophenyl radical **175** which subsequently eliminates nitrogen[202].

The conversion of bicyclic hydrazyls **177** in the presence of oxygen is a further example of N—N scission in the α-position of the nitroxide group. Firstly the oxidation of **177** gives **178**. The latter is assumed to be oxidized via the intermediate **178** to yield **180** (equation 72)[203].

$$\mathbf{(177)} \xrightarrow{O_2} \mathbf{(178)} \rightleftharpoons \mathbf{(179)} \xrightarrow{O_2} \mathbf{(180)} \qquad n = 1 \text{ or } 2 \qquad (72)$$

The low persistence of alkoxy nitroxides is well documented[204,205]. The activation energy for the α-scission of the N—O bond in alkoxy nitroxides has been determined to be 10 kcal/mol[206], indicating that cleavage of alkoxy nitroxides occurs readily.

H. β-Scission of Nitroxides

The oxidation of nitroxides to afford nitrones may be considered as a β-scission. Moreover halogen substituents in the β-position of the nitroxide group can be easily eliminated. Thus it has been shown, that the chlorine atom in β-chlorinated nitroxides is substituted by hydrogen after treatment with tri-*n*-butyltin hydride[207]. The reaction is considered to be a radical-chain reaction, the propagation of the chain involving the steps shown in equation (73).

$$R{-}N(O^{\bullet}){-}CClR^1R^2 + Bu_3Sn^{\bullet} \longrightarrow R{-}N(O){=}CR^1R^2 + Bu_3SnCl$$

$$R{-}N(O){=}CR^1R^2 + Bu_3SnH \longrightarrow R{-}N(O^{\bullet}){-}CHR^1R^2 + Bu_3Sn^{\bullet} \qquad (73)$$

Supposing that intermediate formation of β-chloronitroxides occurs, followed by elimination of chlorine to yield nitrones, several reactions can be reasonably interpreted[208–210] such as the photochemical synthesis of the nitrone **181** from 2-chloro-2-nitrosoadamantane and nitrosobenzene (equation 74)[208].

$$Ad(NO)(Cl) \xrightarrow[-NO]{h\nu} Ad^{\bullet}(Cl) \xrightarrow{PhN{=}O} Ad(Cl){-}N(O^{\bullet}){-}Ph \xrightarrow{-Cl^{\bullet}} Ad{=}N(O)Ph \; \mathbf{(181)} \qquad (74)$$

Ad = adamantane

Similarly, photolysis of 1-nitro-1-nitrosocyclohexane is assumed to occur via the intermediate **182**. The latter eliminates nitrogen dioxide with the formation of **183** which then reacts further (equation 75)[211].

$$\text{(182)} \xrightarrow{-NO_2^{\cdot}} \text{(183)} \longrightarrow \longrightarrow \quad (75)$$

I. Miscellaneous Nitroxide Group Reactions

The most reactive nitroxide, bis(trifluoromethyl) nitroxide, is capable of addition to alkenes. A 2:1 adduct **184** is formed with ethylene in 99% yield (equation 76)[212].

$$2\,(CF_3)_2NO^{\cdot} + CH_2{=}CH_2 \longrightarrow \underset{\textbf{(184)}}{(CF_3)_2N{-}O{-}CH_2CH_2{-}O{-}N(CF_3)_2} \quad (76)$$

Reactions with isobutene, 2-methylbutene-1 and 2-methylbutene-2 also give the corresponding products. On the other hand, hydrogen is abstracted from 3-methylbutene-1, the resulting radical being trapped by a further nitroxide molecule to yield **185** and **186** (equation 77)[173].

$$2\,(CF_3)_2NO^{\cdot} + Me_2CHCH{=}CH_2 \xrightarrow[-(CF_3)_2NOH]{} \underset{\textbf{(185)}}{Me_2C(ON(CF_3)_2)CH{=}CH_2} + \underset{\textbf{(186)}}{Me_2C{=}CH{-}CH_2{-}ON(CF_3)_2} \quad (77)$$

Acetylene and substituted acetylenes can first add two bis(trifluoromethyl) nitroxide molecules, but contrary to the reaction of ethylene very complex reactions occur[213]. In the reaction of bis(trifluoromethyl) nitroxide and benzene, substitution occurs yielding the 1,2,4-tris[*N,N*-bis (trifluoromethyl)] aminoxybenzene[214].

Recently some addition reactions of acyl nitroxides have been reported[215].

Nitroxides of usual reactivity can induce homolytic cleavage of weak σ-bonds. It has been found that in di-*t*-butyl nitroxide as the solvent or in molten 4-oxo-2,2,6,6-tetramethyl piperidine-*N*-oxyl, compounds such as benzoyl peroxide, *N*-bromosuccinimide, allyl and benzyl bromide and even the less reactive secondary and tertiary alkyl bromides and iodides can be cleaved. This process, called homosolvolysis[216], can be formulated as in equation (78). If X is bromine, the almost insoluble oxoammonium bromide precipitates; benzyl and allyl bromides afford the

$$R^1X + 2\,R^2_2N\dot{O} \longrightarrow [\dot{R}^1\ \dot{X}]_{2\,R^2_2NO} \rightleftharpoons [\dot{R}^1\ \dot{O}NR^2_2] + [R^2_2N\dot{O}\ \dot{X}] \quad (78)$$

$$[\dot{R}^1\ \dot{O}NR^2_2] \longrightarrow \dot{R}^1 + \dot{O}NR^2_2$$

$$[\dot{R}^1\ \dot{O}NR^2_2] + [R^2_2N\dot{O}\ \dot{X}] \rightleftharpoons R^1ONR^2_2 + R^2_2NOX \text{ or } R^2_2NO^+X^-$$

corresponding *O*-substituted hydroxylamines, whereas the radicals formed from *N*-bromosuccinimide or benzoyl peroxide abstract hydrogen. Most strikingly, a small amount of di-*t*-butyl nitroxide in an inert solvent is sufficient to induce the dissociation of triphenylmethyl bromide, the triphenylmethyl radical being detected by ESR[217].

Deoxygenation of nitroxides occurs with a variety of reagents such as triethyl phosphite, methyl iodide or iodine, sodium sulphide in dimethyl sulphoxide, tetramethylthiuram disulphide and sulphur dioxide[218]. The mechanism of this reaction has been discussed for triethyl phosphite only[219]. It is assumed that the reaction begins with the formation of a P—O bond; then scission of the N—O bond follows, yielding triethyl phosphate and an aminyl radical which finally abstracts hydrogen from the solvent to form an amine.

VI. REACTIONS WITHOUT DIRECT INVOLVEMENT OF THE NITROXIDE GROUP

The extraordinary persistence of the di-*t*-alkyl nitroxides makes it possible for various reactions to be performed at other reactive positions of the radical molecule, the nitroxide group itself being unaffected[1–4]. Such reactions were first

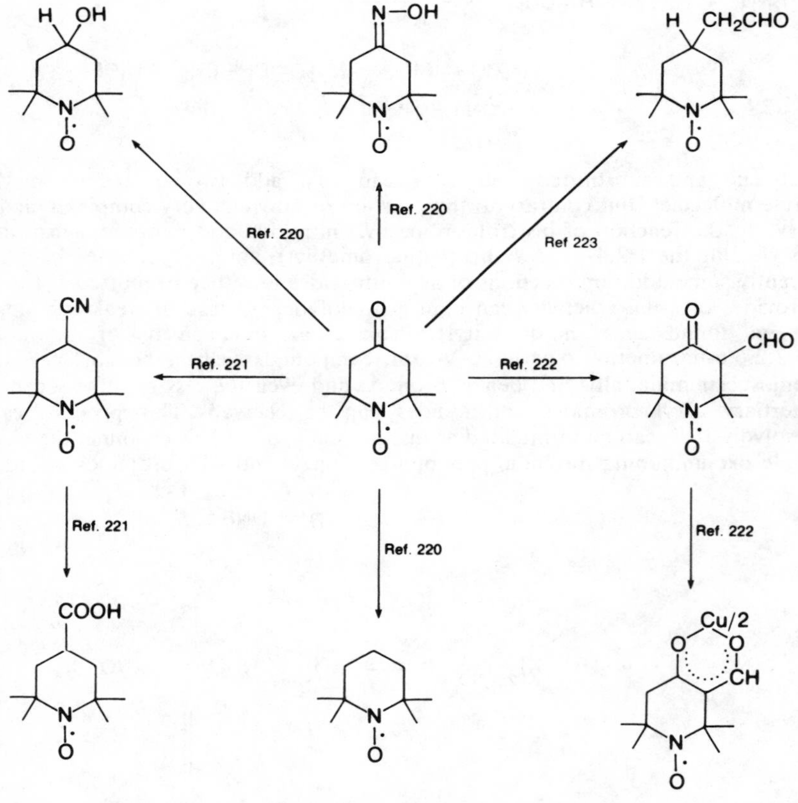

H OH
N—OH
H CH_2CHO
Ref. 220
Ref. 220
Ref 223
CN
O
O
CHO
Ref. 221
Ref. 222
Ref. 221
Ref. 220
Ref. 222
COOH
Cu/2
CH

SCHEME 5

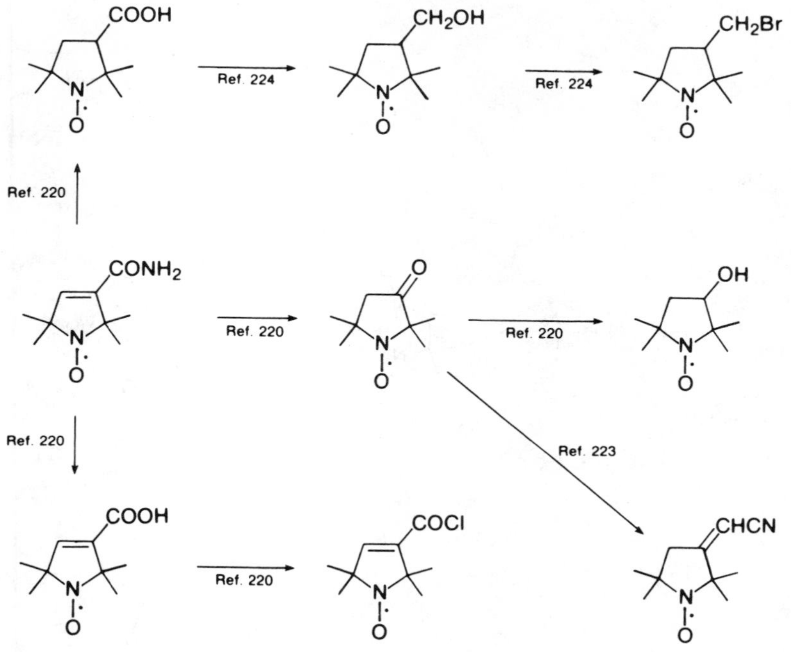

SCHEME 6

carried out by Rozantsev[2,3] using the carbonyl group reactivity of the 4-oxo-2,2,6,6-tetramethylpiperidine-*N*-oxyl. The great numer of possibilities for the application of such reactions is demonstrated in Schemes 5 and 6[220–224]:

The reactions, occurring at the position 2 of 4,4,5,5-tetramethylimidazoline-3-oxide-1-oxyl **187** and its 2-halogeno derivatives, leaving the nitroxide group unchanged (equations 79–81), indicate that the high persistence of the nitroxide group does not originate alone from steric effects[225].

2-Substituted derivatives of **187** can undergo various reactions at the side-chain without loss of the radical properties[225]. Moreover, nitroxides containing chelate-forming groups can form chelate complexes with metal ions[226].

The reactions of nitroxides without direct participation of the nitroxide group have been extensively applied in the field of spin labelling[227,228] during recent years. This technique depends on the fact that a nitroxide which is either covalently or noncovalently incorporated into a biological system can give, via its ESR spectrum, information about its environment. Thus, for instance, the anisotropy of the ESR spectrum of the spin label can give information about the mobility of the biomolecule (see Section III.C.1). Moreover, from the magnitude of the coupling constant a^N, it can be concluded whether the environment of the nitroxide label is polar or nonpolar. Other properties of nitroxides affecting their ESR spectra such as the spin exchange in biradicals, which is dependent on several factors (see Section IV.C) can also be used to gain information about biological systems.

The most important spin labels are 2,2,6,6-tetramethylpiperidine-*N*-oxyls, 2,2,5,5-tetramethylpyrrolidine-*N*-oxyls, 2,2,5,5-tetramethylpyrroline-*N*-oxyls and 4,4-dimethyloxazolidine-*N*-oxyls, substituted by functional groups which are capable

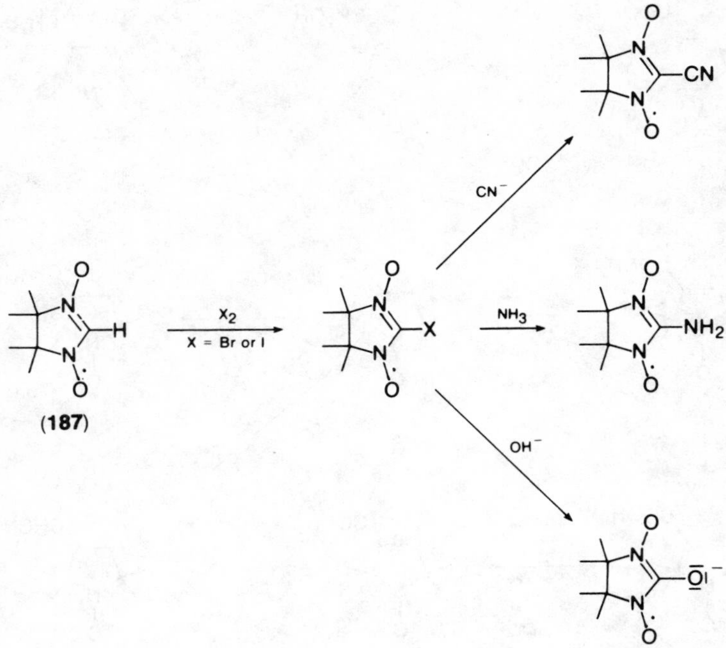

(79)

(80)

(81)

(187)

of undergoing chemical reactions with the reactive sites of biomolecules. A selection of alkylating, acylating, sulphonating and phosphorylating spin labels is shown in formulae **188–199**.

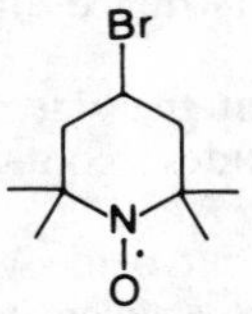

(188) 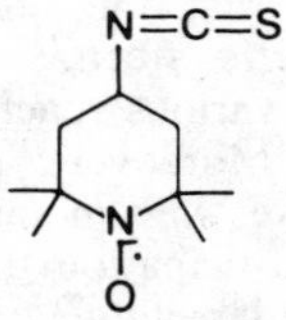 (189) 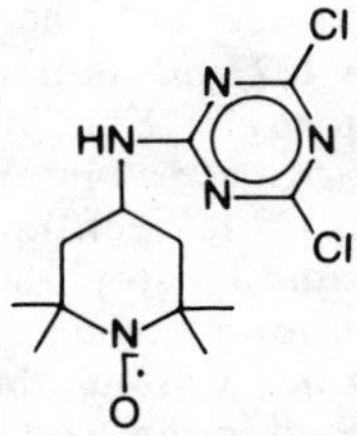 (190) 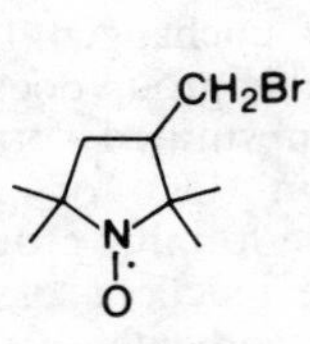 (191)

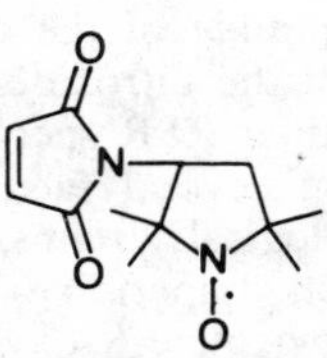

(192) 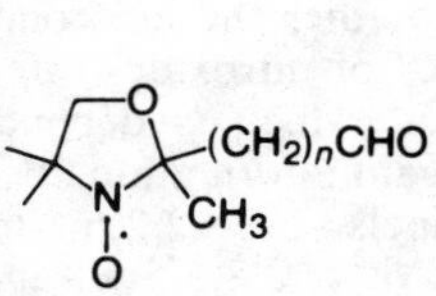 (193) 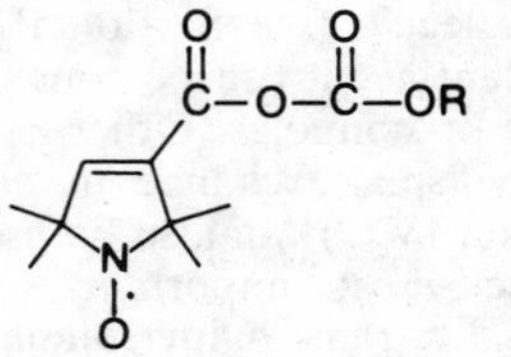 (194)

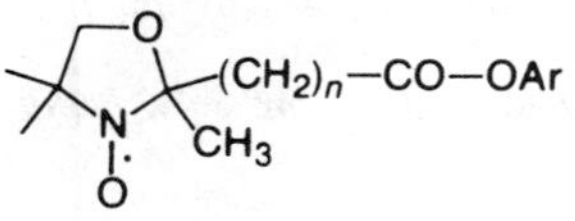

(195) 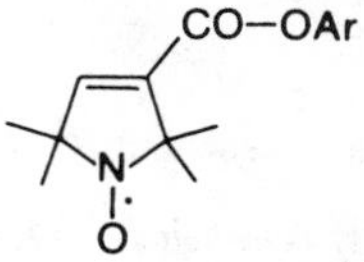 (196) 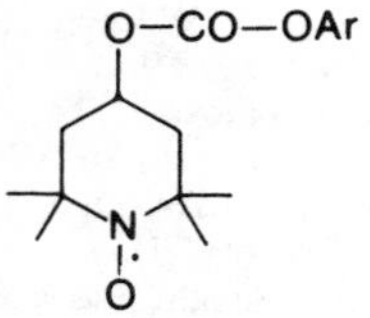 (197)

Ar = $C_6H_4NO_2$-*p*

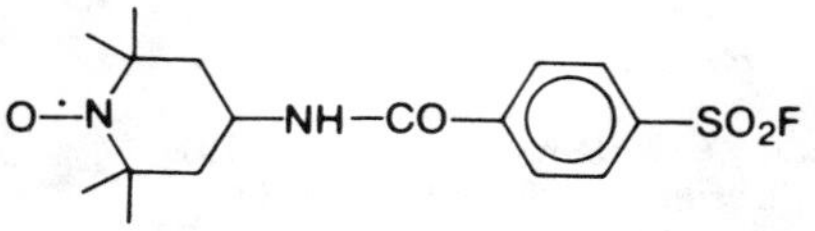

(198) 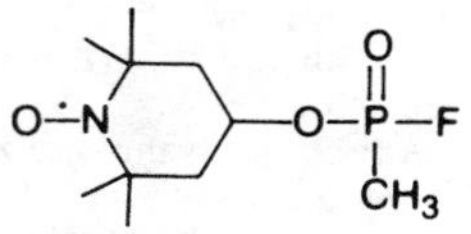 (199)

VII. ACKNOWLEDGEMENTS

The author wishes to thank Dr. E. Wentrup for her help concerning the English language. He thanks Mrs. I. Bublys and Mrs. G. Bach for their help in the preparation of the manuscript.

VIII. REFERENCES

1. A. R. Forrester, J. M. Hay and R. H. Thomson in *Organic Chemistry of Stable Free Radicals*, Academic Press, London, 1968, Chap. 5, pp. 180–246.
2. E. G. Rozantsev, *Free Nitroxyl Radicals*, Plenum Press, New York–London, 1970.
3. E. G. Rozantsev and V. D. Sholle, *Synthesis*, 190, 401 (1971).
4. H. G. Aurich and W. Weiss, *Topics in Current Chemistry (Fortschr. Chem. Forsch.)*, **59**, 65 (1975).
5. Reference 1, p. 191.
6. Reference 2, p. 125.
7. K. U. Ingold, *Accounts Chem. Res.*, **9**, 13 (1976).
8. A. R. Forrester in *Landolt–Börnstein*, Vol. 9, *Magnetic Properties of Free Radicals*, Part c1: *Nitroxide Radicals*, Springer–Verlag, Berlin, 1979, pp. 192–1066.
9. O. Piloty and B. Graf Schwerin, *Ber. Deut. Chem. Ges.*, **34**, 1870, 2354 (1901).
10. Reference 2, pp. 142–160.
11. W. C. Danen and D. D. Newkirk, *J. Amer. Chem. Soc.*, **98**, 516 (1976).
12. B. Maillard and K. U. Ingold, *J. Amer. Chem. Soc.*, **98**, 520 (1976).
13. J. A. Baban and B. P. Roberts, *J. Chem. Soc., Perkin Trans.* 2, 678 (1978).
14. L. R. Mahoney, G. D. Mendenhall and K. U. Ingold, *J. Amer. Chem. Soc. Soc.*, **95**, 8610 (1973).
15. Reference 3, pp. 190–192.
16. Reference 1, pp. 180–181.
17. G. T. Knight and M. J. R. Loadman, *J. Chem. Soc. (B)*, 2107 (1971).
18. G. A. Russeell, E. J. Geels, F. J. Smentowski, K.-Y. Chang, J. Reynolds and G. Kaupp, *J. Amer. Chem. Soc.*, **89**, 3821 (1967).
19. Th. A. J. W. Wajer, A. Mackor and Th. J. deBoer, *Tetrahedron*, **25**, 175 (1969).
20. Reference 3, pp. 192–194.
21. Reference 2, p. 67.
22. Reference 4, p. 72.
23. Reference 1, p. 181.

24. Th. A. J. W. Wajer, H. W. Geluk, J. B. F. N. Engberts and Th. J. deBoer, *Rec. Trav. Chim.*, **89**, 696 (1970).
25. B. Lakatos, B. Turcsanyi and F. Tudos, *Acta Chim. Acad. Sci. Hung.*, **70**, 225 (1971); *Chem. Abstr.*, **76**, 125960p (1972).
26. R. E. Banks, D. J. Edge, J. Freear and R. N. Haszeldine, *J. Chem. Soc., Perkin Trans. 1*, 721 (1974).
27. G. T. Knight and B. Pepper, *Tetrahedron,* **27**, 6201 (1971).
28. A. B. Sullivan, *J. Org. Chem.*, **31**, 2811 (1966).
29. R. W. Layer, *Tetrahedron Letters*, 4413 (1970).
30. W. A. Waters, *J. Chem. Soc., Perkin Trans. 2*, 1078 (1979).
31. Reference 4, p. 76.
32. T. D. Lee and J. F. W. Keana, *J. Org. Chem.*, **41**, 3237 (1976).
33. Reference 4, p. 81.
34. H. G. Aurich and W. Weiss, *Tetrahedron,* **32**, 159 (1976).
35. Reference 3, p. 197.
36. H. G. Aurich, K. Hahn and K. Stork, *Chem. Ber.*, **112**, 2776 (1979).
37. R. West and P. Boudjouk, *J. Amer. Chem. Soc.*, **93**, 5901 (1971).
38. J. R. Roberts and K. U. Ingold, *J. Amer. Chem. Soc.*, **95**, 3228 (1973).
39. H. G. Aurich and H. Czepluch, *Tetrahedron*, **36**, 3543 (1980).
40. E. G. Janzen, *Accounts Chem. Res.*, **4**, 31 (1971).
41. M. J. Perkins in *Essays on Free Radical Chemistry* (Ed. R. O. C. Norman), Special Publication No. 24, Chemical Society, London, 1970, p. 97.
42. A. R. Forrester and S. P. Hepburn, *J. Chem. Soc. (C)*, 701 (1971).
43. W. J. Heilman, A. Rembaum and M. Szwarc, *J. Chem. Soc.*, 1127 (1957).
44. Reference 4, pp. 77–78.
45. Reference 3, pp. 195–196.
46. S. Terabe, K. Kuruma and R. Konaka, *J. Chem. Soc., Perkin Trans. 2*, 1252 (1973).
47. T. Doba, T. Ichikawa and H. Yoshida, *Bull. Chem. Soc. Japan,* **50**, 3158 (1977).
48. Y. Maeda and K. U. Ingold, *J. Amer. Chem. Soc.*, **101**, 4975 (1979).
49. A. Rockenbauer, L. Sümegi, G. Moger, P. Simon, M. Azori and M. Györ, *Tetrahedron Letters*, 5057 (1978).
50. E. G. Janzen and R. V. Shetty, *Tetrahedron Letters*, 3229 (1979).
51. E. G. Janzen, R. L. Dudley and R. V. Shetty, *J. Amer. Chem. Soc.*, **101**, 243 (1979).
52. Reference 4, p. 73.
53. Reference 8, especially pp. 1011–1042.
54. J. A. Maassen, H. Hittenhausen and Th. J. deBoer, *Tetrahedron Letters*, 3213 (1971).
55. A. Rassat and P. Rey, *J. Chem. Soc. (D)*, 1161 (1971).
56. P. Maruthamuthu and J. C. Sciano, *J. Phys. Chem.*, **82**, 1588 (1978).
57. J. A. Howard and J. C. Tait, *Can. J. Chem.*, **56**, 176 (1978).
58. H. Taniguchi and H. Hatano, *Chem. Letters*, 9 (1975).
59. S. Terabe and R. Konaka, *J. Chem. Soc., Perkin Trans. 2*, 369 (1973).
60. W. Ahrens, K. Wieser and A. Berndt, *Tetrahedron,* **31**, 2829 (1975).
61. W. Ahrens, K. Wieser and A. Berndt, *Tetrahedron Letters*, 3141 (1973).
62. J. G. Pacifici and H. L. Browning, Jr., *J. Amer. Chem. Soc.,* **92**, 5231 (1970).
63. R. G. Gasanov, I. I. Kandror and R. Kh. Freidlina, *Tetrahedron Letters*, 1485 (1975).
64. T. Suehiro, M. Kamimori, K. Tokumaru and M. Yoshida, *Chem. Letters*, 531 (1976).
65. M. Kamimori, H. Sakuragi, T. Suehiro, K. Tokumaru and M. Yoshida, *Bull. Chem. Soc. Japan,* **50**, 1195 (1977).
66. P. Schmid and K. U. Ingold, *J. Amer. Soc.,* **100**, 2493 (1978).
67. E. G. Janzen and C. A. Evans, *J. Amer. Chem. Soc.,* **95**, 8205 (1973).
68. Reference 4, pp. 77–79.
69. Reference 4, p. 79.
70. K. Reuter and W. P. Neumann, *Tetrahedron Letters*, 5235 (1978).
71. A. K. Hoffmann, W. G. Hodgson, D. L. Maricle and W. H. Jura, *J. Amer. Chem. Soc.*, **86**, 631 (1964).
72. A. K. Hoffmann, A. M. Feldman, E. Gelblum and W. G. Hodgson, *J. Amer. Chem. Soc.,* **86**, 639 (1964).
73. A. K. Hoffmann, A. M. Feldman and E. Gelblum, *J. Amer. Chem. Soc.,* **86**, 646 (1964).

74. Reference 1, p. 182.
75. D. W. Lamson, R. Sciarro, D. Hryb and R. O. Hutchins, *J. Org. Chem.*, **38**, 1952 (1973).
76. D. Mulvey and W. Waters, *J. Chem. Soc., Perkin Trans. 2*, 1059 (1978).
77. H. G. Aurich and H.-M. Schappert, unpublished results.
78. F. Minisci, R. Galli and A. Quilico, *Tetrahedron Letters*, 785 (1963).
79. H. G. Aurich and U. Watterodt, unpublished results.
80. R. N. Shibaeva, *J. Struct. Chem. USSR (Engl. Transl.)*, **16**, 318 (1975); *Zh. Strukt. Khim.*, **16**, 330 (1975).
81. X. Solans, S. Gali, C. Miravitlles and M. Font-Altaba, *Acta Cryst.*, **B34**, 2331 (1978).
82. J. C. A. Boeyens and G. J. Kruger, *Acta. Cryst.*, **B26**, 668 (1970).
83. D. Bordeaux and J. Lajzèrowicz, *Acta Cryst.*, **B33**, 1837 (1977).
84. J. Morishima, K. Yoshikawa, K. Bekki, M. Kohno and K. Arita, *J. Amer. Chem. Soc.*, **95**, 5817 (1973).
85. A. Rassat and P. Rey, *Tetrahedron,* **29**, 1599 (1973).
86. Y. Ellinger, R. Subra, A. Rassat, J. Douady and G. Berthier, *J. Amer. Chem. Soc.*, **97**, 476 (1975).
87. T. D. Davis, R. E. Christoffersen and G. M. Maggiora, *J. Amer. Chem. Soc.*, **97**, 1347 (1975).
88. Reference 2, p. 126.
89. O. H. Griffith and A. S. Waggoner, *Accounts Chem. Res.*, **2**, 17 (1969).
90. L. J. Libertini and O. H. Griffith, *J. Chem. Phys.*, **53**, 1359 (1970).
91. H. Tanaka and K. Kuwata, *Bull. Chem. Soc. Japan,* **51**, 2451 (1978).
92. Reference 8, p. 193.
93. F. A. Neugebauer in *Landolt–Börnstein*, Vol. 9, *Magnetic Properties of Free Radicals*, Part c1, *Nitrogen-centred Radicals*, Springer–Verlag, Berlin, 1979, pp. 5–180.
94. H. G. Aurich, K. Hahn, K. Stork and W. Weiss, *Tetrahedron,* **33**, 969 (1977).
95. H. G. Aurich, H. Czepluch and K. Hahn, *Tetrahedron Letters*, 4373 (1977).
96. R. W. Kreilick in *Advances in Magnetic Resonance*, Vol. 6 (Ed. J. S. Waugh), Academic Press, New York, 1973, p. 141.
97. F. W. King, *Chem. Rev.*, **76**, 157 (1976), especially pp. 165–166.
98. Y. Kotake, M. Okazaki and K. Kuwata, *J. Amer. Chem. Soc.*, **99**, 5198 (1977).
99. Reference 1, p. 221.
100. C. Morat and A. Rassat, *Tetrahedron,* **28**, 735 (1972).
101. Reference 3, p. 200.
102. P. H. H. Fischer and F. A. Neugebauer, *Z. Naturforsch. (A)*, **19**, 1514 (1964).
103. J. H. Osiecki and E. F. Ullman, *J. Amer. Chem. Soc.*, **90**, 1078 (1968).
104. H. G. Aurich, *Chem. Ber.*, **101**, 1761 (1968).
105. A. Collet, J. Jacques, B. Chion and J. Lajzèrowicz, *Tetrahedron,* **31**, 2243 (1975).
106. J. Morishima, K. Yoshikawa, T. Yonezawa and H. Matsumoto, *Chem. Phys. Letters,* **16**, 336 (1972).
107. A. P. Davies, A. Morrison and M. D. Barratt, *Org. Mass. Spectrum.*, **8**, 43 (1974).
108. B. R. Knauer and J. J. Napier, *J. Amer. Chem. Soc.*, **98**, 4395 (1976).
109. E. G. Janzen, *Accounts Chem. Res.*, **2**, 279 (1969).
110. E. G. Janzen, in *Topics in Stereochemistry*, Vol. 6 (Ed. L. L. Allinger and E. L. Eliel), Wiley–Interscience, New York, 1971, p. 177 (see pp. 184–186).
111. A. H. Cohen and B. M. Hoffman, *J. Phys. Chem.*, **78**, 1313 (1974).
112. H. G. Aurich and H. Czepluch, *Tetrahedron Letters*, 1187 (1978).
113. S. F. Nelsen in *Free Radicals*, Vol. II (Ed. J. Kochi), John Wiley and Sons, New York, 1973, Chap. 21, p. 527 (see pp. 540–542).
114. J. S. Roberts in *Comprehensive Organic Chemistry*, Vol. 2, (Ed. J. O. Sutherland), Pergamon Press, Oxford, 1979, Chap. 6.4.5, p. 205 (see p. 210).
115. T. C. Jenkins, M. J. Perkins and B. Terem, *Tetrahedron Letters*, 2925 (1978).
116. Reference 110, pp. 186–199.
117. A. Rockenbauer, S. A. Abdel-Monem, G. Möger and L. Sümegi, *J. Mol. Struct.*, **46**, 169 (1978).
118. G. Chapelet-Letourneux, H. Lemaire, R. Lenk, M. A. Marechal and A. Rassat, *Bull. Soc. Chim. Fr.*, 3963 (1968).

119. B. C. Gilbert and M. Trenwith, *J. Chem. Soc., Perkin Trans. 2,* 1834 (1973).
120. J. J. Windle, J. A. Kuhnle and B. H. Beck, *J. Chem. Phys.,* **50**, 2630 (1969).
121. A. Hudson and R. G. Luckhurst, *Chem. Rev.,* **69**, 191 (1969).
122. E. G. Janzen and I. Lopp, *J. Mag. Res.,* **7**, 107 (1972).
123. Reference 110, p. 206.
124. Reference 4, p. 89.
125. N. M. Atherton, *Electron Spin Resonance, Theory and Applications*, Halsted Press: a division of John Wiley and Sons, New York, 1973, pp. 182–186.
126. Reference 2, pp. 146–151.
127. E. K. Metzner, L. J. Libertini and M. Calvin, *J. Amer. Chem. Soc.,* **96**, 6515 (1974).
128. J. F. Keana, *Chem. Rev.,* **78**, 37 (1978) (see pp. 60–61).
129. W. B. Gleason and R. E. Barnett, *J. Amer. Chem. Soc.,* **98**, 2701 (1976).
130. Reference 2, pp. 151–157.
131. A. V. Gagua, G. G. Malenkov and V. P. Timofeev, *Chem. Phys. Letters*, **56**, 470 (1978).
132. A. S. Kabankin, G. M. Zhidomirov and A. L. Buchachenko, *J. Mag. Res.,* **9**, 199 (1973).
133. V. Malatesta and K. U. Ingold, *J. Amer. Chem. Soc.,* **95**, 6404 (1973).
134. B. M. Hoffman and Th. B. Eames, *J. Amer. Chem. Soc.,* **91**, 2169 (1969).
135. H. G. Aurich and G. Küttner, unpublished results.
136. E. F. Ullman, L. Call and J. H. Osiecki, *J. Org. Chem.,* **35**, 3623 (1970).
137. J. N. Helbert, P. W. Kopf, E. H. Poindexter and B. E. Wagner, *J. Chem. Soc., Dalton Trans.*, 998 (1975).
138. T. B. Eames and B. M. Hoffman, *J. Amer. Chem. Soc.,* **93**, 3141 (1971).
139. B. M. Hoffman and T. B. Eames, *J. Amer. Chem. Soc.,* **91**, 5168 (1969).
140. A. H. Cohen and B. M. Hoffman, *Inorg. Chem.,* **13**, 1484 (1974).
141. Reference 4, p. 93.
142. Y. Murata and N. Mataga, *Bull. Chem. Soc. Japan,* **44**, 354 (1971).
143. G. M. Burnett, G. G. Cameron and J. Cameron, *J. Chem. Soc., Faraday Trans. 1,* **69**, 864 (1973).
144. Reference 3, p. 406.
145. J. A. Howard and J. C. Tait, *J. Org. Chem.,* **43**, 4279 (1978).
146. T. A. B. M. Bolsman, A. P. Blok and J. H. G. Frijns, *Rec. Trav. Chim.,* **97**, 313 (1978).
147. Reference 3, p. 406.
148. Reference 1, p. 223.
149. Reference 1, p. 224.
150. Reference 4, p. 90.
151. G. D. Mendenhall and K. U. Ingold, *J. Amer. Chem. Soc.,* **95**, 6390 (1973).
152. A. Capiomont, B. Chion and J. Lajzérowicz, *Acta Cryst.,* **B27**, 322 (1971).
153. D. P. Veloso and A. Rassat, *J. Chem. Res. (S)*, 106 (1979).
154. S. H. Glarum and J. H. Marshall, *J. Chem. Phys.,* **62**, 956 (1975).
155. J. Martinie-Hombrouck and A. Rassat, *Tetrahedron,* **30**, 433 (1974).
156. G. E. Ewing, *Angew. Chem.,* **84**, 570 (1972); *Angew. Chem. (Intern. Ed. Engl.),* **11**, 486 (1972).
157. Reference 1, p. 201.
158. R. B. Woodward and C. Wintner, *Tetrahedron Letters*, 2693 (1969).
159. H. G. Aurich, I. Lotz and W. Weiss, *Tetrahedron*, **34**, 879 (1978).
160. D. F. Bowman, J. L. Brokenshire, T. Gillan and K. U. Ingold, *J. Amer. Chem. Soc.,* **93**, 6551 (1971).
161. D. F. Bowman, T. Gillan and K. U. Ingold, *J. Amer. Chem. Soc.,* **93**, 6555 (1971).
162. Reference 161, footnote 23.
163. R. Briere and A. Rassat, *Tetrahedron,* **32**, 2891 (1976).
164. G. D. Mendenhall and K. U. Ingold, *J. Amer. Chem. Soc.,* **95**, 6395 (1973).
165. K. Adamic, F. D. Bowman, T. Gillan and K. U. Ingold, *J. Amer. Chem. Soc., 93, 902* (1971).
166. Reference 4, p. 93.

167. T. Yoshioka, S. Higashida, S. Morimura and K. Murayama, *Bull. Chem. Soc. Japan,* **44**, 2207 (1971).
168. T. C. Jenkins, M. J. Perkins and N. P. Y. Siew, *J. Chem. Soc., Chem. Commun.*, 880 (1975).
169. Reference 3, pp. 403–404.
170. Reference 1, pp. 227–228.
171. T. W. Chan and T. C. Bruice, *J. Amer. Chem. Soc.,* **99**, 7287 (1977).
172. M. A. Schwartz, J. W. Parce and H. M. McConnell, *J. Amer. Chem. Soc.,* **101**, 3592 (1979).
173. R. E. Banks, R. N. Haszeldine and B. Justin, *J. Chem. Soc. (C)*, 2777 (1971).
174. R. E. Banks, D. L. Choudhury and R. N. Hazeldine, *J. Chem. Soc., Perkin Trans. 1*, 1092 (1973).
175. S. A. Hussain, T. C. Jenkins and M. J. Perkins, *Tetrahedron Letters*, 3199 (1977).
176. C. Berti and M. J. Perkins, *Angew. Chem.* **91**, 923 (1979); *Angew. Chem. (Intern. Ed. Engl),* **18**, 864 (1979).
177. J. F. W. Keana, R. J. Dinerstein and F. Baitis, *J. Org. Chem.,* **36**, 209 (1971).
178. A. J. Bogatyreva and A. L. Buchachenko, *Russ. Chem. Rev.,* **44**, 1048 (1975) (see pp. 1057–1058).
179. J. A. Nelson, S. Chou and T. A. Spencer, *J. Amer. Chem. Soc.,* **97**, 648 (1975).
180. L. Call and E. F. Ullman, *Tetrahedron Letters*, 961 (1973).
181. Reference 1, p. 193.
182. D. J. Cowley and W. A. Waters, *J. Chem. Soc. (B),* 96 (1970).
183. Reference 4, p. 96, *see also* H. G. Aurich, S. K. Duggal, P. Höhlein, H.-G. Klingelhöfer and W. Weiss, *Chem. Ber.*, in the press.
184. W. Sümmermann and U. Deffner, *Tetrahedron,* **31**, 593 (1975).
185. Reference 3, pp. 404–405.
186. V. A. Golubev, V. V. Solodova, N. N. Aleinikov, B. L. Korsunskii and E. G. Rozantsev, *Izv. Akad. Nauk SSSR, Ser. Khim.*, 572 (1978); *Chem. Abstr.,* **89**, 43237n (1978).
187. Reference 4, p. 94.
188. B. Ganem, *J. Org. Chem.,* **40**, 1998 (1975).
189. J. A. Cella, J. A. Kelley and E. F. Kenehan, *J. Org. Chem.,* **40**, 1860 (1975).
190. W. Brackman and C. J. Gaasbeek, *Rec. Trav. Chim.,* **85**, 221 (1966).
191. J. A. Cella, J. A. Kelley and E. F. Kenehan, *Tetrahedron Letters*, 2869 (1975).
192. A. Rassat and P. Rey, *Tetrahedron,* **31**, 2673 (1975).
193. A. Rassat and P. Rey, *Tetrahedron,* **30**, 3597 (1974).
194. O. W. Maender and E. G. Janzen, *J. Org. Chem.,* **34**, 4072 (1969).
195. R. L. Craig and J. S. Roberts, *J. Chem. Soc., Chem. Commun.*, 1142 (1972).
196. R. L. Craig, J. Murray-Rust, P. Murray-Rust and J. S. Roberts, *J. Chem. Soc., Chem. Commun.*, 751 (1973).
197. P. J. Carmichael, B. G. Gowenlock and C. A. F. Johnson, *J. Chem. Soc., Perkin Trans. 2*, 1853 (1973).
198. D. R. Anderson and T. H. Koch, *Tetrahedron Letters*, 3015 (1977).
199. E. F. Ullman, L. Call and S. S. Tseng, *J. Amer. Chem. Soc.,* **95**, 1677 (1973).
200. Reference 4, p. 103.
201. Reference 4, p. 104.
202. H. G. Aurich, G. Bach, J. Hahn, G. Küttner and W. Weiss, *J. Chem. Res.,* **122** (S), 1544 (M) (1977).
203. S. F. Nelsen and R. T. Landis, II, *J. Amer. Chem. Soc.,* **95**, 6454 (1973).
204. A. Mackor, Th. A. J. W. Wajer, Th. J. deBoer and J. D. W. van Voorst, *Tetrahedron Letters*, 385 (1967).
205. M. J. Perkins, P. Ward and A. Horsfield, *J. Chem. Soc. (B)*, 395 (1970).
206. D. J. Cowley and H. L. Sutcliffe, *J. Chem. Soc. (B)*, 569 (1970).
207. A. H. M. Kayen, K. deBoer and Th. J. DeBoer, *Rec. Trav. Chim.,* **96**, 1 (1977).
208. A. H. M. Kayen and Th. J. deBoer, *Rec. Trav. Chim.,* **96**, 237 (1977).
209. B. G. Gowenlock, J. Pfab and G. Kresze, *J. Chem. Soc., Perkin Trans. 2*, 511 (1974).
210. B. G. Gowenlock, G. Kresze and J. Pfab, *Tetrahedron Letters*, 593 (1972).

211. T. A. B. M. Bolsman and Th. J. deBoer, *Tetrahedron,* **29**, 3579 (1973).
212. R. E. Banks, R. N. Haszeldine and M. J. Stevenson, *J. Chem. Soc. (C)*, 901 (1966).
213. R. E. Banks, R. N. Haszeldine and T. Myerscough, *J. Chem. Soc. (C)*, 1951 (1971).
214. S. P. Makarov, M. A. Englin, A. F. Videiko, V. A. Tobolin and S. S. Dubov. *Dokl. Akad. Nauk SSSR, Ser. Khim.,* **168**, 344 (1966); *Chem. Abstr.,* **65**, 8742a (1966).
215. S. A. Hussain, T. C. Jenkins, M. J. Perkins and N. P. Y. Siew, *J. Chem. Soc., Perkin Trans. 1*, 2803 (1979).
216. H. Low, I. Paterson, J. M. Tedder and J. Walton, *J. Chem. Soc., Chem. Commun.*, 171 (1977).
217. A. C. Scott and J. M. Tedder, *J. Chem. Soc., Chem. Commun.*, 64 (1979).
218. Reference 4, p. 95.
219. J. I. G. Cadogan and A. G. Rowley, *J. Chem. Soc., Chem. Commun.*, 179 (1974).
220. Reference 3, pp. 408–409.
221. E. J. Rauckman, G. M. Rosen and M. B. Abou-Donia, *J. Org. Chem.,* **41**, 564 (1976).
222. R. Briere, J. -C. Espie, R. Ramasseul, A. Rassat and P. Rey, *Tetrahedron Letters*, 941 (1979).
223. D. J. Kosman and L. H. Piette, *Chem. Commun.*, 926 (1969).
224. B. J. Gaffney in *Spin Labeling, Theory and Applications* (Ed. L. J. Berliner), Academic Press, New York, 1976, Chap. 5, p. 183 (see pp. 208–209).
225. Reference 4, pp. 84–85.
226. S. S. Eaton and G. R. Eaton, *Coord. Chem. Rev.,* **26**, 207 (1978).
227. References 89, 113, 128 and 224.
228. J. D. Morrisett in *Spin Labeling, Theory and Applications* (Ed. L. J. Berliner), Academic Press, New York, 1976, Chap. 8, p. 273.
229. Ch. Reichardt, *Angew. Chem.,* **77**, 30 (1965); *Angew. Chem. (Intern. Ed. Engl.),* **4**, 29 (1965).

Nitrones, Nitronates and Nitroxides,
Edited by S. Patai and Z. Rappoport,

CHAPTER **5**

Appendix to 'Nitroxides'†

HANS GÜNTER AURICH

Fachbereich Chemie, Philipps-Universität Marburg, Marburg, Federal Republic of Germany

†The material in this appendix is divided in the same manner as in the body of the original Chapter 14th Supplement F (see Chapter 4). The section numbers in the appendix are preceded by an asterisk; some section numbers are omitted. The numbers of structures, equations tables and references run continuously in the original chapter and the appendix.

*I. GENERAL AND THEORETICAL ASPECTS

In contrast to an earlier issue the new recommendations of the IUPAC assign the term 'Aminoxyl' instead of 'Aminyl Oxide' to the individual compounds of this class of radicals[230]. 'Nitroxide' should be retained as the group name, however.

The contributions of the First International Symposium on Spin Trapping and Nitroxyl Radical Chemistry have been published[231]. They reflect the widespread interest in nitroxide chemistry. In particular, spin trapping and spin labelling have found extensive application for the study of biochemical, biophysical, pharmacological and medicinal problems. Among the new types of nitroxides, 2,2,5,5-tetramethyl-3-imidazoline-1-oxyls merit special attention.

Throughout the appendix the following abbreviations are used:

TEMPO	2,2,6,6-Tetramethylpiperidine-*N*-oxyl
TEMPOL	4-Hydroxy-2,2,6,6-tetramethylpiperidine-*N*-oxyl (**79**)
TEMPON	4-Oxo-2,2,6,6-tetramethylpiperidine-*N*-oxyl (**9**)
DTBN	Di-*t*-butyl nitroxide (aminoxyl) (**1**)
MNP	2-Methyl-2-nitrosopropane (**28**)
NB	Nitrosobenzene (**32**)
ND	2,3,5,6-Tetramethylnitrosobenzene (nitrosodurene) (**35**)
PBN	α-Phenyl *N*-*t*-butyl nitrone (*N*-*t*-butylbenzylidenamine-*N*-oxide) (**39**)
DMPO	5,5-Dimethyl-1-pyrroline-1-oxide (**40**)

*II. FORMATION OF NITROXIDES

*A. Oxidation of the Hydroxylamino Group

*1. Formation from hydroxylamines

Oxidation of hydroxylamines to afford nitroxides as well as further oxidation products can be favourably performed by cobalt-coordinated peroxyl radicals[232].

*2. Formation from amines

The peroxyl radical assumed to be the intermediate in the conversion of 2,2,6,6-tetramethylpiperidinyl radical into TEMPO can be detected by its ESR signal in a matrix in the temperature range from $-150°$ to $-80\,°C$[233].

The lithium salts of secondary aliphatic amines are oxidized by molecular oxygen to afford the corresponding nitroxides. This reaction proceeds with retention of configuration at the α-carbon atoms as shown by asymmetric substitution. The coupling constants of diastereomeric radicals formed in this way differ considerably. (For instance bis(1-phenylethyl)aminoxyl: racemic mixture *R*, *R*/*S*, *S*, $a^N = 14.9$, $a^H = 9.0$ G; *meso*, $a^N = 14.6$, $a^H = 5.9$ G[234].) In this connection it is remarkable that spin trapping of the 1-phenylethyl radical by 1-phenylnitrosoethane yields only the racemic mixture of the chiral nitroxide.

From trialkylamines with at least one hydrogen atom at the β-position dialkylaminoxyls are formed in aqueous solution in the presence of oxygen by oxidative elimination of one alkyl group[235].

*3. Formation from nitroso compounds

As has been shown in Section II.C.2[45] and II.C.3 (e.g. equations 11[54] and 12[55]) aliphatic and some aromatic nitroso compounds can be photolytically decomposed to give alkyl and aryl radicals which are subsequently trapped by the parent compound affording the corresponding nitroxides. This reaction has been subject to misinterpretation (e.g. equations 26[76] and 27[77]). In fact, some of the nitroxides formed by reactions in which nitroso compounds are involved arise by inadvertent photolysis rather than by molecular-assisted homolysis[236]. However, this should not be true if a hydroxylamine is formed in equilibrium with the nitroso compound causing a redox reaction to occur between these two compounds (see equation 5 for $X = CR_2$ or NR, but not for $X = O$[30]). Thus 2,4,6-trichloronitrosobenzene reacts with *cis*-butene-2[236] and the enamine form of *N*-*t*-butyl-(2-phenylethylidene)amine[237] according to equation (5) to yield the nitroxides **200** and **201**, respectively. **201** is also formed by oxidation of the hydroxylamine intermediate[237].

$$\underset{\textbf{(200)}}{\mathrm{ArN(\dot{O})CH(Me)CH{=}CH_2}} \qquad \underset{\textbf{(201)}}{\mathrm{ArN(\dot{O})CH(Ph)CH{=}NBu\text{-}}t}$$

$$\mathrm{Ar = 2,4,6\text{-}Cl_3C_6H_2\text{—}}$$

$$\mathrm{R^1N{=}O + R^2OOH \rightleftharpoons R^1N(OH){-}OOR^2 \xrightarrow[(R\,N{=}O)]{-H} [R^1N(\dot{O}){-}OOR^2] \longrightarrow R^1NO_2 + {}^{\bullet}OR^2}$$

$$\mathrm{R^1N{=}O + {}^{\bullet}OR^2 \longrightarrow R^1N(\dot{O}){-}OR^2} \qquad (82)$$

$$\mathrm{RN{=}O + MBH_4 \longrightarrow \underset{\textbf{(202)}}{RN(\dot{O}){-}BH_3^-\,M^+}} \qquad (83)$$

$$\mathrm{M = Na\ or\ K}$$

*4. Formation from nitrones

Molecule-induced homolysis of hydroperoxides with nitroso compounds gives initially alkoxynitroxides[238] which should arise by reaction sequence (82). Furthermore, boronitroxides **202** are formed when nitroso compounds are treated with sodium or potassium borohydride in aprotic polar solvents (equation 83)[239].

The molecule-induced homolysis of hydroperoxides with nitrones is also assumed to occur by addition of the hydroperoxide to the nitrone group followed by oxidation of the hydroxylamine as shown in equation (84)[238].

$$\mathrm{R^1\overset{+}{N}(O^-){=}CHR^2 + }t\mathrm{\text{-}BuOOH \rightleftharpoons \underset{\textbf{(203)}}{R^1N(OH){-}CHR^2(OOBu\text{-}}t)}$$

$$\textbf{203} + t\mathrm{\text{-}BuOOH \longrightarrow \underset{\textbf{(204)}}{R^1N(\dot{O}){-}CHR^2(OOBu\text{-}}t)} + \mathrm{H_2O} + t\mathrm{\text{-}BuO^{\bullet}} \qquad (84)$$

$$\textbf{203} + t\mathrm{\text{-}BuO^{\bullet}} \longrightarrow \textbf{204} + t\mathrm{\text{-}BuOH}$$

$$\underset{\textbf{(205)}}{t\text{-BuN}(\dot{O})\text{—CHPh}(\text{C(OH)Me}_2)} + \text{HX} \longrightarrow \underset{\textbf{(206)}}{t\text{-BuN}(\text{OH})\text{—CHPh}(\text{C(OH)Me}_2)} + \text{X}^{\bullet}$$

$$\textbf{(206)} + \underset{\textbf{(207)}}{t\text{-BuN—CHPh (—O—)}} \longrightarrow \textbf{205} \qquad (85)$$

Spin trapping by nitrones performed under irradiation may also take a rather complicated course. For instance, photochemically generated α-hydroxyalkyl radicals are first of all trapped by PBN to give the expected spin adduct **205**. This, however, is photochemically labile and is converted to hydroxylamine **206** under hydrogen abstraction from the solvent. In a competition reaction the nitrone PBN is rearranged to the oxaziridine **207** by the action of light. When irradiation is stopped **207** oxidizes **206** to afford nitroxide **205** in a dark reaction[240].

*C. Spin Trapping

*1. General remarks

Several reviews on spin trapping have appeared[241].

*2. The properties of spin traps

Various novel spin traps adjusted to special requirements and synthesized by introducing specific substituents into the widely used systems, the nitroso compounds and nitrones, are the following: 2,4-dimethyl 3-nitrosobenzenesulphonate (**208**)[242], 4-nitrosopyridine-*N*-oxide (**209**)[243], pentamethoxynitrosobenzene (**210**)[244], trifluoronitrosomethane (**211**)[245] and the compounds **212–216**. **208** and **209** are water-soluble, photostable compounds. **209** and **211** are particularly suited for trapping of O-centred radicals. In the ESR spectra of the spin adducts of **210** the hyperfine splittings are better resolved than in those of other traps, thus providing more information about the trapped radicals.

$$R^1\overset{+}{N}(O^-)\text{=CHR}^2$$

(**212**)[246] $R^1 = t\text{-Bu}, R^2 = 4\text{-}C_6H_4\overset{+}{N}Me_3$
(**213**)[246] $R^1 = t\text{-Bu}, R^2 = 4\text{-}C_6H_4O(CH_2)_{11}Me$
(**214**)[247] $R^1 = t\text{-Bu}, R^2 = 4\text{-}C_5H_4\overset{+}{N}\text{—}O^-$
(**215**)[248] $R^1 = t\text{-Bu}, R^2 = CH\text{=}N(O)Bu\text{-}t$
(**216**)[249] $R^1 = HOCH_2\text{—}C(Me_2), R^2 = Ph$

The well-known DMPO (**40**) has been modified. Thus the lipophilic character of the trap can be successively increased by substituting a methyl group by propyl, hexyl or decyl[250]. Use of 3,3,5,5-tetramethyl-1-pyrroline-1-oxide gives more persistent adducts[251]. There should be a brief comment on a few specific applications of spin traps. Sodium 2-sulphonatophenyl *t*-butyl nitrone (**42**) and **213** were used in sodium decyl sulphate (SDS) micellar solutions[246]. Whereas **42** is located in the aqueous phase, **213** traps the radicals in the lipophilic interior of the micelles. Radicals in the smoke of cigarettes have been trapped by solid-state DMPO which was either placed as a thin film on a glass fibre or was adsorbed at silica gel[252]. By means of its hydroxy group, nitrone **216** could be covalently attached to silaceous surfaces using cyanuric chloride as the linking agent[249].

Under certain circumstances spin traps can act as electron acceptors or even as electron donors. Since the generation of the radicals to be trapped is frequently connected with redox reactions, knowledge of the redox potentials of the traps is necessary in order to know the potential window wherein they are redox inactive. Thus the redox potentials of several nitrones and of MNP were measured in acetonitrile as well as in aqueous solution[247]. It was found that ease of reduction is in the order PBN (**39**) < 3-pyridyl *t*-butyl nitrone < 3-pyridyl-*N*-oxide *t*-butyl nitrone < 3-(*N*-methylpyridinylium) *t*-butyl nitrone, according to the increasing electron-acceptor properties of the aryl group. The oxidation potential was found to increases in the same order. In acetonitrile the oxidation potentials of **42**, **214** and 2-pyridyl-*N*-oxide *t*-butyl nitrone were lower than expected, the potentials of **42** and **214** being even lower than that of PBN. This was ascribed to resonance effects.

Potential measurements in DMF and acetonitrile confirm that reduction of nitrones is more difficult than that of nitroso compounds[253]. The following normalized reduction potentials (Pt electrode, aqueous solution) were determined:

	NB(**32**)	ND(**35**)	MNP(**38**)	PBN(**39**)	DMPO(**40**)
$-E$(V)	0.2	0.44	0.77	1.53	2.18

Comparison of these reduction potentials with the oxidation potentials of various radicals suggests that nitroso compounds can undergo a competition reaction (equation 86). Thus, electron transfer instead of radical trapping can occur with strong electron-donating radicals. This is particularly true for radicals, such as ketyl α-hydroxyalkyl and α-aminoalkyl, with an oxidation potential of less than −0.6. Quantitative trapping by nitroso compounds is only possible for radicals, such as alkyl, aryl or alkoxy, whose oxidation potential are sufficiently positive.

$$R^1N{=}O + R^2\bullet \longrightarrow \begin{cases} R^1N(\dot{O})\text{—}R^2 \\ R^{2+} + R^1\dot{N}\text{—}O^- \xrightarrow{+H^+} R^1N(\dot{O})\text{—}H \end{cases} \qquad (86)$$

In contrast nitrones are reduced only at extremely negative potentials. For example, no electron transfer between nitrones and ketyl radicals occurs. For this reason nitrones can be widely used for spin-trapping experiments. In principle, nitroso compounds can also operate, with strong electron acceptors, as electron donors.

*3. Examples of spin trap reactions

Unambiguous distinction between hydroxyl and peroxyl radicals in spin-trapping experiments is necessary, particularly in view of their application to biological problems. This is possible since the coupling constants found for the adducts of peroxyl and hydroxyl radicals with nitrones are different (cf. **217** and **218**). To confirm this, the hydroxyl spin adducts were formed independently from hydroxyalkyl radicals and nitroso compounds (equation 87)[238].

$$\underset{\textbf{(217)}}{\mathrm{ArN(\dot{O})-CHR(OOH)}} \qquad \underset{\textbf{(218)}}{\mathrm{ArN(\dot{O})-CHR(OH)}} \qquad \mathrm{ArNO} + \mathrm{R\dot{C}HOH} \rightarrow \textbf{218} \quad (87)$$

Ar = duryl

$a^N = 13.12\,G$ $\quad a^N = 13.89\,G$

$a^H_\beta = 4.67\,G$ $\quad a^H = 6.61\,G$

($R = CH_3$, Ar = duryl)

That in fact the O-centred radicals are trapped is also indicated by the spin adducts of the ^{17}O-labelled hydroperoxyl or superoxide radical and the hydroxy radical which exhibit ^{17}O couplings.[254] However, in view of the results with trifluoronitrosomethane[245], the report on the trapping of the hydroperoxyl radical by MNP[57] is presumably an error.

In the same way the spin adducts of alkylperoxyl and alkoxy radicals can be distinguished[238,255]. Since the spin adducts of alkoxy and α-hydroxyalkyl radicals with PBN (**39**) can be distinguished by their ESR parameters, differences in the mechanism of alcohol oxidation by aryloxyls can be revealed. It was found that primary alcohols and *t*-butyl alcohol yield preferably alkoxy radicals on treatment with aroxyls at 100 °C. The alkoxy radicals are assumed to be formed by electron transfer followed by deprotonation. In contrast, secondary alcohols preferentially afford α-hydroxylalkyl radicals which obviously arise by direct hydrogen abstraction[256].

DMPO is one of the most efficient traps. It was used to trap a cobalt–dioxygen complex[257] as well as alkylthiyl radicals[258] and other sulfur-centred radicals derived from thiol drugs and biochemicals[259].

Radicals formed by UV photolysis of peptides and DNA constituents and by photo-induced reactions of benzoyl peroxide with amino acids, peptides, fatty acids and pyrimidines were characterized by spin trapping with MNP[260].

When mixtures of amino acids were γ-irradiated, the radicals formed were first trapped by MNP. Subsequently, the spin adducts were separated by high-performance liquid chromatography and finally the various fractions were identified by ESR[261].

In some cases it is possible to convert the spin adducts in diamagnetic trimethylsilyl derivatives which are stable and volatile and may thus be separated and identified by GLC/mass spectrometry[262].

Ligated boryl radicals **219** were trapped by MNP and other traps. As a competition reaction β-scission of the boryl radicals occurred. Since the rate of the β-scission, k_β, was known, the rate k_{trap} could be determined from the relation k_{trap}/k_β which was estimated from the product ratio **220**/**221**. According to their stronger nucleophilic properties boryl radicals are trapped faster than alkyl radicals [263].

$$\underset{\textbf{(219)}}{i\text{-}Pr_2EtN \rightarrow \dot{B}H_2} \xrightarrow[k_{trap}]{+\,t\text{-}BuN{=}O} \underset{\textbf{(220)}}{t\text{-}BuN(\dot{O})\text{—}BH_2 \rightarrow NEt(Pr\text{-}i)_2} \quad (88)$$

$$\underset{\textbf{(219)}}{i\text{-}Pr_2EtN \rightarrow \dot{B}H_2} \xrightarrow[k_\beta]{-\,i\text{-}PrEtN{=}BH_2} i\text{-}Pr^\bullet \xrightarrow{+\,t\text{-}BuN{=}O} \underset{\textbf{(221)}}{t\text{-}BuN(\dot{O})Pr\text{-}i}$$

Inorganic radicals such as azidyl, cyanatyl, cyanyl and chlorine can be trapped by

PBN[264,265]. However, glyoxal bis(t-butyl) nitrone (**215**) is a superior trap for certain other radicals, e.g. bromine atoms[248]. In the ESR spectra of the adducts of alkyls radicals to this trap even the coupling of γ-protons is frequently resolved[248].

If hydroxy radicals are generated in the presence of 2,4,6-trimethoxyphenyl t-butyl nitrone, the formation of the usual spin adduct is accompanied by hydrogen abstraction from the methoxy group. The 2-oxalkyl radicals (**222**) so formed undergo an intramolecular spin trapping to give nitroxides (**223**)[266].

(222) → (223) (89)

(224) + (225) → (226) ⇌ (227) (90)

(228) → (229) → (230) (91)

Vinyl nitroxides with only one substituent at the β-position (**225**) are extremely reactive. When formed by oxidation of nitrone **224** these nitroxides are immediately trapped by the precursor nitrone to afford **226**, which in turn can undergo an intramolecular spin trap reaction to give **227**[267]. Another intramolecular spin trap reaction is the formation of **230** from **228** in the reaction sequence **228** → **229** → **230**[268].

The formation of spin adducts with nitric oxide is described in a series of papers[269]. In the absence of oxygen, alkyl radicals simply form the corresponding dialkyl nitroxides, whereas in the presence of oxygen acyl alkyl nitroxides are detected. It is assumed that initial formation of alkylperoxyl radicals is followed by several reaction steps in which nitric oxide is involved[269].

*5. *Kinetic studies of spin trapping*

The trapping rates by *para*-substituted phenyl *t*-butyl nitrones were measured for several radicals generated by pulse radiolysis. Whereas no dramatic increase in trapping rates with increasing acceptor properties of X was found for *s*- and *t*-hydroxyalkyl radicals or for primary alkyl radicals, an acceleration by a factor of 17.5 and 28 was observed for hydroxymethyl radicals when X is CN and NO_2, respectively (compared to X = H)[270]. The effect of the strong acceptors CN and NO_2 points to some charge transfer in the transition state. In general, however, the observed substituent effects are mainly due to their influence on the ground state of the nitrone, charge-transfer interaction in the transition state being less important as Schmid and Ingold concluded[66].

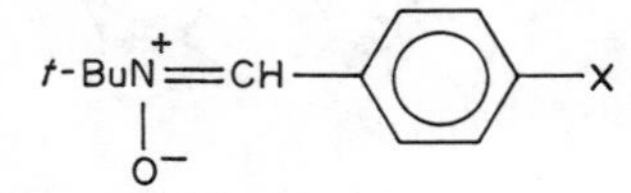

$t\text{-BuN}^+(\text{O}^-)\!=\!\text{CH}-\text{C}_6\text{H}_4-\text{X}$ $\qquad$ $\text{Y}-\text{C}_6\text{H}_4-\text{S}\bullet$

(231)

X = Me, MeO, H, CN, NO_2

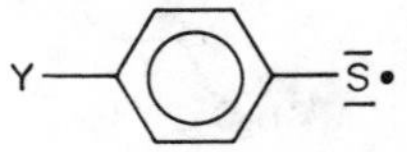

Thiyl radicals **231**, generated by flash photolysis, were trapped by PBN and DMPO[271] and by pentamethyl nitrosobenzene, NB and MNP[272]. A clear increase in trapping rates was found in going from donor to acceptor substituents Y in **231**. The Hammett plots vs. σ^+ were linear with positive slopes. It was concluded that the substituent effect is partly due to a decreasing stabilization of radicals **231** by acceptor substituents Y and partly to a charge transfer in the transition state. This means that, contrary to expectation, the thiyl radicals would act as acceptors and the nitrones and nitroso compounds as donors. The solvent effect on the trapping rate has also been discussed[271].

In the system $\cdot O_2^- / \cdot O_2H$ both radicals are trapped by DMPO in aqueous solution. The values estimated for the trapping rates are far less than those for other O-centred radicals, indicating the low reactivity of $\cdot O_2^-$ and $\cdot O_2H$[273].

In a further kinetic study the rate of rearrangement of the 2,2-dimethyl-3-buten-1-yl radical to yield the 1,1-dimethyl-3-buten-1-yl radical was determined from the ratio of the spin adducts of both radicals with 1-methyl-4-nitroso-3,5-diphenylpyrazole[274].

*III. INVESTIGATION OF NITROXIDES BY PHYSICAL METHODS

A compilation of crystal structure data indicates that in nitroxides with hydroxy or amino groups intermolecular hydrogen bonding involving the nitroxide group exists. Even if other potential acceptors for hydrogen bonding such as amide or carboxyl groups are present, the nitroxide group is always preferred. The geometry of the hydrogen bonding moiety is similar in all structures, corresponding to the σ-model rather than to the π-model. If no hydrogen bonding is possible 'short contacts' between the oxygen atom of the nitroxide group and the carbon atom of a methyl or methylene group of neighbouring molecules exist, resembling the geometric disposition in the hydrogen-bonded molecules[275].

X-ray analysis indicates that the 1:2 complex from copper(II) chloride and 2-phenyl-4,4,5,5-tetramethyl-2-imidazoline-3-oxide-1-oxyl (**90**) consists of structurally discrete molecules with square-planar coordination[276]. The three interacting paramagnetic centres are arranged in a linear fashion. The molecule exists in an electronic doublet ground state. Whereas the metal–oxygen bond is short, the nitrogen–oxygen bond was found to be longer than in the uncomplexed compound. Presumably this is due to transfer of spin density from the nitronyl–nitroxide system to the copper ion.

Discrete nitrosonium cations and perchlorate anions exist in the nitrosonium perchlorate which was obtained from the same nitronyl nitroxide by oxidation with copper perchlorate[277].

The study of a crystal of 2,2,6,6-tetramethyl-4-piperidyl-1-oxyl suberate by neutron diffraction using the classical polarized neutron technique has shown that the unpaired electron is approximately equally distributed between nitrogen and oxygen, at least in this particular nitroxide[278].

The spin density distribution was calculated by the Local Spin Density Theory for the most simple nitroxides. The ratio $\rho^N:\rho^O$ for the unsubstituted aminoxyl was found to be 0.44:0.57, whereas for dimethylaminoxyl a ratio of approximately 1:1 was estimated. Hydrogen bonding by methanol shifts the spin density towards the nitrogen atom[279].

UV spectroscopic studies of DTBN have revealed that the influence of the solvent on the absorption band at 23 000 cm^{-1} parallels the solvent effect on its ESR coupling constants as well as the solvent effect on the IR absorption band ν_{max}(C=O) of acetone. Thus it is concluded that the mode of solvation of the nitroxide group resembles that of the keto group. However, nitroxides are found to be somewhat more basic than ketones[280].

Solvent effects on the visible spectra of TEMPO and TEMPON have also been correlated with the hyperfine coupling constants of DTBN[281]. Microenvironmental effects on energies of visible bands of these nitroxides in electrolyte solutions and in micelles have been studied[282] and their micelle–water distribution coefficients determined[283].

^{15}N-NMR spectroscopy has been applied to studies of nitroxides. Thus, determination of the hyperfine splitting of the nitrogen atom at position 3 in 2,2,5,5-tetramethyl-3-imidazoline-1-oxyls and 3-imidazoline-3-oxy-1-oxyls was possible; this could not be detected by ESR or ENDOR[284].

In addition to ^1H-ENDOR spectroscopy, ^{14}N-ENDOR spectroscopy is being used increasingly. A compilation of the nitroxides studied by ENDOR has been published[285]. The sign of the coupling constants can be determined by such special techniques as CRISP[286] and TRIPLE resonance[285,287]. Superhyperfine coupling of γ- and δ-hydrogen atoms can be found due to the better resolution of the ESR spectra with the aid of the out-of-phase detection method[288].

Changes in the ESR spectra of some diradicals in the plastic phase of cyclohexane at $-10\,°C$ or in the frozen state in toluene at $-150\,°C$ are observed when the orientation of the probe is changed. Under favourable conditions it is thus possible to obtain anisotropic spectra[289].

*IV. SPECIFIC PROPERTIES OF NITROXIDES AS STUDIED BY ESR

*A. Spin Density Distribution

The ^{17}O coupling constant of the ^{17}O-labelled bis(trifluoromethyl)aminoxyl (**2**) was determined ($a^{^{17}O} = 23.6$, $a^N = 9.4$ G)[290]. These values correspond to an increased spin density at oxygen ($\rho^O = 0.67$) and decreased spin density at nitrogen ($\rho^N = 0.28$). This result clearly indicates the effect of electron acceptor substituents on the spin density distribution of the nitroxide group (cf. DTBN (**1**), Table 6). Only recently the ^{17}O coupling constants of a variety of five- and six-membered nitroxides were determined using the nonisotopically enriched compounds[291].

Further discussions concerning the problem of geometry (configuration) of the nitroxide moiety have appeared. In this connection bis(trifluoromethyl)aminoxyl (**2**) and trifluoromethylalkoxyamin-*N*-oxyls were particularly studied[292]. From these studies it was concluded that in **2** the nitroxide group is either planar or quasi-planar; i.e. the inversion barrier is so small that it is overcome by the vibrational energy of most of the molecules, resulting in a time-averaged planar configuration. Other dialkyl nitroxides were considered to be more rigidly planar. In any case temperature-dependent ESR spectra do not constitute strong evidence for the configuration of nitroxides[292].

In general, there seems to be agreement that the energy difference between the planar and the pyramidal configuration of the nitroxide group is extremely small so that conformational or lattice forces are more than sufficient to overcome the inherent geometry of the nitroxide group[293]. The conclusion that an increase of the out-of-plane bonding should occur with increasing temperature was verified by a study of *t*-butylaminoxyl[293].

By MNDO calculations the *transoid* conformation of α-acyl nitroxides was found to be lower in energy than the *cisoid* conformation and to have the higher spin density at nitrogen, in accordance with experiments. In contrast, in calculations for *N*-nitrosonitroxides the *cisoid* conformer turned out to be of lower energy[294].

The nitrogen coupling constants a^{N} for several spin adducts of PBN and for the adduct of the *t*-butoxyl radical to DMPO increase with increasing solvent polarity because of the higher spin density at nitrogen. Due to hyperconjugation the β-hydrogen coupling constants a_{β}^{H} also increase. There is a linear relationship between a^{N} and a_{β}^{H} with the exception of the benzoyl spin adduct of PBN[295].

In contrast, a decrease of a^{N} was found for *t*-butoxy-*t*-butylaminoxyl in aprotic solvents with increasing polarity. Two explanations were given: a^{N} may be diminished by the enhanced delocalization according to mesomeric formula C (equation 92) or by an increase of planarity of the nitroxide moiety[296].

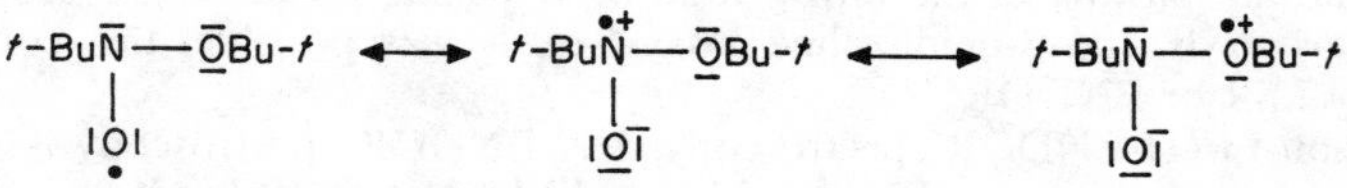

t-BuN̄—ŌBu-t (N–O•) (A) ⟷ t-BuN•+—ŌBu-t (N–O⁻) (B) ⟷ t-BuN̄—O•+Bu-t (N–O⁻) (C) (92)

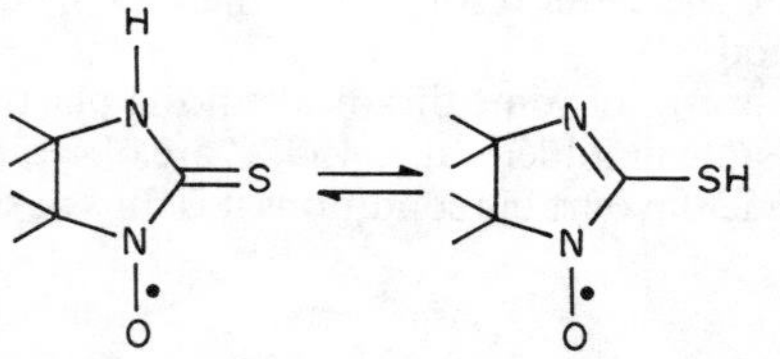

(232a) ⇌ (232b) (93)

The cyclic nitroxide **232** exhibits a special solvent effect in so far as the equilibrium between the thioamide and the iminethiol form is markedly influenced. In apolar solvents, such as benzene, or protic polar solvents, such as ethanol or water, the iminethiol tautomer **232b** is preferred. In aprotic solvents such as DMSO and DMF which are hydrogen bond acceptors the thioamide form **232a** is, however, dominating[297]. A similar influence of the solvent on the tautomerism of the related porphyrexide and its hydrolysis product had been previously observed[298].

An excellent correlation between a^{N} and the σ Hammett parameter was found for *meta*- and *para*-substituted aryl spin adducts of DMPO. Obviously, this correlation is not disturbed by resonance effects because the aryl and nitroxide groups are separated by a tetrahedral carbon atom[299].

As the ESR spectra reveal, diastereomeric spin adducts **233** are formed from alkoxyalkyl and aminoalkyl radicals with PBN[300]. The diastereomeric radicals have different *g*-factors as well as different N and β-H coupling constants. The differences in g-factors and a^N are due to the different orientations of the C—O and C—N dipoles, respectively, with respect to the nitroxide group. The repulsive interactions of the various groups affect the conformation in a different way, giving rise to the differences in a_β^H (cf. Section IV.B).

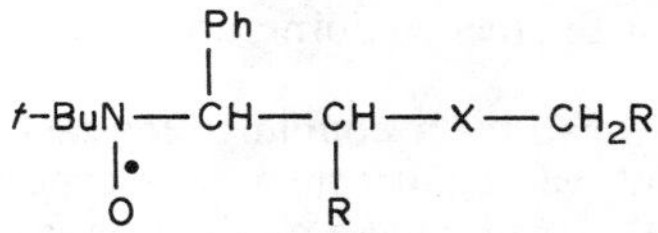

(**233**) X = O, NH

*B. Conformation of Nitroxides

The dependence of the β-hydrogen coupling constant a_β^H (Table 12) on the angle φ (see equation 34) yields interesting conclusions concerning the conformation of β-silyl nitroxides **235**. Their values for a_β^H are larger than those of the corresponding alkyl-substituted nitroxides **234** ($X = CH_3$). They are thus incompatible with conformation C ($a^H \simeq 6.5$ G), which is expected to be favoured by the hyperconjugation effect of the silicon.

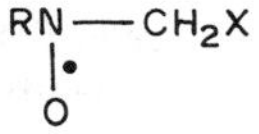

(**234**) X = Me

(**235**) X = $SiMe_3$

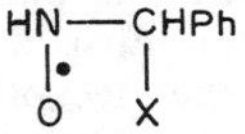

(**236**)

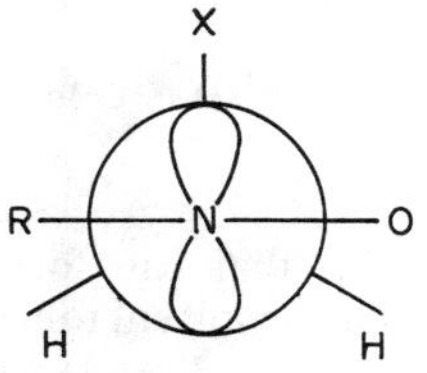

(C) 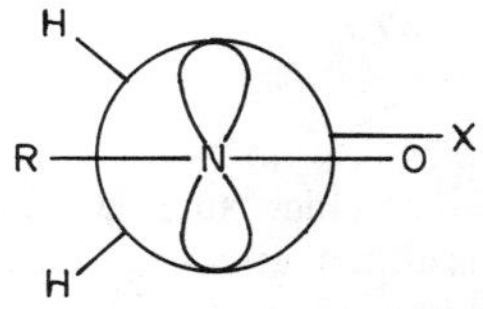 (H) 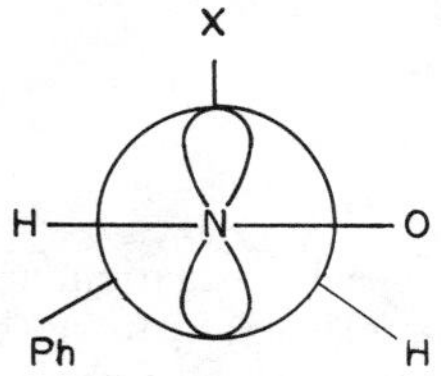 (C′)

On the other hand, conformation H would give rise to values in the order of 19.5 G. Thus it is assumed that a rapid interconversion between the conformers C and H with an averaging of a_β^H takes place. Whereas conformation C is stabilized by π–σ (C–Si) hyperconjugation and/or π–d homoconjugation, conformation H is believed to be

TABLE 12. Comparison of coupling constants a_β^H (G) for nitroxides **234** and **235**[301]

	R = Me	*t*-Bu	Ph
234	10.3	10.3	7.8
235	12.7	14.5	10.8

stabilized by a specific nonbonded attractive interaction between the nitroxide oxygen and the silicon with participation of σ^* or 3d orbitals of silicon[301].

A similar phenomenon was observed on comparison of β-sulphur-, silyl- and germanyl-substituted nitroxides. **236** (X = SBu-*n*, $a_\beta^H = 6.55$ G) is thought to exist exclusively in conformation C′. In contrast, for **236** (X = $SiEt_3$, $a_\beta^H = 14.8$ G) and **236** (X = $GePh_3$, $a_\beta^H = 14.1$ G) an equilibrium with other conformations is assumed to be responsible for the enhancement of a_β^H[302].

On the other hand the β-hydrogen couplings for alkoxyalkyl- and hydroxyalkyl-trifluoromethylaminoxyls are even smaller than those expected for conformation C. It is believed that conformation C is realized as a result of steric and anomeric effects and that additionally the OR group is bent towards the nitroxide π- orbital moving the hydrogen atoms away from this orbital[303]. The selective line-broadening in the ESR spectra of the trimethylamine-ligated boryl-*t*-butylaminoxyl is ascribed to the out-of-phase modulation of the boron and the β-H splitting which is caused by an exchange of two almost equally occupied conformers[263].

In a similar way conclusions may be drawn from the β-^{13}C coupling[304] (Table 13). ^{13}C couplings of about 10 G point to a conformation in which the respective substituent eclipses the nitroxide π-orbital. In nitroxides **81** (R = CN, CCl_3) the substituent R is preferably in the eclipsed position, whereas in nitroxides **81** (R = H, Me) the phenyl group is eclipsed with the π-orbital. In 81 (R = Ph) both phenyl groups form an angle of approximately 30° with the plane of the π-orbital. From these results a stabilizing interaction between the nitroxide group and the phenyl group is concluded which is, however, exceeded by the effect of the electron acceptor substituents CN and CCl_3[285,304].

t-BuN(O•)—CHPh(R) **(81)**

t-BuN(O•)—CH(Ph)—CHR—O—H (O·····H intramolecular) **(237)** + MeOH $\overset{K}{\rightleftharpoons}$ t-Bu—N(O•···H—OMe)—CH(Ph)—CHR(OH) (94)

A special situation arises for molecules such as **237** (R = H) that can form an intramolecular hydrogen bond. a^N is almost unchanged when the aprotic solvent toluene is replaced by the protic solvent methanol ($a^N = 15.0$ and 15.1 G, respectively), but a_β^H changes drastically ($a_\beta^H = 6.6$ and 3.6 G, respectively). Whereas in toluene an intramolecular hydrogen bond exists, formation of an intermolecular hydrogen bond is more favourable in methanol. Thus the conformation of the molecule on which a_β^H depends is quite different in both solvents. Since the nitroxide group is hydrogen bonded in both solvents its direct environment which determines a^N remains, however, almost unchanged[305].

By addition of increasing portions of methanol to a solution of **237** in toluene the equilibrium constant *K* was determined from a_β^H. $\Delta H^0 = -7.4 \pm 0.5$ KJ mol^{-1} and $\Delta S^0 =$

TABLE 13. Coupling constants $a^{^{13}C}{}_\beta$ (G) for nitroxides **81**[285,304]

	R = CN	CCl_3	H	Me	Ph
$a^{^{13}C}{}_{\beta\text{-CR}}$	9.85	9.68	—	3.25	7.38
$a^{^{13}C}{}_{\beta\text{-CPh}}$	—	—	9.82	10.50	7.38

$-19 \pm 2\,J\,mol^{-1}\,K^{-1}$ result from its temperature dependence. ΔH^0 approximately represents the strain energy due to formation of the hydrogen-bonded ring. If $R \neq H$ the two diastereomeric forms of **237** could be detected, the *trans* form being energetically more favourable[305].

The two TEMPO groups attached to the crown ether skeleton in the *syn* position of **238** exhibit only a weak spin–spin interaction. Every change in conformation caused either by the solvent, by variation of pH or complexation with K^+ changes the distance between the nitroxide groups, giving rise to a change in spin–spin interaction[306].

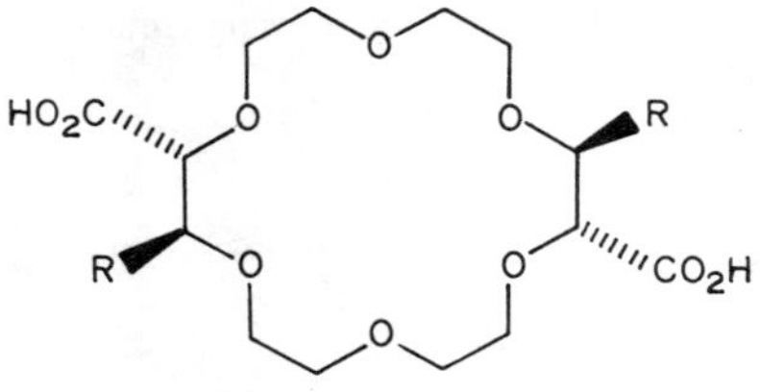

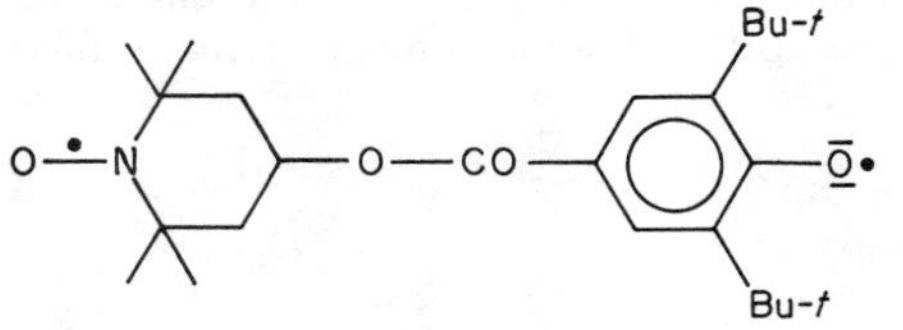

(**239**)

(**238**) R=CONH— (TEMPO)

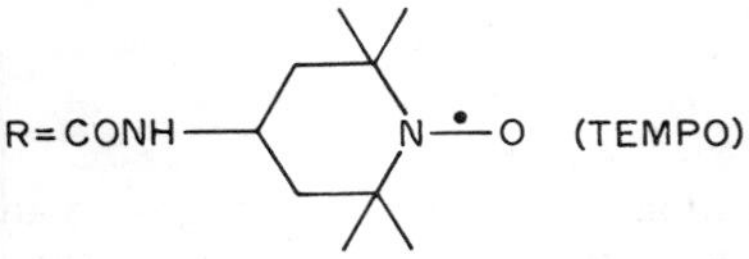

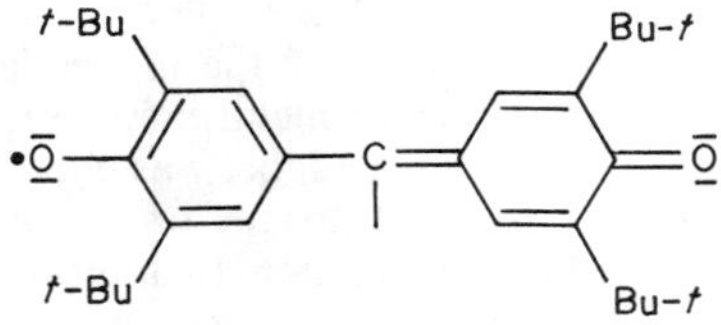

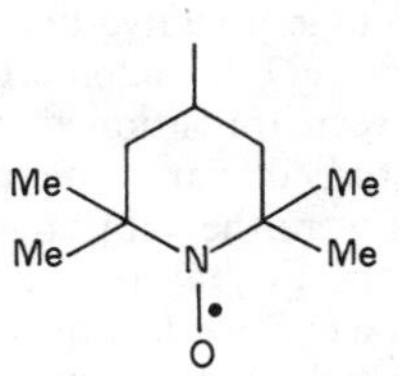

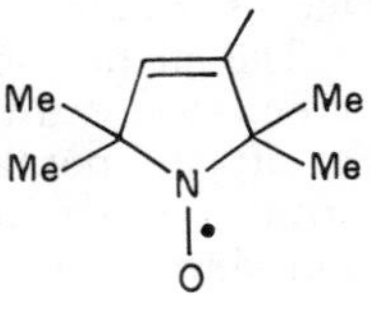

(G) (N_1) (N_2)

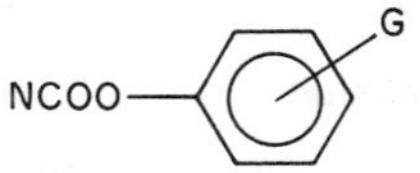

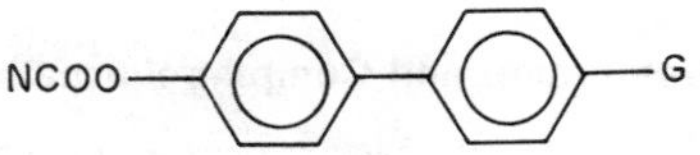

(**240**) (**241**) (**242**)

The ESR spectrum of the hetero diradical **239** reveals that a strong exchange interaction between the two spin systems in the molecule takes place[307]. Hetero diradicals consisting of a six- or five-membered nitroxide (N_1 or N_2) and the galvinoxyl (G) linked either directly (**240**) or by the phenyloxycarbonyl (**241**) or the biphenyloxycarbonyl group (**242**), respectively, were extensively studied by ESR and ENDOR spectroscopy[308]. The magnitude of the exchange integral J relative to the nitrogen hyperfine splitting was found to change from $|J| < a^N|$ to $|J| \gg |a^N|$ depending on the length of the connecting link. The sign of the exchange radical was negative as was shown by ENDOR for three of the biradicals. That means that their ground state is a triplet[308].

Two nitroxide molecules can also be connected by a metal–chelate bridge (**243**). For this purpose cyclic nitroxides were covalently bonded to complexing agents such as β-

dicarbonyl compounds, carboxylic acids, hydroxamic acids, crown ethers and monosubstituted phosphates. Nitrogen-containing heterocycles with a hydroxy group in a favourable position and their corresponding *N*-oxides are also frequently used. In addition to the various piperidine-*N*-oxyls, pyrrolidine-*N*-oxyls and nitronyl nitroxides, particularly 2,2,5,5-tetramethyl-3-imidazoline-1-oxyls and 2,2,5,5-tetramethyl-3-imidazoline-3-oxide-1-oxyls bonded to the complex-forming agents at the 4-position (see for example **244**) were used as nitroxide compounds. Since many of the metal–chelate complexes are soluble in organic solvents the spin-labelled agents can be applied to extract the metal cations from aqueous solutions. The imidazoline compounds are particularly suitable for analytical applications since they make the nitroxide moiety relatively insensitive to an acid medium due to their basicity[309].

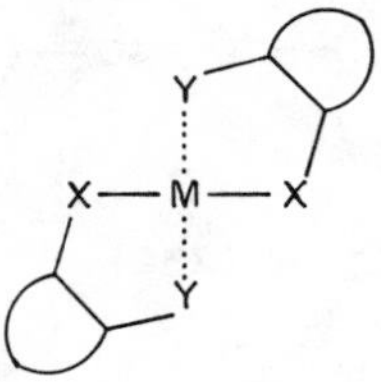

Y, X—M—X, Y

(**243**)

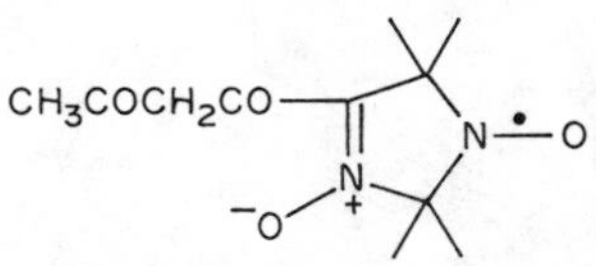

CH_3COCH_2CO—, N—O, N, ^-O, +

(**244**)

With diamagnetic metals and weak interaction of the two nitroxide groups an almost unchanged three-line ESR spectrum is observed. However, if the exchange interaction is strong, the spectrum changes to a five-line pattern (see Section IV.C). With paramagnetic metals the central atom may participate in the exchange interaction[309].

The interaction between transition metals, in various spin states, and the nitroxide group was extensively studied in many metal complexes[310]. In particular, metal complexes of tetraaryl porphyrins were the subject of these studies. The nitroxide label was either covalently bonded to the porphyrin skeleton or to one of the aryl groups or it was introduced by means of a carboxylate or pyridine ligand at the axial position. In addition to porphyrins spin-labelled ethylendiaminetetraacetic acid and β-diketones were used as complex ligands for various metals[310].

*V. REACTIONS INVOLVING THE NITROXIDE GROUP

*A. Protonation and Complex Formation of Nitroxides

Radicals **245** and **246** are protonated at the amino group in acid solution ($pK_A = 4.5$ and 2.0, respectively). Due to the inductive effect of the ammonium moiety a^N decreases on protonation (see Section IV.A: mesomeric formula A is favoured over B). Since proton exchange is fast on the ESR time-scale the coupling constants are weighted averages of the protonated and unprotonated species. Thus, a^N can be used for the determination of pH. (**246**: a^N is reduced from 15.50 G at pH 2.92 to 14.36 G at pH 0.8). **246** was even used to determine the absolute pH value in intravesicular solutions[311].

For nitroxides **247** and **248** fast, slow or moderate proton exchange was observed depending on the substituent R. Fast proton exchange occurs when $pK_A < 3$. **247** (R $= CH_2I$) was connected with a protein and can be used in this way for pH determination in biological systems[312].

A variety of complexes have been described in which the nitroxide group is directly coordinated to the metal[313]. Nitroxides form inclusion complexes with cyclodextrins. With α-cyclodextrin (six glucose units) only the hydrophobic part of DTBN fits into the

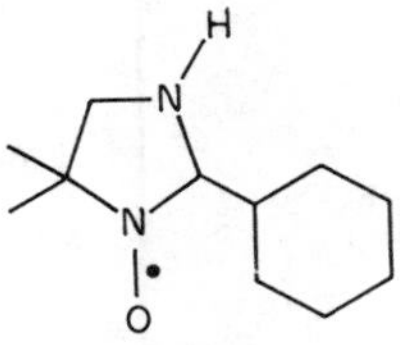

(245) 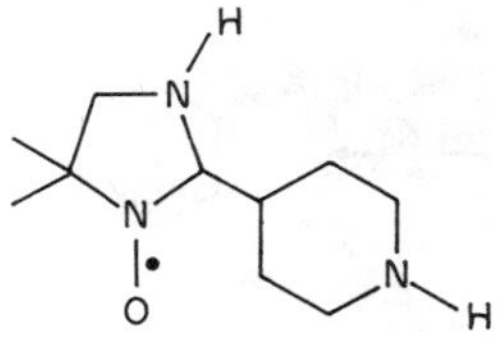 (246) 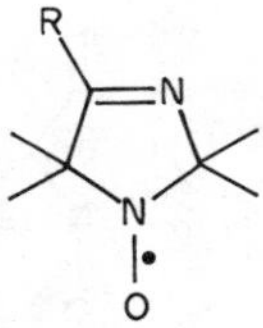 (247) 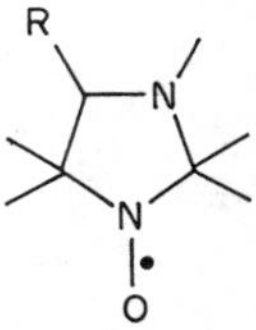 (248)

cavity with the nitroxide moiety remaining outside[314]. In the larger cavity of β-cyclodextrin (seven glucose units) however, the entire molecule is included. However, the nitroxides TEMPO and TEMPOL are only partly enclosed in β-cyclodextrin, the nitroxide group and the hydroxy group, respectively, remaining outside. Furthermore, β-cyclodextrin catalyses the hydrolysis of carbamoyl nitroxides in alkaline medium[315].

The shifts of NMR signals of various substrates in the presence of nitroxides are suggested to be caused by different types of interactions, e.g. hydrogen bonds of the nitroxide lone pair, Lewis acid–base association with the nitroxide lone pair, spin polarization of lone pairs of the substrate or changes of the bulk susceptibility[316].

*B. Addition of Radicals to the Nitroxide Group

1,1,3,3-Tetramethylisoindole-2-oxyl (**249**) reacts readily with alkyl radicals to afford the corresponding *N*-alkoxy compounds **250**. The reaction was used to study a variety of mechanistic problems by identification of the products formed. In this way the following reactions were studied (equation 95): (a) the reaction of *t*-butoxyl radical with methyl acrylate[317], (b) the rearrangement of the radical formed by addition of the *t*-butoxyl radical to norbornadiene[318], (c) the competition between the trapping of the 5-hexenyl radical by **249** and its ring closure to cyclopentylmethyl[319] and (d) the competition between the trapping of the cyclopropylmethyl radical and its ring opening to 3-butenyl[320]. The reaction of *t*-butoxy radical with methyl acrylate, for instance, yielded a number of products **250**, which indicate that apart from addition (63% *t*-$BuOCH_2\dot{C}HCO_2Me$ + 1% $^{\bullet}CH_2CH(OBu\text{-}t)CO_2Me$) other processes take place.

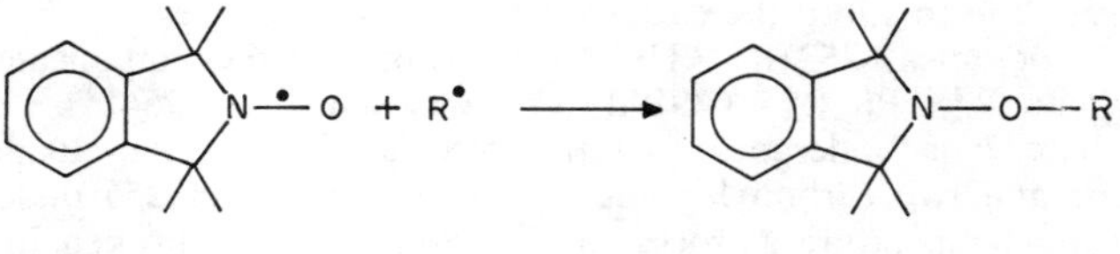

$$\text{N}\!-\!\dot{\text{O}} + \text{R}^{\bullet} \longrightarrow \text{N}\!-\!\text{O}\!-\!\text{R} \quad (95)$$

(249) (250)

R = *t*-$BuOCH_2\dot{C}HCO_2Me$ + others (a)

R = [norbornenyl–OBu-*t*] + [nortricyclyl–OBu-*t*] + [7-OBu-*t* norbornenyl] (b)

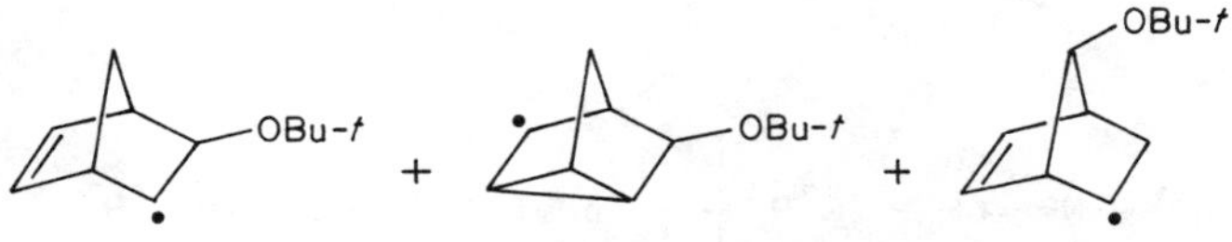

R = $CH_2{=}CH(CH_2)_3\dot{C}H_2$ + *cyclo*-$C_5H_9\dot{C}H_2$ (c)

R = *cyclo*-$C_3H_5\dot{C}H_2$ + $CH_2{=}CHCH_2\dot{C}H_2$ (d)

$$\text{(251)} \xrightarrow[-H]{+X^{\bullet}} \text{(252)} \qquad (96)$$

(a) $X = t\text{-BuO}^{\bullet}$
(b) $X = Ar_2N^{\bullet}$
(c) $X = Ar\dot{N}H$

However, if there is conjugation between the nitroxide group and the aromatic ring as in the indolinone oxyl **251**, substitution of the aromatic ring by the *t*-butoxy radical occurs to give **252a**[321]. Obviously, the first step of the reaction is radical combination, with attack of *t*-butoxyl at the *para* position of **251** followed by oxidation of the intermediate. The same reaction can take place once more at the free *ortho* position. No decomposition of *t*-butoxyl was observed. On the other hand, the methyl radical adds to the nitroxide group, as expected, to afford the *N*-methoxyindolinone. Diarylaminyl radicals react in the same way as *t*-butoxyl, affording the radicals **252b**. With primary arylaminyls further oxidation to a quinonoid intermediate occurs followed by addition of a second arylaminyl and additional oxidation steps[322] (see Section *V.*F).

*C. Dimerization of Nitroxides

As is found for other nitroxides with extended delocalization (equations 39 and 40), the formation of covalently bonded dimers of vinyl nitroxides **253** should be energetically favoured. Such dimers can arise either by C—C bonding (**254**) or C—O bonding (**255**).

In fact, C—C bonding is observed in most cases, but frequently secondary reactions of the dimers **254** take place[323]. Unchanged dimers **254** could be isolated with R^1 = alkyl, R^2 = Ph and R^3 = H[324]. These are relatively stable and do not dissociate. Most of the dimers **254** (R^3 = H) formed under oxidative conditions from their monomers **253** undergo further dehydrogenation to afford the corresponding dehydro dimers[325].

The dimerization of radicals **253** ($R^2 \neq H$, R^3 = acyl) takes a rather complicated course. Thus **253a** ($R^1 = t$-Bu, R^2 = Ph, R^3 = COPh) yields a tricyclic dimer **256**[326]. The C—C-bonded intermediate **254a** undergoes an intramolecular [3 + 2 + 2]cycloaddition in which one nitrone and two carbonyl groups are involved. Dimer **256** dissociates in solution even at room temperature, a process in which four bonds are broken. In contrast,

$$2R^1\text{—}\underset{\underset{O}{|\,\cdot}}{N}\text{—CH=CR}^2R^3 \quad \textbf{(253)}$$

$$\longrightarrow \left[R^1\text{—}\overset{+}{\underset{\underset{O^-}{|}}{N}}\text{=CH—CR}^2R^3\text{—}\right]_2 \textbf{(254)} \quad \text{or} \quad R^1\text{—}\underset{\underset{O\text{—CR}^2R^3\text{—CH=}\overset{+}{\underset{\underset{O^-}{|}}{N}}\text{—R}^1}{|}}{N}\text{—CH=CR}^2R^3 \textbf{(255)} \qquad (97)$$

253a ⇌ [254a] ⇌ (256) (98)

(a) $R^1 = t\text{-Bu}$, $R^2 = Ph$, $R^3 = COPh$

253b ⇌ [255b] ⟶ (257) (99)

(b) $R^1 = R^2 = Ph$, $R^3 = COPh$

dimerization of **253b** ($R^1 = R^2 = Ph$, $R^3 = COPh$) begins with C—O bond formation followed by an intramolecular [3 + 2]cycloaddition to give the bicyclic dimer **257**. The latter does not dissociate up to 150 °C[327]. The effect of substituents on the course of dimerization of the acyl vinyl nitroxides **253** as well as on dissociation of dimers **256** was studied[328].

*D. Disproportionation of Nitroxides

The decomposition of acyl nitroxides **258** affording the hydroxamic acids **259** and the extremely reactive nitrones **260** by disproportionation was studied kinetically[329]. The *A* factors and activation parameters of this bimolecular reaction were found to be unusually low. For this reason it is assumed that in the encounter pair both nitroxide groups are in a rather unfavourable orientation for disproportionation due to their dipole–dipole interaction. Since rearrangement of the unfavourably oriented 'unreactive' complexes into 'reactive' complexes with favourable orientation does not occur readily, only a very few of the encounter complexes afford the disproportionation products, while most of them decompose regenerating the acyl nitroxides. The same assumption may hold for the bimolecular decomposition of dialkyl nitroxides[329].

The acid-catalysed hydrolysis of the decomposition products of nitroxide **261** affords *N*-*t*-butyl-*p*-formylphenylhydroxylamine (**264**) via the intermediate formation of dimers and trimers[330]. The key step of the reaction is disproportionation of **261** giving **262** and **263**.

$$2\ CH_3N(O^{\bullet})COR \longrightarrow CH_3N(OH)COR + CH_2{=}\overset{+}{N}(O^-)COR \qquad (100)$$

(258) (259) (260)

t-BuN(O•)–C6H4–CH2OMe

(261)

⟶ t-BuN⁺(O⁻)=C6H4=CHOMe + t-BuN(OH)–C6H4–CH2OMe ⟶ dimer

(262) (263)

$\xrightarrow{+\ 262}$ trimer $\xrightarrow{H^+/H_2O}$ 2 t-BuN(OH)–C6H4–CH=O + (263) (101)

(264)

*E. Reduction of the Nitroxide Group

The OH bond dissociation energies of a number of *N*-*t*-butylhydroxamic acids and related compounds were found to fall in the range from 74.4 to 79.4 kcal mol^{-1} (cf. Table 10), the bond strength increasing with increasing electron demand in the acyl group[331]. Due to the even higher electron-withdrawing properties of the trifluoromethyl group, the OH bond dissociation energy of bis(trifluoromethyl)hydroxylamine was determined to be 85.3 ± 3 kcal mol^{-1} [332].

As the rate constants for hydrogen abstraction from a number of substrates by bis(trifluoromethyl)aminoxyl (**2**) indicate, this radical, although being the most reactive nitroxide, is still far less reactive than other radicals such as phenyl and *t*-butoxyl. The Arrhenius pre-exponential factors for the reactions of **2** are unusually low. It is assumed that these reactions require transition states that have highly restrictive geometries with a specific orientation of the nitroxide group relative to the bonds which are attacked[332].

2 first abstracts a benzylic hydrogen atom from benzyl cyanide, benzyl chloride or benzyl alcohol. The benzylic radicals so formed rapidly undergo secondary reactions with **2**. Benzyl cyanide reacts in a similar way as toluene (see equation 51)[333].

When cyclic nitroxides were reduced by ascorbic acid it was found that it is usually more difficult to reduce tetramethylpyrrolidine-*N*-oxyls than tetramethylpiperidine-*N*-oxyls. Substituents accelerate the reduction, as is seen in particular in the reaction of 2,2,5,5-tetramethylpyrrolid-3-on-1-oxyl[334].

Whereas nitroxides such as TEMPO and DTBN are not reduced by thiols in the absence of catalysts they are reduced in the presence of superoxide. This is even more surprising since superoxide alone usually oxidizes hydroxylamines to nitroxides. To explain this curiosity, formation of an adduct **265** from TEMPO and $O_2^{\cdot}$ is assumed. Its protonated form is believed to be a stronger oxidizing agent than its components. Thus the thiol can be oxidized by adduct **265** with formation of sulphinic acid and *N*-hydroxypiperidine (**266**). The sulphinic acid is then further oxidized by TEMPO to give sulphonic acid. The reaction rate for this process is about two orders of magnitude higher than for oxidation of hydroxylamines by superoxide[335].

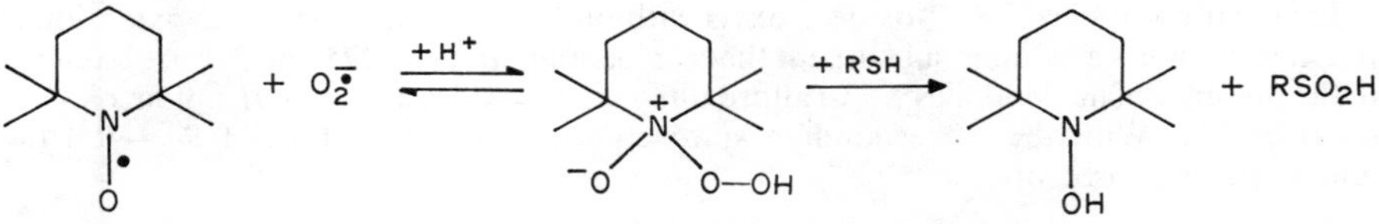

$$\text{TEMPO} + O_2^{\bullet -} \underset{}{\overset{+H^+}{\rightleftharpoons}} \textbf{(265)} \xrightarrow{+\text{RSH}} \textbf{(266)} + RSO_2H$$

$$RSO_2H + 2\ \text{TEMPO} + OH^- \longrightarrow 2\ \textbf{266} + RSO_3^- \qquad (102)$$

Photochemically excited TEMPO is found to be an efficient hydrogen abstractor. It therefore abstracts hydrogen atoms from acetonitrile and toluene under irradiation, affording reaction products as in equation (53). It is assumed that the reaction occurs from an upper excited state[336].

*F. Oxidation of the Nitroxide Group

Nitroxides substituted by primary or secondary alkyl groups are oxidized to form nitrones (see Scheme 4). Since these can easily undergo spin trap reactions or addition of nucleophiles followed by further oxidation, a sequence of reactions is possible adjacent to the nitroxide group (equation 103). Thus, treatment of hydroxylamines substituted by a primary alkyl group with cobalt-complexed peroxyl radicals affords successively the nitroxides **267**, **268** and **269**. In the oxidation of hydroxylamines substituted by two primary alkyl groups nitroxides **270** with two peroxy groups at the α-carbon atoms can be detected. Finally, these peroxyalkyl groups are converted to acyl groups[232].

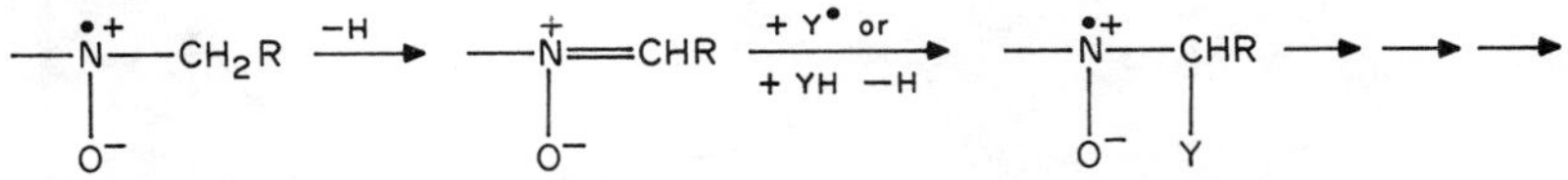

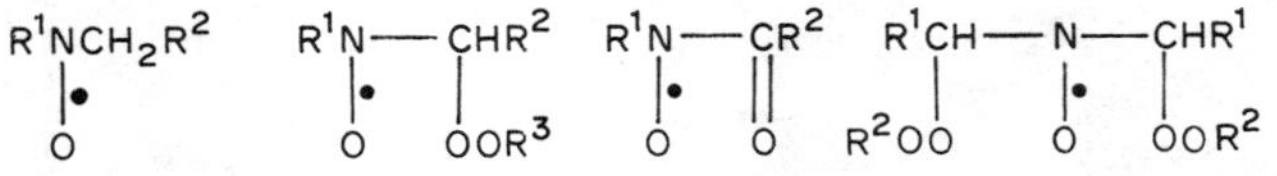

$$-\overset{\bullet +}{N}(O^-)-CH_2R \xrightarrow{-H} -\overset{+}{N}(O^-)=CHR \xrightarrow[+\text{YH}\ -\text{H}]{+\text{Y}^\bullet\ \text{or}} -\overset{\bullet +}{N}(O^-)-CHR(Y) \rightarrow \rightarrow \rightarrow$$

$$\underset{\textbf{(267)}}{R^1N(O^\bullet)CH_2R^2} \quad \underset{\textbf{(268)}}{R^1N(O^\bullet)-CHR^2(OOR^3)} \quad \underset{\textbf{(269)}}{R^1N(O^\bullet)-CR^2(=O)} \quad \underset{\textbf{(270)}}{R^1CH(OOR^2)-N(O^\bullet)-CHR^1(OOR^2)} \qquad (103)$$

When 1-hydroxy-5, 5-dimethyl-3-imidazoline-3-oxides are oxidized in methanol, at first nitroxides **271** and subsequently **272** and **273** can be detected by ESR. The latter have been isolated[337]. They arise from the corresponding nitrone formed by oxidation of **271** (cf. equation 103), subsequent 1, 2- or 1, 4-addition, respectively, of methanol to the nitrone and then more dehydrogenation. Starting from **271** ($R^2 = H$) the oxidation process produces the isolable nitroxides **274**[337].

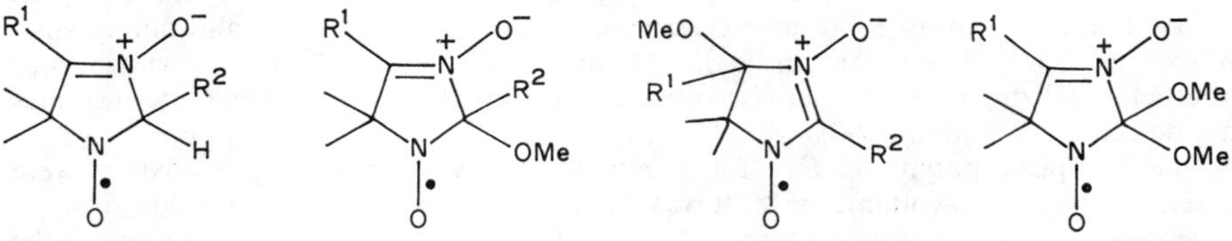

(271) (272) (273) (274)

Even with 3-imidazoline-3-oxide-1-oxyls without hydrogen at position 2, oxidation is possible if there is a suitable substituent that can be split off. Thus **275** and **276** are oxidized in the presence of nucleophiles X^- to afford nitronyl nitroxides **278** via 4-*H*-imidazole-1,3-dioxides **277**. With the corresponding spiro compounds (connection of R^1—R^2) the annexed ring can be opened[338,339].

(275) (276) (277) (278)

$-H^+$, $-e$; $-R^2CHO$; $-R^2COX$; $+X^-$, $-e$

X = OH, OMe, NH_2

$+X^-$, $-e$ (104)

Another example of the reaction sequence oxidation–addition–oxidation is the conversion of **279** to the bicyclic nitroxide **280** in which the oxime group adds to the nitrone intermediate[340]. The same is true for the secondary reaction of the nitroxide **252c** (X = ArNH) formed by oxidative substitution of the indolinone-*N*-oxyl with primary arylaminyls. **252c** is oxidized to the quinoneimine oxide **281**; addition of a second arylamine molecule to **281** and further oxidation yields **282**, which can be oxidized to give again the corresponding quinoneimine oxide[322]. Three oxidation steps without any addition occur when alkyl- or aryl-hydroxylamines substituted at their β-carbon by an electron-acceptor group R^2 form successively nitroxides **283**, nitrones and finally vinyl nitroxides **253**[341]. The latter can be directly detected by ESR if R^3 is an aryl group or another stabilizing group. Otherwise spin adducts from **258** to the nitrone intermediates are detected (see equation 90).

The half-peak potentials $E_{P/2}$ for a number of cyclic di-*t*-alkyl nitroxides were determined by cyclic voltametry[342]. It was found that oxidation of the nitroxide group of piperidine-*N*-oxyls becomes more difficult with increasing electronegativity of the substituent in the piperidine ring. The same is true for substituents in five-membered ring

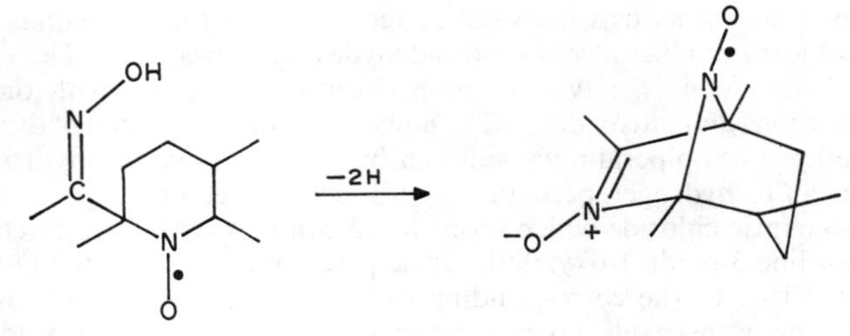

(279) $\xrightarrow{-2H}$ (280) (105)

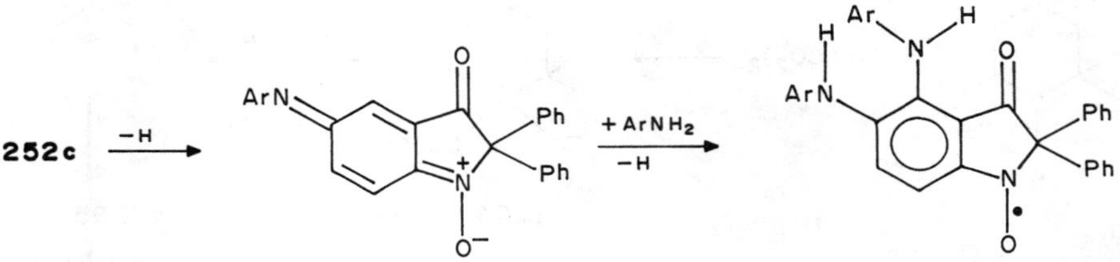

252c $\xrightarrow{-H}$ (281) $\xrightarrow[-H]{+ArNH_2}$ (282)

X=ArNH

(106)

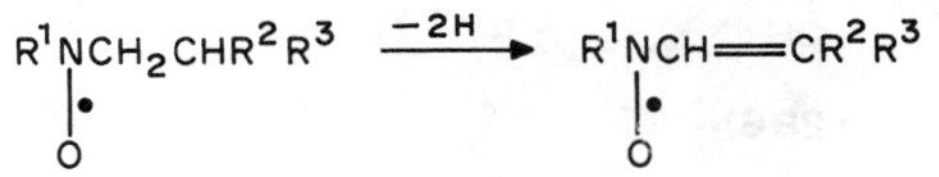

$$R^1N(\dot{O})CH_2CHR^2R^3 \xrightarrow{-2H} R^1N(\dot{O})CH{=}CR^2R^3 \quad (107)$$

(283) (253)

systems. If R is constant in the series, the potential increases due to the inductive effect are in the following order:

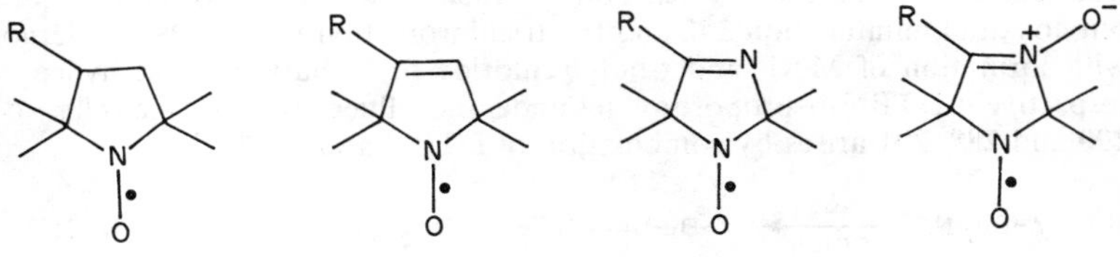

R=CO_2Me 0.42 0.54 0.65 0.92 V

Oxidation of alcohols by oxoammonium salts (see equation 61, last line) was applied in a catalytic cycle. TEMPO or TEMPOL was used as a catalyst in the electrochemical oxidation of primary and secondary alcohols to aldehydes and ketones, respectively[343]. At first the nitroxide was oxidized to the oxoammonium salt which in turn oxidized the alcohol. The hydroxylamine formed by this process reacted with additional oxoammonium salt to give the nitroxide which again entered the catalytic cycle. Primary amines were also oxidized in this way, forming nitriles in acetonitrile or aldehydes in aqueous solution[344].

Cupric ions may also be used as an oxidative agent in the nitroxide-mediated oxidation. Thus it is possible to oxidize alcohols to aldehydes by cupric ions. The oxidation of benzylic and allylic alcohols may even be performed by oxygen with only catalytic amounts of cupric ions and nitroxide[345]. The higher oxidation potential of the 4-methoxy-2,2,2,6-tetramethyl-1-oxopiperidinium salt can be utilized to oxidize hydroxide ions in nonaqueous media to hydrogen peroxide[346] and to cleave ethers[347].

In contrast to cupric chloride which forms a 1:2 complex with 4,4,5,5-tetramethyl-2-phenyl-2-imidazoline-3-oxide-1-oxyl (**90**), cupric perchlorate oxidizes the nitronyl nitroxide **90** in acetonitrile to the corresponding oxoammonium salt[277]. The oxidation of TEMPO by dibenzoyl peroxide takes a rather complicated course. An adduct **284** is possibly initially formed, which then eliminates benzoic acid to give **285**. Further conversion to **287** occurs via **286**[348].

$$\text{TEMPO} + (PhCO_2)_2 \longrightarrow \mathbf{(284)} \xrightarrow{-PhCO_2H} \mathbf{(285)} \longrightarrow \mathbf{(286)} \xrightarrow{-2\,H} \mathbf{(287)} \tag{108}$$

In contrast to the irradiation in pentane yielding products that result from a π, π^* excited state (equation 66) irradiation of DTBN (**1**) in carbon tetrachloride or chloroform affords products **28** and **288–291** which arise via a charge-transfer excited state. The primary photochemical event in the contact charge-transfer excited state is electron transfer from DTBN to carbon tetrachloride with formation of the di-*t*-butyloxammonium cation **292** and the trichloromethyl radical. **292** undergoes cleavage with formation of MNP and *t*-butyl chloride or isobutylene and hydrogen chloride, respectively. DTBN disproportionates under the influence of hydrogen chloride to give **28**, **290** and **288**. **291** arises by combination of DTBN and $^{\bullet}CCl_3$[349].

$$\underset{(1)}{t\text{-}Bu_2NO^{\bullet}} \xrightarrow[CCl_4]{h\nu} \underset{(28)}{t\text{-}Bu\text{-}N{=}O} + \underset{(288)}{t\text{-}Bu_2\overset{+}{N}HOH\ Cl^-} + \underset{(219)}{t\text{-}BuCl} + \underset{(290)}{CH_2{=}CMe_2} + \underset{(291)}{t\text{-}Bu_2N\text{-}O\text{-}CCl_3} \tag{109}$$

$$(t\text{-}Bu_2NO^{+\bullet}\ldots CCl_4^{-\bullet})^* \longrightarrow \underset{(292)}{t\text{-}Bu_2\overset{+}{N}{=}O\ Cl^-} + {}^{\bullet}CCl_3$$

$$t\text{-}Bu_2\overset{+}{N}{=}O\ Cl^- \longrightarrow t\text{-}BuN{=}O + t\text{-}BuCl + CH_2{=}CMe_2 + HCl$$

(292) (28) (289) (290)

$$2\ t\text{-}Bu_2NO^\bullet + HCl \longrightarrow t\text{-}BuN{=}O + CH_2{=}CMe_2 + t\text{-}Bu_2\overset{+}{N}HOH\ Cl^-$$

(1) (28) (290) (288)

$$t\text{-}Bu_2NO^\bullet + {}^\bullet CCl_3 \longrightarrow t\text{-}Bu_2N\text{-}O\text{-}CCl_3$$

(1) (291)

*H. β-Scission of Nitroxides

When the azide radical is trapped by certain nitroso compounds the primary adducts **293** spontaneously eliminate nitrogen. The radical intermediate **294** is subsequently trapped by another molecule of nitroso compound to give **295**[350]. The cyanate radical also affords nitroxides **295** when it is trapped by nitroso compounds. In this reaction carbon monoxide is eliminated from the primary adduct[351]. Since the coupling constants of the spin adduct formed from reaction of diphenyliminyl radical with nitrosodurene are very similar[352], it can be supposed that a radical of structure **295** is also formed in this reaction.

$$\underset{(293)}{RN(O^\bullet)\text{—}\bar{N}\text{—}\overset{+}{N}{\equiv}N} \xrightarrow{-N_2} \underset{(294)}{RN(O^\bullet)\text{—}\bar{N}} \xrightarrow{+\,R\text{—}N{=}O} \underset{(295)}{RN(O^\bullet)\text{—}N{=}\overset{+}{N}(O^-)\text{—}R} \qquad (110)$$

*I. Miscellaneous Nitroxide Group Reactions

Additional examples of homolytic solvolyses (equation 78) have been found, as for instance the reductive cleavage of diazonium salts[353]. The aryl radicals so formed can either combine with additional nitroxide radicals or abstract hydrogen from their alkyl groups. With *o*-allyloxy or -allylamino substituents the aryl radicals undergo an *exo* ring closure to afford radicals **297** which are then trapped by DTBN or TEMPO (equation 111)[354]. When such homosolvolytic reactions initiated by DTBN are performed with a sufficient excess of radical precursor RX, the radical R$^\bullet$ may be trapped by the MNP

$$\underset{(296)}{o\text{-}C_6H_4^\bullet\text{-}XCH_2CH{=}CH_2} \longrightarrow \underset{(297)}{\text{3-}({}^\bullet CH_2)\text{-2,3-dihydrobenzo-X-ole}}$$

X = O or NAc (111)

$$\xrightarrow{R_2NO^\bullet} \text{3-}(CH_2ONR_2)\text{-2,3-dihydrobenzo-X-ole}$$

formed in the course of the reaction to give the corresponding spin adducts **298** (equation 112)[355]. Homosolvolysis occurs extremely fast with compounds that form captodative substituted radicals such as α-bromo-α-methoxy-substituted ketones or methyl acetates[356].

$$\mathrm{RX} + t\text{-}\mathrm{Bu_2NO^\bullet} \rightleftharpoons \mathrm{R^\bullet} + t\text{-}\mathrm{Bu_2\overset{+}{N}{=}O}\ \mathrm{X^-}$$

$$\mathrm{R^\bullet} + t\text{-}\mathrm{Bu_2NO^\bullet} \longrightarrow \mathrm{RH} + \mathrm{CH_2{=}CMe_2} + t\text{-}\mathrm{BuN{=}O} \quad (112)$$

$$\mathrm{R^\bullet} + t\text{-}\mathrm{BuN{=}O} \longrightarrow t\text{-}\mathrm{BuN(O^\bullet)}\text{—}\mathrm{R}$$

(298)

$\mathrm{R = Ph_2CH, R^1CO, Ar}$, Succinimidyl

Diphenylcarbene existing as a triplet molecule in the ground state reacts cleanly with TEMPO with transfer of the oxygen atom (equation 113). The tetramethylpiperidyl radical formed in addition to benzophenone is slowly converted to tetramethylpiperidine by hydrogen abstraction from the solvent[357]. With TEMPOL this reaction proceeds at approximately the same rate as with TEMPO. Insertion into the OH bond takes place as a competition reaction. This reaction is about 10 times slower. Its rate is comparable to that of the carbene insertion into the OH bond of cyclohexanol.

$$\mathrm{Ph_2C}{\uparrow\uparrow} + \mathrm{{}^\bullet O\text{—}N}\langle\text{tetramethylpiperidine}\rangle\text{—X} \longrightarrow \mathrm{Ph_2C{=}O} + \mathrm{{}^\bullet N}\langle\text{tetramethylpiperidine}\rangle\text{—X} \quad (113)$$

X=H or OH

*VI. REACTIONS WITHOUT DIRECT INVOLVEMENT OF THE NITROXIDE GROUP

A remarkable reaction without participation of the nitroxide group is the addition of singlet oxygen to the anthracenyl nitroxide **299** (equation 114). The reaction is accompanied by a striking change in the ESR spectrum and proceeds quantitatively in organic solvents and in the bilayers of vesicles[358]. The 3-imidazoline-3-oxide-1-oxyl **300** undergoes a 1,3-dipoar cycloaddition at its nitrone moiety with suitable dipolarophiles, leaving the nitroxide group intact (equation 115)[359].

$$t\text{-}\mathrm{BuN(O^\bullet)}\text{-}(9,10\text{-di-OMe-anthracenyl}) \xrightarrow{^1\mathrm{O_2}} t\text{-}\mathrm{BuN(O^\bullet)}\text{-}(9,10\text{-di-OMe-9,10-epidioxyanthracenyl}) \quad (114)$$

(299)

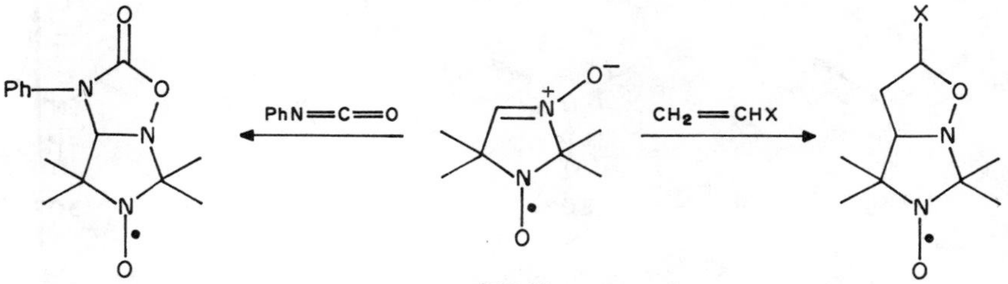

(300)

X=CO_2Me, CN, COMe

(115)

The diradical **302**, the dehydropiperidine rings of which are connected by two acetylenic bridges, arises from coupling of compound **301a** and its iodo derivative **301b** by cuprous chloride[360].

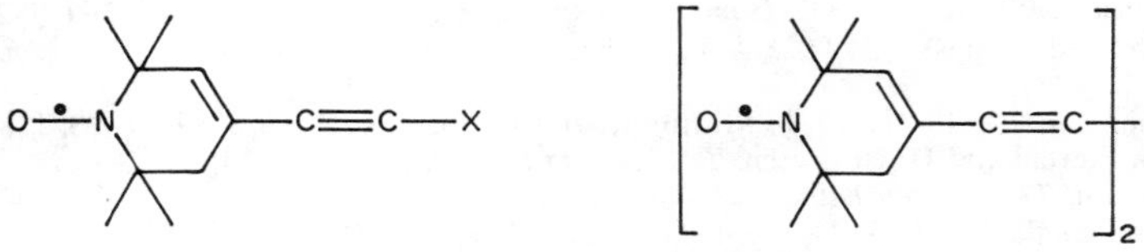

(301)

(a) X=H

(b) X=I

(302)

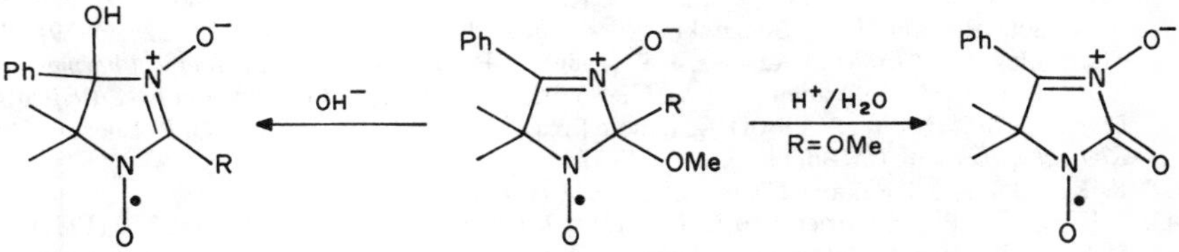

R=Me, Ph

(304)

(303)

(116)

Radicals of type **304** are changed under acidic or basic conditions without direct involvement of the nitroxide group[339] (equation 116). Pyrrolidine-*N*-oxyls used as spin labels (see **191**, **192** and Scheme 6) are modified insofar as the functional groups required for the coupling with biomolecules or other molecules are now also attached to one or both of the alkyl groups at the 2- or the 2- and 5- positions (**305**/**306**), respectively[361].

In addition to nitroxides **188–199**, 3-imidazoline-1-oxyls **247** and 3-imidazoline-3-oxide-1-oxyls (**307**), functionalized at the 4-position, are particularly useful as spin labels[309,359]. A review on the phosphorus compounds labelled by TEMPO has appeared[362]. Spin-labelled metal complexes have been extensively studied[363]. In this connection a number of spin-labelled crown ethers have been prepared[364]. The application of spin labelling to studies of biochemical and biophysical problems has increased considerably[365].

(305) (306) (247) (307)

X	Y
NH_2	CO_2H
OH	CO_2Me

X
CH_2CN
CH_2CO_2H

*VIII. REFERENCES

230. Revised nomenclature for radicals, ions, radical ions and related species. Recommendations 1985, provisional draft. Section C: RC-81.2.4, IUPAC, 1985.
231. J. R. Bolton (Ed.), The 1st International Symposium on Spin Trapping and Nitroxyl Radical Chemistry, July 12–17, 1981, Guelph, Canada, *Can. J. Chem.*, **60**, 1379–1636 (1982).
232. V. Cholvad, A. Staško, A. Tkáč, A. L. Buchachenko and L. Malik, *Coll. Czech. Chem. Commun.*, **46**, 823 (1981).
233. A. Faucitano, A. Buttafava, F. Martinotti and P. Bortulus, *J. Phys. Chem.*, **88**, 1187 (1984).
234. M. B. Eleveld and H. Hogeveen, *Tetrahedron Letters*, **25**, 785 (1984).
235. L. Grossi, *Tetrahedron Letters*, **28**, 3387 (1987).
236. C. Chatgilialoglu and K. U. Ingold, *J. Amer. Chem. Soc.*, **103**, 4833 (1981).
237. H. G. Aurich, J.-M. Heinrich and G. Wassmuth, *J. Chem. Res.*, 222 (S), 3046 (M) (1980).
238. R. Konaka, S. Terabe, T. Mizuta and S. Sakata, *Can. J. Chem.*, **60**, 1532 (1982).
239 M. P. Crozet and P. Tordo, *J. Amer. Chem. Soc.*, **102**, 5696 (1980).
240. J. M. Coxon, B. C. Gilbert and R. O. C. Norman, *J. Chem. Soc., Perkin Trans. 2*, 379 (1981).
241. M. J. Perkins in *Advances in Physical Organic Chemistry*, Vol. 17 (Ed. V. Gold and D. Bethell), Academic Press, New York, 1980, p. 1; D. Rehorek, *Z. Chem.*, **20**, 325 (1980); V. E. Zubarev, V. N. Belevskii and L. T. Bugaenko, *Russ. Chem. Rev.* (Engl. transl.), **48**, 729 (1979); P. J. Thornalley, *Life Chemistry Reports*, **4**, 57 (1986); R. P. Mason in *Spin labeling in Pharmacology* (Ed. J. L. Holtzman), Academic Press, New York, 1984, p. 87; E. G. Janzen in *Free Radicals in Biology*, Vol. 4 (Ed. W. A. Pryor), Academic Press, New York, 1980, p. 115. G. R. Buettner, *Free Radical Biology and Medicine*, **3**, 259 (1987).
242 R. Konaka and S. Sakata, *Chem. Letters*, 411 (1982).
243. J. K. Brown, P. J. Coldrick and E. J. Forbes, *J. Chem. Soc., Chem. Commun.*, 770 (1982).
244. F. Vila, M. Boyer, G. Gronchi, Y. Duccini, O. Santero and P. Tordo, *Tetrahedron Letters*, **25**, 2215 (1984).
245. C. Chatgilialoglu, J. A. Howard and K. U. Ingold, *J. Org. Chem.*, **47**, 4361 (1982).
246. E. G. Janzen and G. A. Coulter, *J. Amer. Chem. Soc.*, **106**, 1962 (1984).
247. G. L. Mc Intire, H. N. Blount, H. J. Stronks, R. V. Shetty and E. G. Janzen, *J. Phys. Chem.*, **84**, 916 (1980).
248. D. Rehorek and E. G. Janzen, *J. Prakt. Chem.*, **327**, 968 (1985); E. G. Janzen, D. Rehorek and H. J. Stronks, *J. Magn. Reson.*, **56**, 174 (1984).
249. E. E. Bankroft, H. N. Blount and E. G. Janzen, *J. Phys. Chem.*, **84**, 557 (1980).
250. D. L. Haire and E. G. Janzen, *Can. J. Chem.*, **60**, 1514 (1982).
251. E. G. Janzen, R. V. Shetty and S. M. Kunanec, *Can. J. Chem.*, **59**, 756 (1981).
252. W. A. Pryor, M. Tamura and D. F. Church, *J. Amer. Chem. Soc.*, **106**, 5073 (1984).
253. I. M. Sosonkin, V. N. Belevskii, G. N. Strogov, A. N. Domarev and S. P. Yarkov, *J. Org. Chem. USSR* (Engl. transl.), **18**, 1313 (1982).
254. C. Mottley, H. D. Connor and R. P. Mason, *Biochem. Biophys. Res. Commun.*, **141**, 622 (1986).
255. E. Niki, S. Yokoi, J. Tsuchiya and Y. Kamiya, *J. Amer. Chem. Soc.*, **105**, 1498 (1983).
256. C. R. H. I. de Jonge, I. V. Khudyakov, S. L. Krenyov, V. A. Kuzmin, A. I. Prokof'ev and M. V. Voevodskaya, *J. Chem. Soc., Perkin Trans. 2*, 347 (1985).
257. D. E. Hamilton, R. S. Drago and J. Telser, *J. Amer. Chem. Soc.*, **106**, 5353 (1984).

258. P. D. Josephy, D. Rehorek and E. G. Janzen, *Tetrahedron Letters*, **25**, 1685 (1984).
259. C. Mottley, K. Toy and R. P. Mason, *Mol. Pharmacology*, **31**, 417 (1987); see also references therein.
260. P. Riesz and I. Rosenthal, *Can. J. Chem.*, **60**, 1474 (1982).
261. N. Iguchi, F. Moriya, K. Makino, S. Rokushika and H. Hatano, *Can. J. Chem.*, **62**, 1722 (1984).
262. K. Abe, H. Suezawa and M. Hirota, *J. Chem. Soc., Perkin Trans. 2*, 29 (1984).
263. V. P. J. Marti and B. P. Roberts, *J. Chem. Soc., Perkin Trans. 2*, 1613 (1986).
264. E. G. Janzen, H. J. Stronks, D. E. Nutter, Jr., E. R. Davis, H. N. Blount, J. L. Poyer and P. B. McCay, *Can. J. Chem.*, **58**, 1596 (1980).
265. W. Kremers and A. Singh, *Can. J. Chem.*, **58**, 1592 (1980).
266. D. Rehorek, C. M. Du Bose and E. G. Janzen, *Z. Chem.*, **24**, 188 (1984).
267. (a) H. G. Aurich, M. Schmidt and T. Schwerzel, *Chem. Ber.*, **118**, 1086 (1985).
(b) H. G. Aurich, J. Eidel and M. Schmidt, *Chem. Ber.*, **119**, 36 (1986).
268. S. Atmaram, A. R. Forrester, M. Gill, R. J. Napier and R. H. Thomson, *Acta Chem. Scand.*, **B36**, 641 (1982).
269. A. Rockenbauer, M. Györ and F. Tüdös, *Tetrahedron Letters*, **27**, 3759, 3763, 3421, 3425, 4795 (1986).
270. C. L. Greenstock and R. H. Wiebe, *Can. J. Chem.*, **60**, 1560 (1982).
271. O. Ito and M. Matsuda, *Bull. Chem. Soc. Japan*, **57**, 1745 (1984).
272. O. Ito and M. Matsuda, *J. Amer. Chem. Soc.*, **105**, 1937 (1983).
273. E. Finkelstein, G. M. Rosen and E. J. Rauckman, *J. Amer. Chem. Soc.*, **102**, 4994 (1980).
274. C. Chatgilialoglu, K. U. Ingold, I. Tse-Sheepy and J. Warkentin, *Can. J. Chem.*, **61**, 1077 (1983).
275. B. Chion and J. Lajzérowicz-Bonneteau, *Acta Cryst.*, **B36**, 998 (1980).
276. J. Laugier, P. Rey, C. Benelli, D. Gatteschi and C. Zanchini, *J. Amer. Chem. Soc.*, **108**, 6931 (1986).
277. A. Caneschi, J. Laugier and P. Rey, *J. Chem. Soc., Perkin Trans. 1*, 1077 (1987).
278. P. J. Brown, A. Capiomont, B. Gillon and J. Schweizer, *Journal of Magnetism and Magnetic Materials*, **14**, 289 (1979).
279. B. Delley, P. Becker and B. Gillon, *J. Chem. Phys.*, **80**, 4286 (1984).
280. M. C. R. Symons and A. S. Pena-Nuñez, *J. Chem. Soc., Faraday Trans. 1*, **81**, 2421 (1985).
281. P. Mukerjee, C. Ramachandran and R. A. Pyter, *J. Phys. Chem.*, **86**, 3189 (1982).
282. C. Ramachandran, R. A. Pyter and P. Mukerjee, *J. Phys. Chem.*, **86**, 3198 (1982).
283. R. A. Pyter, C. Ramachandran and P. Mukerjee, *J. Phys. Chem.*, **86**, 3206 (1982).
284. I. A. Grigor'ev, L. B. Volodarsky, A. Z. Gogolev and R. Z. Sagdeev, *Chem. Phys. Letters*, **122**, 46 (1985).
285. E. G. Janzen, U. M. Oehler, D. L. Haire and Y. Kotake, *J. Amer. Chem. Soc.*, **108**, 6858 (1986).
286. E. G. Janzen and U. M. Oehler, *Chem. Letters*, 1233 (1984).
287. R. Z. Sagdeev, A. Z. Gogolev, I. A. Grigor'ev, G. I. Shchukin, L. B. Volodarsky, W. Möhl and K. Möbius, *Chem. Phys. Letters*, **105**, 223 (1984).
288. M. M. Mossoba, K. Makino, P. Riesz and R. C. Perkins, Jr., *J. Phys. Chem.*, **88**, 4717 (1984).
289. R. Chiarelli, A. Jeunet, J. Michon, P. Michon, C. Morat, A. Rassat and H. U. Sieveking, *Org. Magn. Reson.*, **13**, 216 (1980).
290. C. Chatgilialoglu, V. Malatesta and K. U. Ingold, *J. Phys. Chem.*, **84**, 3597 (1980).
291. A. Rockenbauer, M. Györ, K. Hideg and H. O. Hankovszky, *J. Chem. Soc., Chem. Commun.*, 1651 (1985).
292. C. Chatgilialoglu and K. U. Ingold, *J. Phys. Chem.*, **86**, 4372 (1982); D. A. C. Compton, C. Chatgilialoglu, H. H. Mantsch and K. U. Ingold, *J. Phys. Chem.*, **85**, 3093 (1981).
293. R. Henriquez, M. J. Perkins and D. Griller, *Can. J. Chem.*, **62**, 139 (1984).
294. A. T. Balaban, E.-U. Würthwein and P. v R. Schleyer, *Tetrahedron*, **43**, 405 (1987).
295. E. G. Janzen, G. A. Coulter, U. M. Oehler and J. P. Bergsma, *Can. J. Chem.*, **60**, 2725 (1982).
296. E. G. Janzen and G. A. Coulter, *Tetrahedron Letters*, **22**, 615 (1981).
297. R. Darcy, *J. Chem. Soc., Perkin Trans. 2*, 1089 (1981); *J. Chem. Educ.*, **57**, 907 (1980).
298. H. G. Aurich and J. Trösken, *Tetrahedron*, **30**, 2515, 2519 (1974).
299. D. F. Church, *J. Org. Chem.*, **51**, 1138 (1986).
300. Y. Kotake and K. Kuwata, *Bull. Chem. Soc. Japan*, **54**, 394 (1981).
301. M. Kira, H. Osawa and H. Sakurai, *J. Organomet. Chem.*, **259**, 51 (1983).
302. L. Grossi, L. Lunazzi and G. Placucci, *J. Chem. Soc., Perkin Trans. 2*, 1831 (1983).
303. C. Chatgilialoglu and K. U. Ingold, *Can. J. Chem.*, **59**, 1745 (1981).
304. E. G. Janzen, *Can. J. Chem.*, **62**, 1653 (1984).

305. Y. Kotake and K. Kuwata, *Bull. Chem. Soc. Japan*, **55**, 3686 (1982).
306. H. Dugas, P. Keroack and M. Ptak, *Can. J. Chem.*, **62**, 489 (1984).
307. K. Mukai, H. Yano and K. Ishizu, *Tetrahedron Letters*, **22**, 1903 (1981).
308. B. Kirste, A. Krüger and H. Kurreck, *J. Amer. Chem. Soc.*, **104**, 3850 (1982).
309. Yu. A. Zolotov, O. M. Petrukhin, V. Yu. Nagy and L. B. Volodarsky, *Anal. Chim. Acta*, **115**, 1 (1980).
310. Leading reference: L. Fielding, K. M. More, G. R. Eaton and S. S. Eaton, *Inorg. Chem.*, **26**, 856 (1987).
311. J. F. W. Keana, M. J. Acarregui and S. L. M. Boyle, *J. Amer. Chem. Soc.*, **104**, 827 (1982).
312. V. V. Khramtsov, L. M. Weiner, S. I. Eremenko, O. I. Belchenko, P. V. Schastnev, I. A. Grigor'ev and V. A. Reznikov, *J. Magn. Reson.*, **61**, 397 (1985).
313. Leading reference: see Reference 276; see also C. Benelli, A. Caneschi, D. Gatteschi, J. Laugier and P. Rey, *Angew. Chem.*, **99**, 958 (1987); *Angew. Chem. Intern. Ed. Engl.*, **26**, 913 (1987).
314. M. P. Eastman, B. Freiha, C. C. Hsu, K. C. Lum and C. A. Chang, *J. Phys. Chem.*, **91**, 1953 (1987).
315. M. Okazaki and K. Kuwata, *J. Phys. Chem.*, **88**, 3163, 4181 (1984); **89**, 4437 (1985).
316. D. Draney and C. A. Kingsbury, *J. Amer. Chem. Soc.*, **103**, 1041 (1981).
317. P. G. Griffiths, E. Rizzardo and D. H. Solomon, *Tetrahedron Letters*, **23**, 1309 (1982).
318. W. K. Busfield, I. D. Jenkins, S. H. Thang, E. Rizzardo and D. H. Solomon, *Tetrahedron Letters*, **26**, 5081 (1985).
319. A. L. J. Beckwith, V. W. Bowry, M. O'Leary, G. Moad, E. Rizzardo and D. H. Solomon, *J. Chem. Soc., Chem. Commun*, 1003 (1986).
320. L. Mathew and J. Warkentin, *J. Amer. Chem. Soc.*, **108**, 7981 (1986).
321. L. Greci, *Tetrahedron*, **38**, 2435 (1982).
322. A. Alberti, L. Greci, P. Stipa, P. Sgarabotto and F. Ugozzoli, *Tetrahedron*, **43**, 3031 (1987).
323. H. G. Aurich, *Can. J. Chem.*, **60**, 1414 (1982).
324. Reference 267b.
325. H. G. Aurich, M. Schmidt and T. Schwerzel, *Chem. Ber.*, **118**, 1105 (1985).
326. H. G. Aurich, G. Baum, W. Massa, K.-D. Mogendorf and M. Schmidt, *Chem. Ber.*, **117**, 2615 (1984).
327. H. G. Aurich, K.-D. Mogendorf and M. Schmidt, *J. Org. Chem.*, **49**, 2654 (1984).
328. H. G. Aurich, O. Bubenheim, W. Keßler and K.-D. Mogendorf, *J. Org. Chem.*, **53**, (1988) (in press).
329. D. Griller and M. J. Perkins, *J. Amer. Chem. Soc.*, **102**, 1354 (1980).
330. A. R. Forrester, J. Henderson and S. P. Hepburn, *J. Chem. Soc., Perkin Trans. 1*, 1165 (1981).
331. T. C. Jenkins and M. J. Perkins, *J. Chem. Soc., Perkin Trans. 2*, 717 (1983).
332. T. Doba and K. U. Ingold, *J. Amer. Chem. Soc.*, **106**, 3958 (1984).
333. R. E. Banks, J. M. Birchall, R. N. Haszeldine, R. A. Hughes, S. M. Nona and C. W. Stephens, *J. Chem. Soc., Perkin Trans. 1*, 455 (1981).
334. W. R. Couet, R. C. Brasch, G. Sosnovsky, J. Lukszo, I. Prakash, C. T. Gnewuch and T. N. Tozer, *Tetrahedron*, **41**, 1165 (1985).
335. E. Finkelstein, G. M. Rosen and E. J. Rauckman, *Biochim. Biophys. Acta*, **802**, 90 (1984).
336. L. J. Johnston, M. Tencer and J. C. Scaiano, *J. Org. Chem.*, **51**, 2806 (1986).
337. I. A. Grigor'ev, L. B. Volodarsky, V. F. Starichenko, G. I. Shchukin and I. A. Kirilyuk, *Tetrahedron Letters*, **26**, 5085 (1985).
338. I. A. Grigor'ev, G. I. Shchukin, V. V. Khramtsov, L. M. Weiner, V. F. Starichenko and L. B. Volodarsky, *Bull. Acad. Sci. USSR, Div. Chem. Sci.* (Engl. transl.), 2169 (1985).
339. V. V. Khramtsov, L. M. Weiner, A. Z. Gogolev, I. A. Grigor'ev, V. F. Starichenko and L. B. Volodarsky, *Magn. Reson, Chem.*, **24**, 199 (1986).
340. M. Reinhardt, M. Schulz and J. Römbach, *J. Prakt. Chem.*, **328**, 597 (1986).
341. Reference 267a.
342. G. I. Shchukin, V. A. Ryabinin, I. A. Grigor'ev and L. B. Volodarsky, *J. Gen. Chem. USSR* (Engl. transl.), **56**, 753 (1986).
343. M. F. Semmelhack, C. S. Chou and D. A. Cortés, *J. Amer. Chem. Soc.*, **105**, 4492 (1983).
344. M. F. Semmelhack and C. R. Schmid, *J. Amer. Chem. Soc.*, **105**, 6732 (1983).
345. M. F. Semmelhack, C. R. Schmid, D. A. Cortés and C. S. Chou, *J. Amer. Chem. Soc.*, **106**, 3374 (1984).

346. T. Endo, T. Miyazawa, S. Shiihashi and M. Okawara, *J. Amer. Chem. Soc.*, **106**, 3877 (1984).
347. T. Miyazawa and T. Endo, *Tetrahedron Letters*, **27**, 3395 (1986).
348. G. Moad, E. Rizzardo and D. H. Solomon, *Tetrahedron Letters*, **22**, 1165 (1981).
349. S. Keute, D. R. Anderson and T. H. Koch, *J. Amer. Chem. Soc.*, **103**, 5434 (1981).
350. D. Rehorek and E. G. Janzen, *Z. Chem.*, **24**, 68 (1984).
351. D. Rehorek and E. G. Janzen, *Can. J. Chem.*, **62**, 1598 (1984).
352. T. Suehiro, S. Masuda and N. Motoyama, *Chem. Letters*, 53 (1980).
353. A. C. Scott, J. M. Tedder, J. C. Walton and S. Mhatre, *J. Chem. Soc., Perkin Trans. 2*, 260 (1980); see also J. Smith and J. M. Tedder, *J. Chem. Soc., Perkin Trans. 2*, 895 (1987).
354. A. L. J. Beckwith and G. F. Meijs, *J. Chem. Soc., Chem. Commun.*, 595 (1981).
355. H. Singh, J. M. Tedder and J. C. Walton, *J. Chem. Soc., Perkin Trans. 2*, 1259 (1980).
356. H. Singh and J. M. Tedder, *J. Chem. Soc., Chem. Commun.*, 1095 (1980).
357. H. L. Casal, N. H. Werstiuk and J. C. Scaiano, *J. Org. Chem.*, **49**, 5214 (1984).
358. J. F. W. Keana, V. S. Prabhu, S. Ohmiya and C. E. Klopfenstein, *J. Org. Chem.*, **51**, 3456 (1986).
359. L. B. Volodarsky, V. V. Martin and T. F. Leluch, *Tetrahedron Letters*, **26**, 4801 (1985).
360. V. V. Pavlikov, E. G. Rozantsev, A. B. Shapiro and V. D. Sholle, *Izv. Akad. Nauk SSSR, Ser. Khim.*, 197 (1980).
361. J. F. W. Keana, G. S. Heo and G. T. Gaughan, *J. Org. Chem.*, **50**, 2346 (1985); J. F. W. Keana, S. E. Seyedrezai and G. T. Gaughan, *J. Org. Chem.*, **48**, 2644 (1983); J. W. F. Keana and V. S. Prabhu, *J. Org. Chem.*, **51**, 4300 (1986).
362. M. Konieczny and G. Sosnovsky, *Synthesis*, 682 (1981).
363. Leading references 309 and 310.
364. J. F. W. Keana, J. Cuomo, L. Lex and S. E. Seyedrezai, *J. Org. Chem.*, **48**, 2647 (1983); see also Reference 306.
365. See for example: H. M. Swartz, M. Sentjurc and P. D. Morse, II, *Biochim. Biophys. Acta*, **888**, 82 (1986); E. G. Ankel, C.-S. Lai, L. E. Hopwood and Z. Zivkovic, *Life Science*, **40**, 495 (1987); H. Dugas and A. Rodriguez, *Can. J. Chem.*, **60**, 1421 (1982).

Author index

This author index is designed to enable the reader to locate an author's name and work with the aid of the reference numbers appearing in the text. The page numbers are printed in normal type in ascending numerical order, followed by the reference numbers in parentheses. The numbers in *italics* refer to the pages on which the references are actually listed.

Subject index